Hard Rock Mine Reclamation

Hard Rock Mine Reclamation

From Prediction to Management of Acid Mine Drainage

Edited By

Bruno Bussière and
Marie Guittonny

CRC Press is an imprint of the
Taylor & Francis Group, an **informa** business

First edition published 2021
by CRC Press
6000 Broken Sound Parkway NW, Suite 300, Boca Raton, FL 33487-2742

and by CRC Press
2 Park Square, Milton Park, Abingdon, Oxon, OX14 4RN

CRC Press is an imprint of Taylor & Francis Group, LLC

ISBN: 978-1-138-05451-6 (hbk)
ISBN: 978-1-315-16669-8 (ebk)

Typeset in Times
by SPi Global, India

Contents

Acknowledgments

The editors would first like to thank their institution, the Université du Québec en Abitibi-Témiscamingue (Quebec, Canada), for supporting teaching, training, research, and services related to mines and the environment for over 30 years. This support was essential for the creation of the Research Institute on Mines and the Environment (RIME) in 2012. RIME creates and shares interdisciplinary knowledge to solve environmental problems associated with abandoned and active mine sites in Quebec, Canada, and abroad. This book is the logical continuity of this knowledge development and transfer.

The editors would also like to acknowledge our copyeditors, Dr Joanne Muzak, for 13 chapters, and Gary Schudel for three chapters. Their exceptional contributions have significantly improved the quality of the book.

Some of the figures have been developed in collaboration with Marine Malfoy, Marie Bois-Joyal, and Faneva Rarison. The editors would like to thank them for their precious contribution.

This book would not have been possible without the engagement of all the contributing authors. This four-year journey included many meetings, always held in collegiality and with enthusiasm. The authors worked hard to meet various deadlines. The editors thank all of them for their collaboration and dedication.

Finally, the editors and authors would like to thank their research partners, particularly governmental ministries (especially the Quebec Ministry of Energy and Natural Resources), funding agencies (Natural Sciences and Engineering Research Council of Canada, and Fonds de recherche du Québec—Nature et Technologies), consulting and mining companies (too numerous to be listed here), for all these years of precious collaboration. They gave access to their resources and their sites, and their constructive interactions allowed the editors and authors to develop their expertise and applied knowledge.

Editors

Professor Bruno Bussière is the Scientific Director of the Research Institute on Mines and the Environment (RIME UQAT-Polytechnique) at Université du Québec en Abitibi-Témiscamingue (UQAT), where he also holds the Industrial NSERC-UQAT Chair on Mine Site Reclamation. Trained as a mining engineer (1991), he holds a master's degree (1993) and a PhD (1999) in hydrogeology and mining environment from École Polytechnique de Montréal. His teaching and research activities mostly relate to mining geoenvironment and hydrogeology, including constitutive and numerical modeling of unsaturated flow in soils and mine wastes, characterization of tailings and waste rocks, passive mine water treatment, hydrogeotechnical aspects of mine wastes disposal, mine water quality prediction, mineral separation in tailings, and reclamation methods for surface disposal sites including control of acid mine drainage and contaminated neutral drainage. In addition to publishing numerous technical papers in refereed journals and conference proceedings, Professor Bussière has trained many specialists in the field of mines and the environment.

Marie Guittonny is a professor in mine site revegetation since 2013 at the Research Institute on Mines and the Environment at the Université du Québec en Abitibi-Témiscamingue. Trained as a biologist (1999), she holds a master's degree (2001) and a PhD (2004) in Biosciences of the Environment from Université d'Aix-Marseille (France). Her research focuses on rehabilitating degraded sites, and her teaching and research activities revolve around plant and mine substrates' relationships to ensure the long-term success of mine site reclamation. In particular, Dr Guittonny works on overcoming the limitations of mine wastes to establish vegetation and on studying plant effects on the performance of mine reclamation methods, especially engineered covers, including root colonization effects on the water budget, oxygen consumption, and material properties.

Authors

Dr Akué Sylvette Awoh
Université du Québec en Abitibi-Témiscamingue, Research Institute on Mines and the Environment (RIME), 445 boul. de l'Université, Rouyn-Noranda (QC), J9X 5E4, Canada

Mostafa Benzaazoua
Université du Québec en Abitibi-Témiscamingue, Research Institute on Mines and the Environment (RIME), 445 boul. de l'Université, Rouyn-Noranda (QC), J9X 5E4, Canada

Pr Vincent Boulanger-Martel
Université du Québec en Abitibi-Témiscamingue, Research Institute on Mines and the Environment (RIME), 445 boul. de l'Université, Rouyn-Noranda (QC), J9X 5E4, Canada

Pr Bruno Bussière
Université du Québec en Abitibi-Témiscamingue, Research Institute on Mines and the Environment (RIME), 445 boul. de l'Université, Rouyn-Noranda (QC), J9X 5E4, Canada

Jean Côté
Université Laval, Faculté des sciences et de génie, Département de génie civil et de génie des eaux, pavillon Adrien-Pouliot, 1065, av. de la Médecine, Québec (Québec), G1V 0A6, Canada

Pr Isabelle Demers
Université du Québec en Abitibi-Témiscamingue, Research Institute on Mines and the Environment (RIME), 445 boul. de l'Université, Rouyn-Noranda (QC), J9X 5E4, Canada

Adrien Dimech
Université du Québec en Abitibi-Témiscamingue, Research Institute on Mines and the Environment (RIME), 445 boul. de l'Université, Rouyn-Noranda (QC), J9X 5E4, Canada

Pr Marie Guittonny
Université du Québec en Abitibi-Témiscamingue, Research Institute on Mines and the Environment (RIME), 445 boul. de l'Université, Rouyn-Noranda (QC), J9X 5E4, Canada

Pier-Luc Labonté-Raymond
Polytechnique Montréal, Civil Geological and Mining engineering department, C.P. 6079, succ. Centre-Ville, Montréal (QC), H3C 3A7, Canada

Pr Abdelkabir Maqsoud
Université du Québec en Abitibi-Témiscamingue, Research Institute on Mines and the Environment (RIME), 675 1re avenue, Val-d'Or (QC), J9P 1Y3, Canada

Pr Mamert Mbonimpa
Université du Québec en Abitibi-Témiscamingue, Research Institute on Mines and the Environment (RIME), 445 boul. de l'Université, Rouyn-Noranda (QC), J9X 5E4, Canada

Pr Carmen Mihaela Neculita
Université du Québec en Abitibi-Témiscamingue, Research Institute on Mines and the Environment (RIME), 445 boul. de l'Université, Rouyn-Noranda (QC), J9X 5E4, Canada

Pr Thomas Pabst
Polytechnique Montréal, Civil Geological and Mining engineering department, Bureau A-360 C.P. 6079, succ. Centre-Ville, Montréal (QC), H3C 3A7, Canada

Pr Benoît Plante
Université du Québec en Abitibi-Témiscamingue, Research Institute on Mines and the Environment (RIME), 445 boul. de l'Université, Rouyn-Noranda (QC), J9X 5E4, Canada

Gary Schudel
Université du Québec en Abitibi-Témiscamingue, Research Institute on Mines and the Environment (RIME), 445 boul. de l'Université, Rouyn-Noranda (QC), J9X 5E4, Canada

G. Ward Wilson
University of Alberta, Faculty of Engineering - Civil and Environmental Engineering Department, 7-203 Donadeo Innovation Centre for Engineering, 9211 - 116 Street NW, Edmonton (Alberta), T6G 1H9, Canada

Gérald J. Zagury
Polytechnique Montréal, Civil Geological and Mining engineering department, Bureau A-355 C.P. 6079, succ. Centre-Ville, Montréal (QC), H3C 3A7, Canada

Introduction

Marie Guittonny and Bruno Bussière

CONTEXT

As the mining industry continues to grow, so too do its environmental impacts and the need for efficient reclamation methods.

The mining industry produces metals that are essential for global economic development. To do so, mining operations exploit ore deposits. For these ore bodies, both those exploited by open pits and underground mines, the proportion of the economic value compared to the whole excavated rock is very low, which implies that large quantities of solid wastes are produced annually, particularly waste rocks and mine tailings (Lottermoser 2007). Waste rocks are economically nonviable material excavated to reach ore-bearing rock. They are usually stored in waste rock piles at the surface. Tailings are ground rock particles resulting from ore processing. Tailings are usually pumped or transported into tailings storage facilities (Bussière 2007). Both waste types must be stored in a manner that assures their long-term physical and chemical stability. Many environmental challenges are associated with waste rock and tailings storage areas, the main one being the generation of acid mine drainage (AMD). AMD can occur if meteoric water and atmospheric oxygen come into contact with sulfide minerals that can be contained in both types of mine wastes. When the problem is poorly controlled, it can cause significant impacts on ecosystems located near mining operations (e.g., Wolkersdorfer and Bowell 2004; Moncur et al. 2005; Bussière 2009; Salvarredy-Aranguren et al. 2008).

Despite the potential negative impacts of mining operations, metal needs are continuously increasing in our society, and recycling does not provide enough raw materials to satisfy those needs. Moreover, with the development of large-scale mining equipment, the global trend is more toward the exploitation of large and low-grade ore deposits via open-pit mines that generate larger amounts of wastes. Thus, to fulfill the needs of both developed and developing countries, more and more metal mines will be opened, more and more mine wastes will be generated, and the potential to generate more and more AMD will also grow.

In addition to managing technical challenges, new mines will have to obtain the social acceptance of communities. Because the management of AMD problems associated with hard rock mines started only in the 1970s, a lot of old mine sites generate environmental contamination and degrade the quality of life in adjacent communities. Furthermore, the costs of reclamation for abandoned mine sites are transferred to taxpayers, which increases communities' circumspection of any proposals for new mine projects. To accept non-renewable resource extraction on their land, communities have increasingly elevated reclamation requirements. Consequently, stakeholders attach a growing importance to social and environmental responsibilities of mine companies when selecting the projects they fund. In particular, mine companies must demonstrate to financiers and communities at the design stage that negative environmental impacts such as AMD will be minimized during and after mining operation and that there will be efficient reclamation techniques at the post-closure stage.

Within this context, this book provides fundamental background about the main bio-chemico-geotechnical mechanisms underlying mine site reclamation and describes practical methods to reclaim hard rock mine wastes facilities and control their AMD contamination potential. The book focuses on mine site reclamation from an engineering point of view. The authors identify new reclamation issues and propose well-tested as well as innovative approaches to addressing them.

The research on AMD and hard rock mine site reclamation is 50 years old, and the accumulated knowledge is ripe to be gathered and integrated.

AMD is not a recent problem in the mining industry. Georgius Agricola (1556 [1912]), in his famous book *De re metallica*, mentions water contamination from ore extractions: "[W]hen the ores are washed, the water which has been used poisons the brooks and streams, and either destroys the fish or drives them away".

In many countries that produce mineral products, the problem of AMD was (and still is) mainly linked to hard rock mines. Initially, this environmental problem was not considered as important in hard rock mines as it was in U.S. coal mines, which may explain the relatively small volume of research on the subject before the 1970s. In the early 1970s, a few publications in Canada rang the alarm by identifying serious environmental problems related to AMD (e.g., Hawley and Shikaze 1971; Rivett and Oko 1971; Hawley 1972). At the time, reclamation of tailings impoundments (as for coal mine land; e.g., Bramble and Ashley 1955) was done mainly by stabilizing the surface with vegetation combined to lime amendments (e.g., Leroy 1973; Ludeke 1973). However, research on acid-generating tailings impoundments conducted in Canada and elsewhere showed that revegetation success and the neutralization effects of the amendment were temporary and that AMD was only slightly reduced, and only for a short period of time (e.g., Moffet et al. 1977). Since then, most engineering research has been devoted to finding solutions to this problem with an emphasis on controlling AMD formation at the source. Nonetheless, the science of AMD control for the reclamation of hard rock mine sites is young, and the information available is disseminated in journal articles, conference papers and reports, lacking an integrative effort.

This book aims to gather the available information in one contemporary, interdisciplinary, and integrated document. It allows geoenvironmental engineers and professionals to get an overview of reclamation methods to control AMD and to update their knowledge to design efficient mine site reclamation methods.

CONTENT AND AUTHORS

A book on mine site reclamation that takes an interdisciplinary approach is highly needed by the scientific and mining community.

Since mine site reclamation is a young science, the training needs for professionals and graduate students working in this field are huge. **Thus, this book targets students in graduate programs focused on mines and the environment as well as professionals who are already working in departments related to mine site reclamation.** It is written for an audience who is already familiar with the subject of mine site reclamation.

The book focuses on the reclamation of waste storage facilities from hard rock mines operations. These facilities constitute the main source of AMD pollution during and after hard rock mine operations. Reclamation aims to return a mine site to a satisfactory state, which means that the site should not threaten human health or security, should not generate in the long term any contaminant that could significantly affect the surrounding environment, and should be aesthetically acceptable to communities. Moreover, long-term control of the contamination must be done without continuous maintenance, which excludes the collect and treat option to reclaim a mine site. The establishment of vegetation on the site is also favored to control erosion and restore the site's natural appearance. Finally, when possible, reclamation that allows the site a second life after the mine closure is preferred.

The book is composed of 14 chapters, divided into four main themes (see Figure 0.1): fundamentals, reclamation approaches, tools to design and validate, and performance improvement.

Fundamentals: The book begins with an explanation of the biogeochemical processes underlying the problem of mine water contamination from AMD, followed by a chapter that reviews the fundamental concepts linked to water, gas, and heat movement in geological materials used in covers. These chapters prepare the reader for the following chapters, which describe six main reclamation techniques for hard rock mines.

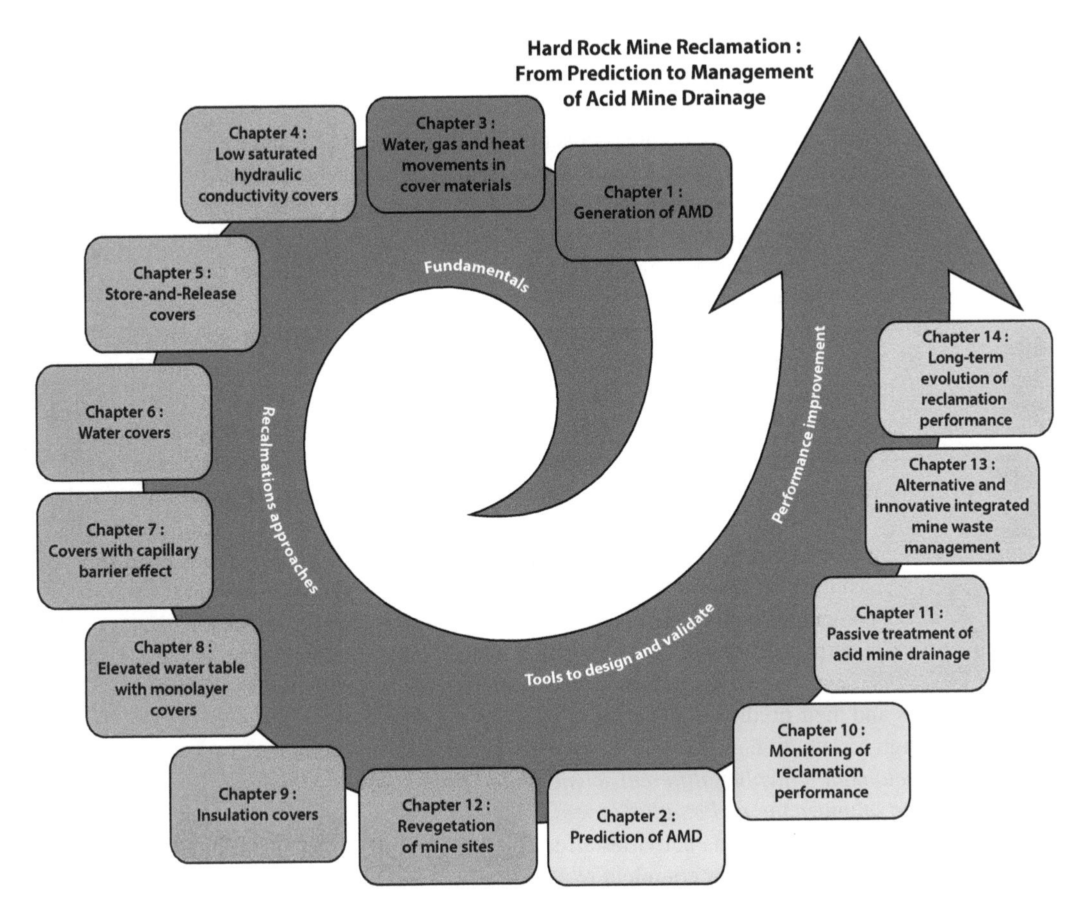

FIGURE 0.1 Content and structure of the book.

Reclamation approaches: Chapters 4 to 9 provide historical context for the development of a specific reclamation method, conceptual and technical descriptions of the method, and the factors that influence its design and performance. Special attention is paid to providing design approaches for each main reclamation method. Finally, the main advantages and limits of the different reclamation techniques are specified to promote comparative analysis, and research and development needs are identified. Thus, the reader will understand the remaining work to be done to reach an optimal application of the different reclamation methods.

The two first reclamation techniques presented aim to control water infiltration. Two types of impervious covers are described: low saturated hydraulic conductivity covers (LSHCC) (Chapter 4) and store-and-release (SR) covers (Chapter 5). Then, oxygen barrier techniques are described, including water covers (Chapter 6), covers with capillary barrier effects (CCBE) (Chapter 7), and elevated water table (EWT) with monolayer covers (Chapter 8). Chapter 9 discusses insulation covers, which control mine waste temperature in order to avoid contaminant generation; this type of cover, however, can only be used under Arctic climates. To finalize the reclamation after implementation of AMD control methods, Chapter 12 presents ecological knowledge and a general design approach related to the revegetation of mine sites, a mandatory step in most recent regulations on mine site reclamation.

Tools to design and validate: Designing a reclamation plan is a long process that starts with the environmental characterization of the mine wastes and the validation of preliminary scenarios in the laboratory. Chapter 2 presents the different tools available to characterize the AMD potential of mine wastes and how we can simulate reclamation scenarios at the laboratory or field scale. Chapter 10 presents the different parameters that must be monitored after reclamation implementation and the equipment available to measure these parameters. The objective is not to describe in detail all technologies but to identify the different approaches and their advantages and limits.

Performance improvement: Since interstitial waters in mine wastes sometimes have had the time to be contaminated before a reclamation method is applied, water needs to be temporarily treated. In a reclamation context, passive treatment techniques of water are more appropriate due to the reduced maintenance needed for their functioning and because they do not rely on human presence to operate the systems; these techniques are the focus of Chapter 11. Most of the time, reclamation is not integrated at the development stage of a mine project even if the concept of design for closure could significantly help mining companies reduce their long-term liabilities and reclamation costs. In Chapter 13, the main waste management modes are presented with an emphasis on how each of them can affect the final reclamation, from both a technical and economical point of view. This chapter connects the different stages of mine life cycle (exploration, operations, and ore treatment) to reclamation. Subjects such as co-disposal of waste rocks and tailings, tailings desulfurization, waste segregation and reduction at the source, and progressive reclamation are discussed. Finally, the book dedicates an entire chapter to the long-term performance of reclamation methods by considering that reclaimed sites are integrated into a changing environment. In particular, exposure to organisms' colonization from the surrounding environment, as well as changes in the properties of materials, are tackled because they can reduce the long-term performance of reclamation methods and eventually lead to the contamination of the environment. Similarly, the need to conceive covers by using climate data that integrate climate changes and their prediction uncertainty is discussed. These considerations will improve our capacity to design reclamation methods that ensure an acceptable long-term performance.

Mine site reclamation involves different sciences, and professionals with different expertise must work together to optimize the reclamation process.

Environmental impacts of mine activities affect different land components, such as rocks, soils, water, air, and organisms. Thus, a complete reclamation plan requires the knowledge and integration of several disciplines, including hydrology, geology, chemistry, biology, pedology, geotechnics, engineering, and ecology, among others. This interdisciplinarity represents an important challenge because each discipline has its own vocabulary, studied objects, problem-solving approaches, and spatiotemporal frame. **To address this challenge, the book involved ten main authors who agreed to work collectively and interdisciplinarily.** The editors have paid particular attention to the structure of each chapter and to the terminology used to guarantee the consistency of the content.

Most authors and the two editors belong to the Research Institute on Mines and the Environment (RIME) UQAT-Polytechnique. RIME UQAT-Polytechnique was created in 2013, following over ten years of applied research collaboration, by the Université du Québec en Abitibi-Témiscamingue (UQAT) and Polytechnique Montréal. Focused on environment and mine waste management, RIME aims to develop environmental solutions for the entire mine lifecycle. During our years spent working together, the mining industry, regulations authorities, and the research team members have shared disciplinary skills to create a wholistic vision of reclamation projects and their associated challenges. Each reclamation project integrates contamination prediction, waste management, contamination mitigation and treatment, to end with storage facilities' revegetation. Over the 20 last years of collaboration, the RIME team has had to apply and combine developed techniques on sites managed by industrial and government partners; these real case studies, mainly Canadian, are used throughout the book to concretely illustrate its interdisciplinary approach.

FINAL REMARKS

The book prioritizes the chemical stability of mine wastes over their physical stability, even if both must be evaluated at the design stage of any reclamation plan. We consider physical stability of mine waste storage facilities a subject of its own and beyond the scope of the book. The subject has also been covered by others in relatively recent books (e.g., Blight, 2010).

The book presents six mine site reclamation methods, with complementary chapters (Chapters 11 to 14) essential to perform a complete and successful work. The editors consider these methods the most relevant in the context of hard rock mines. However, other methods have been suggested in the literature to reclaim AMD mine sites, such as organic covers (Ribeta et al. 1995), use of bactericide (Kleinmann 1998), passivation or coating of sulfide minerals (Zhang and Evangelou 1998), and amendment with neutralizing agents (Hakkou et al. 2009). However, they are not described in this book mainly because of one or more of these reasons: the low material availability, the incapacity to maintain AMD control with time, the relatively low efficiency to control AMD, and the costs associated with implementation.

The book focuses on the control of AMD generated by mine wastes from hard rock mines, especially on base and precious metal mines. More recently, some of these mines faced a new type of contamination called contaminated neutral drainage (sometimes referred to as metal leaching; Plante et al. 2014). This aspect is not directly treated in the book. The authors are also aware that the same and other challenges are related to mine site reclamation from other types of hard rock mine ores. For example, uranium mines can generate radioactive contamination. The editors decided not to focus on radioactive contaminants. However, if uranium mine wastes are AMD generating, the methods presented in the book could be applied (Peacey et al. 2002). Also, the development of electronics and energy storage technologies pushes for the exploitation of rare earth elements and lithium mines. The biogeochemistry and environmental effects of mine drainage from wastes coming from these new types of mines are still poorly documented and may need specific reclamation methods (Edahbi et al. 2019). The AMD reclamation methods described in the book may constitute a useful knowledge basis for the adaptation and development of specific reclamation methods for these strategic metals.

Despite its interdisciplinary approach, the book doesn't include the contribution of social sciences specialists to deal with social aspects of reclamation plans. Given the need to obtain a social license to operate for new mine projects (Moffat and Zhang 2014), however, we think that special attention should be given to integrate the social aspects by co-designing the reclamation plans with the communities impacted by mining activities. Communities should also be involved in the long-term monitoring, and efforts should be paid to maximizing the economic impacts of reclamation for communities.

REFERENCES

Agricola, G. 1556. *De re metallica.* Translated by H.C. Hoover and L.H. Hoover in 1912. Reprinted 1950. New York: Dover Publications.

Blight, G. 2010. *Geotechnical engineering for mine waste storage facilities.* Boca Raton, FL: CRC Press/Balkema.

Bramble, W. C., and R. H. Ashley. 1955. Natural revegetation of spoil banks in central Pennsylvania. *Ecology* 36:417–423.

Bussière, B. 2007. Colloquium 2004: Hydro-geotechnical properties of hard rock tailings from metal mines and emerging geo-environmental disposal approaches. *Canadian Geotechnical Journal* 44, no. 9:1019–1052. https://doi.org/10.1139/T07-040.

Bussière, B. 2009. Acid mine drainage from abandoned mine sites: problematic and reclamation approaches. In *Proceedings of International Symposium on Geoenvironmental Engineering, ISGE2009, Advances in Environmental Geotechnics*, 8–10 September 2009, Hangzhou, China, ed. Y. Chen, X. Tang, and L. Zhan, 111–125. Berlin: Springer.

Edahbi, M., B. Plante, and M. Benzaazoua. 2019. Environmental challenges and identification of the knowledge gaps associated with REE mine wastes management. *Journal of Cleaner Production* 212:1232–1241.

Hakkou, R., M. Benzaazoua, and B. Bussière. 2009. Laboratory evaluation of the use of alkaline phosphate wastes for the control of acidic mine drainage. *Mine Water and the Environment* 28:206–218.

Hawley, J. R. 1972. *The problem of acid mine drainage in the province of Ontario.* Ministry of the Environment, Special Projects Section (Mining), Industrial Wastes Branch.

Hawley, J. R., and K. H. Shikaze. 1971. The problem of acid mine drainage in Ontario. *Canadian Mining Journal* 92, no. 6:82–93.

Kleinmann, R. L. P. 1998. Bactericidal control of acidic drainage. In *Coal mine drainage prediction and pollution prevention in Pennsylvania*, ed. K. C. Brady, M. W. Smith, and J. Schueck, 15-1–15-6. Harrisburg, PA: PA DEP.

Leroy, J.-C. 1973. How to establish and maintain growth on tailings in Canada – Cold winters and short growing seasons. In *Proceeding of Tailings Disposal Today*, ed. C. L. Aplin and G.O. Argall, 411–449. San Francisco: J.R. Miller Freeman Publications Inc.

Lottermoser, B. 2007. *Mine wastes: Characterization, treatment and environmental impacts*. 2nd ed. Berlin: Springer.

Ludeke, K. L. 1973. Vegetative stabilization of copper mine tailings disposal berms of Pima Mining Company. In *Proceeding of Tailings Disposal Today*, ed. C. L. Aplin and G. O. Argall, 377–410. San Francisco: J.R. Miller Freeman Publications Inc.

Moffat, K., and A. Zhang. 2014. The paths to social licence to operate: An integrative model explaining community acceptance of mining. *Resources Policy* 39:61–70.

Moffet, D., G. Zahary, M. C. Campbell, and J. C. Ingles. 1977. *CANMET's environmental and process research on uranium*. CANMET Report 77–53.

Moncur, M. C., C. J. Ptacek, D. W. Blowes, and J. L. Jambor. 2005. Release, transport and attenuation of metals from an old tailings impoundment. *Applied Geochemistry* 20:639–659.

Peacey, V., E. K. Yanful, and R. Payne. 2002. Field study of geochemistry and solute fluxes in flooded uranium mine tailings. *Canadian Geotechnical Journal* 39, no. 2:357–376.

Plante, B., B. Bussière, and M. Benzaazoua. 2014. Lab to field scale effects on contaminated neutral drainage prediction from the Tio mine waste rocks. *Journal of Geochemical Exploration* 137:37–47.

Ribeta, I., C. J. Ptacek, D. W. Blowes, and J. F. L. Jambor. 1995. The potential for metal release by reductive dissolution of weathered mine tailings. *Journal of Contaminant Hydrology* 17, no. 3:239–273.

Rivett, L. S., and U. M. Oko. 1971. Tailings disposal, generation of acidity from pyrrhotite and limestone neutralization of wastewater at Falconbridge's Onaping mines. *CIM Bulletin* 74:186–191.

Salvarredy-Aranguren, M. M., A. Probst, M. Roulet, and M.-P. Isaure. 2008. Contamination of surface waters by mining wastes in the Milluni Valley (Cordillera Real, Bolivia): Mineralogical and hydrological influences. *Applied Geochemistry* 23:1299–1324.

Wolkersdorfer, C., and R. Bowell. 2004. Contemporary reviews of mine water studies in Europe, part 1. *Mine Water and the Environment* 23:162–182.

Zhang, Y. L., and V. P. Evangelou. 1998. Formation of ferric hydroxide-silica coatings on pyrite and its oxidation behavior. *Soil Science* 163:53–62.

1 Generation of Acid Mine Drainage

Benoît Plante, Gary Schudel, and Mostafa Benzaazoua

The aim of this chapter is to provide an overview of basic geochemical knowledge regarding the generation of acid mine drainage (AMD). Chapter 2 then describes the most common procedures used to predict the onset of AMD in order to comprehend the objectives and mechanisms of the different reclamation methods discussed in the subsequent chapters (Chapters 4 to 9). This chapter focuses on the fundamentals of the acid-generating and acid-consuming geochemical reactions, the most significant factors that influence these reactions, and the associated current research and development need. To concentrate on the most common reactions and most significant factors affecting reaction rates, we have omitted some details on related subjects. Readers are directed to the most relevant studies should they want to supplement the information provided here.

1.1 HISTORICAL PERSPECTIVE AND OVERVIEW OF THE PROBLEM

Contamination of the environment surrounding mining activities has been long recognized. In his well-known 1556 book, *De re metallica*, German mineralogist and metallurgist Georgius Agricola described what is perhaps one of the first accounts of environmental preoccupation related to mining operations. Agricola (1556/1912) states that the people recognized that mining and smelting activities not only contaminated streams' receiving process waters but also had a significant impact on nearby forests and lands, which, in turn, affected the associated fauna and flora:

> But besides this, the strongest argument of the detractors is that the fields are devastated by mining operations, for which reason formerly Italians were warned by law that no one should dig the earth for metals and so injure their very fertile fields, their vineyards, and their olive groves. Also they argue that the woods and groves are cut down, for there is need of an endless amount of wood for timbers, machines, and the smelting of metals. And when the woods and groves are felled, then are exterminated the beasts and birds, very many of which furnish a pleasant and agreeable food for man. Further, when the ores are washed, the water which has been used poisons the brooks and streams, and either destroys the fish or drives them away. Therefore the inhabitants of these regions, on account of the devastation of their fields, woods, groves, brooks and rivers, find great difficulty in procuring the necessaries of life, and by reason of the destruction of the timber they are forced to greater expense in erecting buildings. Thus it is said, it is clear to all that there is greater detriment from mining than the value of the metals which the mining produces.

Despite Agricola's assessment that mine waters seemed to be poisonous to receiving streams, the first formal research on AMD began in the 1920s with studies of coal mine wastes from the Appalachian region of the United States (for an overview of early research on AMD, see Paine 1987). Fundamental studies on AMD, particularly pyrite oxidation, have also been performed from the 1920s onward (Colmer and Hinkle 1947), and the involvement of microorganisms in AMD generation, especially in ferrous to ferric iron oxidation, was assessed beginning in the late 1940s (Colmer and Hinkle 1947; Temple and Colmer 1951; Singer and Stumm 1970). The first AMD prediction studies, which still focused on US coal mine wastes and were mainly qualitative, occurred in the 1940s and 1950s (Braley 1949, 1960). The first quantitative AMD prediction studies started in the 1970s at West Virginia University (1971). These studies lead to the famous United States Environmental Protection Agency report by Sobek et al. (1978), which described the procedures for determining the neutralization and acid generation potentials of mine wastes that would later become the basis for modern acid–base accounting tests (see Chapter 2). The first modern research on the reclamation of AMD-generating mine waste disposal areas was initiated in the 1980s. The International Conference on Acid Rock Drainage series was also established in the late 1980s and early 1990s. The conference, which is still held every three years, is attended by specialists from all fields related to AMD mine site reclamation.

The wide variety of geological settings in which AMD may arise leads to a similarly wide array of AMD compositions. The generally low-sulfide sulfur contents of coal mine wastes, which are also often poor in neutralizing minerals, leads to AMD containing relatively low concentrations of metals and sulfate. On the other hand, hard rock mines of base and precious metals operate in a wide array of mineralogical contexts, from low-sulfur/low-neutralizing minerals to high-sulfur/high-neutralizing mineral. This can lead to drainage waters with metal and sulfate concentrations ranging from several parts per million to hundreds or thousands of grams per liter. Table 1.1 compares typical AMD water quality data from base and precious metal mines.

Although there are no official, widely recognized definitions for the different types of mine drainages, the classification system suggested by Nordstrom et al. (2015), which is based on the *Global Acid Rock Drainage Guide* (INAP 2009), is used in this book (Figure 1.1). Under this system, AMD refers to any mine water sample with pH values below 6, where sulfate is the dominant ion. Neutral mine drainage, which is also called contaminated neutral drainage (CND) or more simply metal leaching, typically ranges in pH from 6.0 to 9.0. Metal leaching at near-neutral conditions can occur in a variety of geologic environments where the acid generated from sulfide oxidation is neutralized by sufficient acid buffering within the source materials but still allows for some metals, such as Ni and Zn, to leach into the drainage waters at problematic concentrations. CND can also develop in the lag time before the onset of AMD, which is essentially the period in which the neutralizing minerals present in a material are still able to buffer the pH of the drainage waters to near-neutral values. Finally, alkaline mine drainage is defined as having a pH ranging from 9.0 to 12.0. At present, there are very few examples of alkaline mine drainage in the literature, and the conditions and processes leading to its formation are not yet well understood.

TABLE 1.1

Examples of Acid Mine Drainage Occurrences from Metal Mines

	Lowest pH	SO_4 (mg/L)	Fe (mg/L)	Cu (mg/L)	Zn (mg/L)
Richmond mine, Iron Mountain, California, USA[1]	–3.6	Up to 760,000	Up to 141,000	Up to 4,760	Up to 23,500
Sherridon, Manitoba, Canada[2]	0.67	280,000	129,000	1,600	55,000
Manitou mine, Quebec, Canada[3]	2	2,500	10,000	235	350
Genna Luas, Sardinia, Italy[4]	0.6	203,000	77,000	220	10,800

Source: Data taken from
[1] Nordstrom et al. (2000);
[2] Moncur et al. (2003, 2005);
[3] Ethier (2018);
[4] Frau (2000).

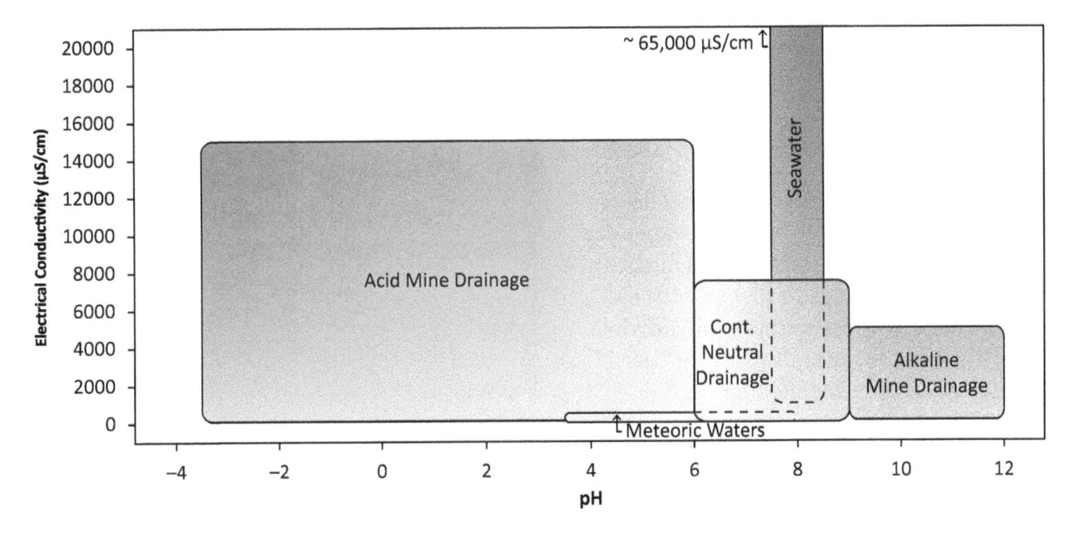

FIGURE 1.1 pH vs. electrical conductivity plot characterizing the different types of mine drainage waters, as well as meteoric waters and seawater. The pH values for AMD, CND, and alkaline mine drainage were based on Nordstrom et al. (2015) and INAP (2009). Data were gathered from various sources: AMD (Ball and Nordstrom 1989; Nieto et al. 2007; Søndergaard et al. 2008; Pope et al. 2010; Cruz et al. 2013), CND (Ball and Nordstrom 1989; Church et al. 2007; Conesa et al. 2006; Heikkinen et al. 2009), alkaline mine drainage (Nordstrom et al. 1989; Azzie 2002; Dahrazma and Kharghani 2012), meteoric waters (Zunckel et al. 2003; Topçu et al. 2002; Santos et al. 2011; Zhang et al. 2012), and seawater (Chester and Jickells 2012; Tyler et al. 2017).

1.2 MINERALOGICAL ASPECTS OF AMD GENERATION

The generation of acidic drainage waters relies on a combination of acid generation from sulfide oxidation and insufficient neutralization provided by other minerals. The resulting water quality of the mine drainage waters is influenced by a combination of geochemical phenomena, as represented in Figure 1.2. This section describes the geochemical reactions of the main minerals involved in AMD, as well as secondary mineral precipitation and sorption. This chapter covers only the basics

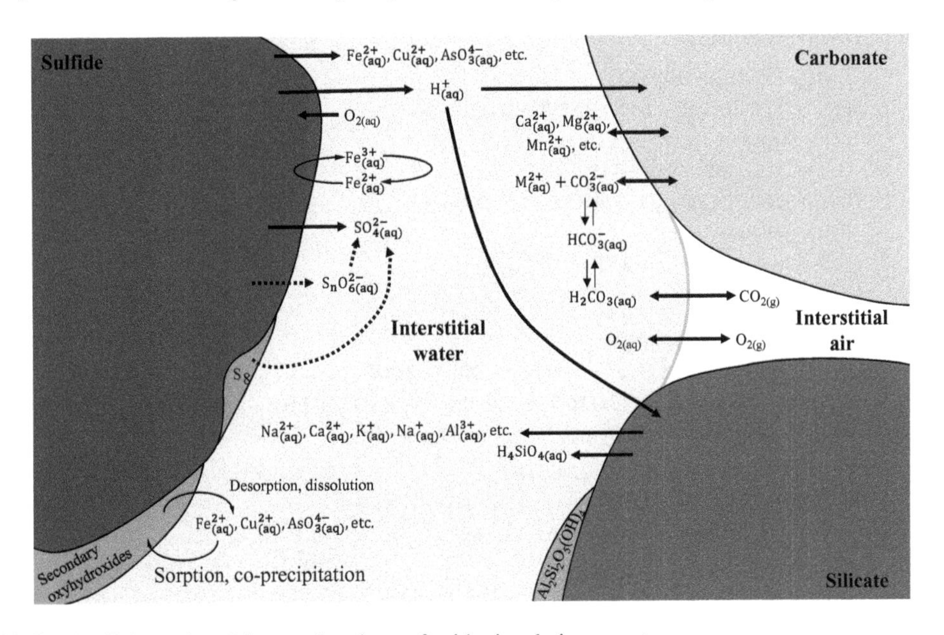

FIGURE 1.2 Schematics of the geochemistry of acid mine drainage waters.

of geochemical reactions behind AMD; the interested reader can find more exhaustive coverage in studies by Nordstrom et al. (2015), Lindsay et al. (2015), and Blowes et al. (2014).

1.2.1 Sulfide Oxidation: The Main Source of Acid Generation

The main source of acid in mine wastes is the oxidation of sulfide minerals. The amount of acid (H^+) generated by sulfide oxidation varies depending on the mechanisms involved (Table 1.2) and is controlled by the kinetics of the reactions. Pyrite oxidation by dissolved oxygen generates acid and ferrous iron (Fe^{2+}; Equation 1.1), which may undergo oxidation (Equation 1.2) and hydrolysis under certain conditions to generate oxyhydroxides such as ferrihydrite (Equation 1.3). Adding equations 1.1 to 1.3 gives the overall oxidation reaction of pyrite by dissolved oxygen (Equation 1.4), where 4 moles of H^+ are generated for each mole of pyrite oxidized. Pyrite can also be oxidized by Fe^{3+} (Equation 1.5) instead of oxygen if the geochemical conditions are favorable to Fe^{3+} solubility (i.e., unfavorable to its hydrolysis), which is typically in low pH waters (approximately 3.5–4.5) and oxidized environments (typically Eh >300 mV). Pyrite oxidation by Fe^{3+} generates Fe^{2+} and 16 moles of H^+ per mole of pyrite. Therefore, the acid generated by pyrite oxidation in acidic conditions is four times greater than that generated at higher pH values that promote iron oxyhydroxide precipitation (approximately at pH values >4.5, depending on the redox conditions and dissolved Fe concentrations). In addition, the 15 moles of ferrous iron produced can, in turn, undergo oxidation following Equation 1.2 (favored by the involvement of iron-oxidizing bacteria) and oxidize other pyrite molecules, therefore acting as a catalyst for pyrite oxidation.

Pyrrhotite can be oxidized by O_2 (Equations 1.6 and 1.7) or Fe^{3+} (Equation 1.8), just like pyrite. As can be deduced from Equation 1.6, the iron deficiency in the pyrrhotite crystal lattice (i.e., the value of "x") makes the stoichiometry of its reaction slightly different depending on the extent of the deficiency, producing between ¼ mole of H^+ (x=0.125, $Fe_{0.875}S$ or Fe_7S_8) to none (x=0, FeS). However, the Fe^{2+} released upon pyrrhotite oxidation can also undergo oxidation and hydrolysis (Equation 1.2 and 1.3) and contribute to the generation of further acid; combining these reactions

TABLE 1.2
Pyrite and Pyrrhotite Oxidation Reactions

Pyrite Oxidation

Pyrite oxidation by oxygen

$FeS_2 + 7/2\ O_2 + H_2O \rightarrow Fe^{2+} + 2\ SO_4^{2-} + 2\ H^+$ (1.1)

$Fe^{2+} + 1/4\ O_2 + H^+ \rightarrow Fe^{3+} + 1/2\ H_2O$ (1.2)

$Fe^{3+} + 3\ H_2O \rightarrow Fe(OH)_3 + 3\ H^+$ (1.3)

$FeS_2 + 15/4\ O_2 + 7/2\ H_2O \rightarrow Fe(OH)_3 + 2\ SO_4^{2-} + 4\ H^+$ (1.4)

Pyrite oxidation by Fe^{3+}

$FeS_2 + 14\ Fe^{3+} + 8\ H_2O \rightarrow 15\ Fe^{2+} + 2\ SO_4^{2-} + 16\ H^+$ (1.5)

Pyrrhotite

Pyrrhotite oxidation (complete)

$Fe_{(1-x)}S + (2-1/2x)\ O_2 + x\ H_2O \rightarrow (1-x)\ Fe^{2+} + SO_4^{2-} + 2x\ H^+$ (1.6)

$Fe_{(1-x)}S + 3/4\ (3-x)\ O_2 + 1/2\ (5-3x)\ H_2O \rightarrow (1-x)\ Fe(OH)_3 + SO_4^{2-} + 2\ H^+$ (1.7)

$Fe_{(1-x)}S + (8-2x)\ Fe^{3+} + 4\ H_2O \rightarrow (9-3x)\ Fe^{2+} + SO_4^{2-} + 8\ H^+$ (1.8)

Pyrrhotite oxidation (partial—polysulfide)

$Fe_{(1-x)}S + 1/2(1-x)\ O_2 + 2(1-x)\ H^+ \rightarrow (1-x)\ Fe^{2+} + S^0 + (1-x)\ H_2O$ (1.9)

$Fe_{(1-x)}S + (2-2x)\ Fe^{3+} + 4\ H_2O \rightarrow (3-3x)\ Fe^{2+} + S^0$ (1.10)

Pyrrhotite dissolution (non-oxidative)

$FeS + 2\ H^+ \rightarrow Fe^{2+} + H_2S$ (1.11)

gives the overall oxidation reaction of pyrrhotite by dissolved oxygen (Equation 1.7), in which 2 moles of H^+ are generated per mole of pyrrhotite, regardless of its iron stoichiometry. Like pyrite, pyrrhotite generates four times more acid when Fe^{3+} acts as the oxidant (Equation 1.8).

In some circumstances, sulfur oxidation may not proceed to the sulfate state and may generate more reduced states such as elemental sulfur or polysulfides with O_2 (Equation 1.9) or Fe^{3+} (Equation 1.10) as the oxidant. The non-oxidative dissolution of pyrrhotite (Equation 1.11) occurs in acidic conditions and is characterized by the release of Fe^{2+} in solution, followed by sulfur release in the form of HS^- (Thomas et al. 1998, 2001; Janzen et al. 2000). More details on the oxidation of pyrrhotite can be found in studies by Belzile et al. (2004) and Blowes et al. (2014), and references therein. The oxidation of other metal sulfides, such as arsenopyrite (FeAsS), chalcopyrite ($CuFeS_2$), and sphalerite (ZnS), is also described by numerous authors (Rimstidt et al. 1994; Janzen et al. 2000; Chopard et al. 2017; see Blowes et al. 2014 for a compilation of other relevant references) but will not be covered explicitly here. Nonetheless, the readers should be aware that other metal sulfides can also be oxidized by O_2 and Fe^{3+} and keep in mind that some sulfides generate less acid than pyrite and pyrrhotite or no acid at all (such as sphalerite and galena), while others may generate more acid, such as for arsenopyrite (e.g., Rimstidt et al. 1994) and gersdorffite (e.g., Chopard et al. 2017).

The mechanistic details of sulfide oxidation are still the subject of intense debate among scientists after decades of studies (e.g., Lowson 1982; Schippers and Sand 1999; Belzile et al. 2004; see Blowes et al. 2014 for additional key references). However, the crucial aspect to remember from the stoichiometry of sulfide oxidation is that both the amount of acid and the rate at which it is generated increase when conditions become acidic and oxidizing enough, which explains why AMD is a chain reaction that is difficult to stop once it has started.

It is well known that microbial processes significantly influence the cycling of Fe and S in AMD-generating mine tailings. It is now generally accepted that microbes participate mainly through an indirect mechanism in which they facilitate and significantly accelerate the oxidation of Fe^{2+} to Fe^{3+} (Singer and Stumm 1970; Rohwerder et al. 2003). Therefore, given that the rate of abiotic Fe^{2+} oxidation is up to five orders of magnitude lower than that of biotic Fe^{2+} oxidation below pH 3 (Nordstrom 2003; Nordstrom and Southam 1997), iron-oxidizing acidophiles ultimately control the rate of sulfide oxidation under acidic conditions (Rohwerder et al. 2003; Lindsay et al. 2015).

The contact between two sulfide minerals in a conductive media (such as AMD) leads to a voltaic cell promoting galvanic interactions, where one mineral is preferentially oxidized and transfers electrons to the other, which is therefore protected from oxidation (e.g., Holmes and Crundwell 1995; Kwong et al. 2003; Cruz et al. 2005; Chopard et al. 2017; Qian et al. 2018). Therefore, galvanic interactions do not require oxygen or Fe^{3+} and can proceed despite the use of oxygen barriers in reclamation techniques (see Chapters 6, 7, and 8).

Pyrite oxidation rates have been studied extensively, and many of these rates were compiled by Holmes and Crundwell (2000) and reported in the study by Blowes et al. (2014), where the rates are expressed as a function of dissolved O_2, pH, as well as concentrations of Fe^{2+} and Fe^{3+}. The pyrrhotite oxidation rate has not been studied as extensively as for pyrite, but studies by Nicholson and Scharer (1994) and Janzen et al. (2000) provide both oxidative and non-oxidative rate data for several pyrrhotite samples that highlight that pyrrhotite oxidation occurs at rates between 20 and 100 times faster than pyrite (Blowes et al. 2014). However, most of these rate data were obtained in closed systems under controlled environments and are difficult to upgrade to field conditions. Chopard et al. (2017) compared the oxidation rates of sulfides in laboratory kinetic tests, which showed that pyrrhotite oxidizes approximately eight times faster than pyrite. Thus, although pyrrhotite oxidation generates half as much acid as pyrite (see Equations 1.4 and 1.5 for pyrite and Equations 1.7 and 1.8 for pyrrhotite), it does so significantly faster than pyrite. Therefore, the presence of significant pyrrhotite in mine wastes requires fast, effective neutralizers in order to buffer this acid and prevent or delay AMD generation.

The next section describes acid neutralization processes in mine wastes.

1.2.2 Acid Neutralization

While acid generation in mine wastes results primarily from the oxidation of iron-sulfide minerals, acid neutralization occurs through reactions between H^+ and a wide variety of rock-forming minerals, essentially carbonates, soluble silicates, and oxyhydroxides (Table 1.3; Blowes et al. 2014; Dubrovsky et al. 1985). Principally, acid neutralization involves the dissolution of carbonates, silicates, and oxyhydroxides, but acid may also be neutralized through exchanges with cations on the surface of clay minerals (Lottermoser 2010). The most effective acid-neutralizing minerals are carbonates (e.g., Sherlock et al. 1995; Blowes et al. 2014) because of their relatively high dissolution rate in comparison to silicates. Calcite (Equations 1.12 and 1.13) and dolomite (Equations 1.14 and 1.15), two carbonate minerals, consume different amounts of acid depending on the pH to which they are submitted.

Siderite is also able to consume acid (Equation 1.16). However, because of Fe^{2+} oxidation (Equation 1.2) and subsequent Fe^{3+} hydrolysis (Equation 1.3), which release as many hydrogen ions as consumed by siderite dissolution (Equation 1.16), the whole process is not globally acid neutralizing (Equation 1.17). In addition to Fe, metals that have the potential to undergo hydrolysis within acidic to near-neutral pH values, such as Al and Mn, will proportionally decrease the neutralization potential of their host mineral. The hydrolysable metal content of neutralizing minerals needs to be taken into account if their host minerals constitute a significant proportion of the overall neutralization potential of a given material (e.g., Paktunc 1999b; Plante et al. 2012; Bouzahzah et al. 2014; Elghali et al. 2019b).

Various silicate minerals, such as chlorite, biotite, olivine, pyroxene, and plagioclase feldspars (e.g., anorthite; Equation 1.18), are also able to neutralize acid. Numerous studies assess the contribution of silicate minerals to acid neutralization (e.g., Sherlock et al. 1995; Johnson et al. 2000; Jurjovec et al. 2002; Moncur et al. 2005; Jambor et al. 2002, 2007). Many silicates possess similarly elevated neutralization capacities relative to calcite and, as a group, silicates theoretically represent the greatest source of acid neutralization in most wastes. For example, considering only stoichiometry, the dissolution of 1 mole calcite is required to neutralize 1 mole of sulfuric acid (Equations 1.12 and 1.13), whereas only 0.5 mole of anorthite (Equation 1.18) is required to neutralize the same amount of acid.

However, as stated earlier, the ability of silicate minerals to buffer mine waters is significantly limited by their slow dissolution kinetics relative to sulfide oxidation rates (e.g., Jamieson et al. 2015; Blowes et al. 2014). In determining which minerals will realistically contribute to buffering the pH of acid drainages on relevant timescales, both the stoichiometry of neutralization reactions

TABLE 1.3

Reactions of Acid-neutralizing Minerals

Calcite

$$CaCO_3 + H^+ \rightarrow Ca^{2+} + HCO_3^- \quad (1.12)$$

$$CaCO_3 + 2\ H^+ \rightarrow Ca^{2+} + H_2CO_3\ (pH < 6.4) \quad (1.13)$$

Dolomite

$$CaMg(CO_3)_2 + 2\ H^+ \rightarrow Ca^{2+} + Mg^{2+} + 2\ HCO_3^- \quad (1.14)$$

$$CaMg(CO_3)_2 + 4\ H^+ \rightarrow Ca^{2+} + Mg^{2+} + 2\ H_2CO_3\ (pH < 6.4) \quad (1.15)$$

Siderite

$$FeCO_3 + 2\ H^+ \rightarrow Fe^{2+} + H_2CO_3 \quad (1.16)$$

$$FeCO_3 + 1/4\ O_2 + 5/2\ H_2O \rightarrow Fe(OH)_3 + H_2CO_3 \quad (1.17)$$

Anorthite

$$CaAl_2Si_2O_8 + 2\ H^+ + H_2O \rightarrow Ca^{2+} + Al_2Si_2O_5(OH)_4 \quad (1.18)$$

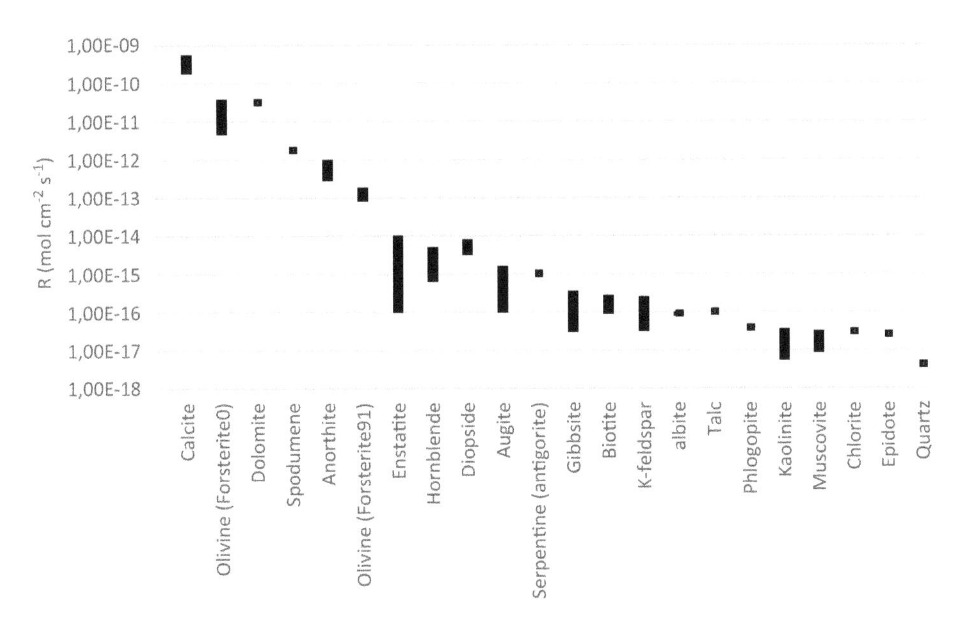

FIGURE 1.3 Dissolution rates at pH 5 (Adapted from Paktunc 1999a).

and their rates under given environmental conditions must be considered (Jamieson et al. 2015). Dissolution rates were compiled by Paktunc (1999a and references therein) and are shown in Figure 1.3. While calcite dissolution proceeds quite rapidly, rates of silicate dissolution are several orders of magnitude slower than typical rates for calcite (McKibben and Barnes 1986; Wollast 1990; Paktunc 1999a).

1.2.3 Secondary Minerals

Secondary minerals are formed by combining reaction products of primary minerals in mining wastes upon their metallurgical treatment and/or exposure to atmospheric conditions in waste rock piles and tailings storage facilities. In potentially AMD-generating materials, the geochemical reactions are often driven by sulfide oxidation and the subsequent response of acid-neutralizing minerals. Therefore, many anionic and cationic products are leached in mine drainage waters. They may form a wide range of secondary minerals in situ, as their constituting ions are made available by reaction of their parent primary minerals (Table 1.4). The most significant secondary minerals encountered in mine drainage waters are sulfates, oxides/hydroxides (often named oxyhydroxides), and carbonates (e.g., Alpers et al. 1994; Blowes et al. 2014). Their crystallization may have a significant effect on the composition of the drainage waters and eventually on the hydrogeological properties of the materials themselves. Indeed, since secondary precipitates may partially fill the voids between the grains, a phenomenon that leads to the formation of a hardpan typical of highly weathered AMD-generating tailings (e.g., Blowes et al. 1991; McGregor and Blowes 2002; Moncur et al. 2005, 2009; Graupner et al. 2007; DeSisto et al. 2011; Elghali et al. 2018), their type and occurrence need to be assessed in mine water studies.

The iron released from sulfide oxidation leads to the formation of a variety of iron oxyhydroxides, the most thermodynamically stable and most widely encountered being goethite (α-FeOOH; Equation 1.19), which is responsible for the ocher color typical of highly oxidized AMD-generating mine wastes (Blowes et al. 2014). Ferrihydrite (Equation 1.20), often reported as amorphous $Fe(OH)_3$, is a poorly crystalline secondary mineral that is also widely reported in oxidized AMD-generating mine wastes (Jambor and Dutrizac 1998; Blowes et al. 2014). Aluminum oxyhydroxides such as gibbsite (Equation 1.21) and boehmite (Equation 1.22) are also

TABLE 1.4
Examples of Secondary Mineral Formation

Oxides and Hydroxides		
Goethite:	$Fe^{3+} + 2\ H_2O \rightarrow FeOOH + 3\ H^+$	(1.19)
Ferrihydrite:	$Fe^{3+} + 3\ H_2O \rightarrow Fe(OH)_3 + 3\ H^+$	(1.20)
Gibbsite:	$Al^{3+} + 3\ H_2O \rightarrow Al(OH)_3 + 3\ H^+$	(1.21)
Boehmite:	$Al^{3+} + 2\ H_2O \rightarrow AlOOH + 3\ H^+$	(1.22)
Sulfates		
Jarosite:	$K^+ + 3\ Fe^{3+} + 2\ SO_4^{2-} + 6\ H_2O \rightarrow KFe_3(SO_4)_2(OH)_6 + 6\ H^+$	(1.23)
Schwertmannite:	$8\ Fe^{3+} + SO_4^{2-} + 14\ H_2O \rightarrow Fe_8O_8(OH)_6(SO_4) + 22\ H^+$	(1.24)
Gypsum:	$Ca^{2+} + SO_4^{2-} + 2\ H_2O \rightarrow CaSO_4.2H_2O$	(1.25)
Barite:	$Ba^{2+} + SO_4^{2-} \rightarrow BaSO_4$	(1.26)
Carbonates		
Siderite:	$Fe^{2+} + CO_3^{2-} \rightarrow FeCO_3$	(1.27)

often encountered in AMD-generating mine sites where the dissolution of aluminosilicate minerals is significant (e.g., Blowes et al. 2014; Lindsay et al. 2015; and references therein). Iron and aluminum oxyhydroxides are also known to sorb and co-precipitate a wide variety of metals, and it has often been demonstrated that they exert a significant control on the concentrations of metals in mine drainage in tailings (e.g., Al et al. 2000; Heikkinen and Räisänen, 2008, 2009; Hakkou et al. 2008; Blowes et al. 2014; Lindsay et al. 2015) as well as in waste rocks (Stockwell et al. 2006; Blackmore et al. 2018).

Most sulfate minerals are hydrated and quite soluble. They will only be encountered in dry periods as efflorescent salts and will disappear upon re-wetting of the material (e.g., Nordstrom and Alpers 1999; Nordstrom 2011; Blowes et al. 2014). Typical examples of soluble hydrated sulfates comprise iron sulfate $FeSO_4.nH_2O$ ($1 < n < 7$) and magnesium sulfates such as epsomite $MgSO_4.7H_2O$. Although jarosites are often composed of its K end-member ($KFe_3(SO_4)_2(OH)_6$; Equation 1.23), they will incorporate other cations as they are formed. More rarely, other end-member jarosites can also be encountered depending on the geochemical environment in which they are formed. Schwertmannite (Equation 1.24) may also be encountered in acid-generating tailings (Blowes et al. 2014) but is metastable and will transform to goethite (Schwertmann and Carlson 2005). However, some sulfate minerals are partly soluble, such as gypsum $CaSO_4{\cdot}2H_2O$ (Equation 1.25), or highly insoluble, such as barite $BaSO_4$ (Equation 1.26), and may exert a significant control over the aqueous concentrations of their constituents. Gypsum is by far the most common secondary sulfate encountered in acid-generating mine wastes (Blowes et al. 2014) because of the prevalence of sulfate (from sulfide oxidation) and calcium (from carbonate and silicate dissolution) ions in mine waters.

Secondary carbonates can form in waters having high alkalinity values. Secondary siderite (Equation 1.27) has been reported in the Kidd Creek tailings as coatings on ankerite-dolomite grains in association with iron oxyhydroxides (Al et al. 2000), as well as at Elliott Lake (Paktunc and Davé 2002), where secondary calcite was also detected. However, these secondary carbonates will be temporary in AMD-generating tailings since they will dissolve as the pH is lowered by continued acid generation.

Microorganisms are also known to have an effect on the stability of secondary iron oxyhydroxides by facilitating the reductive dissolution of iron (such as the acidophilic bacteria *Acidiphilium* spp. and *Ferrimicrobium acidiphilum*, and the mixotrophic acidophiles *Sulfobacillus* spp. and Am. ferrooxidans; Blowes et al. 2014; Bridge and Johnson 1998, 2000; Johnson et al. 2009) or sulfur (such as the autotrophic acidophiles *Acidithiobacillus* spp.; Blowes et al. 2014; Hallberg 2010;

Ohmura et al. 2002). Many specific examples of the microbiological influence on secondary mineral stability can be found in studies by Blowes et al. (2014), Lindsay et al. (2015), Nordstrom et al. (2015), and references therein.

1.3 FACTORS INFLUENCING THE DEVELOPMENT OF AMD CONDITIONS

Numerous factors influence the development of AMD conditions by affecting the kinetics of the reactions involved. Some of these factors are intrinsic to the materials themselves, such as their mineralogical composition (e.g., modal composition, grain size distribution, specific surface area, and degree of liberation) and its evolution over time (e.g., depletion of certain minerals, buildup of a passivating layer of secondary minerals over sulfides and neutralizers), as well as their hydrogeological properties (e.g., hydraulic conductivity, water retention curve, and permeability function). In addition, extrinsic factors can also have a significant influence on the rates of reactions, such as climatic conditions (temperature, atmospheric humidity, etc.) and geochemical environments (pH, microbiological activity, degree of saturation, redox conditions, oxygen availability, etc.), which are closely linked to the management and reclamation techniques used. Consequently, the factors influencing the development of AMD conditions are discussed parallel to the hydrogeological and geochemical environments typically encountered in tailings storage facilities (TSFs) and waste rock piles. The reader is directed to Chapter 3 for a comprehensive description of fluid flow control (gas and water), which is responsible for these conditions.

1.3.1 MINERALOGICAL AND HYDROGEOLOGICAL PROPERTIES

The mineralogical composition of the materials, as well as their physical and hydrogeological properties, influence their reactivity. Obviously, the grain size distribution and degree of liberation of the reactive minerals within the materials, concomitant to their specific surface area and textures, have a significant impact on sulfide oxidation; finer grain sizes will present greater surface areas available to oxidation (e.g., Nicholson and Scharer 1994; Parbhakar-Fox et al. 2013; Kalyoncu Erguler et al. 2014; Erguler and Kalyoncu Erguler 2015; Elghali et al. 2018, 2019a, 2019b, 2019c). More particularly, the degree of liberation of minerals plays a key role in reactivity, as surface processes such as sulfide oxidation or acid neutralization will only be possible on exposed surfaces (e.g., Elghali et al. 2018, 2019c). Thus, efforts to predict the onset of acidic drainages from mine wastes (Chapter 2) may need to consider the differential segregation and degree of liberation of sulfides and neutralizing minerals within different grain sizes (Elghali et al. 2018, 2019a) in order to avoid inaccurate predictions, particularly for waste rocks.

Similarly, if the mine waste management scenario involves conditions favorable to a differential distribution of grain sizes (such as end- or push-dumping for waste rock piles and spigoting for TSF), the creation of local heterogeneities in material properties will induce a heterogeneous geochemical behavior (e.g., Bussière 2007). Furthermore, the differentiation of grain sizes will also have an effect on the hydrogeological properties of the materials governing gas and water movement through them (see Chapter 3 for a description of hydrogeological properties and their relationship with fluid movement).

The next section of this chapter discusses the influence of typical storage conditions of mine tailings and waste rocks on the development of acidic drainage. Chapters 4 to 9 detail on how to successfully design and implement reclamation scenarios using the materials' properties to control fluid movement and sulfide oxidation in tailings ponds and waste rock piles.

1.3.2 TAILINGS

The storage conditions of acid-generating mine wastes will have a significant effect on their geochemical behavior. Conventional TSFs are constructed by pumping a slurry that typically contains between

25% and 45% of solids retained within an impoundment (e.g., Bussière 2007). The tailings are often discharged from the crest of the surrounding dikes, and the natural segregation of the particles create beaches made of coarser and denser particles close to the discharge points, whereas the finer and lighter particles are transported further down the middle of the impoundment (e.g., Blight 2003).

The grain size distribution of hard rock mine tailings typically shows a D_{10} from 1 to 4 μm and a D_{60} ranging from 10 to 50 μm, whereas their proportion passing 80 μm vary between 70% and 97% (e.g., Bussière 2007). Upon consolidation, these tailings show typical porosity values between 0.33 and 0.5 (Bussière 2007). In conventional TSFs and under temperate climates, capillary forces between the grains can keep the water table close to the surface of the tailings at a variable degree of saturation, depending on the storage conditions. However, the water table can be much deeper in certain circumstances, for example, under arid climates or at the surface of high TSFs (approximately higher than 30 m). For more details, see the water covers, CCBEs, and EWTs in Chapters 6, 7, and 8, respectively.

Figure 1.4 illustrates the hydrogeological and geochemical environments in a typical profile of acid-generating tailings within a conventional TSF located under typical Canadian conditions (i.e., a wet climate). The degree of saturation is typically lowest at the surface and increases deeper within the vadose zone (unsaturated) to complete saturation below the capillary fringe, where the water table cycles up and down. Since oxygen can only migrate from the surface of the tailings downward (within the vadose zone), sulfide oxidation and acid generation mainly proceed from the top down. As the weathering progresses, the sulfides toward the surface are depleted and/or passivated with a rim of secondary oxyhydroxides, which allows the oxygen to migrate further beneath the surface and oxidize the sulfides deeper within the impoundment (gaseous O_2 profiles in Figure 1.4).

As the oxidation front migrates inward, a neutralization sequence develops within the tailings. Indeed, the different rates of acid neutralization shown in Figure 1.3 engender a series of sequential neutralization reactions upon weathering, which result in the progressive depletion of specific neutralizing phases, as observed in numerous tailings impoundments (Dubrovsky et al. 1985; Johnson et al. 2000; Blowes and Jambor 1990; Blowes et al. 1991; Lindsay et al. 2015). Blowes et al. (2003) outlined a sequence involving the following reactions with decreasing pH plateaus, as illustrated in Figure 1.4:

- pH 6.5–7.5: buffering by calcite and dolomite dissolution, precipitation of siderite, gibbsite $Al(OH)_3$, and ferrihydrite $Fe(OH)_3$;
- pH 4.8–6.3: buffering by siderite, precipitation of $Al(OH)_3$ and $Fe(OH)_3$;

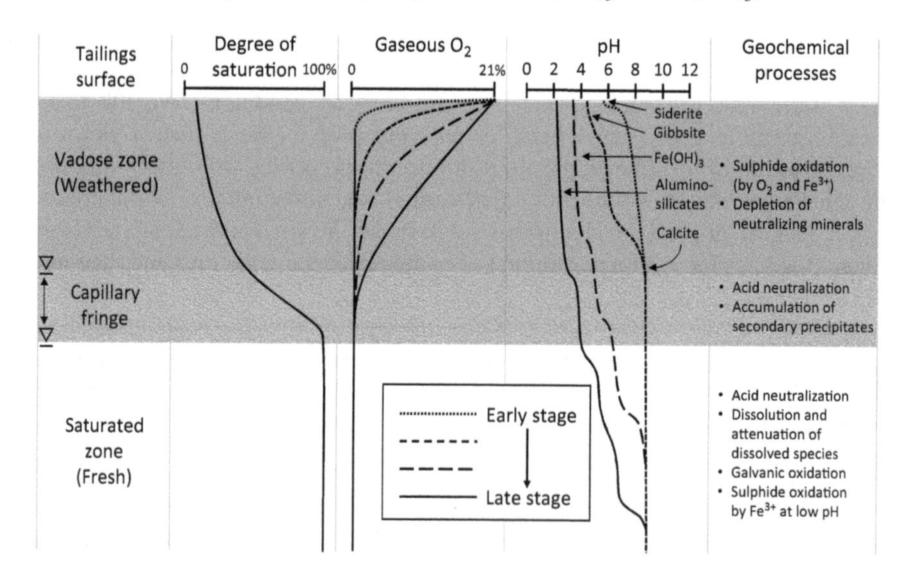

FIGURE 1.4 Hydrogeological and geochemical processes within AMD-generating tailings ponds.

- pH 4.0–4.3: buffering by $Al(OH)_3$;
- pH 2.5 < 3.5: buffering by $Fe(OH)_3$;
- pH < 3.0: buffering by aluminosilicates.

The soluble reaction products (from sulfide oxidation and acid neutralization) are transported by the infiltrating water toward the capillary fringe, where the water table height oscillates up and down, making this specific area particularly prone to secondary precipitations. As a result, secondary precipitates tend to accumulate near this capillary fringe and may cement the grains together into a hardpan (see Section 1.2.3), which significantly affects fluid movement through the tailings and, potentially, the effectiveness of reclamation methods based on fluid flow control (Gilbert et al. 2003; Ethier et al. 2018; Pabst et al. 2017, 2018; Ethier 2018; Elghali et al. 2019b).

Oxygen migrates significantly slower below the capillary fringe, where the tailings are saturated (see Chapter 3). Therefore, it is often hypothesized that the saturated zone mostly involves acid neutralization, dissolution of neutralization products, and attenuation of metals leached from the vadose zone. However, galvanic mechanisms are possible within the saturated zone, if two different sulfide particles are in contact within acidic and conductive waters, even if these conditions are limited to a microenvironment surrounding the grains involved. The iron pathway of sulfide oxidation is also possible where iron-rich waters are in contact with available sulfide surfaces, with the contribution from microorganisms.

The secondary minerals formed upon tailings and waste rock weathering can accumulate in acid-generating materials. Changes in microenvironment conditions will affect the stability of these minerals. For example, when submitted to wet–dry cycles, they will release part of their constituents (e.g., Fe, H^+, SO_4^{2-}, and K^+), as well as the adsorbed and co-precipitated metals associated with them, back into drainage waters (e.g., Johnson et al. 2000; Moncur et al. 2005; Lindsay et al. 2015; Blackmore et al. 2018; Elghali et al. 2018). In addition to the acid released by the dissolution of acidic salts, the released Fe will undergo a second round of hydrolysis (Equation 1.3) to generate tertiary minerals, generating acid in the process (e.g., Moncur et al. 2005, 2009; Blowes et al. 2014). For this reason, highly oxidized mine wastes (such as in abandoned tailings facilities) will continue to generate acidic leachates even after complete depletion of the available sulfides (e.g., Elghali et al. 2019b).

This Fe release will also be increased when highly oxidized wastes are placed in more reduced conditions—for example, if an oxygen barrier–type cover is installed over tailings (e.g., see Chapters 6, 7, and 8). As a result, covering weathered AMD-generating tailings in a manner that prevents oxygen ingress will not instantly improve the water quality. Instead, a gradual improvement is to be expected, as a new equilibrium is established within the more reduced, induced conditions (e.g., Bussière et al. 2009; Ethier 2018; Ethier et al. 2018; Pabst et al. 2017, 2018). An increase in acid generation and metals release can even be anticipated within the first months or years after reclamation, which will gradually decrease upon depletion of the accumulated secondary iron oxyhydroxides (e.g., Ribeta et al. 1995; Ethier et al. 2018). In the meantime, it can be appropriate to install a passive treatment system in order to meet the water quality criteria (see Chapter 11 for a review of passive treatment methods).

1.3.3 Waste Rocks

Waste rock piles contain the non-economical rocks that were blasted to reach the ore. They consist of a wide range of particle sizes from fine, micron-scale grains to boulders of several meters, stockpiled either in a single bench or successive benches. The waste rock pile construction method influences geotechnical and hydrogeological properties by creating heterogeneities within the piles in terms of porosity and particle segregation (e.g., Smith et al. 2013; Dawood and Aubertin 2014; Amos et al. 2015; Lahmira et al. 2016, 2017). Typically, the surfaces where machinery circulates

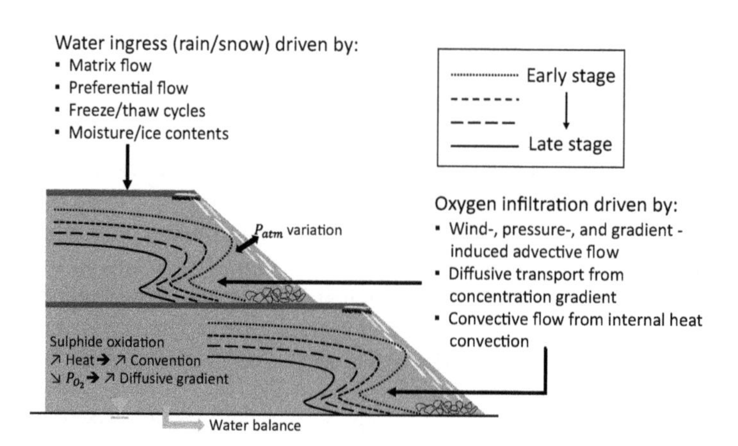

FIGURE 1.5 Hydrogeological and geochemical processes within AMD-generating waste rock piles; darker areas are made of denser, smaller particles (Modified from Amos et al. 2015; Lefebvre et al. 2001; Ritchie 1994).

will be composed of denser, smaller particles, while the extremities of the pile or benches show a segregation of particles, as the biggest boulders roll down the slope, while the smaller particles tend to remain upslope.

When waste rock piles are constructed in successive benches, each bench will show these heterogeneous properties, and the piles will show a succession of those heterogeneities. Figure 1.5 illustrates a conceptual model of a waste rock pile made of two successive benches (for more details, see also Chapter 13 on integrated mine waste management methods). As a result, fluid circulation within waste rock piles, and consequently the geochemical behavior of the materials within, will be driven by a complex arrangement of heterogeneous hydrogeological and geotechnical properties (e.g., Dawood and Aubertin 2014; Lahmira et al. 2016, 2017), as illustrated in Figure 1.5.

Water circulation in waste rock piles will be driven by matrix flow in areas with smaller particles and smaller pores exerting capillary forces, and by preferential flow paths where larger boulders offer bigger, more connected voids in which capillary forces are negligible (see Chapter 3 for details). Oxygen movement through the piles will be much easier than within tailings, as the bigger voids enable air to infiltrate all sides that are open to the atmosphere and move toward the core of the dump. In addition, wind- and gradient-driven advective flows will drive air inside the pile, as will conductive flows from the heat generated by sulfide oxidation and diffusive transport caused by the concentration gradient resulting from oxygen consumption inside the reactive zone of the pile (e.g., Amos et al. 2015; Lorca et al. 2016; Vriens et al. 2018). Consequently, waste rock piles will tend to be weathered from the exterior toward the interior of the benches, and as weathering progresses, oxygen is allowed to migrate further inside the pile, as illustrated in Figure 1.5, where the advancement of the weathering front is outlined (Ritchie 1994; Lefebvre et al. 2001; Amos et al. 2015). Finally, the results obtained by Blackmore et al. (2018) show that water quality in acid-generating waste rock piles is controlled by a combination of (hydro-)geochemical and microbiological mechanisms.

1.3.4 Open Pits

Open pits and underground galleries walls are also weathered upon exposure to atmospheric conditions until they are flooded and/or refilled. The soluble weathering products from the walls are transported down to the bottom of the pit and/or galleries (and eventually to the groundwater), where secondary precipitates are likely to accumulate (e.g., Blowes et al. 2014; Castendyk et al.

2015a, 2015b; Gammons and Duaime 2006). When the pits or galleries are flooded, the secondary precipitates formed upon weathering may be solubilized under the more reduced conditions induced by flooding, which releases their constituents back into the flood water or the groundwater (e.g., Blowes et al. 2014) along with the sorbed and co-precipitated species associated with them. An extensive review of pit lake water quality can be found in the study by Castendyk et al. (2015a). Therefore, the presence of secondary precipitates in open pits and underground workings need to be assessed when planning the water management upon refilling. However, the sulfide minerals that remain over the surface of the flooded pit will continue to oxidize after flooding. In addition, if the pit lake is acidic enough and contains dissolved Fe^{3+}, the iron pathway of sulfide oxidation can continue even if O_2 is low within the pit lake (e.g., Pellicori et al. 2005; Gammons and Duaime 2006). More details about pit backfilling are presented in Chapter 13.

1.3.5 Temperature

The temperature to which the mine wastes are exposed has a significant effect on their reactivity. Since sulfide oxidation is an exothermic reaction, the rate of oxidation increases with temperature, which increases heat production, feeding a chain reaction that can lead to high temperatures, especially within waste rock piles (e.g., Amos et al. 2015). In turn, this chain reaction induces pressure, oxygen concentration, and temperature gradients that promote gas convection through waste rocks, further promoting AMD. Indeed, temperatures over 65°C were recorded within the Doyon waste rock pile (Lefebvre et al. 2001).

Inversely, the decrease in the sulfide oxidation rate with decreasing temperature is often reported to follow an Arrhenius behavior (Equations 1.28 and 1.29; e.g., Coulombe et al. 2012; Elberling 2001, 2005; Meldrum et al. 2001).

$$k = Ae^{\frac{-E_a}{RT}} \tag{1.28}$$

$$\ln\frac{k_2}{k_1} = -\frac{E_a}{R}\left(\frac{1}{T_2} - \frac{1}{T_1}\right) \tag{1.29}$$

where k_1 and k_2 are the reaction rates at temperatures T_1 and T_2, respectively; E_a is the activation energy; A is a pre-exponential factor, and R is gas constant. The different iron oxidizers and sulfur oxidizers identified in the previous section are mesophiles that are optimally active within temperatures that mostly range between 20°C and 45°C (Blowes et al. 2014; Lindsay et al. 2015). Thus, lowering the temperature will decrease the rate of microbial processes involved in sulfide oxidation (e.g., Elberling 2001, 2005). In addition, since the different mechanisms of sulfide oxidation require aqueous media, freezing the wastes will significantly decrease the amount of liquid water and, consequently, the extent of oxidation that might occur. Therefore, the mine waste management and reclamation scenarios for many mines operating in cold environments are aimed at keeping their wastes frozen (see Chapter 9). However, the sulfides may oxidize below the freezing point, as demonstrated using oxygen consumption tests in mine tailings at temperatures between 0°C and –4°C (Elberling 2001; Meldrum et al. 2001; Coulombe et al. 2012) and as low as −11°C (Elberling 2005). Hence, for each site located in cold environments, it is crucial to determine the temperature at which the mine waste can be considered non-acid-generating (T_{target}; see Chapter 9 for more details).

Since acid generation accelerates when the pH reaches acidic values (as described in Section 1.2.1), sulfide oxidation needs to be slowed down as early as possible upon surface deposition or at least during the lag time before AMD is clearly established. When tailings are stored in such a way that the oxygen ingress is significantly slowed down, then acid generation will be equally decreased. The same reasoning applies to freezing acid-generating tailings in order to diminish sulfide oxidation through liquid water availability and temperature control.

1.4 RESEARCH NEEDS

Despite the extensive research performed since the 1970s on the geochemistry of AMD generation, many knowledge gaps remain and require additional research. This section describes current research needs.

- *The influence of management and reclamation methods on AMD generation and water quality*
  The impacts of novel mine waste management and reclamation methods (e.g., desulfurization, progressive reclamation, segregation, co-disposal, backfilling, densified tailings, etc.) on the generation of AMD need to be investigated. More specifically, the impact of gas and water flow through the mine waste disposal areas on the geochemical response of the materials need to be understood in order to determine their short- and long-term effects on drainage water quality.
- *Geochemical behavior of weathered AMD-generating mine wastes after reclamation*
  More research efforts need to be invested toward understanding the time needed for reclaimed sites containing weathered mine wastes to stop generating AMD, including through the use of numerical reaction transport tools. Further research on the long-term effectiveness of reclamation methods based on freezing the mine wastes should also be conducted, particularly to understand the geomicrobial mechanisms that occur below the freezing point.
- *Geometallurgical tools and geoenvironmental modeling*
  There is an increasing interest in the use of automated mineralogical tools to identify potential contaminants and their source terms (e.g., Parbhakar-Fox et al. 2013; Benzaazoua et al. 2017; Elghali et al. 2018). However, automated mineralogical analyses are used on a case-to-case basis, and results interpretation relies on the sole expertise of the geochemists analyzing the data. Therefore, more research is needed to develop and test systematic approaches that would enable the identification of potential contaminants as early as possible in the mining cycle (e.g., during exploration and feasibility studies) while taking into account the factors that influence their potential release in mine waters (e.g., their texture, degree of liberation, and mineralogical associations).
- *Microbial processes involved in the geochemistry of mine wastes*
  Recent advances in culture-independent techniques, which are based on the analysis of DNA or RNA extracted directly from environmental samples without the need for prior cultivation, have exponentially increased knowledge on microbial communities in mine wastes. These techniques have revealed that the microbiomes of mine wastes and acidic drainages are comprised of complex consortiums of bacteria, archaea, and Eukaryotes (e.g., Auld et al. 2013; Baker and Banfield 2003; Hallberg 2010; Hua et al. 2015; Johnson 2007; and references therein). Hence, the development and democratization of these genomic techniques will help fulfill research needs in microbiological involvement in the development of contaminated mine waters. Among other aspects, there needs to be additional research into the influence of pH, salinity, dissolved oxygen, and organic carbon availability on the microbiological involvement in S and Fe cycling in mine wastes, as well as on identifying the microbial involvement in sub-zero sulfide oxidation.
- *Water quality prediction using numerical reactive transport modeling tools*
  Numerical reactive transport modeling tools, such as MIN3P (Mayer et al. 2002), have been demonstrated to adequately reproduce the geochemical and hydrogeological processes in a wide variety of settings, from mine tailings reclamation scenarios (e.g., Molson et al. 2008; Pabst et al. 2017) to water treatment using reactive barriers (e.g., Mayer et al. 2001; Weber et al. 2013), including isotopic fractionation during groundwater treatment processes (e.g., Gibson et al. 2011; Jamieson-Hanes et al. 2017). More recently, numerical reactive transport modeling tools have been used to successfully predict the water quality of intermediate-scale field lysimeters containing over 9000 kg of waste rocks using a conceptual model based on

laboratory-derived rates of reaction (Wilson et al. 2018a, 2018b). However, more research is needed to better define the proxies necessary to scale the laboratory results up to intermediate and full scales.

- *CND or metal leaching from mine wastes*
  As environmental regulations related to mining become more restrictive, and as the mining industry seeks to continually decrease its environmental impacts, increasing research and development efforts are driven toward understanding CND development and its prevention, control, and treatment. Although CND is not discussed in this chapter, the mechanisms leading to its development are the same as for AMD, and many of the research needs identified for AMD are also applicable to CND.

REFERENCES

Agricola, G. (1556/1912). *De re metallica.* Translated from the first Latin edition by H. Hoover and L. H. Hoover. New York: Dover Publications.

Al, T. A., C. J. Martin, and D. W. Blowes. 2000. Carbonate-mineral/water interactions in sulfide-rich mine tailings. *Geochimica et Cosmochimica Acta* 64, no. 23: 3933–3948.

Alpers, C. N., D. W. Blowes, D. K. Nordstrom, and J. L. Jambor. 1994. Secondary minerals and acid mine water chemistry. In *Short course handbook on environmental geochemistry of sulfide mine-wastes*, ed. J. L. Jambor and D. W. Blowes, 247–270. Waterloo, ON: Mineralogical Association of Canada.

Amos, R. T., D. W. Blowes, B. L. Bailey, D. C. Sego, L. Smith, and A. I. M. Ritchie. 2015. Waste-rock hydrogeology and geochemistry. *Applied Geochemistry* 57: 140–156.

Auld, R. R., M. Myre, N. C. S. Mykytczuk, L. G. Leduc, and T. J. S. Merritt. 2013. Characterization of the microbial acid mine drainage microbial community using culturing and direct sequencing techniques. *Journal of Microbiological Methods* 93, no. 2: 108–115.

Azzie, B. A.-M. 2002. Coal mine waters in South Africa: Their geochemistry, quality and classification. PhD diss., University of Cape Town.

Baker, B. J., and J. F. Banfield. 2003. Microbial communities in acid mine drainage. *FEMS Microbiology Ecology* 44, no. 2: 139–152.

Ball, J. W., and D. K. Nordstrom. 1989. Final revised analyses of major and trace elements from acid mine waters in the Leviathan Mine drainage basin, California and Nevada; October 1981 to October 1982. In *Water-Resources Investigations Report.* U.S. Geological Survey, https://pubs.er.usgs.gov/publication/wri894138

Belzile, N., Y.-W. Chen, M.-F. Cai, and Y. Li. 2004. A review on pyrrhotite oxidation. *Journal of Geochemical Exploration* 84, no. 2: 65–76.

Benzaazoua, M., H. Bouzahzah, Y. Taha, et al. 2017. Integrated environmental management of pyrrhotite tailings at Raglan Mine: Part 1 challenges of desulphurization process and reactivity prediction. *Journal of Cleaner Production* 162, Supplement C: 86–95.

Blackmore, S., B. Vriens, M. Sorensen, et al. 2018. Microbial and geochemical controls on waste rock weathering and drainage quality. *Science of the Total Environment* 640–641: 1004–1014.

Blight, G. E. 2003. Quantified comparisons of disposal of thickened and unthickened tailings. Paper presented at *Tailings and Mine Waste '03*, Fort Collins, CO.

Blowes, D. W., and J. L. Jambor. 1990. The pore-water geochemistry and the mineralogy of the vadose zone of sulfide tailings, Waite Amulet, Quebec, Canada. *Applied Geochemistry* 5, no. 3: 327–346.

Blowes, D. W., E. J. Reardon, J. L. Jambor, and J. A. Cherry. 1991. The formation and potential importance of cemented layers in inactive sulfide mine tailings. *Geochimica et Cosmochimica Acta* 55, no. 4: 965–978.

Blowes, D. W., C. J. Ptacek, and J. Jurjovec. 2003. Mill tailings hydrogeology and geochemistry. In *Environmental aspects of mine wastes*, ed. J. L. Jambor, W. D. Blowes, and A. I. M. Ritchie, 95–116. Ottawa, ON: Mineralogical Association of Canada.

Blowes, D. W., C. J. Ptacek, J. L. Jambor, et al. 2014. 11.5 - The geochemistry of acid mine drainage. In *Treatise on geochemistry*, 2nd ed., ed. K. Turekian and H. Holland, 131–190. Oxford: Elsevier.

Bouzahzah, H., M. Benzaazoua, B. Bussiere, and B. Plante. 2014. Prediction of acid mine drainage: Importance of mineralogy and the test protocols for static and kinetic tests. *Mine Water and the Environment* 33, no. 1: 54–65.

Braley, S. A. 1949. *Annual summary report of Commonwealth of Pennsylvania Department of Health*, Industrial Fellowship No. B.3. Melton Institute, Pittsburgh, PA.

Braley, S. A. 1960. *Special report on the oxidation of pyrite conglomerate: Coal.* Research Project No. 370-6, Advisory Committee on the Ohio River Valley Water Sanitation Commission.

Bridge, T. A. M., and D. B. Johnson. 1998. Reduction of soluble iron and reductive dissolution of ferric iron-containing minerals by moderately thermophilic iron-oxidizing bacteria. *Applied and Environmental Microbiology* 64, no. 6: 2181–2186.

Bridge, T. A. M., and D. B. Johnson. 2000. Reductive dissolution of ferric iron minerals by acidiphilium SJH. *Geomicrobiology Journal* 17, no. 3: 193–206.

Bussière, B. 2007. Colloquium 2004: Hydrogeotechnical properties of hard rock tailings from metal mines and emerging geoenvironmental disposal approaches. *Canadian Geotechnical Journal* 44, no. 9: 1019–1052.

Bussière, B., R. Potvin, A.-M. Dagenais, M. Aubertin, A. Maqsoud, and J. Cyr. 2009. Restauration du site minier Lorraine, Latulipe, Québec : Résultats de 10 ans de suivi. *Déchets Sciences et Techniques* 54, no. 2: 49–64.

Castendyk, D. N., L. E. Eary, and L. S. Balistrieri. 2015a. Modeling and management of pit lake water chemistry 1: Theory. *Applied Geochemistry* 57: 267–288.

Castendyk, D. N., L. S. Balistrieri, C. Gammons, and N. Tucci. 2015b. Modeling and management of pit lake water chemistry 2: Case studies. *Applied Geochemistry* 57: 289–307.

Chester, R., and T. Jickells. 2012. *Marine geochemistry*. 3rd ed. New York: Wiley-Blackwell.

Chopard, A., B. Plante, M. Benzaazoua, H. Bouzahzah, and P. Marion. 2017. Geochemical investigation of the galvanic effects during oxidation of pyrite and base-metals sulfides. *Chemosphere* 166: 281–291.

Church, S. E., P. von Guerard, and S. E. Finger. 2007. Integrated investigations of environmental effects of historical mining in the Animas River watershed, San Juan County, Colorado. 2 vol., U.S. Geological Survey Professional Paper 1651.

Colmer, A. R., and M. E. Hinkle. 1947. The role of microorganisms in acid mine drainage: A preliminary report. *Science* 106, no. 2751: 253–256.

Conesa, H. M., A. Faz, and R. Arnaldos. 2006. Heavy metal accumulation and tolerance in plants from mine tailings of the semiarid Cartagena–La Unión mining district (SE Spain). *Science of the Total Environment* 366, no. 1: 1–11.

Coulombe, V., B. Bussière, J. Côté, and P. Garneau. 2012. Performance of insulation covers to control acid mine drainage in cold environment. In *Cold regions engineering 2012: Sustainable infrastructure development in a changing cold environment: Proceedings of the International Conference on Cold Regions Engineering, Québec*, Québec, Canada, August 19–22, 2012, ed. B. Morse and Guy Doré, 789–799. https://ascelibrary.org/doi/book/10.1061/9780784412473.

Cruz, L. S., A. Garralón, A. Escribano, et al. 2013. Chemical characteristics of acid mine drainage from an As-W mineralized zone in western Spain. *Procedia Earth and Planetary Science* 7: 284–287.

Cruz, R., R. M. Luna-Sánchez, G. T. Lapidus, I. González, and M. Monroy. 2005. An experimental strategy to determine galvanic interactions affecting the reactivity of sulfide mineral concentrates. *Hydrometallurgy* 78, no. 3–4: 198–208.

Dahrazma, B., and M. Kharghani. 2012. The impacts of alkaline mine drainage on Ba, Cr, Ni, Pb and Zn concentration in the water resources of the Takht coal mine, Iran. *Earth Sciences Research Journal* 16: 109–112.

Dawood, I., and M. Aubertin. 2014. Effect of dense material layers on unsaturated water flow inside a large waste rock pile: A numerical investigation. *Mine Water and the Environment* 33, no. 1: 24–38.

DeSisto, S. L., H. E. Jamieson, and M. B. Parsons. 2011. Influence of hardpan layers on arsenic mobility in historical gold mine tailings. *Applied Geochemistry* 26, no. 12: 2004–2018.

Dubrovsky, N. M., J. A. Cherry, E. J. Reardon, and A. J. Vivyurka. 1985. Geochemical evolution of inactive pyritic tailings in the Elliot Lake uranium district. *Canadian Geotechnical Journal* 22, no. 1: 110–128.

Elberling, B. 2001. Environmental controls of the seasonal variation in oxygen uptake in sulfidic tailings deposited in a permafrost-affected area. *Water Resources Research* 37, no. 1: 99–107.

Elberling, B. 2005. Temperature and oxygen control on pyrite oxidation in frozen mine tailings. *Cold Regions Science and Technology* 41, no. 2: 121–133.

Elghali, A., M. Benzaazoua, H. Bouzahzah, B. Bussière, and H. Villarraga-Gómez. 2018. Determination of the available acid-generating potential of waste rock, part I: Mineralogical approach. *Applied Geochemistry* 99: 31–41.

Elghali, A., M. Benzaazoua, B. Bussière, and H. Bouzahzah. 2019a. Determination of the available acid-generating potential of waste rock, part II: Waste management involvement. *Applied Geochemistry* 100: 316–325.

Elghali, A., M. Benzaazoua, B. Bussière, C. Kennedy, R. Parwani, and S. Graham. 2019b. The role of hardpan formation on the reactivity of sulfidic mine tailings: A case study at Joutel mine (Québec). *Science of the Total Environment* 654: 118–128.

Elghali, A., M. Benzaazoua, B. Bussière, and T. Genty. 2019c. Spatial mapping of acidity and geochemical properties of oxidized tailings within the former Eagle/Telbel mine site. *Minerals* 9, no. 3. https://doi.org/10.3390/min9030180

Erguler, Z. A., and G. Kalyoncu Erguler. 2015. The effect of particle size on acid mine drainage generation: Kinetic column tests. *Minerals Engineering* 76: 154–167.

Ethier, M.-P. 2018. Évaluation de la performance d'un système de recouvrement monocouche avec nappe surélevée pour la restauration d'un parc à résidus miniers abandonné. PhD diss., Université du Québec en Abitibi-Témiscamingue.

Ethier, M.-P., B. Bussière, M. Aubertin, A. Maqsoud, I. Demers, and S. Broda. 2018. In situ evaluation of performance of reclamation measures implemented on abandoned reactive tailings disposal site. *Canadian Geotechnical Journal* 55, no. 12: 1742–1755.

Frau, F. 2000. The formation-dissolution-precipitation cycle of melanterite at the abandoned pyrite mine of Genna Luas in Sardinia, Italy: Environmental implications. *Mineralogical Magazine* 64, no. 6: 995–1006.

Gammons, C. H., and T. E. Duaime. 2006. Long term changes in the limnology and geochemistry of the Berkeley Pit Lake, Butte, Montana. *Mine Water and the Environment* 25, no. 2: 76–85.

Gibson, B. D., R. T. Amos, and D. W. Blowes. 2011. 34S/32S fractionation during sulfate reduction in groundwater treatment systems: Reactive transport modeling. *Environmental Science and Technology* 45, no. 7: 2863–2870.

Gilbert, S. E., D. R. Cooke, and P. Hollings. 2003. The effects of hardpan layers on the water chemistry from the leaching of pyrrhotite-rich tailings material. *Environmental Geology* 44, no. 6: 687–697.

Graupner, T., A. Kassahun, D. Rammlmair, et al. 2007. Formation of sequences of cemented layers and hardpans within sulfide-bearing mine tailings (mine district Freiberg, Germany). *Applied Geochemistry* 22, no. 11: 2486–2508.

Hakkou, R., M. Benzaazoua, and B. Bussière. 2008. Acid mine drainage at the abandoned Kettara mine (Morocco): 2. mine waste geochemical behavior. *Mine Water and the Environment* 27, no. 3: 160–170.

Hallberg, K. B. 2010. New perspectives in acid mine drainage microbiology. *Hydrometallurgy* 104, no. 3: 448–453.

Heikkinen, P. M., and M. L. Räisänen. 2008. Mineralogical and geochemical alteration of Hitura sulphide mine tailings with emphasis on nickel mobility and retention. *Journal of Geochemical Exploration* 97, no. 1: 1–20.

Heikkinen, P. M., and M. L. Räisänen. 2009. Trace metal and As solid-phase speciation in sulphide mine tailings – Indicators of spatial distribution of sulphide oxidation in active tailings impoundments. *Applied Geochemistry* 24, no. 7: 1224–1237.

Heikkinen, P., M. Räisänen, and R. Johnson. 2009. Geochemical characterisation of seepage and drainage water quality from two sulphide mine tailings impoundments: Acid mine drainage versus neutral mine drainage. *Mine Water and the Environment* 28, no. 1: 30–49.

Holmes, P. R., and F. K. Crundwell. 1995. Kinetic aspects of galvanic interactions between minerals during dissolution. *Hydrometallurgy* 39, no. 1: 353–375.

Holmes, P. R., and F. K. Crundwell. 2000. The kinetics of the oxidation of pyrite by ferric ions and dissolved oxygen: An electrochemical study. *Geochimica et Cosmochimica Acta* 64, no. 2: 263–274.

Hua, Z. S., Y. J. Han, L. X. Chen, et al. 2015. Ecological roles of dominant and rare prokaryotes in acid mine drainage revealed by metagenomics and metatranscriptomics. *ISME Journal* 9, no. 6: 1280–1294.

INAP (International Network for Acid Prevention). 2009. *Global acid rock drainage guide (GARD Guide).* http://www.gardguide.com.

Jambor, J. L., and J. E. Dutrizac. 1998. Occurrence and constitution of natural and synthetic ferrihydrite, a widespread iron oxyhydroxide. *Chemical Reviews* 98: 2549–2585.

Jambor, J. L., J. E. Dutrizac, L. A. Groat, and M. Raudsepp. 2002. Static tests of neutralization potentials of silicate and aluminosilicate minerals. *Environmental Geology* 43, no. 1–2: 1–17.

Jambor, J. L., J. E. Dutrizac, and M. Raudsepp. 2007. Measured and computed neutralization potentials from static tests of diverse rock types. *Environmental Geology* 52, no. 6: 1173–1185.

Jamieson, H. E., S. R. Walker, and M. B. Parsons. 2015. Mineralogical characterization of mine waste. *Applied Geochemistry* 57: 85–105.

Jamieson-Hanes, J. H., H. K. Shrimpton, H. Veeramani, C. J. Ptacek, A. Lanzirotti, M. Newville, and D. W. Blowes. 2017. Evaluating zinc isotope fractionation under sulfate reducing conditions using a flow-through cell and in situ XAS analysis. *Geochimica et Cosmochimica Acta* 203: 1–14.

Janzen, M. P., R. V. Nicholson, and J. M. Scharer. 2000. Pyrrhotite reaction kinetics: Reaction rates for oxidation by oxygen, ferric iron, and for nonoxidative dissolution. *Geochimica et Cosmochimica Acta* 64, no. 9: 1511–1522.

Johnson, D. B. 2007. Physiology and ecology of acidophilic microorganisms. In *Physiology and Biochemistry of Extremophiles*, ed. C. Gerday and N. Glansdorff, 257–270. doi:10.1128/9781555815813.ch20.
Johnson, D. B., P. Bacelar-Nicolau, N. Okibe, A. Thomas, and K. B. Hallberg. 2009. Ferrimicrobium acidiphilum gen. nov., sp. nov. and Ferrithrix thermotolerans gen. nov., sp. nov.: Heterotrophic, iron-oxidizing, extremely acidophilic actinobacteria. *International Journal of Systematic and Evolutionary Microbiology* 59, no. 5: 1082–1089.
Johnson, R. H., D. W. Blowes, W. D. Robertson, and J. L. Jambor. 2000. The hydrogeochemistry of the Nickel Rim mine tailings impoundment, Sudbury, Ontario. *Journal of Contaminant Hydrology* 41, no. 1–2: 49–80.
Jurjovec, J., C. J. Ptacek, and D. W. Blowes. 2002. Acid neutralization mechanisms and metal release in mine tailings: A laboratory column experiment. *Geochimica et Cosmochimica Acta* 66, no. 9: 1511–1523.
Kalyoncu Erguler, G., Z. A. Erguler, H. Akcakoca, and A. Ucar. 2014. The effect of column dimensions and particle size on the results of kinetic column test used for acid mine drainage (AMD) prediction. *Minerals Engineering* 55: 18–29.
Kwong, Y. T. J., G. W. Swerhone, and J. R. Lawrence. 2003. Galvanic sulphide oxidation as a metal-leaching mechanism and its environmental implications. *Geochemistry: Exploration, Environment, Analysis* 3, no. 4: 337–343.
Lahmira, B., R. Lefebvre, M. Aubertin, and B. Bussière. 2016. Effect of heterogeneity and anisotropy related to the construction method on transfer processes in waste rock piles. *Journal of Contaminant Hydrology* 184: 35–49.
Lahmira, B., R. Lefebvre, M. Aubertin, and B. Bussière. 2017. Effect of material variability and compacted layers on transfer processes in heterogeneous waste rock piles. *Journal of Contaminant Hydrology* 204: 66–78.
Lefebvre, R., D. Hockley, J. Smolensky, and P. Gélinas. 2001. Multiphase transfer processes in waste rock piles producing acid mine drainage. 1: Conceptual model and system characterization. *Journal of Contaminant Hydrology* 52, no. 1–4: 137–164.
Lindsay, M. B. J., M. C. Moncur, J. G. Bain, J. L. Jambor, C. J. Ptacek, and D. W. Blowes. 2015. Geochemical and mineralogical aspects of sulfide mine tailings. *Applied Geochemistry* 57: 157–177.
Lorca, M. E., K. U. Mayer, D. Pedretti, L. Smith, and R. D. Beckie. 2016. Spatial and temporal fluctuations of pore-gas composition in sulfidic mine waste rock. *Vadose Zone Journal* 15, no. 10.
Lottermoser, B. 2010. *Mine wastes: Characterization, treatment and environmental impacts*. Berlin: Springer.
Lowson, R. T. 1982. Aqueous oxidation of pyrite by molecular oxygen. *Chemical Reviews* 82, no. 5: 461–497.
Mayer, K. U., D. W. Blowes, and E. O. Frind. 2001. Reactive transport modeling of an in situ reactive barrier for the treatment of hexavalent chromium and trichloroethylene in groundwater. *Water Resources Research* 37, no. 12: 3091–3103.
Mayer, K. U., E. O. Frind, and D. W. Blowes. 2002. Multicomponent reactive transport modeling in variably saturated porous media using a generalized formulation for kinetically controlled reactions. *Water Resources Research* 38, no. 9: 13-1–13-21.
McGregor, R. G., and D. W. Blowes. 2002. The physical, chemical and mineralogical properties of three cemented layers within sulfide-bearing mine tailings. *Journal of Geochemical Exploration* 76, no. 3: 195–207.
McKibben, M. A., and H. L. Barnes. 1986. Oxidation of pyrite in low temperature acidic solutions: Rate laws and surface textures. *Geochimica et Cosmochimica Acta* 50, no. 7: 1509–1520.
Meldrum, J. L., H. E. Jamieson, and L. D. Dyke. 2001. Oxidation of mine tailings from Rankin Inlet, Nunavut, at subzero temperatures. *Canadian Geotechnical Journal* 38, no. 5: 957–966.
Molson, J., M. Aubertin, B. Bussière, and M. Benzaazoua. 2008. Geochemical transport modelling of drainage from experimental mine tailings cells covered by capillary barriers. *Applied Geochemistry* 23, no. 1: 1–24.
Moncur, M. C., C. J. Ptacek, D. W. Blowes, and J. L. Jambor. 2003. Fate and transport of metals from an abandoned tailings impoundment after 70 years of sulphide oxidation. In *Proceedings of Sudbury 03, Mining and the Environment III*, ed. S. Graeme, P. Beckett, and H. Conroy, 238–247. Sudbury, Ontario: Laurentian University.
Moncur, M. C., C. J. Ptacek, D. W. Blowes, and J. L. Jambor2005. Release, transport and attenuation of metals from an old tailings impoundment. *Applied Geochemistry* 20: 639–659.
Moncur, M. C., J. L. Jambor, C. J. Ptacek, and D. W. Blowes. 2009. Mine drainage from the weathering of sulfide minerals and magnetite. *Applied Geochemistry* 24: 2362–2373.
Nicholson, R. V., and J. M. Scharer. 1994. Laboratory studies of pyrrhotite oxidation kinetics. In *Environmental geochemistry of sulfide oxidation*, ed. C. M. Alpers and D. W. Blowes, 14–30. Washington, DC: American Chemical Society.

Nieto, J. M., A. M. Sarmiento, M. Olías, et al. 2007. Acid mine drainage pollution in the Tinto and Odiel Rivers (Iberian Pyrite Belt, SW Spain) and bioavailability of the transported metals to the Huelva Estuary. *Environment International* 33, no. 4: 445–455.

Nordstrom, D. K. 2003. Effects of microbiological and geochemical interactions in mine drainage. *Environmental Aspects of Mine Wastes* 31: 227–238.

Nordstrom, D. K. 2011. Hydrogeochemical processes governing the origin, transport and fate of major and trace elements from mine wastes and mineralized rock to surface waters. *Applied Geochemistry* 26: 1777–1791.

Nordstrom, D. K., and C. N. Alpers. 1999. Negative pH, efflorescent mineralogy, and consequences for environmental restoration at the Iron Mountain Superfund site, California. *Proceedings of the National Academy of Sciences* 96: 3455–3462.

Nordstrom, D. K., and G. Southam. 1997. Geomicrobiology of sulphide mineral oxidation. *Reviews in Mineralogy* 35: 381–390.

Nordstrom, D. K., J. W. Ball, R. J. Donahoe, and D. Whittemore. 1989. Groundwater chemistry and water-rock interactions at Stripa. *Geochimica et Cosmochimica Acta* 53, no. 8: 1727–1740.

Nordstrom, D. K., C. N. Alpers, C. J. Ptacek, and D. W. Blowes. 2000. Negative pH and extremely acidic mine waters from Iron Mountain, California. *Environmental Science and Technology* 34: 254–258.

Nordstrom, D. K., D. W. Blowes, and C. J. Ptacek. 2015. Hydrogeochemistry and microbiology of mine drainage: An update. *Applied Geochemistry* 57: 3–16.

Ohmura, N., K. Sasaki, N. Matsumoto, and H. Saiki. 2002. Anaerobic respiration using Fe(3+), S(0), and H(2) in the chemolithoautotrophic bacterium Acidithiobacillus ferrooxidans. *Journal of Bacteriology* 184, no. 8: 2081–2087.

Pabst, T., J. Molson, M. Aubertin, and B. Bussière. 2017. Reactive transport modelling of the hydro-geochemical behaviour of partially oxidized acid-generating mine tailings with a monolayer cover. *Applied Geochemistry* 78: 219–233.

Pabst, T., B. Bussiere, M. Aubertin, and J. Molson. 2018. Comparative performance of cover systems to prevent acid mine drainage from pre-oxidized tailings: A numerical hydro-geochemical assessment. *Journal of Contaminant Hydrology* 214: 39–53.

Paine, P. J. 1987. An historic and geographic overview of acid mine drainage. In *Proceedings of the Acid Mine Drainage Seminar/Workshop*, Halifax, NS, 23–26 March 1987, 1–45. Minister of Supply and Services Canada, Catalogue No. En 40–11–7/1987.

Paktunc, A. D. 1999a. Characterization of mine wastes for prediction of acid mine drainage. In *Environmental impacts of mining activities: Emphasis on mitigation and remedial measures*, ed. J. M. Azcue, 19–40. Berlin: Springer.

Paktunc, A. D. 1999b. Mineralogical constraints on the determination of neutralization potential and prediction of acid mine drainage. *Environmental Geology* 39: 103–112.

Paktunc, A. D., and N. K. Davé. 2002. Formation of secondary pyrite and carbonate minerals in the Lower Williams Lake tailings basin, Elliot Lake, Ontario, Canada. *American Mineralogist* 87: 593–602.

Parbhakar-Fox, A., B. Lottermoser, and D. Bradshaw. 2013. Evaluating waste rock mineralogy and microtexture during kinetic testing for improved acid rock drainage prediction. *Minerals Engineering* 52: 111–124.

Pellicori, D. A., C. H. Gammons, and S. R. Poulson. 2005. Geochemistry and stable isotope composition of the Berkeley pit lake and surrounding mine waters, Butte, Montana. *Applied Geochemistry* 20: 2116–2137.

Plante, B., B. Bussière, and M. Benzaazoua. 2012. Static tests response on 5 Canadian hard rock mine tailings with low net acid-generating potentials. *Journal of Geochemical Exploration* 114: 57–69.

Pope, J., N. Newman, D. Craw, D. Trumm, and R. Rait. 2010. Factors that influence coal mine drainage chemistry West Coast, South Island, New Zealand. *New Zealand Journal of Geology and Geophysics* 53, no. 2–3: 115–128.

Qian, G., R. Fan, M. D. Short, R. Schumann, J. Li, R. S. C. Smart, and A. R. Gerson. 2018. The effects of galvanic interactions with pyrite on the generation of acid and metalliferous drainage. *Environmental Science & Technology* 52, no. 9: 5349–5357.

Ribeta, I., C. J. Ptacek, D. W. Blowes, and J. L. Jambor. 1995. The potential for metal release by reductive dissolution of weathered mine tailings. *Journal of Contaminant Hydrology* 17: 239–273.

Rimstidt, J. D., A. Chermak John, and M. Gagen Patrick. 1994. Rates of reaction of galena, sphalerite, chalcopyrite, and arsenopyrite with Fe(III) in acidic solutions. In *Environmental geochemistry of sulfide oxidation*, ed. C. M. Alpers and D. W. Blowes, 2–13. Washington, DC: American Chemical Society.

Ritchie, A. I. M. 1994. Sulfide oxidation mechanisms: Controls and rates of oxygen transport. In *Short course handbook on environmental geochemistry of sulfide mine-wastes*, ed. J. L. Jambor and D. W. Blowes, 201–246. Waterloo, ON: Mineralogical Association of Canada.

Rohwerder, T., T. Gehrke, K. Kinzler, and W. Sand. 2003. Bioleaching review part A. *Applied Microbiology and Biotechnology* 63, no. 3: 239–248.

Santos, P. S. M., M. Otero, E. B. H. Santos, and A. C. Duarte. 2011. Chemical composition of rainwater at a coastal town on the southwest of Europe: What changes in 20 years? *Science of the Total Environment* 409, no. 18: 3548–3553.

Schippers, A., and W. Sand. 1999. Bacterial leaching of metal sulfides proceeds by two indirect mechanisms via thiosulfate or via polysulfides and sulfur. *Applied and Environmental Microbiology* 65, no. 1: 319–321.

Schwertmann, U., and L. Carlson. 2005. The pH-dependent transformation of schwertmannite to goethite at 25°C. *Clay Minerals* 40: 63–66.

Sherlock, E. J., R. W. Lawrence, and R. Poulin. 1995. On the neutralization of acid rock drainage by carbonate and silicate minerals. *Environmental Geology* 25: 43–54.

Singer, P. C., and W. Stumm. 1970. Acidic mine drainage: The rate determining step. *Science* 167: 1121–1123.

Smith, L. J. D., M. C. Moncur, M. Neuner, et al. 2013. The Diavik Waste Rock Project: Design, construction, and instrumentation of field-scale experimental waste-rock piles. *Applied Geochemistry* 36: 187–199.

Sobek, A. A., W. A. Schuller, J. R. Freeman, and R. M. Smith. 1978. *Field and laboratory methods applicable to overburden and minesoils*. EPA 600/2-78-054, 203pp.

Søndergaard, J., B. Elberling, and G. Asmund. 2008. Metal speciation and bioavailability in acid mine drainage from a high Arctic coal mine waste rock pile: Temporal variations assessed through high-resolution water sampling, geochemical modelling and DGT. *Cold Regions Science and Technology* 54, no. 2: 89–96.

Stockwell, J., L. Smith, J. L. Jambor, and R. Beckie. 2006. The relationship between fluid flow and mineral weathering in heterogeneous unsaturated porous media: A physical and geochemical characterization of a waste-rock pile. *Applied Geochemistry* 21: 1347–1361.

Temple, K. L., and A. R. Colmer. 1951. The autotrophic oxidation of iron by a new bacterium: Thiobacillus ferrooxidans. *Journal of Bacteriology* 62: 605–611.

Thomas, J. E., C. F. Jones, W. M. Skinner, and R. S. C. Smart. 1998. The role of surface sulphur species in the inhibition of pyrrhotite dissolution in acid conditions. *Geochimica et Cosmochimica Acta* 62: 1555–1565.

Thomas, J. E., W. M. Skinner, and R. S. C. Smart. 2001. A mechanism to explain sudden changes in rates and products for pyrrhotite dissolution in acid solution. *Geochimica et Cosmochimica Acta* 65: 1–12.

Topçu, S., S. Incecik, and A. T. Atimtay. 2002. Chemical composition of rainwater at EMEP station in Ankara, Turkey. *Atmospheric Research* 65, no. 1: 77–92.

Tyler, R. H., T. P. Boyer, T. Minami, M. M. Zweng, and J. R. Reagan. 2017. Electrical conductivity of the global ocean. *Earth, Planets and Space* 69, no. 1, 156.

Vriens, B., M. St. Arnault, L. Laurenzi, L. Smith, K. U. Mayer, and R. D. Beckie. 2018. Localized sulfide oxidation limited by oxygen supply in a full-scale waste-rock pile. *Vadose Zone Journal* 17.

Weber, A., A. S. Ruhl, and R. T. Amos. 2013. Investigating dominant processes in ZVI permeable reactive barriers using reactive transport modeling. *Journal of Contaminant Hydrology* 151: 68–82.

West Virginia University. 1971. Mine spoil potentials for water quality and controlled erosion. 14010 EJE 12/71. U.S. Environmental Protection Agency (USEPA), Washington, DC.

Wilson, D., R. T. Amos, D. W. Blowes, J. B. Langman, C. J. Ptacek, L. Smith, and D. C. Sego. 2018a. Diavik Waste Rock Project: A conceptual model for temperature and sulfide-content dependent geochemical evolution of waste rock – Laboratory scale. *Applied Geochemistry* 89: 160–172.

Wilson, D., R. T. Amos, D. W. Blowes, J. B. Langman, L. Smith, and D. C. Sego. 2018b. Diavik Waste Rock Project: Scale-up of a reactive transport model for temperature and sulfide-content dependent geochemical evolution of waste rock. *Applied Geochemistry* 96: 177–190.

Wollast, R. 1990. Rate and mechanism of dissolution of carbonates in the system $CaCO_3$-$MgCO_3$. In *Aquatic chemical kinetics: Reaction rates of processes in natural waters*, ed. W. Stumm, 431–445. New York: Wiley.

Zhang, X., H. Jiang, Q. Zhang, and X. Zhang. 2012. Chemical characteristics of rainwater in northeast China, a case study of Dalian. *Atmospheric Research* 116: 151–160. https://doi.org/10.1016/j.atmosres.2012.03.014.

Zunckel, M., C. Saizar, and J. Zarauz. 2003. Rainwater composition in northeast Uruguay. *Atmospheric Environment* 37, no. 12: 1601–1611. https://doi.org/10.1016/S1352-2310(03)00007-4.

2 Prediction of Acid Mine Drainage

Benoît Plante, Gary Schudel, and Mostafa Benzaazoua

The potential for acid mine drainage (AMD) generation from wastes can be predicted through static and kinetic tests. While static tests examine only the balance between acid-generating and neutralizing reactions, kinetic tests attempt to create accelerated weathering conditions in order to determine the relative rates of these reactions as well as the concentrations and release rates of potential contaminants in long-term drainages (e.g., Bouzahzah et al. 2014a, 2014b). However, the predictions made using these tests are driven by the representativeness of the samples and their relevance to the desired goals of the prediction work, which makes sampling as important as the testing itself.

This chapter describes a general strategy for AMD prediction, discusses the challenges of sampling for prediction purposes, presents an overview of the most used static and kinetic tests, outlines factors that must be considered in the operation of kinetic tests, presents the application of numerical tools to AMD prediction, and identifies critical research needs in AMD prediction. Although the static and kinetic tests will only be briefly described in this section, the interested reader can refer to other review papers and reports on the subject such as studies by Bouzahzah et al. (2014a), Bouzahzah et al. (2014b), Plante et al. (2015), the *MEND report 1.20.1* (MEND 2009), as well as to Dold's (2017) critical review of acid rock drainage prediction to learn more about the details of each method presented here.

2.1 AMD PREDICTION STRATEGY AND SAMPLING CONSIDERATIONS

The general strategy to predict the onset of AMD is to examine each lithology of a given deposit separately (e.g., SRK 1989). A sampling campaign should be established with the goal of identifying the risk of acid generation from the different lithologies. Ideally, each lithology should be sampled in a way that ensures the best spatial coverage possible of the rocks that will be mined (or exposed on pit walls) according to the mining plan. This is often a challenge, particularly for mine projects, since exploration campaigns will mostly aim for ore material and will not necessarily cross each of the non-economic lithologies destined to end up in waste rock piles. Generally, the number of samples should be proportional to the mass of each geological unit considered, based on the rule of thumb first proposed by SRK in 1989 (Equation 2.1), which is illustrated in Figure 2.1.

$$N = 0.026\,M^{0.5} \tag{2.1}$$

where N is the number of samples, and M is the mass of a given geologic unit.

The SRK rule of thumb (Figure 2.1) was not designed to be followed rigorously but should serve as a first approximation (SRK 1989; Plante et al. 2015), especially in the case of larger geological units that would require hundreds of samples. Lithologies can be demonstrated to be homogeneous with less than the recommended number of samples, while some others may be highly heterogeneous

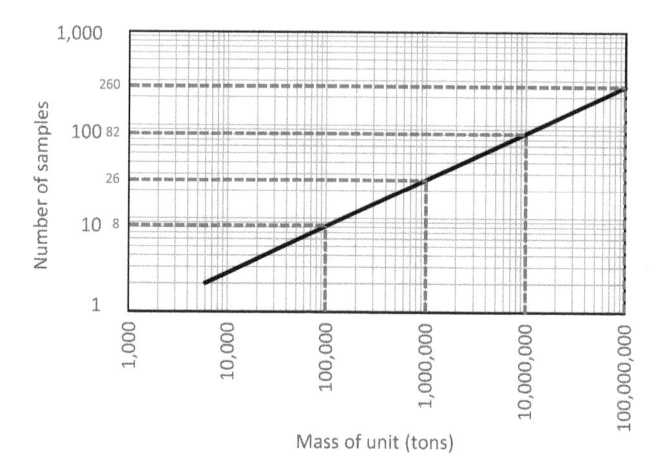

FIGURE 2.1 Illustration of the SRK (1989) rule of thumb for the number of samples required.

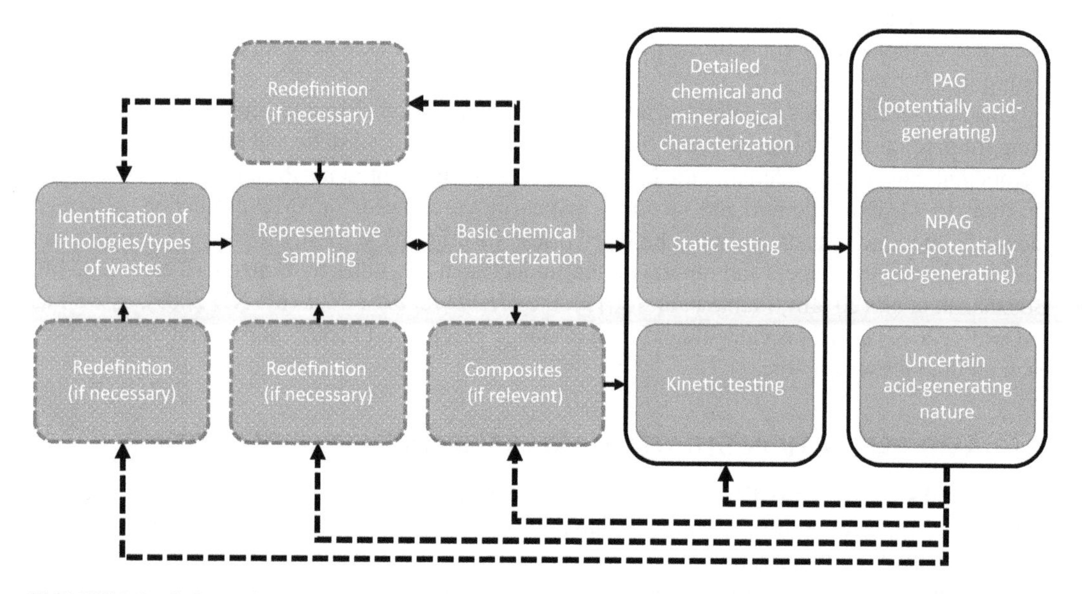

FIGURE 2.2 Schematics of the AMD prediction methodology (modified from Plante et al. 2015).

and require more than what is recommended by the rule of thumb. The proposed sampling strategy is shown in Figure 2.2, which illustrates that upon characterizing a set of samples, it can be decided to redefine the lithologies or the number of representative samples required. Therefore, the sampling strategy should be iterative and adjusted accordingly upon the evolution of the mining plan and refinements to the geologic units, as well as based on the statistical demonstration of the representativeness of the samples analyzed thus far.

In order to reduce the costs associated with prediction analyses, it is recommended to submit the samples to basic chemical characterization, which consists of analyzing simple parameters, such as whole-rock chemistry and sulfur/carbon analyses, to appreciate the heterogeneity of the different lithologies and to identify potential problematic areas of the deposit. At this preliminary step, it is also imperative to have as much knowledge of the mineralogical composition of the materials as possible, which enables linking some of the simple parameters to the mineralogical composition, and, therefore, to estimate the potential for acid generation and metals leaching (see Section 2.2.1). This does not necessarily mean going for very detailed mineralogical analyses; often, project geologists can easily provide the necessary information. The irony is that project geologists will often concentrate on the economic minerals in the deposit and might not know that some of the additional knowledge they acquire while looking through the core samples (such as the presence of carbonates that could provide buffering capacity or the presence of minerals that can easily dissolve and release problematic elements, like arsenic sulfosalts) are important to the mine planners dealing with future mine waste management and water quality. Consequently, having a team of people from all the different aspects of mine development share information is a simple key to successful mine waste management and water quality prediction.

Once the first set of samples are analyzed for their basic properties, the results can be used (according to the mineralogical information gathered) to select representative samples of a given lithology or area to be submitted to more advanced characterizations and testing. For instance, samples representing the mean composition of a lithology or the 75th percentile of a given problematic parameter (e.g., sulfur or arsenic content) for that lithology could be chosen or made from a composite, depending on the objectives of the chosen tests (e.g., MEND 2009; Plante et al. 2015).

The detailed mineralogical composition of the representative samples can then be determined using a combination of different methods, such as X-ray diffraction, optical and scanning electron microscopy, and automated mineralogical analyses such as Qemscan or MLA (e.g., Benzaazoua et al. 2017; Pooler and Dold 2017; Elghali et al. 2018). Particular attention needs to be given to the identification of all minerals that are involved in acid generation or acid consumption, as well as to their associations and degree of liberation (e.g., Strömberg and Banwart 1999; Parbhakar-Fox et al. 2013; Jamieson et al. 2015; Parbhakar-Fox and Lottermoser 2015; Benzaazoua et al. 2017; Dold 2017; Brough et al. 2017; Elghali et al. 2018).

The proposed prediction strategy (Figure 2.2) not only identifies the lithologies that are potentially acid generating (PAG) or non-acid generating (NPAG) and, therefore, provides insights about the risks associated with each but also enables design management scenarios that will help control acid generation (see Chapter 13). For instance, it may be beneficial to separate the waste rock lithologies in order to store acid-generating lithologies over acid-neutralizing ones that will buffer the pH and precipitate a significant part of the metals generated by sulfide oxidation (e.g., MEND 2009; Plante et al. 2015). Similarly, it may be possible to mine NPAG ores prior to acid-generating ones in order to have a first layer of acid-consuming materials at the base of the tailings pond (see Chapter 14). The Meadowbank (Agnico Eagle Mines Ltd.) mine waste rock management plan is a very good example of planning based on the acid-generating properties of the different lithologies (Voyer and Robert 2015).

Sampling for waste rock and tailings will require a different approach, depending on the development stage of a mine and the amount of material needed. Table 2.1 summarizes the sampling possibilities, while the following sections provide more detail about waste rock and tailings sampling.

2.1.1 Sampling of Waste Rocks

Sampling waste rocks of a mine project that is yet to enter production can be done either through sampling the core samples (or trenches) from exploration campaigns, from the rock excavated for exploration ramps, or from bulk samples. The main challenge here will reside in obtaining samples from all waste rock lithologies covering the entire portion to be mined as much as possible, as the core program principally aims for the ores rather than for uneconomical waste rocks. Of course, bulk samples and exploration ramps may facilitate the collection of a higher volume of waste rock samples than the cores, but only for the lithologies that are either available for bulk sampling or encountered during ramp construction, which will obviously be representative of the local area of the lithology but not necessarily the whole of it.

Sampling waste rocks in existing piles also poses a particular challenge. Indeed, the highly heterogeneous nature of waste rock piles, as well as their height, makes them extremely hard to sample

TABLE 2.1
Overview of Waste Rocks and Tailings Sampling Options

	Sampling Options	Challenges
Waste rocks		
Mine projects	Core samples Bulk samples Exploration ramp material	Limited quantity of materials Limited spatial covering of all lithologies
Operating or closed mines	Regular sampling upon production Existing waste rock piles	Heterogeneity in size and composition Difficult sampling at depth
Tailings		
Mine projects	Pilot testing materials	Limited quantity of materials Limited representativeness of pilot testing materials
Operating or closed mines	Regular sampling at the plant Tailings storage facility sampling	Heterogeneity in tailings ponds composition and grain size distribution Difficult access inward of ponds

at depth. Indeed, not only are waste rocks highly heterogeneous (grain sizes from µm to meter scale, variable composition) but the pile construction method adds heterogeneities throughout the structure (more or less compacted/fine areas, preferential segregation along slopes) (e.g., Anterrieu et al. 2010; Lahmira et al. 2017). Many authors have attempted to define sampling strategies for waste rock piles (e.g., MEND 1994; Price and Kwong 1997, Smith et al. 2000). For instance, Smith et al. (2000) proposed to divide a pile surface into at least 30 equal areas, each to be sampled for at least 100 g of their <1 cm fraction, air dried, and sieved to <2 mm, and to make a composite of these subsamples. Although this approach was proposed as a screening strategy to characterize and compare waste rock piles, they argue that it can be adapted to target other populations such as waste rock pile lifts or benches, geologic units, or any other operational units. If possible, waste rocks from operating mines should be sampled regularly as the piles are being constructed, at a frequency that will be deemed relevant based on tonnage and expected heterogeneities. Finally, the waste rock sampling should also consider the scale of the planned kinetic tests. Indeed, larger-scale kinetic tests might enable larger particles; Section 2.3 will provide more details regarding sample volume and maximum particle size requirements.

2.1.2 Sampling of Tailings

Tailings samples for a mine project can be obtained from the pilot-scale ore recovery tests that are performed in the feasibility studies. The main challenges associated with such samples are the representativeness of future operations and quantity. Indeed, the pilot metallurgical testing can produce tailings that can end up being quite different from those generated in the full-scale production, especially in terms of grain size distribution (and degree of liberation of minerals) and mineralogical composition. In addition, small-scale metallurgical testing is often limited to kg-sized batches, making larger-scale kinetic testing impossible.

Sampling tailings in existing ponds or other storage facilities can also be a challenge, essentially because of preferential segregation of particle sizes and densities across the site, mineralogical heterogeneities in the tailings, and health and safety issues. Indeed, conventional tailings deposition methods can induce preferential segregations of particles in a tailings ponds (Blight 2010), which depends on the deposition method and tailings properties (see Chapter 13 and Bussière 2007 for more details), and such heterogeneities can, in turn, have a significant impact on their geochemical behavior. In addition, mineralurgical plants often treat ores from various types of ores within a given mine over years and even process ores from different mines, which leads to important heterogeneities within the tailings management facilities. Taking these heterogeneities into account requires sampling the tailings at various depths and, if available, historical accounts of the tailings deposition. Finally, it can be very hazardous to access the tailings ponds, and some sampling campaigns might only be feasible when the tailings are frozen (of course, given that the facility is within an area that freezes over long enough). Otherwise, safety measures need to be taken to ensure a safe access to the tailings. As for waste rocks, tailings from operating mines should be sampled regularly at the plant, with a frequency that is deemed relevant based on tonnage and expected heterogeneities.

2.2 STATIC TESTS

Once the representative samples are chosen, they can be submitted to AMD prediction tools. The first set of tools available for water quality prediction are static tests, which are presented and discussed in the following.

2.2.1 The Sobek et al. (1978) Method and Its Derivatives

Static tests aim to compare the acid generation and acid neutralization potentials (NPs) of mine wastes. Sobek et al. (1978) detailed the first static prediction test upon its development from the team of Professor R.M. Smith's team at West Virginia University during the 1970s (Skousen 2017). What is now known as the "Sobek" method consists of converting the total S (%S_t) content to maximum potential acidity (AP) by multiplying by a conversion factor of 31.25 (Equation 2.2), expressed as the amount of calcite that would be necessary to neutralize the acid expected in kg/t equivalents. This method considers that all the sulfur is present within pyrite and that all that pyrite will oxidize to generate acid.

$$AP\left(\frac{kgCaCO_3}{t}\right) = \%S_t \frac{1000kg\,/\,t}{100\%}\frac{M_{CaCO_3}}{M_S} = 31.25 * \%S_t \tag{2.2}$$

In the Sobek method, the NP is calculated by back titration of a sample aliquot submitted to a 24-h acid attack (0.1 mol/L HCl). The amount of acid to add is determined using a preliminary assessment of carbonate reactivity called the "fizz test," which consists of qualifying the amount of bubbling (or fizz) generated when the sample is in contact with 25% HCl. As for the AP, the NP is expressed in kg/t of calcite equivalents. Comparison between the AP and NP, either by their difference (Net NP, or NNP = NP – AP) or by their ratio (NP ratio or NPR = NP/AP), determines whether the material will be acid generating or a net neutralizer. The criteria generally used are illustrated in Figure 2.3 and are as follows: the material is considered NPAG if NPR > 2 or NNP > 20 kg $CaCO_3$/t; it is considered PAG if NPR < 1 or NNP < –20 kg $CaCO_3$/t; and its acid-generating nature is considered uncertain if the results fall between these numbers (MEND 2009). However, different criteria are available depending on the legislation; site-specific criteria can also be defined.

Although Sobek tests are widely used as a screening assessment tool, they have certain limits, including the following:

- incomplete dissolution of some minerals such as dolomite and ankerite, which leads to an under-estimation of the NP (e.g., Kwong and Ferguson 1997; Plante et al. 2012);
- incomplete time for ferrous iron oxidation and hydrolysis upon siderite (or other Fe-bearing neutralizing mineral) dissolution (e.g., Skousen et al. 1997; Jambor et al. 2003), which overestimates the NP;
- more silicate dissolution than in typical storage conditions due to heating, which overestimates the NP (e.g., MEND 1991, 1996).

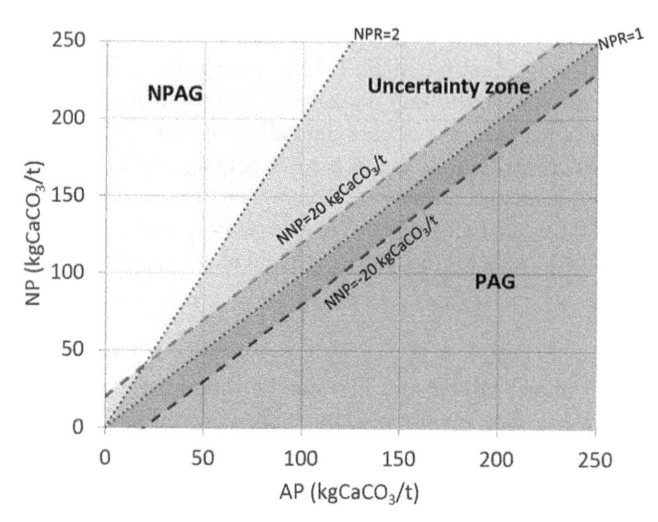

FIGURE 2.3 Criteria used to interpret Sobek test results.

Knowledge about the mineralogical composition of the materials allows for the interpretation of the Sobek test results, taking those limitations into account. In addition, many variations have been proposed to the Sobek procedure over the years in order to bypass these potential over- or under-estimation issues related to the AP or NP calculation, such as:

- performing the NP acid attack at room temperature instead of at 85°C (MEND 1989);
- subtracting the sulfate sulfur from the total sulfur to consider only the sulfide sulfur content for AP calculation (MEND 1989);
- using different concentrations of dilute HCl and a different end-point pH (Lawrence 1990; Lawrence and Wang 1997);
- considering the siderite content (Skousen et al. 1997);
- using a calculated fizz based on the carbon content (Frostad et al. 2003; Bouzahzah et al. 2015).

All variations of the Sobek procedure are relevant and can be used, but because these protocol differences can have significant effects on the NP results (e.g., Jambor et al. 2007), it is imperative to know which variation is used by the laboratory and to interpret the results accordingly, taking into account the composition of the materials.

A series of pure minerals was submitted to standard Sobek tests, and the results provided an interesting overview of the NP provided by each one (Jambor et al. 2007). Some of these results are shown in Table 2.2. This compilation highlights the orders of magnitude difference between the NP of carbonates versus that of the available NP from other minerals. Although the NP of the best neutralizing silicates is at least 25 times lower than that of carbonates, they can nevertheless provide enough neutralization to buffer the pH to near-neutral values.

Another simple static test consists of using the bulk carbon content to calculate the carbonate NP (or CNP) by multiplying the carbon content ($\%C_t$) by a factor of 83.3 in order to express the result in kg/t equivalents of calcite (Equation 2.3). The AP is calculated with the sulfur content as in the Sobek method (Equation 2.2). Since carbon and sulfur analyses with induction furnaces are fast and relatively cheap, this method enables a simple estimation of the net acid generation (NAG) potential of materials (MEND 2009).

$$CNP\left(\frac{kgCaCO_3}{t}\right) = \%C_t \frac{1000kg/t}{100\%} \frac{M_{CaCO_3}}{M_C} = 83.3 * \%C_t \tag{2.3}$$

TABLE 2.2
Sobek NP values of Selected Minerals (Jambor et al. 2007 and References Therein)

Mineral	Sobek NP Values (kg $CaCO_3$/t)
Calcite	1000
Olivine/forsterite	38
Serpentine	32
Nepheline	25
Albite–anorthite series	1–14 (linear trend from albite to anorthite)
Analcime	11
Apatite	8
Chlorite/clinochlore	6
Pyroxene	5
K-feldspar	1

Obviously, this method is relevant only if the carbon content is related to carbonate minerals. Other forms of carbon found in waste rocks and tailings, such as graphitic carbon or organic carbon, must be taken into account. The same goes for siderite, a carbonate mineral that is not a net neutralizer (see Chapter 1, Equation 1.17). In these cases, the total carbon content should be replaced by the neutralizing carbonate carbon content. The results from carbonate NP can be interpreted using the same criteria as for the Sobek tests (Figure 2.3).

2.2.2 NAG Tests

The NAG test was developed as a stand-alone prediction test, while the kinetic NAG test was developed to estimate the lag time before the development of acid drainage (Miller et al. 1997). These tests are based on the forced oxidation of the sulfides using hydrogen peroxide and to allow some time (up to 2 to 3 days) for the neutralizers to buffer the pH. The final pH of the solution (NAG_{pH}) determines whether the material is acid generating, while the acidity titration estimates the intensity of acid generation. Materials are considered PAG if their $NAG_{pH} < 4.5$ and NPAG if $NAG_{pH} > 4.5$. In addition, analyses of the soluble species at the end of a NAG test can provide insight as to the potential contaminants to watch during kinetic testing and in the field.

Three different variations of the NAG test are available: the single NAG, the sequential NAG, and the kinetic NAG test. The single NAG test consists of a single peroxide addition and is relevant for materials containing up to approximately 1% sulfide sulfur. For materials containing more sulfide sulfur, a single peroxide addition is often not enough to oxidize all the available sulfides; in those cases, sequential NAG tests are necessary, which are simply consecutive single peroxide additions in the same sample until no more reactions are observed upon peroxide addition. Finally, the kinetic NAG tests consist of single or sequential NAG tests in which the pH and temperature are continuously monitored. The intensity of heat production and pH decrease upon peroxide additions (and sulfide oxidation), and the speed at which the pH increases back upon reaction of the neutralizing minerals, provide insight as to the reactivity of the minerals involved in AMD generation.

Combining NAG and Sobek test results can help refine the predictions. For instance, NAG_{pH} can be plotted against the NPR results for the samples, as illustrated in Figure 2.4 (the same could be done for NNP).

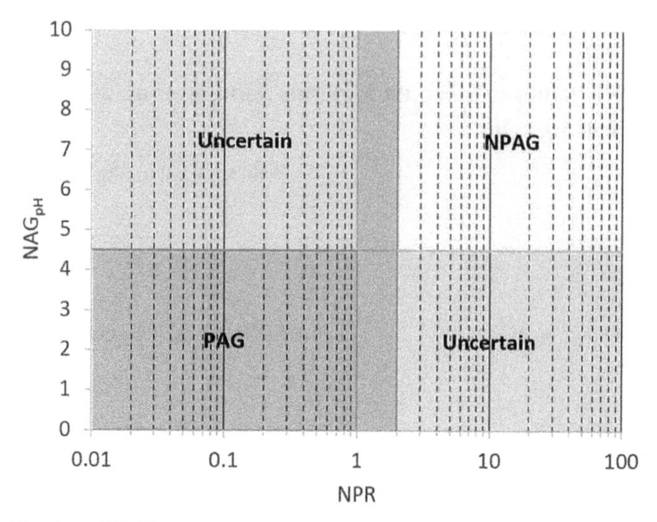

FIGURE 2.4 Combination of NAG and NPR test results for static test interpretation.

2.2.3 Mineralogical Static Tests

In parallel to the Sobek and NAG static tests, mineralogical calculations of the AP and NP were proposed by several authors, including Lawrence and Scheske (1997) and Paktunc (1999a, 1999c), based on the sum of the individual contribution of each mineral to the NP and AP. Although these methods were not used often outside research facilities at first, their use by mining practitioners is constantly increasing with the popularization of reliable and accurate mineralogical characterizations at decreasing costs from various commercial laboratories. The Lawrence and Scheske (1997) method is shown in Equation 2.4, where M_{CaCO3} is the calcite molar mass, while C_{Mi}, M_{Mi}, and R_i are, respectively, the concentration (wt%), molar mass, and relative reactivity (unitless) of the "i" mineral.

$$NP\left(\frac{kgCaCO_3}{t}\right) = \frac{1000kg}{t} M_{CaCO_3} \sum_{i=1}^{n} \frac{C_{Mi}R_i}{M_{Mi}} \tag{2.4}$$

Indeed, the calculation involves the relative reactivity as classified by MEND (1993) based on a compilation of reaction rates made by Sverdrup (1990). The method received criticism upon its release; Paktunc (1999b) pointed out that the Lawrence and Scheske (1997) paper contained discrepancies in the compilation of the relative rates and that the mineralogical composition of the samples was calculated using a normative calculation that was not relevant for the samples studied. However, it was later demonstrated that its results compare relatively well with other NP methods (e.g., Plante et al. 2012) when used properly and interpreted considering its limitations.

The Paktunc method calculates a mineralogical NP and AP (Equations 2.5 and 2.6, respectively). The Paktunc NP calculation takes into account the proportion of hydrolyzable cations (such as Fe) within the neutralizing minerals and the stoichiometry of neutralization, while the AP considers the different stoichiometry of sulfide oxidation of pyrite and pyrrhotite.

$$NP\left(\frac{kg\ H_2SO_4}{t}\right) = \sum_{i=1}^{n} \frac{10X_i\omega_a c_i}{n_{M,i}\omega_i} \tag{2.5}$$

$$AP\left(\frac{kg\ H_2SO_4}{t}\right) = \sum_{i=1}^{n} \frac{10n_{M,a}X_i\omega_a}{\omega_i} \tag{2.6}$$

where 10 is the conversion factor (1000 kg/t divided by 100%), X_i is the concentration of mineral "i" (mass %), ω_a et ω_i is the molar mass of H_2SO_4 and "i" mineral (g/mol), c_i is the proportion of hydrolyzable cations in neutralizing mineral "i," $n_{M,i}$, moles of mineral "i" necessary to neutralize 1 mole of H_2SO_4, and $n_{M,a}$ represents the moles of H_2SO_4 generated by the oxidation of 1 mole of the sulfide mineral (2 for pyrite and 1 for pyrrhotite).

The NP and AP results from the Paktunc method are in kg H_2SO_4/t, but converting to equivalents of $CaCO_3$ is easily done by multiplying the results by a factor of 1.02 (ratio of the molar masses of $CaCO_3$ and H_2SO_4, respectively, 100.09 and 98.08 g/mol, respectively).

2.2.4 Static Test Comparison and Selection

Each static test procedure has its own advantages and limitations, which are all related to the experimental procedure and the response of minerals to the testing conditions. Therefore, as much knowledge as possible is needed in order to select the most appropriate static tests and interpret the results. Table 2.3 compares the static tests and can be used to determine (1) the best procedure given the mineralogical composition of the samples to be tested and (2) which parameters are important to consider when interpreting the results.

TABLE 2.3
Static Test Comparison

Advantages	Limits
Sobek et al. (1978)	
- Widely used since its release - Fast procedure	- NP overestimation in presence of Fe/Mn-bearing neutralizers - NP overestimation because of high temperatures - AP overestimation of other sulfur species than sulfide - Fizz test is subjective
Sobek modified by Lawrence and Wang (1997)	
- Lower temperature minimizes NP overestimation compared to original procedure	- Insufficient acid addition to measure NP values higher than 125kg$CaCO_3$/t - NP overestimation in presence of Fe/Mn-bearing neutralizers
Sobek modified by Jambor et al. (2003), Skousen et al. (1997), Stewart et al. (2006)	
- Peroxide addition enables to bypass NP overestimation in presence of Fe/Mn-bearing neutralizers	- Same limits as original procedure
Sobek modified by Bouzahzah (2015)	
- Fizz determination based on carbonate contents instead of objective evaluation - Same advantages as Sobek mod. by Lawrence and Wang (1997)	- Same limits as Sobek mod. by Lawrence and Wang (1997)
Carbonate NP	
- Simple, fast, cheap	- Overestimation of carbon species other than carbonates - Overestimation in presence of Fe/Mn-bearing neutralizers
Single and sequential NAG tests (Miller et al. 1997)	
- Provides a NNP analog in a single test	- Lack of separate NP and AP
Kinetic NAG tests (Miller et al. 1997)	
- Same as the single procedure - Provides clues on speeds of acid generation and neutralization	- Same as the single procedure

(Continued)

TABLE 2.3 (*Continued*)
Static Test Comparison

Advantages	Limits
Lawrence and Scheske (1997)	
- Considers silicate NP	- Controversy over relative reactivity ratings - Overestimation in presence of Fe/Mn-bearing neutralizers
Paktunc (1999c)	
- Considers Fe/Mn contents of neutralizers - Considers the different stoichiometry of pyrite and pyrrhotite oxidation	- Overestimation when considering silicates; better suited when carbonates are the main neutralizers

2.3 KINETIC TESTS

The primary goal of kinetic tests, such as humidity cell and column tests, is to reproduce the weathering processes that would be expected in the field at an accelerated pace in order to determine dissolution and drainage characteristics with respect to time (Paktunc 1999a). These tests essentially submit geologic materials to repeated weathering cycles, which consist of rinsing samples with deionized water, then collecting and analyzing the leachates for relevant geochemical parameters such as pH, redox potential, electrical conductivity, acidity, alkalinity, and concentration of dissolved species. By tracking changes in the geochemical properties of leachates over time, conclusions can be made, for example, about whether samples will generate acid (i.e., comparing the rates of acid generation vs. neutralization) and when, which contaminants can be expected in drainages and at what concentrations. Kinetic tests can be conducted at the laboratory scale or at the field scale. They can, therefore, be used to answer diverse questions about the behavior of mine wastes in a variety of environmentally relevant contexts.

With respect to static tests, the major challenges common to most kinetic tests are that they are longer in duration (weeks to years), more expensive, and more challenging to interpret. However, kinetic testing is essential when

- insufficient information is provided by static tests about whether samples will generate AMD;
- there is a potential for the development of contaminated neutral drainage (CND);
- expected contaminants and their concentrations in drainage waters need to be determined;
- site-specific field conditions need to be considered and reproduced;
- the lag time before AMD generation needs to be evaluated;
- or the efficiency of reclamation scenarios needs to be assessed and compared.

2.3.1 Overview of Kinetic Tests

Although many kinetic test procedures are available, their general principles are similar, and they can be tackled using a generalized approach, as illustrated in Figure 2.5. Indeed, materials to be submitted to kinetic testing need to be characterized before performing the tests; upon the finalization of the tests, the leachate data are interpreted; and post-testing materials characterizations provide additional information about the geochemical behavior of the materials.

The most relevant initial characterizations necessary for kinetic testing are listed in Figure 2.5. These characterizations enable the installation of materials for kinetic testing in conditions as close as possible to the expected field conditions in terms of porosity, volumetric water content (VWC), and suction. They also enable a more robust interpretation of the results.

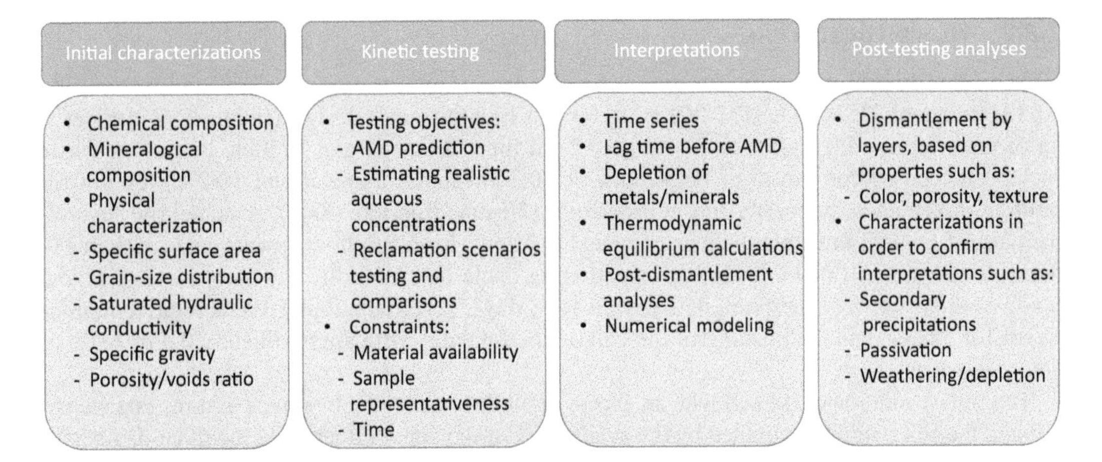

FIGURE 2.5 Characteristics of the main steps involved in kinetic testing.

The type of kinetic procedure must be chosen based on the pursued objectives, time constraints, and quantity of material. For instance, humidity cells are best suited to determine the long-term onset of AMD but less appropriate to estimate the water quality in terms of contaminant concentrations or to simulate and compare reclamation scenarios. On the other hand, column tests often provide more realistic leachate qualities and enable comparisons of reclamation scenarios but require longer test durations and needing more materials. Field tests (in situ barrels or field experimental cells) submit the materials to the most realistic environmental conditions but involve even more materials and require more expertise for installation, maintenance, and interpretation.

Interpretation of kinetic testing is done by looking at the time series of the different parameters measured in the leachates. The depletion of the source terms of the different species analyzed can also be deduced (e.g., Benzaazoua et al. 2004). Thermodynamic equilibrium calculations can be used to evaluate the speciation of the chemical species in solution, as well as saturation indices of secondary minerals (e.g., Hakkou et al. 2009; El Adnani et al. 2016). Finally, numerical modeling can support the interpretation by identifying the main factors of influence on water quality and predicting the long-term geochemical behavior of the materials (e.g., Molson et al. 2008; Ouangrawa et al. 2009; Pabst et al. 2018).

Post-testing characterizations of the physical and mineralogical properties of the materials are also essential to complement leachate data from kinetic testing and aid in interpretation of the weathering behavior. They allow identifying which minerals will provide short- and long-term buffering capacities. Whenever possible, dismantlement of the materials in layers can help identify the weathering profile and its characteristics (e.g., Bussière et al. 2004; Demers et al. 2008). It can also enable to verify the presence of secondary minerals suggested to precipitate by thermodynamic equilibrium calculation. The profiles of porosity, grain size distribution, and VWC can also provide useful information on the evolution of the materials, as well as validate the results of some of the sensors installed at the beginning of the test (e.g., Ouangrawa et al. 2010). Some of the post-dismantlement physical and mineralogical characterization results are also needed for numerical modeling.

The following section will cover the fundamentals of the most commonly used kinetic tests in AMD prediction: humidity cells, weathering cells, column tests, and field procedures. It will also explain how to interpret kinetic test data and describe the initial and post-testing characterizations necessary.

2.3.2 Humidity Cell Tests

Unlike other kinetic tests, the designs and procedures for humidity cells are standardized by ASTM standard D5744 (ASTM 2013). A typical humidity cell design (Figure 2.6) utilizes 1 kg of material in a Plexiglass cell with a 10.2 cm inner diameter and 20.3 cm height for waste rocks (crushed and/or sieved to <6.3 mm), or 20.3 cm inner diameter and 10.2 cm height for tailings (taken as is, generally in the order of <150 μm; Bussière 2007). A geotextile layer at the base of the cell prevents the loss of fines but allows for air to flow upward and leachates to drain downward (MEND 1991). Each weathering cycle in a humidity cell lasts for 7 days and involves three steps: (1) aeration with dry air for 3 days, (2) circulation of 100% relative humidity air for 3 days, and (3) flushing of the cell on the final day with approximately 0.5 to 1.0 L of deionized water.

The forced aeration of the cell with an excess of air during the first two steps ensures that weathering in the cell will not be limited by the availability of oxygen. Furthermore, the three-day period of dry air leads to evaporation of some of the pore water in the cell, which increases oxygen diffusion rates by several orders of magnitude (see Chapter 3) as well as the concentrations of oxidation products in pore waters, thus resulting in increased weathering and acid generation. The circulation of humid air helps to maintain the level of pore water in the cell following the period of drainage and evaporation and allows for the re-dissolution and diffusion of secondary precipitates, which may have formed during the dry phase. The final phase (the flushing of the cell) ensures that the weathering products that were accumulated during each cycle are leached out of the cell (MEND 1991; ASTM 2013; Bouzahzah et al. 2015). The leachate volume is recorded, and the leachate is analyzed for the relevant parameters.

The ASTM norm mentions that humidity cell tests should be run for at least 20 weekly cycles. In practice, humidity cells are often run for 30 to 40 weeks and more. Stopping the tests should be done when the objective of the test is attained, when the results are stabilized over the last cycles, and/or when no subsequent changes in the parameters are expected.

Because of the size of the humidity cell apparatus, using instrumentation such as VWC or suction probes is almost impossible. Whenever such information is important, other larger types of kinetic tests need to be considered.

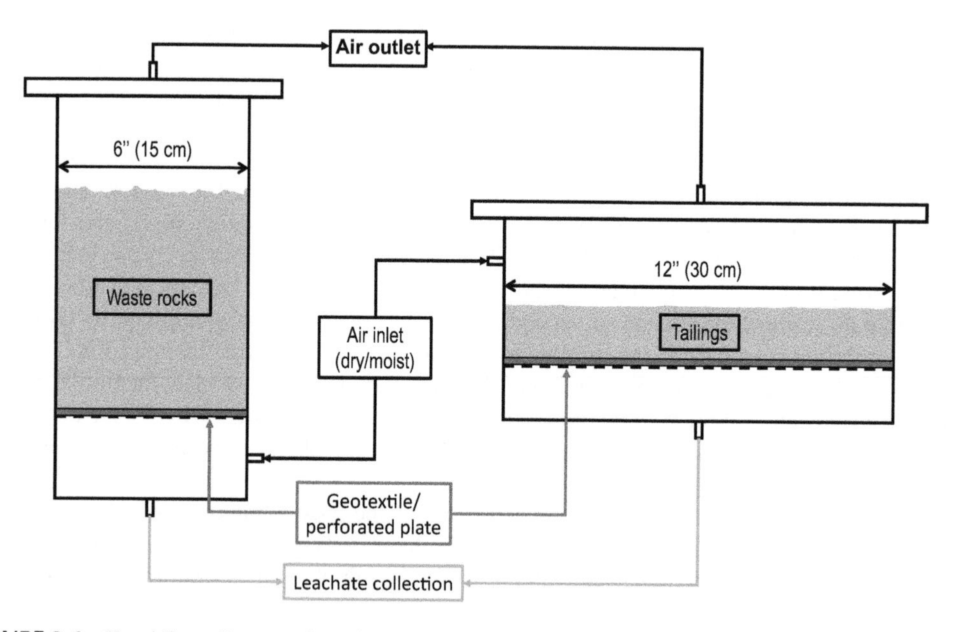

FIGURE 2.6 Humidity cell test settings for waste rocks and tailings.

2.3.3 Weathering Cell Tests

Weathering cell tests are a small-scale version of the humidity cell test, developed by Villeneuve et al. (2003) based on the work of Cruz et al. (2001). It consists of 9 cm Büchner funnels equipped with a 0.45-μm glass fiber filter containing approximately 67 g of materials, which are leached bi-weekly (for instance, on Mondays and Thursdays) with 50 mL deionized water (Figure 2.7). The water is left in contact with the materials for approximately three hours before being vacuum filtered for leachate collection and analyses. The two weekly leachates can either be analyzed separately or combined for weekly analysis.

As for humidity cells, the tests should be stopped only when the objective of the test is attained, when the results are stabilized over the last cycles, and/or when no subsequent changes in the parameters are expected. Villeneuve et al. (2009) demonstrated that the weathering cell test results are comparable to those of the humidity cell tests despite conditions being slightly more aggressive (e.g., more oxidizing conditions). Therefore, weathering cells can be useful in cases when there is not enough material to perform humidity cells, as well as when less time is available to reach the objectives of kinetic testing.

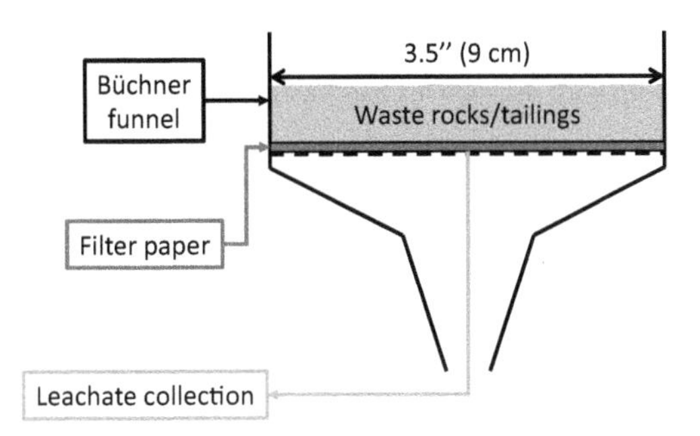

FIGURE 2.7 Weathering cell apparatus.

2.3.4 Column Tests

While the design and procedures of column tests are not standardized in practice, recommendations are provided in ASTM standard D4874 (ASTM 2014) and various other documents (e.g., MEND 1991; Prediction Workgroup of the Acid Drainage Technology Initiative 2000; Demers et al. 2011). As a rule of thumb, the diameter of a column should be at least six to ten times greater than the largest particle size in the tested material; no guidelines are set for column height (MEND 1991; ASTM 2014). In practice, columns are often made of 15 and 30 cm tubing (6 or 12 inches; Figure 2.8), requiring particles <2.3 and 5 cm, respectively, but can also be of other diameters. Columns are typically designed with either a porous ceramic disk or a geotextile at the base, both of which are used to prevent the loss of fines (Bouzahzah et al. 2014b). In addition, the porous ceramic enables control over the suction in the materials. Generally, a porous ceramic plate is used for tests involving fine-grained materials, such as tailings, and geotextile-covered perforated disks are used for coarse-grained materials, such as waste rocks. At the beginning of testing, the level of the water table is adjusted to the desired level (usually in accordance with what is expected in the field) by adding water and applying the necessary suction (Bussière et al. 2004).

Columns are flushed with deionized water, generally in a continuous single pass, using peristaltic pumps or gravity-assisted flow (MEND 1991). The volume of leachate is adjusted based on the objective of the test and can vary from a trickle leach from the top all the way to a complete saturation from the bottom up, corresponding to volumes anywhere from less than a liter up to tens of liters, depending on the size of the column and the voids ratio of the materials. The volume of leachate can be chosen to reflect the local climatic conditions to be tested. Leachates are collected

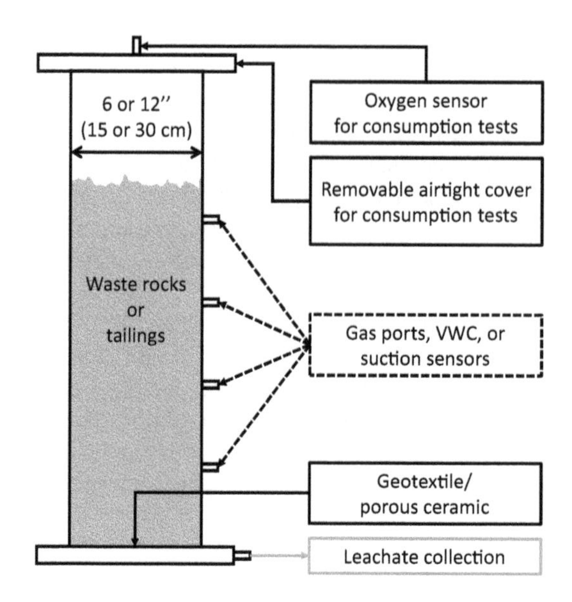

FIGURE 2.8 Column test setup with instruments and main components.

from the base of the column, usually once per cycle, which generally lasts 15 to 30 days; however, leachates may be sampled more frequently if needed.

The column tests can be large enough to enable monitoring of several parameters using different types of sensors and/or sampling ports at relevant depths. For instance, VWC and suction probes can be relevant especially when testing reclamation scenarios (Bussière et al. 2004; Kalonji-Kabambi et al. 2017; Larochelle et al. 2019).

Furthermore, the rate of sulfide oxidation can be described through oxygen dynamics in the columns. Indeed, oxygen consumption tests evaluate the flux of oxygen diffusing into and being consumed by the tailings (Elberling et al. 1994; Elberling and Nicholson 1996; Mbonimpa et al. 2003, 2011). They consist of sealing the air chamber at the top of the column and monitoring the decrease in oxygen concentration over a few hours. The oxygen flux and consumption can also be estimated through an analysis of interstitial gas composition using oxygen sensors or gas sampling ports at different depths into the materials. Solving the Fick's laws (Chapter 3) enables the estimation of the oxygen flux at the surface (mol/m^2/year), which can be used to assess the efficiency of oxygen barrier-type covers.

2.3.5 Field Procedures

The acid-generating potential of a material can be assessed in the field using a variety of methods, including field columns or barrels, field test pads, and experimental waste rock piles. These tests can be held in settings that mimic the real conditions (more realistically than in laboratory tests) to which the wastes are going to be exposed, in terms of wet–dry cycles, freeze–thaw cycles and temperature variations, rainwater leachings, volume and frequencies (instead of deionized water), as well as liquid to solid ratio, to name a few. As for laboratory tests, they can be instrumented in order to follow the relevant hydrogeological and/or geochemical parameters, they can be submitted to oxygen consumption tests, and their interstitial oxygen profiles can be monitored. These setups also allow for test management and reclamation scenarios.

Field columns or barrels are relatively simple to set up and consist of installing columns filled with the material to assess in a field setting as close as possible to the expected conditions of exposition, ideally at the mine site. The setups are submitted to natural wettings by rainwater, and the leachates are collected and measured from the bottom for analyses of the parameters described for laboratory kinetic tests, as needed. The schematics are similar to those of the laboratory column tests (Figure 2.7). Typical examples of field barrels can be found in a study by Hirsche et al. (2017) for the Antamina mine waste rocks and in a study by Bossé et al. (2013) for reclamation scenarios of the Kettara mine site.

Field test pads are made of an impervious geomembrane (with protective sand layers if needed) over which a few cubic meters of the mine tailings or waste rocks are deposited, and the leachates are collected and measured at the bottom for analyses. Typically, field test pads are either aboveground (Figure 2.9) or buried (Figure 2.10). Aboveground cells are perfect for self-supporting materials (such as waste rock or filtered tailings), which are simply placed on top of a geomembrane (without the use of dikes; Figure 2.9), while buried cells are best suited for loose materials such as conventional and thickened tailings that simply fill a cavity lined with the geomembrane (Figure 2.10). Typical examples of aboveground cells can be found in a study by Bussière et al. (2011), which were used to study the Lac Tio mine waste rocks, while examples of buried field test pads can be found in a study by Bussière et al. (2007), where various cover scenarios for tailings facilities are compared to an uncovered control test pad.

Experimental waste rock piles are also used (Figure 2.11), where lysimeters are installed at strategic locations for leachate collection, measurement, and analyses (e.g., Bailey et al. 2016; Martin et al. 2017; Poaty et al. 2018). The most heavily documented case is perhaps that of the Diavik mine site (e.g., Smith et al. 2013) that was used to study the geochemical behavior and microbiology of its AMD-generating waste rocks (e.g., Bailey et al. 2015, 2016), as well as water/gas transport and temperature regimes (e.g., Chi et al. 2013; Neuner et al. 2013; Amos et al. 2015) within the pile.

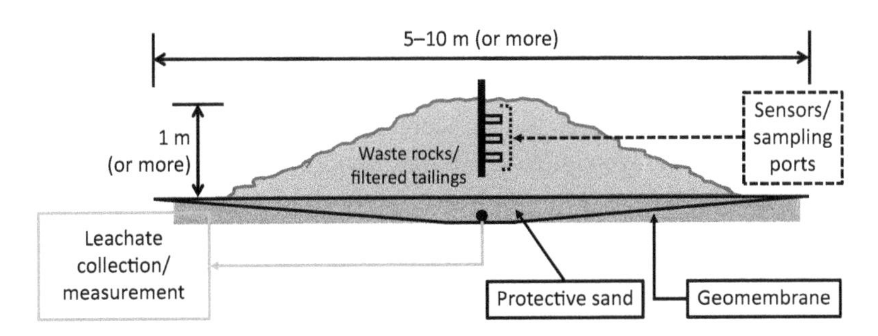

FIGURE 2.9 Field test pad for waste rocks or self-supporting tailings.

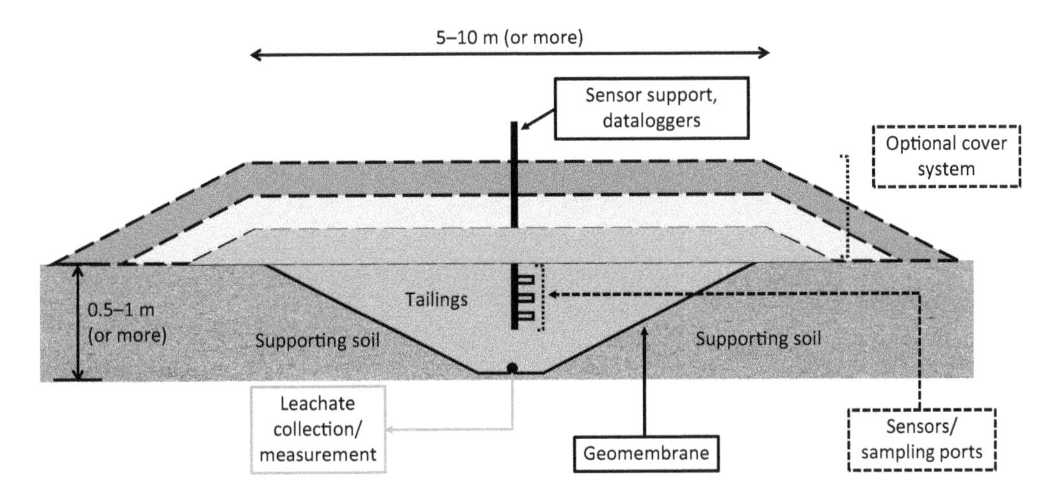

FIGURE 2.10 Field test pad for tailings, including an optional cover system.

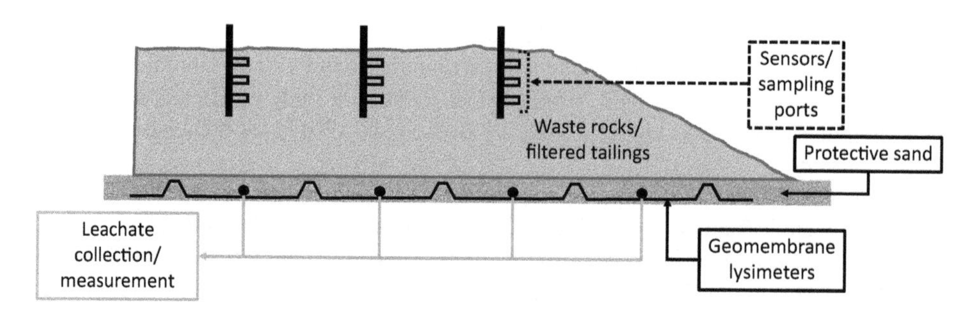

FIGURE 2.11 Schematic representation of field experimental pile of waste rocks or self-supporting tailings.

Reclamation scenarios for waste rock piles can also be tested using an experimental pile. For instance, insulation covers were tested at the Diavik mine (e.g., Smith et al. 2013), while a flow control layer was tested at the Lac Tio mine site (e.g., Martin et al. 2017). As is the case for laboratory kinetic tests, dismantlement (or deconstruction) of experimental field cell tests enables the characterization of the post-testing materials, provides useful information about the weathering profile, and validates some of the data provided by the sensors.

2.3.6 Kinetic Test Interpretation

Kinetic test results may demonstrate that the materials are acid-generating by generating acidic leachates during the actual test. However, some materials might not generate AMD during the tests, and therefore, more advanced, long-term interpretation methods are required to assess their acid generation potential. Three such methods will be compared in the following: the ASTM recommended method (ASTM 2013), the oxidation–neutralization method (Benzaazoua et al. 2004), and the chemical/mineralogical depletion method (Villeneuve et al. 2009). All methods consider the same hypotheses – namely, that (1) the geochemical conditions will remain the same during the tests (pH, Eh, temperature, and the release ratio of oxidation and neutralization rates) and (2) depletion of one of the minerals will not significantly affect the reaction rates of the other minerals. The interested reader is invited to look at the references provided for more details.

The humidity cell procedure (ASTM 2013) suggests a method proposed by White and Jeffers (1994) to compare the time for NP depletion, based on the calcium and magnesium generation rates in the leachates, and for AP depletion, based on the sulfate production rate. The calculation of the time needed for AP and NP depletion are shown in Equations 2.7 and 2.8, where AP and NP are expressed as $kgCaCO_3/t$, and R_{SO4}, R_{Ca}, and R_{Mg} are the release rates of sulfate, calcium, and magnesium, respectively (converted in kg $CaCO_3/t$).

$$time\ for\ AP\ depletion = \frac{AP}{R_{SO_4}} \tag{2.7}$$

$$time\ for\ NP\ depletion = \frac{NP}{R_{Ca} + R_{Mg}} \tag{2.8}$$

Using these equations, the materials for which time for AP depletion is greater than that of NP depletion are considered acid-generating in the long term. An example is shown in Figure 2.12, where the experimental results are extrapolated, illustrating that the NP will likely be depleted before the AP; this material is, therefore, considered PAG.

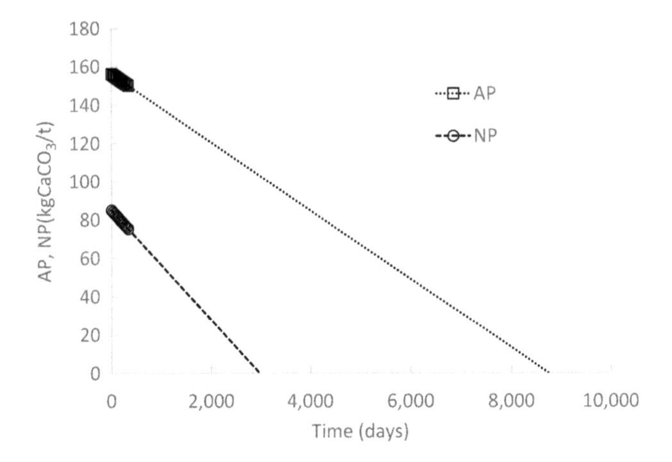

FIGURE 2.12 Example of the ASTM interpretation method.

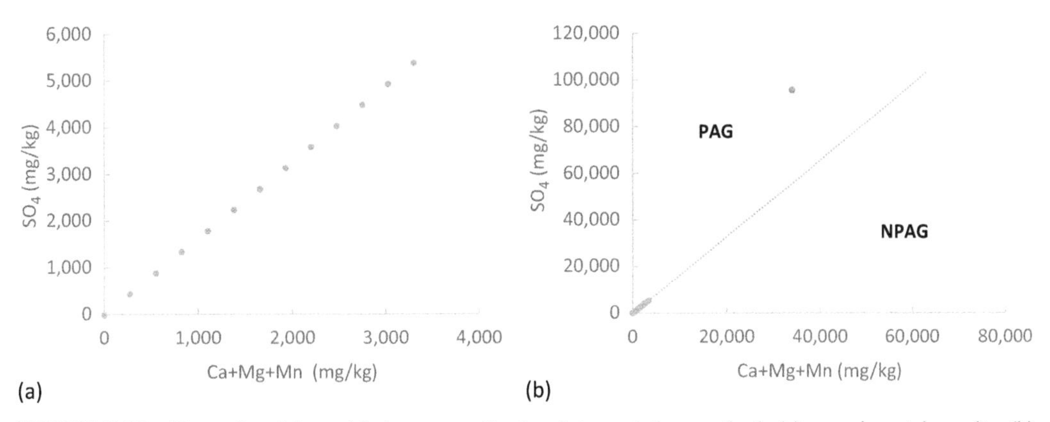

FIGURE 2.13 Example of the oxidation–neutralization interpretation method; (a) experimental results; (b) extrapolation of the results shown in (a) and comparison with the initial composition.

The oxidation–neutralization curves method (Benzaazoua et al. 2004) consists of plotting the cumulative loads of calcium, magnesium, and manganese (representing neutralizing mineral dissolution) leached out of kinetic tests against that of sulfate loads (resulting from sulfide oxidation). Figure 2.13 illustrates a typical example of the use of this interpretation method. The long-term assessment of AMD generation is done by comparing the point corresponding to the initial contents of these elements with an extrapolation of the relationship obtained (usually linear). If the point falls on the sulfur side of the relationship, the neutralizing minerals will deplete first, and it suggests that the material is PAG. Inversely, the point will fall on the Ca + Mg + Mn side of the relationship for non-acid-generating materials. This method was successfully applied with various materials (e.g., Awoh et al. 2014; Benzaazoua et al. 2004; Bouzahzah et al. 2014a; El Adnani et al. 2016; Pabst et al. 2014; Villeneuve et al. 2009).

The chemical/mineralogical depletion method (Villeneuve et al. 2009) is essentially very similar to the ASTM method, but instead of calculating the times for AP and NP depletion, it calculates the time for depletion of the minerals responsible for acid generation and its neutralization, such as sulfides and carbonates. In order to do this, sulfate is used to calculate the depletion of sulfide minerals, while elements such as Ca, Mg, and Mn are used to calculate the depletion time of carbonate minerals, using Equations 2.9 and 2.10, where R_{SO4}, R_{Ca}, R_{Mg}, and R_{Mn} (mg/kg/days) are the release rates of sulfate, calcium, magnesium, and manganese, respectively. S_{init} and $Carbonate_{init}$ are the initial sulfide and carbonate contents (mg/kg), respectively, and M_{SO4} and M_S are the molar masses of sulfate (96.06 g/mol) and sulfur (32.065 g/mol), respectively.

$$time\ for\ sulfide\ depletion = \frac{M_{SO_4} S_{init}}{M_S R_{SO_4}} = \frac{2.9958 * S_{init}}{R_{SO_4}} \tag{2.9}$$

$$time\ for\ carbonate\ depletion = \frac{Carbonate_{init}}{R_{Ca} + R_{Mg} + R_{Mn}} \tag{2.10}$$

Given that the initial composition of the materials is known with enough precision, the depletion of separate minerals can be calculated using one of its components as a proxy (e.g., Cu for chalcopyrite, Zn for sphalerite, Ca for calcite, Mg for dolomite, etc.). Other elements, such as Na and Al, can be used to calculate the time to depletion of aluminosilicates.

The literature contains countless studies involving kinetic tests. However, the manner in which the results are interpreted varies a lot between studies, and major issues can be found in many of them. To provide more guidance on AMD prediction, Maest and Nordstrom (2017) provided a series of suggestions regarding kinetic testing, including the following:

- Perform a complete chemical and mineralogical characterization of the materials
- Conduct complete weekly leachate analyses of the major, minor, and trace elements
- Pursue the kinetic tests as long as possible (i.e., > 1 year for humidity cell tests) when the leachates do not become acidic during the tests in order to provide extensive geochemical data to be interpreted in the long term
- Vary the climatic and hydrogeological conditions in parallel studies at larger scales to more closely match the expected variation of field conditions and to take into account the effects of available surface area and total mass being tested.

These recommendations are consistent with others in the literature, such as in studies by Villeneuve et al. (2009), Parbhakar-Fox and Lottermoser (2015), Parbhakar-Fox et al. (2013), and those in this book.

2.3.7 Numerical Tools

Many numerical tools may be used to aid in kinetic test interpretation, from simple geochemical speciation mass transfer models, such as PHREEQC (Parkhurst and Appelo, 2013) and Visual MINTEQ (USEPA 2006), to advanced reactive transport models, such as MIN3P (Mayer et al. 2002). An extensive review of numerical models was conducted by Alpers and Nordstrom (1999) and updated by Blowes et al. (2014).

Geochemical speciation mass transfer models calculate the activities of every ion in solution and use their speciation to calculate, for instance, the saturation indexes of potential minerals and the exchange between aqueous, gaseous, and solid phases at different temperatures. These calculations are based on processes such as geochemical speciation, acid–base and redox equilibrium, precipitation/dissolution, and sorption phenomena, as well as some reaction kinetics and simple one-dimensional solute transport (Blowes et al. 2014). These types of models are used for three main types of modeling: speciation, inverse modeling (also called mass balance modeling), and forward modeling (also called reaction path modeling). The main characteristics and examples of application of these models are compared in Table 2.4.

All these different models can be used, for instance, to interpret kinetic test results or field data by calculating the probable precipitation of secondary phases or sorption phenomena that are occurring

TABLE 2.4
Comparison of the Main Geochemical Speciation Mass Transfer Model Types

Model Type	Characteristics	Examples of Applications
Speciation	Calculates the speciation of all ions in solution (e.g., ratio of As^{3+}/As^{5+})	Saturation index calculations
Inverse (or mass balance) modeling – mass transfer	Calculates the transfer from a solid phase in which a liquid traveled, when the initial and final water qualities are known.	Interpretation of groundwater transport in a known geological setting
Inverse (or mass balance) modeling – mixing	Enables to calculate the result of mixing two waters of known qualities	Water quality of a stream receiving AMD
Forward (or reaction path) modeling	Enables to calculate the evolution of a known water quality upon mass transfer reactions with another phase (solid, liquid, and gaseous)	AMD treatment

within tailings (e.g., Blowes et al. 1998; Moncur et al. 2005; Gunsinger et al. 2006). In that particular case, the input parameters are the water quality, specifically the species concentrations, pH, redox potential, temperature, acidity, alkalinity, and the presence of a gas phase (e.g., atmosphere with 400 ppm CO_2).

Reactive transport models combine the equations that describe the geochemical reactions between liquid, solid, and gas phases (as those in geochemical speciation mass transfer models previously described) with equations that describe the physical transport of gas and liquids (such as laws governing saturated and unsaturated flow, advection, and dispersion; Blowes et al. 2014). Usually, the transport part of the model works separately from the geochemical portion of the code (mostly using forward modeling). Thus, for each increment, the model will start by calculating the transport of fluids and will resolve the forward modeling geochemistry equations, taking into account the movement of fluids it just calculated. Examples of such models are MIN3P (Mayer et al. 2002), MINTRANT (Walter et al. 1994), and PHAST (Parkhurst et al. 2004).

For instance, the MIN3P solves the Richards equations of unsaturated flow for the movements of fluids (water and gas) and solves the mass balance equations for advective-diffusive solute transport and diffusive gas transport (Mayer et al. 2012). The model includes the MINTEQA2 (Allison et al. 1991) and WATEQf4 (Ball and Nordstrom 1991) thermodynamic databases. The MIN3P code was demonstrated to adequately simulate the oxidation of pyrite and associated release of metals in mining wastes, as well as the subsequent attenuation processes (Mayer et al. 2012). It was also used to study the influence of heterogeneities in waste rock piles on the uncertainty related to AMD predictions (Pedretti et al. 2020).

These models can be calibrated using the initial physical properties (e.g., porosity, water retention curve, and hydraulic function) and mineralogical characterization, as well as the leachate quality time series of a given laboratory or field kinetic test. Once the model is calibrated on experimental data, key factors of influence on the water quality can be identified by performing parametric studies. It can also be used to predict the future water quality.

Reactive transport modeling was frequently applied to AMD cases in order to better understand the geochemical behavior of mine wastes, based on laboratory tests or field test pad results (Jurjovec et al. 2004; Brookfield et al. 2006; Pedretti et al. 2017; Wilson et al. 2018). Numerical modeling was also used to assess the performance of reclaimed mine sites with covers with capillary barrier effects (Molson et al. 2008) and the performance of monolayer covers with elevated water table (Ouangrawa et al. 2009; Pabst et al. 2017, 2018). Reactive transport modeling was also used to model the development of CND in waste rocks (Demers et al. 2013).

2.4 RESEARCH AND DEVELOPMENT NEEDS

Although many AMD prediction tools are already available and widely used, significant improvements are still possible in order to reduce the uncertainties associated with the predictions. Examples are listed as follows:

- *Including the degree of liberation in ABA assessment and kinetic test interpretation*
  Recently, attempts have been made to interpret ABA and kinetic test results in conjunction with advanced mineralogical characterizations that not only quantify the spatial distribution of minerals but also look at mineralogical variations in different grain sizes and describe the mineralogical associations and degree of liberation (e.g., Benzaazoua et al. 2017; Brough et al. 2017; Parbhakar-Fox et al. 2013). Most efforts related to the degree of liberation were aimed at the sulfide minerals, but improvements are still needed to characterize the neutralizing minerals. Results published thus far are promising, but the exact manner in which to take these data into consideration when interpreting the static and kinetic test data still needs to be generalized, as it is mostly on a case-to-case basis for now. More research is also required to include the information regarding the degree of liberation of minerals in reactive transport models.
- *Using metal isotope geochemistry in kinetic testing*
  Many different geochemical reactions, such as secondary precipitations and sorption, can induce isotope fractionation (the relative partitioning of the heavier and lighter isotopes between two coexisting phases) in dissolved metals. Therefore, the isotope fractionation analysis can provide insights as to which phenomena are mainly responsible for the water quality of the leachates (e.g., Skierszkan et al. 2016, 2019; Veeramani et al. 2015; Salifu et al. 2018, 2019) and enable its prediction over the long term. As metal isotopic analyses are becoming more available, there is a great opportunity to improve our knowledge of the geochemistry of mine wastes.
- *Modeling other types of contaminated drainages*
  As tough AMD remains the major challenge of the mining industry, there is an increasing interest in reducing the impact of mining operations on the environment. Consequently, mines generating other types of contaminated drainage (e.g., CND, radioactive drainage, rare earth elements leaching from mine wastes, and alkaline drainage) are being constantly challenged to improve their environmental performance, which calls for a better understanding of the processes governing water quality in these settings.
- *Understanding the microbiological influence on the geochemistry of mining wastes*
  It has been long known that microbes influence the geochemistry of mining wastes (e.g., Nordstrom et al. 2015). However, the vast majority of the microbiological investigations were limited to semi-quantitative evaluation of the size of populations (e.g., most probable number analyses) and culture-based techniques. As genomic analysis techniques are being constantly developed, there is an opportunity to discover new species of bacteria and gain a better understanding of how the microbiota interacts with mining wastes and influences its geochemistry (e.g., Bruneel et al. 2005; Hallberg 2010; Auld et al. 2013; Park et al. 2018). Moreover, there is a great opportunity to understand the microbiological influence on the geochemistry and water quality of the materials submitted to kinetic tests at various scales compared to in situ conditions, which will greatly improve the predictive capabilities.

REFERENCES

Allison, J. D., D. S. Brown, and K. J. Novo-Gradac. 1991. *MINTEQA2/PRODEFA2, A Geochemical Assessment Model for Environmental Systems*. EPA/600/3–91/021.

Alpers, C. N., and D. K. Nordstrom. 1999. Geochemical modeling of water-rock interactions in mining environments. In *Environmental geochemistry of mineral deposits*, eds. G. S. Plumlee and M. J. Logsdon, Littleton, Colorado. Published by The Society of Economical Geologists, pp. 289–323.

Amos, R. T., D. W. Blowes, B. L. Bailey, D. C. Sego, L. Smith, and A. I. M. Ritchie. 2015. Waste-rock hydrogeology and geochemistry. *Applied Geochemistry* 57 (0):140–156.

Anterrieu, O., M. Chouteau, and M. Aubertin. 2010. Geophysical characterization of the large-scale internal structure of a waste rock pile from a hard rock mine. *Bulletin of Engineering Geology and the Environment* 69, no. 4: 533–548.

ASTM. 2013. *ASTM International D5744–13: Standard Test Method for Laboratory Weathering of Solid Materials Using a Humidity Cell*. ASTM International.

ASTM. 2014. *Astm International D4874: Standard Test Method for Leaching Solid Material in a Column Apparatus*. ASTM International.

Auld, R. R., M. Myre, N. C. S. Mykytczuk, L. G. Leduc, and T. J. S. Merritt. 2013. Characterization of the microbial acid mine drainage microbial community using culturing and direct sequencing techniques. *Journal of Microbiological Methods* 93, no. 2: 108–115.

Awoh, A., M. Mbonimpa, B. Bussière, B. Plante, and H. Bouzahzah. 2014. Laboratory study of highly pyritic tailings submerged beneath a water cover under various hydrodynamic conditions. *Mine Water and the Environment* 33, no. 3: 241–255.

Bailey, B. L., D. W. Blowes, L. Smith, and D. C. Sego. 2015. The Diavik Waste Rock project: Geochemical and microbiological characterization of drainage from low-sulfide waste rock: active zone field experiments. *Applied Geochemistry* 62:18–34.

Bailey, B. L., D. W. Blowes, L. Smith, and D. C. Sego. 2016. The Diavik Waste Rock project: geochemical and microbiological characterization of drainage from low-sulphide waste rock: active zone field experiments. *Applied Geochemistry* 62:18–34.

Ball, J. W., and D. K. Nordstrom. 1991. User's manual for WATEQ4F, with revised thermodynamic data base and test cases for calculating speciation of major, trace, and redox elements in natural waters. *USGS Open File Report* 183:91–183.

Benzaazoua, M., H. Bouzahzah, Y. Taha, et al. 2017. Integrated environmental management of pyrrhotite tailings at Raglan Mine: Part 1 challenges of desulphurization process and reactivity prediction. *Journal of Cleaner Production* 162:86–95.

Benzaazoua, M., B. Bussière, A. M. Dagenais, and M. Archambault. 2004. Kinetic tests comparison and interpretation for prediction of the Joutel tailings acid generation potential. *Environmental Geology* 46, no. 8: 1086–1101.

Blight, G. E. 2010. *Geotechnical Engineering for Mine Waste Storage Facilities*. 1st ed. London: CRC Press.

Blowes, D. W., J. L. Jambor, C. J. Hanton-Fong, L. Lortie, and W. D. Gould. 1998. Geochemical, mineralogical and microbiological characterization of a sulphide-bearing carbonate-rich gold-mine tailings impoundment, Joutel, Quebec. *Applied Geochemistry* 13, no. 6: 687–705.

Blowes, D. W., C. J. Ptacek, J. L. Jambor, et al. 2014. 11.5—The geochemistry of acid mine drainage. In *Treatise on geochemistry*, 2nd ed., ed. K. Turekian and H. Holland, 131–190. Oxford: Elsevier.

Bossé, B., B. Bussière, R. Hakkou, A. Maqsoud, and M. Benzaazoua. 2013. Assessment of phosphate limestone wastes as a component of a store-and-release cover in a semiarid climate. *Mine Water and the Environment* 32, no. 2:152–167.

Bouzahzah, H., M. Benzaazoua, B. Bussière, and B. Plante. 2014a. Prediction of acid mine drainage: Importance of mineralogy and the test protocols for static and kinetic tests. *Mine Water and the Environment* 33, no. 1: 54–65.

Bouzahzah, H., M. Benzaazoua, B. Bussière, and B. Plante. 2014b. Revue de littérature détaillée sur les tests statiques et les essais cinétiques comme outils de prédiction du drainage minier acide. *Déchets Sciences et Techniques* 66:14–31.

Bouzahzah, H., M. Benzaazoua, B. Bussière, and B. Plante. 2015. ASTM normalized humidity cell kinetic test: protocol improvements for optimal sulphide tailings reactivity. *Mine Water and the Environment* 34, no. 3: 242–257.

Bouzahzah, H., M. Benzaazoua, B. Plante, and B. Bussiere. 2015. A quantitative approach for the estimation of the "fizz rating" parameter in the acid-base accounting tests: A new adaptations of the Sobek test. *Journal of Geochemical Exploration* 153:53–65.

Brookfield, A. E., D. W. Blowes, and K. U. Mayer. 2006. Integration of field measurements and reactive transport modelling to evaluate contaminant transport at a sulphide mine tailings impoundment. *Journal of Contaminant Hydrology* 88:1–22.

Brough, C., J. Strongman, R. Bowell, et al. 2017. Automated environmental mineralogy; the use of liberation analysis in humidity cell testwork. *Minerals Engineering* 107:112–122.

Bruneel, O., R. Duran, K. Koffi, et al. 2005. Microbial diversity in a pyrite-rich tailings impoundment (Carnoulès, France). *Geomicrobiology Journal* 22, no. 5: 249–257.

Bussière, B. 2007. Colloquium 2004: hydrogeotechnical properties of hard rock tailings from metal mines and emerging geoenvironmental disposal approaches. *Canadian Geotechnical Journal* 44:1019–1052.

Bussière, B., M. Aubertin, M. Mbonimpa, J. W. Molson, and R. P. Chapuis. 2007. Field experimental cells to evaluate the hydrogeological behaviour of oxygen barriers made of silty materials. *Canadian Geotechnical Journal* 44, no. 3: 245–265.

Bussière, B., M. Benzaazoua, M. Aubertin, and M. Mbonimpa. 2004. A laboratory study of covers made of low-sulphide tailings to prevent acid mine drainage. *Environmental Geology* 45, no. 5: 609–622.

Bussière, B., I. Demers, I. Dawood, et al. 2011. Comportement géochimique et hydrogéologique des stériles de la mine Lac Tio. *Proceedings of the Symposium 2011 sur l'environnement et les mines, Rouyn-Noranda,* Québec, Canada. Published by the Canadian Institute of Mining, Metallurgy and Petroleum (CIM)

Chi, Xiaotong, Richard T. Amos, Marek Stastna, David W. Blowes, David C. Sego, and Leslie Smith. 2013. The Diavik Waste Rock project: implications of wind-induced gas transport. *Applied Geochemistry* 36 (0):246–255.

Cruz, R., B. A. Mendez, M. Monroy, and I. Gonzalez, 2001. Cyclic voltammetry applied to evaluate reactivity in sulfide mining residues. *Applied Geochemistry* 16:1631–1640.

Demers, I., B. Bussière, M. Aachib, and M. Aubertin. 2011. Repeatability evaluation of instrumented column tests in cover efficiency evaluation for the prevention of acid mine drainage. *Water, Air, and Soil Pollution* 219:113–128.

Demers, I., B. Bussière, M. Benzaazoua, M. Mbonimpa, and A. Blier. 2008. Column test investigation on the performance of monolayer covers made of desulphurized tailings to prevent acid mine drainage. *Minerals Engineering* 21 (4):317–329.

Demers, I., J. Molson, B. Bussière, and D. Laflamme. 2013. Numerical modeling of contaminated neutral drainage from a waste-rock field test cell. *Applied Geochemistry* 33: 346–356.

Dold, B. 2017. Acid rock drainage prediction: a critical review. *Journal of Geochemical Exploration* 172:120–132.

El Adnani, M., B. Plante, M. Benzaazoua, R. Hakkou, and H. Bouzahzah. 2016. Tailings weathering and arsenic mobility at the abandoned Zgounder Silver Mine, Morocco. *Mine Water and the Environment* 35:508–524.

Elberling, B., and R. V. Nicholson. 1996. Field determination of sulphide oxidation rates in mine tailings. *Water Resources Research* 32:1773–1784.

Elberling, B., R. V. Nicholson, E. J. Reardon, and P. Tibble. 1994. Evaluation of sulphide oxidation rates: a laboratory study comparing oxygen fluxes and rates of oxidation product release. *Canadian Geotechnical Journal* 31:375–383.

Elghali, A., M. Benzaazoua, H. Bouzahzah, B. Bussière, and H. Villarraga-Gómez. 2018. Determination of the available acid-generating potential of waste rock, part i: mineralogical approach. *Applied Geochemistry* 99:31–41.

Frostad, S. R., W. A. Price, and H. Bent. 2003. Operational NP determination – accounting for iron manganese carbonates and developing a site-specific fizz rating. *Proceedings of the Sudbury 2003 Mining and the environment Conference, Sudbury 2003*, ed. G. Spiers, P. Beckett, and H. Conroy, 231–237. Sudbury, ON, Canada. Published by Laurentian University, Sudbury, ON, Canada.

Gunsinger, M. R., C. J. Ptacek, D. W. Blowes, J. L. Jambor, and M. C. Moncur. 2006. Mechanisms controlling acid neutralization and metal mobility within a Ni-rich tailings impoundment. *Applied Geochemistry* 21:1301–1321.

Hakkou, R., M. Benzaazoua, and B. Bussière. 2009. Laboratory evaluation of the use of alkaline phosphate wastes for the control of acidic mine drainage. *Mine Water and the Environment* 28, no. 3: 206–218.

Hallberg, K. B. 2010. New perspectives in acid mine drainage microbiology. *Hydrometallurgy* 104, no. 3–4: 448–453.

Hirsche, D. T., R. Blaskovich, K. U. Mayer, and R. D. Beckie. 2017. A study of Zn and Mo attenuation by waste-rock mixing in neutral mine drainage using mixed-material field barrels and humidity cells. *Applied Geochemistry* 84:114–125.

Jambor, J. L., J. E. Dutrizac, and M. Raudsepp. 2007. Measured and computed neutralization potentials from static tests of diverse rock types. *Environmental Geology* 52:1173–1185.

Jambor, J. L., J. E. Dutrizac, M. Raudsepp, and L. A. Groat. 2003. Waste management: effect of peroxide on neutralization-potential values of siderite and other carbonate minerals. *Journal of Environmental Quality* 32:2373–2378.

Jamieson, H. E., S. R. Walker, and M. B. Parsons. 2015. Mineralogical characterization of mine waste. *Applied Geochemistry* 57:85–105.

Jurjovec, J., D. W. Blowes, C. J. Ptacek, and K. U. Mayer. 2004. Multicomponent reactive transport modeling of acid neutralization reactions in mine tailings. *Water Resources Research* 40:W1120201–W1120217.

Kalonji Kabambi, A., B. Bussière, and I. Demers. 2017. Hydrogeological behaviour of covers with capillary barrier effects made of mining materials. *Geotechnical and Geological Engineering* 35, no. 3: 1199–1220.

Kwong, Y. T. J., and K. T. Ferguson. 1997. Mineralogical changes during NP determinations and their implications. *Proceedings of the 4th International Conference on Acid Rock Drainage (ICARD)*, ed. Mine Environment Neutral Drainage (MEND) Program, CANMET, 435–437. Vancouver, Canada. Published by Natural Resources Canada, Ottawa, Canada.

Lahmira, B., R. Lefebvre, M. Aubertin, and B. Bussière. 2017. Effect of material variability and compacted layers on transfer processes in heterogeneous waste rock piles. *Journal of Contaminant Hydrology* 204:66–78.

Larochelle, C. G., B. Bussière, and T. Pabst. 2019. Acid-generating waste rocks as capillary break layers in covers with capillary barrier effects for mine site reclamation. *Water, Air, and Soil Pollution* 230, no. 3: 1–16.

Lawrence, R. W. 1990. Prediction of the behaviour of mining and processing wastes in the environment. In *Proceedings of the Western regional symposium on mining and mineral processing wastes*, ed. F.M. Doyle, pp.115–121. Littleton, CO, U.S.A. Published by the Society for Mining, Metallurgy, and Exploration.

Lawrence, R. W., and M. Scheske. 1997. A method to calculate the neutralization potential of mining wastes. *Environmental Geology* 32:100–106.

Lawrence, R. W., and Y. Wang. 1997. Determination of neutralisation potential in the prediction of acid rock drainage, *Proceedings of the 4th International Conference on Acid Rock Drainage (ICARD)*, ed. Mine Environment Neutral Drainage (MEND) Program, CANMET, 451–464. Vancouver, Canada. Published by Natural Resources Canada, Ottawa, Canada.

Maest, A. S., and D. K. Nordstrom. 2017. A geochemical examination of humidity cell tests. *Applied Geochemistry* 81:109–131.

Martin, V., B. Bussière, B. Plante, et al. 2017. Controlling water infiltration in waste rock piles: design, construction, and monitoring of a large-scale in-situ pilot test pile. *Proceedings of the Canadian Geotechnical Conference GeoOttawa*, ed. Canadian Geotechnical Society, 1–8. Ottawa, ON, Canada. Published by the Canadian Geotechnical Society, Ottawa, ON, Canada.

Mayer, K. U., R. T. Amos, S. Molins, and F. Gérard. 2012. Reactive transport modeling in variably saturated media with MIN3P: basic model formulation and model enhancements. In *Groundwater reactive transport models*, eds. Zhang, F., Yeh, G.T., Parker, J.C.,186–211. Published by Bentham Science Publishers, Oak Park, IL, U.S.A.

Mayer, K. U., E. O. Frind, and D. W. Blowes. 2002. Multicomponent reactive transport modeling in variably saturated porous media using a generalized formulation for kinetically controlled reactions. *Water Resources Research* 38, no.9: 13-1–13-21.

Mbonimpa, M., M. Aubertin, M. Aachib, and B. Bussiere. 2003. Diffusion and consumption of oxygen in unsaturated cover materials. *Canadian Geotechnical Journal* 40:916–932.

Mbonimpa, M., M. Aubertin, and B. Bussière. 2011. Oxygen consumption test to evaluate the diffusive flux into reactive tailings: interpretation and numerical assessment. *Canadian Geotechnical Journal* 48:878–890.

MEND. 1988. *MEND Report 5.5.1 Reactive Acid Mine Tailings Stablization (RATS) Research Plan*, ed. Canmet. Published by Canmet, Ottawa, ON, Canada.

MEND. 1989. *MEND report 1.16.1a: Investigation of predictive techniques for acid mine drainage*, ed. Lawrence, R.W., Poling, G.P. and Marchant, P.B. Published by Canmet, Ottawa, ON, Canada.

MEND. 1991. *MEND report 1.16.1b: Acid rock drainage prediction manual*, ed. Marchant, P. B.; Lawrence, R. W. Published by Canmet, Ottawa, ON, Canada.

MEND. 1993. *MEND report 1.32.1: Prediction and prevention of acid rock drainage from a geological and mineralogical perspective*, ed. Kwong, Y.J.T. Published by Canmet, Ottawa, ON, Canada.

MEND. 1994. *MEND report 4.5.1: Review of waste rock sampling techniques*. Ed. SENES Consultants Ltd., Golder Associates Ltd., and Laval University. Published by Canmet, Ottawa, ON, Canada.

MEND. 1996. *MEND report 1.16.3: Determination of Neutralization Potential for Acid Rock Drainage Prediction*, ed. Lawrence, R. W. and Wang, Y. Published by Canmet, Ottawa, ON, Canada.

MEND. 2009. *MEND report 1.20.1: Prediction manual for drainage chemistry from sulphidic geologic materials*, ed. Price, W.A. Published by Canmet, Ottawa, ON, Canada.

Miller, A., A. Robertson, and T. Donahue. 1997. *Advances in Acid Drainage Prediction Using the NAG test, Proceedings of the 4th International Conference on Acid Rock Drainage (ICARD)*, Vancouver, Canada, ed. Mine Environment Neutral Drainage (MEND) Program, CANMET, 533–549. Vancouver, Canada. Published by Natural Resources Canada, Ottawa, Canada.

Molson, J., M. Aubertin, B. Bussiere, and M. Benzaazoua. 2008. Geochemical transport modelling of drainage from experimental mine tailings cells covered by capillary barriers. *Applied Geochemistry* 23:1–24.

Moncur, M. C., C. J. Ptacek, D. W. Blowes, and J. L. Jambor. 2005. Release, transport and attenuation of metals from an old tailings impoundment. *Applied Geochemistry* 20:639–659.

Neuner, M., L. Smith, D. W. Blowes, D. C. Sego, L. J. D. Smith, N. Fretz, and M. Gupton. 2013. The Diavik waste rock project: water flow through mine waste rock in a permafrost terrain. *Applied Geochemistry* 36 (0):222–233.

Nordstrom, D. K., D. W. Blowes, and C. J. Ptacek. 2015. Hydrogeochemistry and microbiology of mine drainage: an update. *Applied Geochemistry* 57:3–16.

Ouangrawa, M., M. Aubertin, J. W. Molson, B. Bussière, and G. J. Zagury. 2010. Preventing acid mine drainage with an elevated water table: Long-term column experiments and parameter analysis. *Water, Air, and Soil Pollution* 213, no. 1–4: 437–458.

Ouangrawa, M., J. Molson, M. Aubertin, B. Bussière, and G. J. Zagury. 2009. Reactive transport modelling of mine tailings columns with capillarity-induced high water saturation for preventing sulfide oxidation. *Applied Geochemistry* 24, no. 7: 1312–1323.

Pabst, T., M. Aubertin, B. Bussière, and J. Molson. 2014. Column tests to characterise the hydrogeochemical response of pre-oxidised acid-generating tailings with a monolayer cover. *Water, Air, and Soil Pollution* 225, no. 2: 1–21.

Pabst, T., B. Bussiere, M. Aubertin, and J. Molson. 2018. Comparative performance of cover systems to prevent acid mine drainage from pre-oxidized tailings: A numerical hydro-geochemical assessment. *Journal of Contaminant Hydrology* 214:39–53.

Pabst, T., J. Molson, M. Aubertin, and B. Bussière. 2017. Reactive transport modelling of the hydro-geochemical behaviour of partially oxidized acid-generating mine tailings with a monolayer cover. *Applied Geochemistry* 78:219–233.

Paktunc, A. D. 1999a. Characterization of mine wastes for prediction of acid mine drainage. In *Environmental impacts of mining activities: Emphasis on mitigation and remedial measures*, ed. J. M. Azcue, 19–40. Berlin: Springer.

Paktunc, A. D. 1999b. Discussion of "A method to calculate the neutralization potential of mining wastes" by Lawrence and Scheske (1997). *Environmental Geology* 38:82–84.

Paktunc, A. D. 1999c. Mineralogical constraints on the determination of neutralization potential and prediction of acid mine drainage. *Environmental Geology* 39:103–112.

Parbhakar-Fox, A., and B. G. Lottermoser. 2015. A critical review of acid rock drainage prediction methods and practices. *Minerals Engineering* 82:107–124.

Parbhakar-Fox, A., B. Lottermoser, and D. Bradshaw. 2013. Evaluating waste rock mineralogy and microtexture during kinetic testing for improved acid rock drainage prediction. *Minerals Engineering* 52:111–124.

Park, Jin Hee, Bong-Soo Kim, and Chul-Min Chon. 2018. Characterization of iron and manganese minerals and their associated microbiota in different mine sites to reveal the potential interactions of microbiota with mineral formation. *Chemosphere* 191 (Supplement C):245–252.

Parkhurst, D. L., and C. A. J. Appelo. 2013. *Description of Input and Examples for PHREEQC Version 3—A Computer Program for Speciation, Batch-Reaction, One-Dimensional Transport, and Inverse Geochemical Calculations Book 6, Chapter 43 of section A*. US Geologiecal Survey Techniques and Methods. https://pubs.usgs.gov/tm/06/a43/pdf/tm6-A43.pdf.

Parkhurst, D. L., K. L. Kipp, O. Engesgaard, and S. R. Charlton. 2004. *PHAST-A program for simulating ground-water flow, solute transport, and multicomponent geochemical reactions*. US Geological Survey. https://pubs.usgs.gov/tm/2005/tm6A8/pdf/tm6a8.pdf.

Pedretti, D., K. U. Mayer, and R. D. Beckie. 2017. Stochastic multicomponent reactive transport analysis of low quality drainage release from waste rock piles: controls of the spatial distribution of acid generating and neutralizing minerals. *Journal of Contaminant Hydrology* 201:30–38.

Pedretti, D., K. U. Mayer, and R. D. Beckie. 2020. Controls of uncertainty in acid rock drainage predictions from waste rock piles examined through Monte-Carlo multicomponent reactive transport. *Stochastic Environmental Research and Risk Assessment* 34, no. 1: 219–233.

Plante, B., B. Buessière, H. Bouzahzah, M. Benzaazoua, I. Demers, and E. H. B. Kandji. 2015. *Revue de littérature en vue de la mise à jour du guide de caractérisation des résidus miniers et du minerai. Report from Université du Québec en Abitibi-Témiscamingue (UQAT)/Unité de recherche et de service en technologie minérale (URSTM)*, Rouyn-Noranda, Québec, Canada.

Plante, B., B. Bussière, and M. Benzaazoua. 2012. Static tests response on 5 Canadian hard rock mine tailings with low net acid-generating potentials. *Journal of Geochemical Exploration* 114:57–69.

Poaty, B., B. Plante, B. Bussière, M. Benzaazoua, T. Pabst, V. Martin, M. Thériault, and P. Nadeau. 2018. *Geochemical Behavior of Different Waste Rock Configurations from the Lac Tio mine: comparison between column Tests and Experimental Waste Rock Pile Results. Tailings and mine wastes 2018*, Keystone, Colorado.

Pooler, R., and B. Dold. 2017. Optimization and quality control of automated quantitative mineralogy analysis for acid rock drainage prediction. *Minerals* 7, no. 1, 12.

Prediction Workgroup of the Acid Drainage Technology Initiative. 2000. *Prediction of water quality at surface coal mines*. Morgantown, VA: National Mine Land Reclamation Center, West Virginia University.

Price, W. A., and Y. T. J. Kwong. 1997. Waste Rock Weathering Sampling and Analysis: Observations From the British Columbia Ministry of Employment and Investment Database. *Proceedings of the 4th International Conference on Acid Rock Drainage (ICARD), Vancouver, Canada*, ed. Mine Environment Neutral Drainage (MEND) Program, CANMET, 31–45. Vancouver, Canada. Natural Resources Canada, Ottawa, Canada.

Salifu, M., T. Aiglsperger, L. Hällström, et al. 2018. Strontium (87Sr/86Sr) isotopes: a tracer for geochemical processes in mineralogically-complex mine wastes. *Applied Geochemistry* 99:42–54.

Salifu, M., T. Aiglsperger, C.- M. Mörth, and L. Alakangas. 2019. Stable sulphur and oxygen isotopes as indicators of sulphide oxidation reaction pathways and historical environmental conditions in a Cu–W–F skarn tailings piles, south-central Sweden. *Applied Geochemistry* 110: 1–10.

Skierszkan, E. K., K. U. Mayer, D. Weis, and R. D. Beckie. 2016. Molybdenum and zinc stable isotope variation in mining waste rock drainage and waste rock at the Antamina mine, Peru. *Science of the Total Environment* 550:103–113.

Skierszkan, E. K., K. U. Mayer, D. Weis, J. Roberston, and R. D. Beckie. 2019. Molybdenum stable isotope fractionation during the precipitation of powellite ($CaMoO_4$) and wulfenite ($PbMoO_4$). *Geochimica et Cosmochimica Acta* 244:383–402.

Skousen, J. 2017. A methodology for geologic testing for land disturbance: acid-base accounting for surface mines. *Geoderma* 308:302–311.

Skousen, J., J. Renton, H. Brown, et al. 1997. Neutralization potential of overburden samples containing siderite. *Journal of Environmental Quality* 26, no. 3: 673–681.

Smith, L. J. D., M. C. Moncur, M. Neuner, et al. 2013. The Diavik Waste rock project: design, construction, and instrumentation of field-scale experimental waste-rock piles. *Applied Geochemistry* 36:187–199.

Smith, K. S., C. A. Ramsey, and P. L. Hageman. 2000. Sampling Strategy for the Rapid Screening of Mine-Waste Dumps on Abandoned Mine Lands. *Proceedings of the 5th International Conference on Acid Rock Drainage (ICARD)* , ed. Society for Mining and Metallurgy, 1453–1461. Littleton, CO. The U.S. Geological Survey, Denver, CO.

Sobek, A. A., W. A. Schuller, J. R. Freeman, and R. M. Smith. 1978. *Field and Laboratory Methods Applicable to Overburden and Minesoils*. EPA 600/2-78-054.

SRK (Steffen, Robertson and Kirsten, Inc.). 1989. *Draft Acid Rock Drainage Technical Guide*. Vol. 1. Vancouver: British Columbia Acid Mine Drainage Task Force.

Stewart, W.A., S.D. Miller, and R. Smart. 2006. Advances in Acid Rock Drainage (ARD) Characterization of Mine Wastes. *Proceedings of the 7th International Conference on Acid Rock Drainage (ICARD)*, ed. Barnhisel, R.I., 2098–2119. St. Louis, MO Published by the American Society of Mining and Reclamation (ASMR), Lexington, KY.

Strömberg, B., and S. A. Banwart. 1999. Experimental study of acidity-consuming processes in mining waste rock: Some influences of mineralogy and particle size. *Applied Geochemistry* 14, no. 1: 1–16.

Sverdrup, H. U. 1990. *The Kinetics of Base Cation Release due to Chemical Weathering*. Lund, Sweden: Lund University Press.

USEPA. 2006. *MINTEQA2, Metal Speciation Equilibrium Model for Surface and Ground Water*, version 3.1. https://vminteq.lwr.kth.se/.

Veeramani, H., J. Eagling, J. H. Jamieson-Hanes, L. Kong, C. J. Ptacek, and D. W. Blowes. 2015. Zinc isotope fractionation as an indicator of geochemical attenuation processes. *Environmental Science & Technology Letters* 2, no. 11: 314–319.

Villeneuve, M., B. Bussière, and Benzaazoua, M. 2009. Assessment of Interpretation Methods for Kinetic Tests Performed on Tailings Having a Low Acid Generating Potential, Proceedings of the 8th International Conference on Acid Rock Drainage (ICARD) Securing the Future: Mining, Metals & the Environment in a Sustainable Society, ed. Swedish Association of Mines, Mineral and Metal Producers, 200–210. Skelleftea, Sweden. Curran Associates, Inc., Red Hook, NY

Villeneuve, M., B. Bussière, M. Benzaazoua, M. Aubertin, and M. Monroy. 2003. The influence of kinetic test type on the geochemical response of low acid generating potential tailings. In *Proceedings of the Tailings and mine waste '03 conference*, 269–279. Vail, CO: Sweets & Zeitlinger.

Voyer, E., and S. Robert. 2015. Waste rock management and closure planning in northern climate: The Meadowbank mine, Nunavut. *Proceedings of the Symposium 2015 sur l'environnement et les mines. Rouyn-Noranda, Québec*, 1–15. Rouyn-Noranda, Québec, Canada. Published by the Canadian Institute of Mining, Metallurgy and Petroleum (CIM), Ottawa, ON, Canada.

Walter, A. L., E. O. Frind, D. W. Blowes, C. J. Ptaceck, and J. W. Molson. 1994. Modeling of multicomponent reactive transport in groundwater: 1. Model development and evaluation. *Water Resources Research* 30, no. 11: 3137–3148.

White, W. W., and T. H. Jeffers. 1994. Chemical predictive modeling of acid mine drainage from metallic sulphide-bearing waste rock. In *Environmental geochemistry of sulphide oxidation*, 608–630. ACS Symposium Series.

Wilson, D., R. T. Amos, D. W. Blowes, J. B. Langman, L. Smith, and D. C. Sego. 2018. Diavik Waste Rock Project: Scale-up of a reactive transport model for temperature and sulfide-content dependent geochemical evolution of waste rock. *Applied Geochemistry* 96:177–190.

3 Water, Gas, and Heat Movement in Cover Materials

Mamert Mbonimpa, Vincent Boulanger-Martel, Bruno Bussière, and Abdelkabir Maqsoud

3.1 INTRODUCTION

Chapters 1 and 2 highlight the mechanisms contributing to acid mine drainage (AMD) generation, in which water and atmospheric oxygen are two main components required for the direct oxidation of sulfide minerals contained in tailings and waste rocks. Consequently, most reclamation covers seek to limit AMD generation by controlling water infiltration and/or oxygen migration to the mine wastes beneath the covers. The performance of such covers lies in limiting fluids (water and oxygen) ingress to the sulfide tailings. The movement of these fluids may occur under saturated or unsaturated conditions and under frozen or unfrozen conditions in inert or oxygen-reactive materials. Details on different types of covers and their performances are given in the subsequent chapters (i.e., Chapters 4 to 9). In order to design covers, water and oxygen movement in porous media must be well understood for site-specific climatic conditions, taking into account soil-atmosphere exchanges. Chapter 1 also discusses how low temperatures can decrease the kinetics of the oxidation reactions of sulfide minerals. Accordingly, insulation covers can be used in cold regions with the objective to favor the aggradation of permafrost into the reactive mine wastes and cover materials. The design of such covers involves heat transfer equations (see Chapter 9).

This chapter details the basic soil index properties used to describe the different phases in porous media, as well as the main theoretical aspects related to water flow, gas diffusion, and heat transfer through water-saturated and unsaturated cover materials. The properties required for addressing issues related to the water balance (i.e., precipitation, evapotranspiration, water storage, infiltration, lateral diversion, and runoff) of covers used for mine site reclamation are also recalled. Measurement techniques of the pertinent parameters are also given along with predictive equations to estimates these parameters. The use of such predictive models should, however, be restricted to preliminary phases of projects. The methodological approaches to design each cover system and evaluate their performances through physical modeling (lab and field tests) at different scales are described in Chapters 4 to 9, without neglecting the long-term monitoring program of the installations (Chapter 10).

3.2 BASIC PROPERTIES

Unfrozen and unsaturated porous materials, such as tailings, waste rocks, and cover materials, are generally considered as triphasic media, with solids, water, and air present in different proportions. The medium becomes biphasic in the absence of water (dry soil) or air (saturated material). Based on the volume, mass, and weight of these phases (neglecting the mass of air), several properties can be defined, including the gravimetric water content, the specific gravity, the wet and dry density (and the corresponding unit weight), the porosity, the void ratio, the degree of saturation, the volumetric water content, and the volumetric air content. These parameters are defined below, and the relationships that link them during soil cover design are presented where relevant. The grain-size distribution (GSD) and the associated characteristics are also presented.

3.2.1 Definitions

The abovementioned three-phase system of porous materials can be presented schematically in Figure 3.1. In this Figure, M, W, and V stand for mass, weight, and volume, respectively, while the subscripts a, s, t, v, and w stand for air, solids, total, voids, and water, respectively. In addition to the mass, M, of the different phases, the weight, W, can also be used (with $W = M \times g$, where g is the gravitational constant, $g = 9.81\ m/s^2$). Various relationships between these parameters can be found in geotechnical textbooks (e.g., Holtz and Kovacs 1981; McCarthy 2007). The main index properties defined from the relative phase properties are given in Table 3.1.

If solid particles are assumed to have a constant volume, V_s, any volume change in a soil affects the volume of the voids, V_v. Volume increases can be induced by expansion (e.g., freezing, expansive soils) and volume decrease, which occurs mainly by shrinkage, settlement, and compaction (see Section 3.2.5 for details on compaction).

3.2.2 Relevant Relationships Between Soil Index Properties

The most relevant equations applicable to the cover design and monitoring are noted in Equations (3.2) to (3.8), where all variables are defined in Table 3.1.

$$e = \frac{n}{1-n} \text{ and } n = \frac{e}{1+e} \tag{3.1}$$

$$e = \frac{wG_S}{S_r} \tag{3.2}$$

$$\rho_{wet} = \left(\frac{1+w}{1+e}\right) G_s \rho_w \tag{3.3}$$

$$\gamma_{wet} = g \times \rho_{wet} \tag{3.4}$$

$$\rho_{dry} = \frac{\rho_{wet}}{1+w} \tag{3.5}$$

$$\theta = nS_r = w(1-n)G_s = w\frac{\rho_{dry}}{\rho_w} \tag{3.6}$$

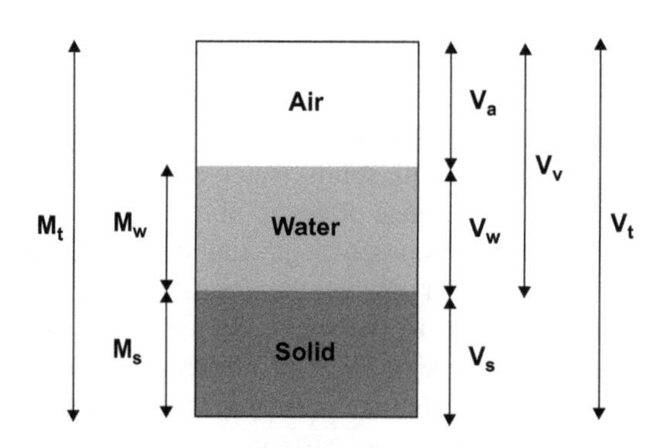

FIGURE 3.1 Three-phase diagram of porous materials.

TABLE 3.1
Definitions of Basic Index Properties Derived from the Soil Phases

Parameter	Symbol	Units	Equation	Definition of Variables
Gravimetric water content	w	kg.kg^{-1}	$w=\frac{M_w}{M_s}$	M_w: mass of water ($=M_t - M_s$) M_s: mass of solids (or dry soil)
Wet density	ρ_{wet}	kg.m^{-3}	$\rho_{wet}=\frac{M_t}{V_t}$	M_t: total mass V_t: total volume
Dry density	ρ_{dry}	kg.m^{-3}	$\rho_{dry}=\frac{M_s}{V_t}$	M_s: mass of solids (or dry soil) V_t: total volume
Wet unit weight	γ_{wet}	N.m^{-3}	$\gamma_{wet}=\frac{W_t}{V_t}$	W_t: weight of wet soil V_t: total volume
Dry unit weight	γ_{dry}	N.m^{-3}	$\gamma_{dry}=\frac{W_s}{V_t}$	W_s: weight of solids (or dry soil) V_t: total volume
Density of solids grains	ρ_s	N.m^{-3}	$\rho_s=\frac{M_s}{V_s}$	M_s: mass of solids (or dry soil) V_s: volume of solid grains
Specific gravity	G_s	–	$G_s=\frac{\rho_s}{\rho_w}$	ρ_s: density of solids grains ρ_w: density of water
Void ratio	e	m^3.m^{-3}	$e=\frac{V_v}{V_s}$	V_v: volume of voids V_s: volume of solid grains
Porosity	n	m^3.m^{-3}	$n=\frac{V_v}{V_t}$	V_v: volume of voids V_t: total volume
Degree of saturation	S_r	m^3.m^{-3}	$S_r=\frac{V_w}{V_v}$	V_w: volume of water V_v: volume of voids
Volumetric water content	θ	m^3.m^{-3}	$\theta=\frac{V_w}{V_t}$	V_w: volume of water V_t: total volume
Volumetric air content	θ_a	m^3.m^{-3}	$\theta_a=\frac{V_a}{V_t}$	V_a: volume of air V_t: total volume

$$\theta_a = n - \theta = (1 - S_r)n \tag{3.7}$$

$$S_r = \frac{w}{\frac{\rho_w}{\rho_{dry}} - \frac{1}{G_s}} \tag{3.8}$$

3.2.3 Grain-Size Distribution

The solid phase in a soil is characterized by its physical, mineralogical, and chemical characteristics. GSDs represent the proportions by dry mass of the different particle sizes in a soil over a specified grain size range. It is one of the most commonly used physical properties for the classification of different soil types for engineering purposes, particularly with respect to water drainage and mechanical compaction. As described in the sections below, many soil properties depend on GSD and the related soil type, including hydrogeological properties (e.g., bulk density, porosity, water retention, residual water content, and relative permeability), flow properties (e.g., tortuosity and intrinsic permeability), frost heave, and thermal conductivity. These thermal properties also depend on the mineralogical characteristics of the particles.

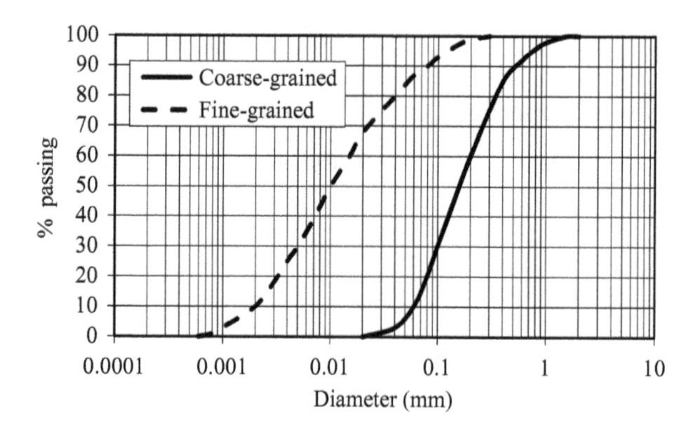

FIGURE 3.2 Examples of grain-size distributions for coarse- and fine-grained soils.

The main GSD parameters used to characterize a material include the diameters D_{10}, D_{30}, and D_{60}, where D_x is the diameter at x% of grains passing on the cumulative GSD curve. Specific D values are used to calculate the coefficients of uniformity $C_U = (D_{60}/D_{10})$ and of curvature $C_C = [(D_{30})^2/(D_{10} \times D_{60}]$. The percentage of particles passing for the grain diameter of ~80 μm or 0.08 mm (Sieve # 200), P_{80}, is also an important parameter used for the soil classification. According to the Unified Soil Classification System (ASTM 2017a) soils are coarse-grained (i.e., sands or gravels) when $P_{80} < 50\%$ and fine-grained (i.e., silts or clays) when $P_{80} > 50\%$. Figure 3.2 shows examples of GSDs for a coarse-grained soil ($P_{80} \approx 20\%$) and a fine-grained soil ($P_{80} \approx 90\%$). For the coarse-grained soil, $D_{10} = 0.06$ mm, $D_{30} = 0.1$ mm, and $D_{60} = 0.2$ mm; therefore, $C_U = 3.3$ and $C_C = 0.8$. For the fine-grained soil, $D_{10} = 0.002$ mm, $D_{30} = 0.005$ mm, and $D_{60} = 0.015$ mm; therefore, $C_U = 7.5$ and $C_C = 0.8$. The D_{10}, which is often referred to as the effective diameter, and C_U will be used later for the prediction of hydraulic properties.

3.2.4 Atterberg's Limits for Fine-Grained Soils

Some engineering properties of fine-grained soils, such as compressibility, permeability, and strength, depend strongly on the water content, which define the consistency of fine-grained soils. With decreasing water content, a fine-grained soil may appear at the liquid, plastic, semi-solid, or solid states. Consistency limits, or Atterberg's limits, correspond to the water content at which the fine-grained soil changes from one state to another. Therefore, there are three limits: the liquid limit (LL or w_L), the plastic limit (PL or w_P), and the shrinkage limit (w_S) (e.g., Holtz and Kovacs 1981; McCarthy 2007).

The w_L is defined as the water content at which a fine-grained soil changes from the liquid state to the plastic state. That is, the water content at which the soil no longer flows like a liquid. The w_p is the water content separating the plastic and semi-solid states or the point at which a fine-grained soil can no longer be remolded without cracking. The w_S is the water content at which a fine-grained soil changes from the semi-solid to solid state or the water content at which the soil no longer changes volume upon drying. At the w_S, any loss of water is compensated by the entry of air into the pores.

The procedures to determine the most frequently used limits (w_L and w_p) are described in ASTM Standard D2487-17 (ASTM 2017b). These limits can be used to define the plasticity index PI = ($w_L - w_p$). When the actual water content (w) of a fine-grained soil is known, the consistency index CI = (($w_L - w$)/PI). The practical significance and use of these indexes are described in geotechnical textbooks (e.g., Holtz and Kovacs 1981; McCarthy 2007). For example, the PI and w_L are used for the classification of fine-grained soils, while the CI is used to indicate a soil's consistency: liquid when CI = 0 for $w = w_L$; very soft when $0 < CI \leq 0.5$; soft when $0.5 < CI \leq 0.75$, etc. The w_L and

PI are also used for suggesting appropriate, natural clayey materials to be used in low saturated hydraulic conductivity covers (see Chapter 4).

3.2.5 Compaction of Cover Materials

Bulk materials used in each layer of a soil cover need to be compacted for reaching the specified design properties (e.g., ρ_{wet} or ρ_{dry}, n, hydraulic conductivity). Generally, materials are placed in layers of specified thicknesses (lifts), which are compacted (via surface compaction) until the desired dry density is obtained. The main objective of compaction is to reduce the volume of voids filled with air, V_a, and, therefore, the porosity (see Figure 3.1). Several types of compactors and compaction equipment can be used to achieve compaction (see Holtz and Kovacs 1981; Kutzner 1997; and Blight 2010). The in-situ ρ_{wet} can be controlled using techniques such as nuclear gauge, sand replacement, balloon densometer, and core cutter (see details in a book by Blight 2010). The in-situ water content can be controlled using oven drying, microwave drying, and nuclear gauge (see Blight 2010). Knowing ρ_{wet} and w, ρ_{dry} can be calculated using Equation 3.5. Continuous compaction control helps to optimize the number of passes required (generally between four and eight for dam materials; Kutzner 1997) to achieve the desired ρ_{dry}.

Ralph Roscoe Proctor showed in 1933 that the ρ_{dry} achieved for a given compaction effort depends on the water content of the soil and, thus, developed the Proctor compaction test. This test is still widely used to determine the optimal water content at which a given soil type reaches its maximum dry density. The mold sizes, number of layers, and blows per layer; hammer weight and drop height; and the largest particle size of the soil are fixed in the standard effort (ASTM 2012a) and modified effort (ASTM 2012b) Proctor tests. A graphical plot of the $w - \rho_{dry}$ relationship provides the compaction or Proctor curve. Figure 3.3 illustrates a Proctor curve obtained with the modified effort method for a till having a specific gravity (G_s) of 2.808. When water is added to a relatively dry soil, it acts as a lubricant that causes the soil particles to slide against one another more easily and to be oriented into a densely packed state during compaction. This means that dry density increases with increasing water content until reaching a maximum value. When additional water is added, the pore water pressure pushes the soil particles apart, thereby decreasing friction; the dry density then decreases. For the tested till, the maximum dry density ($\rho_{dry\text{-}max}$) is 2285 kg.m^{-3}, and the corresponding optimal water content (w_{opt}) is 5.5%. The corresponding porosity (n) and degree of saturation (S_r) can be calculated using Equations 3.6 and 3.8, respectively: n = 0.19 and S_r = 67%. The Proctor compaction curve is divided into the dry side for $w < w_{opt}$ and the wet side for $w > w_{opt}$.

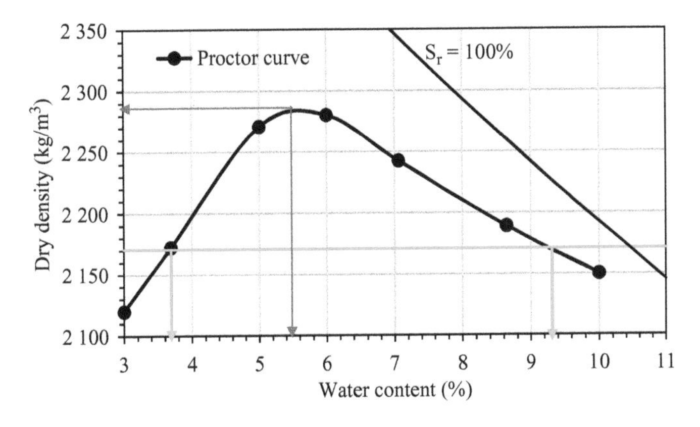

FIGURE 3.3 Proctor curve obtained with the modified effort Proctor test for a till and curve of the dry density of the till compacted up to saturation.

For a given soil, the $\rho_{dry\text{-}max}$ increases and the w_{opt} decreases with increasing compaction effort (Holtz and Kovacs 1981).

Equation 3.8 can be used to calculate how the dry density changes with respect to the water content for a given degree of saturation. For S_r = 100%, the obtained curve is presented in Figure 3.3. This curve gives the dry density that will be reached if the soil is compacted up to saturation; that is, until the voids are completely filled with water ($V_a = 0$). In other words, the Proctor curve cannot cross the density curve at saturation unless water is lost from the sample during compaction.

In practice, project specifications generally provide the required soil density to ensure adequate soil compaction is achieved. This density is often expressed in terms of degree of compaction or the compaction ratio (CR), which is defined as the ratio of the in-situ dry density and the maximum dry density ($CR = \rho_{dry}/\rho_{dry-max}$). For the data presented in Figure 3.3, CR = 95%, meaning that the in-situ soil will be compacted to 95% of the maximum dry density (ρ_{dry} = 2171 kg.m^{-3}). As indicated in Figure 3.3, this target is achieved when the till is compacted with w values ranging between 3.6% and 9.2%. However, several studies have shown that, in the case of clayey soils, lower hydraulic conductivity values are reached when compaction is performed on the wet side of the Proctor curve (Mitchell et al. 1965; Holtz and Kovacs 1981). This issue is especially important for the compaction of natural clayey materials that are used in hydraulic barrier covers (see Chapter 4).

As previously mentioned, the largest particle size of the soil is limited for each type of Proctor test. In order to meet standard practices for coarser-grained soils, the fraction that is coarser than the largest particle size must be removed by sieving. However, because this coarser fraction is present in the field, the materials tested in the laboratory may not be fully representative of the materials compacted in situ. For this reason, corrections of dry density and water content for soils containing oversized particles can be done following ASTM Standard D4318-17e1 (ASTM 2017a).

3.3 WATER FLOW IN SATURATED COVER MATERIALS

Cover materials and the reactive mine wastes below can be under saturated or unsaturated conditions. Therefore, it is necessary to first introduce the fundamental equations governing water flow in saturated soils. Darcy's and continuity equations and the usual experimental and predictive methods to assess the saturated hydraulic conductivity are discussed.

3.3.1 Governing Equations

Equations describing water flow in saturated porous media are extensively documented in the literature (e.g., Hillel 1998; Freeze and Cherry 1979). Darcy (1856) was the first to observe that the volumetric flow rate (volume of water per unit of time) through a saturated, homogeneous sand placed in a pipe was proportional to the head loss and the cross-sectional area and inversely proportional to the flow distance (or length of the sample). This observation is expressed as Equation (3.9).

$$Q = -k_{sat} \frac{\Delta h}{\Delta L} A \tag{3.9}$$

where Q is the flow rate (m^3.s^{-1}), k_{sat} is the saturated hydraulic conductivity (m.s^{-1}), h is the total hydraulic head (m), Δh is the head loss along the sample (m), ΔL is the flow distance or length of the sample (m), and A is flow cross-sectional area (m^2)

The difference in Δh is the flow driving potential. The ratio between the head loss and flow distance ($\Delta h/\Delta L$) is called hydraulic gradient (i). The sign – in Equation (3.9) indicates that water flows in the direction of decreasing hydraulic head. When the kinetic flow energy is neglected (due to slow groundwater velocities), h, which is expressed as an energy per unit of weight in Bernouilli's equation, is defined as follows:

$$h = h_p + z \tag{3.10}$$

where h_p (m) is the pressure head and z (m) is the elevation head from a reference elevation; z is positive or negative above or below this reference elevation, respectively. Equation (3.9) can be expressed as follows in Equation (3.11):

$$\frac{Q}{A} = q_w = k_{sat}\frac{\Delta h}{\Delta L} = k_{sat}\, i \tag{3.11}$$

In this case, q_w ($m.s^{-1}$) represents the specific flow rate of water, which is commonly called Darcy's flow velocity. The saturated hydraulic conductivity (k_{sat}) is constant for a given saturated porous media at a given porosity or void ratio (see Section 3.3.3).

In some cases, the intrinsic permeability (K_i), which is defined in Equation (3.12), is used to represent the void capacity through which a fluid (gas or liquid) can flow without taking into account the fluid properties.

$$K_i = k_f\frac{\eta_f}{\gamma_f} \tag{3.12}$$

where K_i is the intrinsic permeability (m^2), k_f is the saturated hydraulic conductivity (k_{sat} for water) or saturated air conductivity (of the dry porous material), η_f is the dynamic viscosity of the fluid (1×10^{-3} Pa.s or $N.s.m^{-2}$ or $kg.m^{-1}.s^{-1}$ for water at 20°C), and γ_f is the unit weight of the fluid (9.81 $kN.m^{-3}$ for water).

For heterogeneous and anisotropic materials where the coordinate system is oriented along the principal axes x, y, and z, the 3-D components of Darcy's flow velocities become as follows:

$$q_{w-x} = k_{sat-x}\frac{dh}{dx} \tag{3.13}$$

$$q_{w-y} = k_{sat-y}\frac{dh}{dy} \tag{3.14}$$

$$q_{w-z} = k_{sat-z}\frac{dh}{dz} \tag{3.15}$$

where k_{sat-x}, k_{sat-y}, and k_{sat-z} are the saturated hydraulic conductivities in the x-, y-, and z-axes, respectively.

Solving Equations (3.13), (3.14), and (3.15) requires knowledge of the spatial distribution of h(x,y,z) and k_{sat}(x,y,z). As presented in Section 3.2.2, k_{sat} can be determined in the laboratory. Once k_{sat} is known, h(x,y,z) is obtained from the continuity equation. In the Cartesian tridimensional domain of x, y, and z, a differential soil element (i.e., a representative elementary volume) with the dimensions Δx, Δy, and Δz is generally used to develop the continuity equation (Figure 3.4).

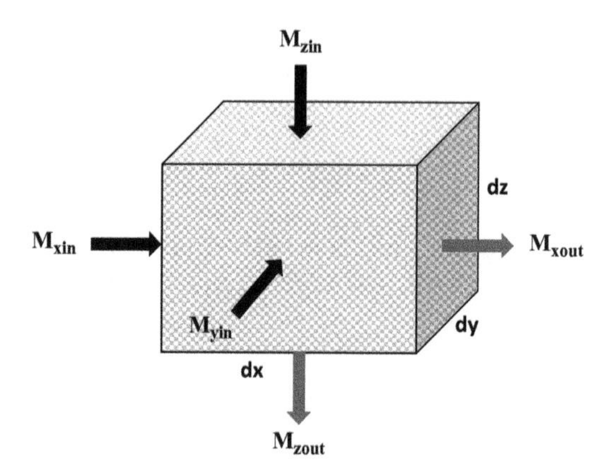

FIGURE 3.4 Balance of water fluxes through a representative elementary volume.

In this Figure 3.4, M represents the mass flux of water (kg.s^{-1}) in the x, y, and z directions, while the subscripts "in" and "out" stand for inflow and outflow, respectively. The mass of inflow given by Equation (3.16) is the water entering along x-axis per unit of time:

$$M_{xin} = \rho_w q_{w-x} dydz \tag{3.16}$$

The mass of outflow water along x-axis is given by:

$$M_{xout} = \left[\rho_w q_{w-x} + \frac{\partial \left(\rho_w q_{w-x} \right) dx}{\partial x} \right] \rho_w q_{w-x} dydz \tag{3.17}$$

Therefore, the net mass accumulation resulting from flow along the x-axis is:

$$\Delta M_x = M_{xin} - M_{xout} = -\frac{\partial \left(\rho_w q_{w-x} \right) dxdydz}{\partial x} \tag{3.18}$$

Net mass accumulations for ΔM_y and ΔM_z can be obtained similarly.

For steady-state flow in incompressible porous media (i.e., where the porosity remains constant), the law of mass conservation gives:

$$\Delta M_x + \Delta M_y + \Delta M_z = 0 \tag{3.19}$$

$$-\left[\frac{\partial \left(\rho_w q_{w-x} \right)}{\partial x} + \frac{\partial \left(\rho_w q_{w-y} \right)}{\partial y} + \frac{\partial \left(\rho_w q_{w-z} \right)}{\partial z} \right] dxdzdy = 0 \tag{3.20}$$

Assuming that the density of water is invariable along the axes, one obtains:

$$\frac{\partial q_{w-x}}{\partial x} + \frac{\partial q_{w-y}}{\partial y} + \frac{\partial q_{w-z}}{\partial z} = 0 \tag{3.21}$$

Substituting the specific flow rates $q_{w\text{-}x}$, $q_{w\text{-}y}$, and $q_{w\text{-}z}$ in the previous equation with the Darcy's law gives:

$$\frac{\partial}{\partial x}\left(k_{sat-x}\frac{\partial h}{\partial x}\right)+\frac{\partial}{\partial x}\left(k_{sat-y}\frac{\partial h}{\partial y}\right)+\frac{\partial}{\partial z}\left(k_{sat-z}\frac{\partial h}{\partial z}\right)=0 \tag{3.22}$$

If $k_{sat\text{-}x}$, $k_{sat\text{-}y}$, and $k_{sat\text{-}z}$ do not vary along the corresponding axes, Equation (3.22) becomes:

$$k_{sat-x}\frac{\partial^2 h}{\partial^2 x}+k_{sat-y}\frac{\partial^2 h}{\partial^2 y}+k_{sat-z}\frac{\partial^2 h}{\partial^2 z}=0 \tag{3.23}$$

For isotropic conditions ($k_{sat\text{-}x} = k_{sat\text{-}y} = k_{sat\text{-}z}$), the continuity equation becomes the Laplace equation:

$$\frac{\partial^2 h}{\partial^2 x}+\frac{\partial^2 h}{\partial^2 y}+\frac{\partial^2 h}{\partial^2 z}=0 \tag{3.24}$$

With known values of $k_{sat\text{-}x}$, $k_{sat\text{-}y}$, and $k_{sat\text{-}z}$, solving the continuity Equation (3.23) allows for the determination of the hydraulic head h(x,y,z), which is used to obtain the hydraulic gradient in Darcy's law. The numerical methods used to solve these equations are implemented in different numerical codes.

It is sometimes assumed that $k_{sat\text{-}x} = k_{sat\text{-}y}$ to distinguish the vertical and horizontal water flow. In this case, the ratio between $k_{sat\text{-}x}$ and $k_{sat\text{-}z}$ represents the anisotropy ratio.

3.3.2 Methods for Measuring Saturated Hydraulic Conductivity in the Laboratory

Saturated hydraulic conductivity is required to predict water flow into saturated porous media and is generally determined on undisturbed or remolded samples in the laboratory, where the influence of factors such as porosity and freeze-thaw cycles can be controlled. Furthermore, vertical values of k_{sat} are generally determined. Standardized permeability tests, including constant head and variable head tests, are mainly used depending on the expected k_{sat} value or the type of soil. For granular materials with an expected $k_{sat} > 10^{-5}$ m.s^{-1}, constant head tests in rigid- or flexible-wall permeameters according to ASTM Standard D2434-19 (ASTM 2019a) are recommended. Variable head tests in rigid-wall permeameters (ASTM 2015) or in flexible-wall permeameters (triaxial cells; ASTM 2016a) are recommended for fine-grained materials with an expected $k_{sat} < 10^{-5}$ m.s^{-1}.

The 14 most important measurement errors occurring during tests in rigid-wall or flexible-wall permeameters were presented by Chapuis (2012). Among the most important actions that could be taken to improve the quality of such tests are:

- Reaching full saturation ($S_r = 1$);
- Measuring the hydraulic head loss within the tested specimen for rigid-wall permeameters;
- Using flexible-wall permeameters for plastic materials (for which both the LL w_L and PL w_P can be determined);
- Avoiding preferential leakage along permeameter walls by using rigid permeameters for non-plastic materials (for which it is not possible to determine w_p or both w_L and w_P);
- Avoiding clogging of porous stones when testing mixtures of fine and coarse materials;
- Avoiding the movement of fine particles into the spaces between coarser particles; that is, internal erosion or suffosion.

In temperate climates, cover materials can be submitted to freeze-thaw cycles, which, in some cases, can affect negatively the hydrogeological properties of cover materials, including the k_{sat}. The effect of freeze-thaw cycles on k_{sat} can be assessed in the laboratory using flexible-wall permeameters (ASTM 2019b) and rigid-wall permeameters (ASTM 2016b) by performing permeability tests on samples before and after being exposed to freeze-thaw cycles.

3.3.3 Predictive Methods for Determining Saturated Hydraulic Conductivity

In the preliminary phases of projects, the saturated hydraulic conductivity can be estimated using various relationships given in the literature. Chapuis (2012) compared the performance of 45 methods for predicting k_{sat}. The accuracy of such predictive equations depends highly on the quality of the experimental results used for calibrating these equations. Plastic (clayey) materials are generally avoided in mine covers because of problems linked to construction (difficulty of installation) and to the evolution of their properties. More specifically, in the contexts where shrinkage and freeze-thaw cycles can induce drying cracks and changes in a material's hydrogeotechnical properties (n, k_{sat}) (see details in Chapter 4). Consequently, most cover materials are non-plastic, and emphasis is put here on the predictive model developed by Mbonimpa et al. (2002) for such types of soils.

The k_{sat} (m.s^{-1}) is defined by Kozeny-Carman as following (Aubertin et al. 1996):

$$k_{sat} = C_1 \frac{g}{\rho_w \eta_w} \frac{1}{G_s^2 S_m^2} \tag{3.25}$$

where C_1 is a material constant, ρ_w (kg.m^{-3}) and η_w (Pa.s or kg.m^{-1}s^{-1}) are the density and dynamic viscosity of water (both depend on the water temperature), and S_m is the specific surface area per unit mass of material (m^2.kg^{-1}).

Mbonimpa et al. (2002) modified this equation by expressing S_m using GSD characteristics for non-plastic materials (Equation 3.26).

$$S_m = \frac{f_s}{\rho_s D_h} \tag{3.26}$$

where f_s is a shape factor (six for idealized spheres) and D_h is the equivalent diameter that can be estimated from a GSD divided into different fractions with an average diameter of D_i. D_h. is calculated using Equation 3.27 in which each size fraction D_i has a mass percentage $p_{m,i}$ (%).

$$D_h = \frac{100}{\sum_i \frac{p_{m,i}}{D_i}} \tag{3.27}$$

It was observed that D_h can be estimated using D_{10} and C_U as follows:

$$D_h = C_U^{1/6} D_{10} \tag{3.28}$$

The saturated hydraulic conductivity at 20°C (temperature correction is required with deviations in water temperature) is given by:

$$k_{sat}\left(m.s^{-1}\right) = 9.80 \frac{e^5}{1+e} C_U^{1/3} D_{10}^2 \left(cm\right) \tag{3.29}$$

These equations were validated for granular and low-plasticity materials ($w_L \leq 20\%$) with 4.0×10^{-8} m.s^{-1} $\leq k_{sat} \leq 3.0$ m.s^{-1} ($0.35 \leq e \leq 1.27$; $1 \leq C_U \leq 227$; 4×10^{-6} cm $\leq D_{10} \leq 1.5$ cm; and $w_L \leq 20\%$).

3.4 WATER FLOW IN UNSATURATED COVER MATERIALS

Soil covers used for mine site reclamation are generally located in the vadose zone. The saturation conditions may control the movement of fluids (gas and water) through the cover materials, which, consequently, can affect the performance of the covers that seek to limit water and/or oxygen infiltration. The water retention curve (WRC) and unsaturated hydraulic conductivity function (k_u) are the main parameters required for determining the water flow (or percolation) in unsaturated materials. In the following, the water balance that controls the saturation conditions, the governing Richard's equations, and the common experimental and predictive methods used to assess WRCs and k_u functions are presented. The concept of capillary barrier effects, which are created when layers with contrasting WRCs are superimposed, is described. This concept is used in store-and-release (SR) covers (see Chapter 5) and covers with capillary barrier effects (see Chapter 7). Oxygen migration in unsaturated soils is described in Section 3.5. Water exchanges due to soil-atmosphere interactions (i.e., evaporation and evapotranspiration) are addressed in Section 3.6.

3.4.1 Water Balance

The water balance helps to determine the extent of unsaturated conditions, which depends on water storage and the volumetric water content within the cover materials. It is also used to quantify the performance of covers, particularly for low-k_{sat} covers (see Chapter 4). The water balance can be expressed using Equation (3.30).

$$P = \Delta S - R - Pe - AE - T \tag{3.30}$$

where P is the precipitation falling on the surface of the cover (m^3.m^{-2} or m); R is the surface runoff over the cover (m^3.m^{-2} or m); Pe is the percolation from the cover (m^3.m^{-2} or m); AE is the actual evaporation at the surface of the cover (m^3.m^{-2} or m); T is the transpiration, which is applicable when plants have grown on the cover (m^3.m^{-2} or m); and ΔS is the variation in water storage, S (m^3.m^{-2} or m).

The determination of the water percolation from the cover is based on the water flow (in and out) in unsaturated cover materials (see Section 3.4.2). In the following, more details related to unsaturated flow are presented since it affects the different components of the water balance. The water balance equation can be slightly modified in the presence of additional inflow and outflow; this is the case for the water covers (see Chapter 6).

3.4.2 Governing Equations

Equations describing water flow in unsaturated porous media (that determine Pe) are well documented in the literature (Fredlund and Rahardjo 1993; Hillel 1998). In the case of unsaturated soils, the flow driving potential (h) is given by:

$$h = \psi + z \tag{3.31}$$

where ψ (m) is the pressure head (negative value in opposite to h_p used in Equation (3.10), which is positive) called matric suction, and z (m) is the elevation head from a reference elevation (positive values along the z-axis are oriented downward).

The suction (ψ) depends on the water retained in the porous media and is expressed through the WRC, which can be defined in terms of $\theta(\psi)$, $w(\psi)$, or $S_r(\psi)$ (see relationships between suction and θ, w, or S_r in Equation 3.6).

Richards (1931) showed that Darcy's law (see Equation 3.9) also applies in unsaturated soils, where the hydraulic head is defined by Equation (3.31) and where the proportionality factor, k_u, represents the unsaturated hydraulic conductivity. Equation (3.32) presents the Richards equation. In contrast to k_{sat}, which is a constant, k_u is a function of suction ($k_u(\psi)$) or of θ, w, or S_r.

$$q_w = -k_u \frac{\Delta h}{\Delta L} = -k_u \frac{\Delta(\psi + z)}{\Delta L} \tag{3.32}$$

The law of mass conservation is also applied to develop the continuity equation. However, in this case, the difference between the entering and exiting fluxes corresponds to the water storage or the accumulation of water in the elementary volume per unit of time. Equations (3.21) and (3.22) can then be generalized as:

$$\frac{\partial \theta}{\partial t} = -\left(\frac{\partial q_{w-x}}{\partial x} + \frac{\partial q_{w-y}}{\partial y} + \frac{\partial q_{w-z}}{\partial z}\right) \tag{3.33}$$

$$\frac{\partial \theta}{\partial t} = \left(\frac{\partial}{\partial x}\left[k_{u-x}(h)\frac{\partial h}{\partial x}\right] + \frac{\partial}{\partial y}\left[k_{u-y}(h)\frac{\partial h}{\partial y}\right] + \frac{\partial}{\partial z}\left[k_{u-z}(h)\frac{\partial h}{\partial z}\right]\right) \tag{3.34}$$

Since $\partial z/\partial x$ and $\partial z/\partial y$ are equal to 0 and $\partial z/\partial z$ is equal to 1, Equation (3.34) can also be rewritten as the following considering Equation (3.31):

$$\frac{\partial \theta}{\partial t} = \frac{\partial}{\partial x}\left(k_{u-x}(\psi)\frac{\partial \psi}{\partial x}\right) + \frac{\partial}{\partial y}\left(k_{u-y}(\psi)\frac{\partial \psi}{\partial y}\right) + \frac{\partial}{\partial z}\left(k_{u-z}(\psi)\frac{\partial \psi}{\partial z}\right) + \frac{\partial k(\psi)}{\partial z} \tag{3.35}$$

For having an equation with one variable, ψ, the water capacity function, $C(\psi)$, which is approximated by the slope of the WRC ($C = \partial\theta/\partial\psi$) is used. Equation (3.35) then becomes:

$$C(\psi)\frac{\partial \psi}{\partial t} = \frac{\partial}{\partial x}\left(k_{u-x}(\psi)\frac{\partial \psi}{\partial x}\right) + \frac{\partial}{\partial y}\left(k_{u-y}(\psi)\frac{\partial \psi}{\partial y}\right) + \frac{\partial}{\partial z}\left(k_{u-z}(\psi)\frac{\partial \psi}{\partial z}\right) + \frac{\partial k(\psi)}{\partial z} \tag{3.36}$$

This 3-D Equation (3.36) is solved using numerical tools to predict the pressure distribution $\psi(x, y, z)$ in unsaturated media. In most cases, this is reduced to a 2-D equation. More details on water flow through unsaturated media can be found in the literature (e.g., Fredlund and Rahardjo 1993, Hillel 1998). As previously mentioned, the two main functions that are used to describe the unsaturated behavior of a granular material are the WRC and k_u. The methods used to obtain these two functions are described in the next sections.

3.4.3 Experimental Methods for Measuring the WRC

The WRC is usually expressed as the volumetric water content, θ, with respect to ψ. The $\theta(\psi)$ relationship can be measured in the laboratory under controlled drainage or wetting paths and in situ under natural drainage and wetting using various techniques. The WRCs obtained under drainage and wetting processes are different due to hysteresis effects that are represented by a considerable difference in the θ values for a given suction. Hysteresis effects are extensively explained in the literature (Poulovassilis 1962, Davis et al. 2009, Mualem and Beriozkin 2009; Maqsoud et al. 2012).

A number of techniques have been developed to evaluate the WRC in the laboratory. The general principle of the techniques listed below is to measure the suction applied to the specimen (the suction is increased once equilibrium reached), while the gravimetric or volumetric water content is calculated or measured.

- The axis translation technique (Tempe Cell, pressure plate, and membrane extractor) for which the suction corresponds to the gas pressure applied to the sample that is placed on a saturated ceramic disk or membrane with an air entry value (AEV) greater than the maximum suction being applied (e.g., Dane and Topp 2002; Marihno et al. 2008).
- Different types of sensors can be used to measure suction including, among others, (e.g., Fredlund and Rahardjo 1993; Dane and Topp 2002; Maqsoud et al. 2007):
  - Tensiometers equipped with a porous cup that is installed in the soil, a water reservoir, and a pressure gauge. Water moves in the direction of decreasing potential until equilibrium is reached inside and outside the cup.
  - Thermocouple psychrometers that measure the relative humidity in the air phase of the soil, which is used, in turn, to estimate the total suction through Kelvin's equation.
- The filter paper method (ASTM 2016c; Fredlund and Rahardjo 1993; Dane and Topp 2002; Noguchi et al. 2012; Almeida et al. 2015; Oleszczuk et al. 2018): A calibration equation linking the suction and the gravimetric water content in the filter is first determined (using a psychrometer for example). The filter is then placed in direct contact with a wet soil so that water is exchanged. Once equilibrium is reached, the gravimetric water content of filter is determined and used with the calibration equation to determine the suction.
- The vapor equilibrium technique for very high suctions (e.g., Romero 1999; Tang and Cui 2005), which is based on Kelvin's law relating the relative humidity generated by salt solutions to the suction.

Figure 3.5 shows typical drying and wetting WRCs with their main characteristics. The volumetric water content at saturation, θ_s, ($\psi = 0$) corresponds to the porosity (for non-shrinking materials).

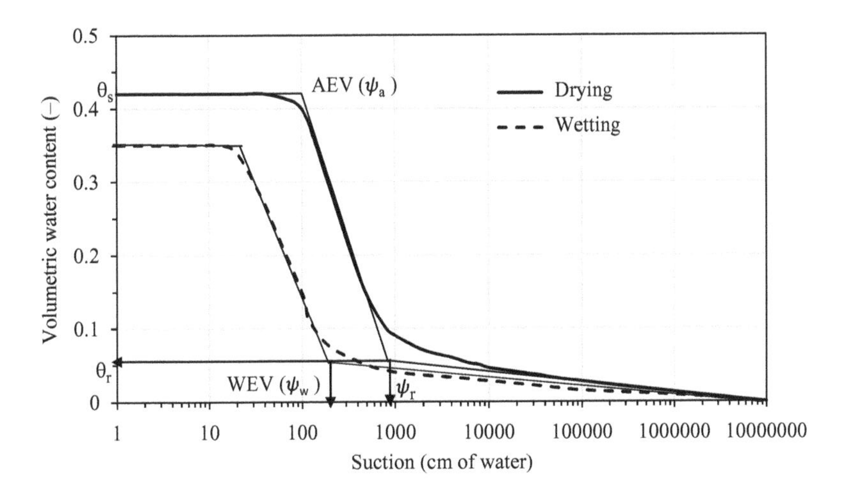

FIGURE 3.5 Definition of the main soil characteristics for typical drying and wetting WRCs.

The AEV corresponds to the matric suction value that must be exceeded before air enters into the soil pores. The AEV is generally determined using the tangent method (Fredlund and Xing 1994). The residual water content (θ_r) corresponds to the amount of water retained by adhesive forces in disconnected pores and immobile films so that it cannot be easily drained from the soil. The corresponding suction on the drying WRC is the residual suction (ψ_r). When a dry soil is wetted, water starts to displace air at the water entry value (ψ_w) of the soil.

When a tested material shrinks under increasing suction, the selected method must be able to account for variations in n or e at each suction level; that is, the shrinkage curve $n(\psi)$ or $e(\psi)$, is required. This is possible by testing many samples simultaneously so that a specimen can be removed at each suction level to determine its volume and deduce the actual value of n or e. Otherwise, considering a constant porosity (the initial porosity of the specimen) results in inaccurate WRCs (Mbonimpa et al. 2006; Liu et al. 2012a).

Field methods to determine WRCs generally use probes to measure the volumetric water content and the suction simultaneously at the same location. Direct and indirect methods can be used to assess volumetric water content. The direct method involves determining the w and the ρ_{wet} on undisturbed samples, as well as the G_s, in order to determine n, θ, and S_r. In contrast to the direct method, the indirect method uses probes to determine θ. The main advantage of the indirect method is that θ can be measured continuously and at a predefined frequency. The main techniques that can be used for in-situ ψ and θ measurements are presented in Chapter 10.

3.4.4 Descriptive Equations for the WRC

Experimentally obtained ψ and θ are fitted with analytical models to obtain the WRC. The parameters of these models are often used in different analytical and numerical solutions. The Gardner (1958), Brooks and Corey (1964), van Genuchten (1980) and Fredlund and Xing's (1994) models are the most commonly used as described below:

Gardner (1958)'s model

$$\theta = \theta_r + \frac{\theta_s - \theta_r}{1 + a\psi^{n_G}} \tag{3.37}$$

where θ is the volumetric water content, θ_s is the volumetric water content at saturation ($\approx n$) in soils that are non-deformable under increasing suction, θ_r is the residual volumetric water content, a is a parameter related to the inverse of the AEV of the soil, and n_G is a parameter related to the pore size distribution of the soil.

Brooks and Corey (1964)

$$\theta = \theta_s \text{ for } \psi \le \psi_a \text{ and } \theta = \theta_r + (\theta_s - \theta_r)\left(\frac{\psi_a}{\psi}\right)^{\lambda} \text{ for } \psi > \psi_a \tag{3.38}$$

where ψ_a is a parameter related to the AEV of the soil, and λ is a pore size index that is related to the pore size distribution of the soil.

van Genuchten (1980)

$$\theta = \theta_r + \frac{(\theta_s - \theta_r)}{\left[1 + (\alpha\psi)^{n_v}\right]^{m_v}} \tag{3.39}$$

where α is a model parameter (m^{-1}) that is generally approximated as the inverse of the AEV, ψ is the matric suction or pressure head (m), n_v is a dimensionless model adjusting parameter (-), and m_v is a dimensionless model fitting parameter (-).

The Gardner, Brooks and Corey, and van Genuchten's models are sometimes expressed in terms of reduced volumetric water content (θ_e) with:

$$\theta_e = \frac{\theta - \theta_r}{\theta_s - \theta_r} \tag{3.40}$$

Fredlund and Xing (1994)

$$\theta = \left[1 - \frac{\ln\left(1 + \frac{\psi}{\psi_r}\right)}{\ln\left(1 + \frac{10^6}{\psi_r}\right)}\right] \times \left[\frac{\theta_s}{\left\{\ln\left[e + \left(\frac{\psi}{a}\right)^n\right]\right\}^m}\right] \tag{3.41}$$

where ψ_r is the residual suction, a is a fitting parameter related to the AEV (kPa), n is a fitting parameter related to the slope of the WRC, m is a fitting parameter related to θ_r, and e is the Euler number (2.71828).

The first factor in Equation (3.41) forces volumetric water content to zero at a suction of 10^6 kPa (or $\approx 10^7$ cm). This condition is not satisfied for the Gardner, Brooks and Corey, and van Genuchten's models, as this leads to $\theta = \theta_r$ for any $\psi \geq \psi_r$.

3.4.5 Methods for Predicting the WRC

Experimental techniques for obtaining WRCs are quite complex, and thus, predictions are often used, especially in the preliminary phases of projects. Many pedo-transfer functions based on empirical equations or on mathematical regressions have been proposed in recent years for specific applications in soil sciences. For engineering applications, predictive models that aim at linking WRCs with their physical properties, such as particle size and porosity, have also been developed (Arya and Paris 1981; Kovács 1981; Haverkamp and Parlange 1986; Aubertin et al. 1998, 2003; Arya et al. 1999; Fredlund et al. 2002; Aubertin et al. 1998, 2003).

One of these models is the modified Kovács (MK) model (Aubertin et al. 1998, 2003). The main advantage of the MK model is that it considers the two phenomena of water retention by capillarity and by adhesion, which are quantified from material characteristics such as the void ratio, the coefficient of uniformity, and the effective diameter (Aubertin et al. 1998, 2003). The total degree of saturation is expressed by combining the capillary (S_c) and adhesion (S_a) components, as shown in Equation (3.42):

$$\mathrm{S_r} = \frac{\theta}{n} = 1 - \langle 1 - \mathrm{S_a} \rangle (1 - \mathrm{S_c}) \tag{3.42}$$

The Macaulay brackets ⟨ ⟩ used in equation (3.42) give $\langle y \rangle = 0.5(y + |y|)$, meaning that $\langle y \rangle = y$ for $y \geq 0$ and $\langle y \rangle = 0$ for $\langle y \rangle < 0$. The S_c and S_a components are given by (Aubertin et al. 2003):

$$S_c = 1 - \left[\left(h_{co} / \psi\right)^2 + 1\right]^m exp\left[-m\left(h_{co} / \psi\right)^2\right] \tag{3.43}$$

$$S_a = a_c \left(1 - \frac{ln\left(1 + \psi / \psi_r\right)}{ln\left(1 + \psi_o / \psi_r\right)}\right) \frac{\left(h_{co} / \psi_n\right)^{2/3}}{e^{1/3}\left(\psi / \psi_n\right)^{1/6}} \tag{3.44}$$

where h_{co} is the equivalent capillary height (cm), m is a pore size distribution parameter ($m = 1/C_U$), a_c is an adhesion coefficient (-) that is approximately constant ($a_c = 0.010$ when ψ is expressed in cm of water), ψ_n is a normalization parameter introduced for unit consistencies ($\psi_n = 1$ cm when ψ is given in cm), ψ_0 is the suction (cm) corresponding approximately to complete dryness ($\theta = 0$ at $\psi = \psi_0 = 10^7$ cm of water), *and* ψ_r is the suction at residual water content (cm).

The second factor in Equation (3.44) was inspired by Equation (3.41). The equivalent capillary height, h_{co} (cm), is related to the equivalent pore diameter and depends on the solid surface area. The value of h_{co} is obtained using Equation (3.45):

$$h_{co} = \frac{0.75}{\left[1.17 log\left(C_U\right)+1\right] e D_{10}} \tag{3.45}$$

The residual suction can be estimated using Equation (3.46).

$$\psi_r = 0.86 h_{co}^{1.2} \tag{3.46}$$

An example is given for the coarse- and fine-grained soils presented in Figure 3.2. For the coarse-grained soil with $D_{10} = 0.006$ cm, $C_U = 3.3$, and assuming $n = 0.40$, the MK model leads to θ_w values of 0.349 and 0.100 for suctions of 40 cm and 100 cm, respectively. For the fine-grained soil with $D_{10} = 0.0002$ cm, $C_U = 7.5$, and assuming $n = 0.40$, θ_w equals 0.344 and 0.178 for suctions of 1000 cm and 10,000 cm, respectively.

The MK model, which was initially developed for non-plastic materials, has been extended to make predictions for coherent non-compressible materials (Aubertin et al. 2003), shrinking materials for which the void ratio depends on the actual suction (Mbonimpa et al. 2006a), and soils with significant hysteresis effects (Maqsoud et al. 2012).

3.4.6 Assessing the Unsaturated Hydraulic Conductivity

Knowledge of the unsaturated hydraulic conductivity function, $k_u(\psi)$, as well as the WRC, $\theta_w(\psi)$, is required to solve equations describing flow under unsaturated conditions. The $k_u(\psi)$ of porous materials can be determined in the laboratory using techniques including the outflow method, the steady-state method, and the instantaneous profile (Watson 1966; Baker et al. 1974; Klute and Dirksen 1986; Fredlund and Rahardjo 1993; Hillel 1998; Askarinejad et al. 2012). The $k_u(\psi)$ function can also be determined in the field (Baker et al. 1974; Hillel 1998). However, these techniques are difficult to apply and very expensive. Therefore, it is more practical to estimate $k_u(\psi)$ in terms of relative unsaturated hydraulic conductivity ($k_r = k_u/k_{sat}$) from the WRC using empirical, macroscopic, and statistical models. The most used statistical models are the Childs and Collis-George (1950), Burdine (1953), Mualem (1976), and Fredlund et al.'s (1994) models, which are described by Equations 3.47, 3.48, 3.49, and 3.50, respectively.

$$k_r = \frac{k_u\left(\theta_e\right)}{k_{sat}} = \theta_e^{\xi} \frac{\int_0^{\theta_e} \left(\theta_e - \gamma\right)\psi^{-2}\left(\gamma\right) d\gamma}{\int_0^{l1} \left(1-\gamma\right)\psi^{-2}\left(\gamma\right) d\gamma} \tag{3.47}$$

$$k_r = \frac{k_u\left(\theta_e\right)}{k_{sat}} = \theta_e^{\xi} \frac{\int_0^{\theta_e} \psi^{-2}\left(\gamma\right) d\gamma}{\int_0^{1} \psi^{-2}\left(\gamma\right) d\gamma} \tag{3.48}$$

$$k_r = \frac{k_u(\theta_e)}{k_{sat}} = \theta_e^{\xi}\left[\frac{\int_0^{\theta_r}\psi^{-1}(\gamma)d\gamma}{\int_0^1\psi^{-1}(\gamma)d\gamma}\right]^2 \quad (3.49)$$

$$k_r = \frac{k_u(\psi)}{k_{sat}} = \frac{\int_{\psi}^{\psi_0}\frac{\theta(v)-\theta(\psi)}{v^2}\theta'(v)dv}{\int_{\psi_a}^{\psi_0}\frac{\theta(v)-\theta_s}{v^2}\theta'(v)dv} \quad (3.50)$$

where θ_e is the reduced volumetric water content (see Equation 3.40), ξ is a pore connectivity parameter that accounts for the presence of tortuous flow paths ($\xi = 0$ for the Childs et Collis-Georges's (1950) model, 2 for the Burdine's (1953) model, and 0.5 for the Mualem's (1976) model), γ is a dummy variable associated with the water content, v is a dummy variable representing the suction, ψ_s is the suction associated with θ_s ($\psi_s = 0$, but a small value, 0.01 cm, is used to avoid a singularity in the integration), ψ_a is the AEV, ψ_0 is the suction at dryness (10^6 kPa), and θ' is the derivative of the volumetric water content function ($\theta' = \partial\theta/\partial\psi$), which corresponds to the water capacity function C used in Equation 3.36.

Due to their mathematical formulations, the Child and Collis-George (1950), Burdine (1953), and Mualem's (1976) models lead to $k_r = 0$ for $\theta_e = 0$, for $\theta_w \geq \theta_r$, or $\psi \geq \psi_r$. However, the Fredlund et al.'s (1994) model allows for calculation of k_r values beyond ψ_r and gives $k_r = 0$ at $\psi = 10^6$ kPa (or $\approx 10^7$ cm).

All the statistical models presented earlier can be solved with numerical methods when the WRC is known. The measured WRC can be defined using the WRC model parameters (see Equations 3.37, 3.38, 3.39, 3.40, and 3.41). Closed-form analytical solutions for k_r have been developed using some specific WRC model parameters. The most well-known equation was derived from the van Genuchten (Equation 3.39) and Mualem's (Equation 3.49) models. Equations (3.51) and (3.52) present the simplest analytical solutions in terms of water content and suction, respectively, when $m_v = 1 - 1/n_v$ is assumed (van Genuchten et al. 1980).

$$k_r(\theta_e) = \theta_e^{\xi}\left[1-\left(1-\theta_e^{1/m_v}\right)^{m_v}\right]^2 \quad (3.51)$$

$$k_r(\psi) = \frac{\left\{1-(\alpha\psi)^{m_v n_v}\left[1+(\alpha\psi)^{n_v}\right]^{-m_v}\right\}^2}{\left[1+(\alpha\psi)^{n_v}\right]^{m_v\xi}} \quad (3.52)$$

When experimental WRC data are not available, predicted WRCs can also be combined with the statistical models to derive preliminary functions of k_r. WRCs predicted with the MK model for granular (non-plastic) materials have been used for these purposes with satisfactory results (Mbonimpa et al. 2006b).

3.4.7 Capillary Barrier Effects

A capillary barrier effect is an unsaturated phenomenon that occurs above the water table in layered covers when a fine-grained material is placed over a coarse-grained material due to contrasts in their unsaturated hydraulic properties. Numerous publications on this phenomenon can be found

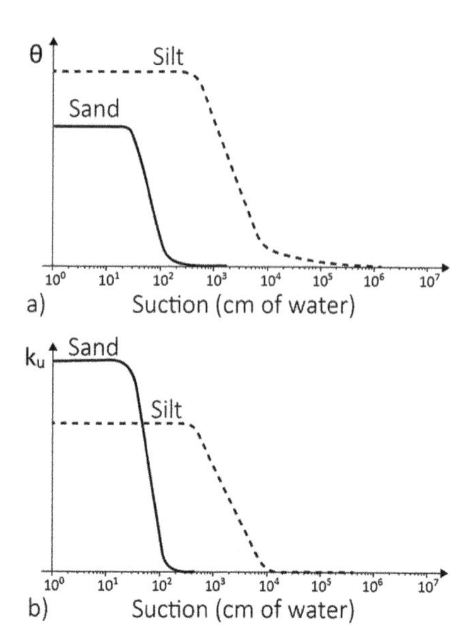

FIGURE 3.6 Explaining capillary barrier effects using (a) water retention curves and (b) unsaturated hydraulic conductivity curves of coarse- and fine-grained materials.

in the literature (e.g., Nicholson et al. 1989; Akindunni et al. 1991; Morel-Seytoux 1992; Aubertin et al. 1995; O'Kane et al. 1998; Bussière et al. 2003). Plotting the WRCs and the k_u curves of the two different materials (e.g., a sand and a silt) as shown in Figure 3.6 can help to understand this phenomenon.

The WRCs in Figure 3.6a indicate that the AEV and the ψ_r of the sand are lower than those of the silt. The corresponding $k_u(\psi)$ functions shown in Figure 3.6b indicate that $k_u \approx k_s$ for $\psi \leq \psi_a$ and k_u begins to decrease for $\psi > \psi_a$. For suctions lower than the AEV of the sand, the k_u is lower for the silt than for the sand. With increasing suctions, k_u decreases for the sand until it becomes very low, while the silt remains almost saturated. Therefore, if a silt layer is placed over a sand layer that drains easily, the hydraulic conductivity of the sand layer rapidly decreases to levels that impede vertical water flow from the fine-grained material. Consequently, the fine-grained material accumulates water and remains close to saturation. This phenomenon is an underlying principle in the designs of both covers with capillary barrier effects and SR covers (see Chapters 5 and 7).

Capillary barrier effects can also be created when a coarse-grained material is placed over a fine-grained material. When the coarse-grained material is at the surface and is exposed to soil-atmosphere interactions (see Section 3.6), the material may reduce significantly evaporation from the underlying material. A drained, coarse-grained material has a very low k_u that limits the movement of water from the fine-grained soil to the surface. The use of coarse-grained materials to protect against evaporation is particularly interesting for the elevated water table reclamation technique (see Chapter 8).

3.5 MOLECULAR DIFFUSION OF OXYGEN IN INERT AND REACTIVE COVER MATERIALS

Understanding oxygen transport mechanisms (i.e., advection and molecular diffusion) is critical for the design of oxygen barrier covers, which limit oxygen availability in reactive tailings. Oxygen transport by advection can result from water infiltration (dissolved oxygen), pressure gradients induced by barometric variations, wind flow, and natural air convection. Advection is predominant in coarse materials, while molecular diffusion becomes the main oxygen transport mode through relatively fine cover materials and tailings. Molecular diffusion is driven by concentration gradients. The governing equations and the methods to assess the required parameters are presented below, focusing primarily on Fickian diffusion. Non-Fickian diffusion of oxygen is not addressed here since it is not considered to significantly influence oxygen fluxes passing through covers (Demers, 2008).

3.5.1 Governing Equations

Similar to water flow, diffusion of oxygen from the atmosphere through porous materials at the Earth's surface may occur in three dimensions. However, for simplification purposes, 1-D diffusion is considered in the following. Fick's first law is used to evaluate the oxygen flux. In this case, the oxygen flux $F(z,t)$ at depth z in a porous material and time t is given by Equation (3.53) (Hillel 1998; Mbonimpa et al. 2003, 2011; Aachib et al. 2004).

$$F(z,t) = -\theta_{eq} \times \frac{\partial C(z,t)}{\partial z} = -D_e \frac{\partial C(z,t)}{\partial z} \tag{3.53}$$

where $F(z,t)$ is the diffusive flux of O_2 ($kg.m^{-2}s^{-1}$); θ_{eq} is the equivalent (diffusion) porosity (m^3m^{-3}); D^* and D_e are the bulk and effective diffusion coefficients, respectively ($m^2.s^{-1}$) ($D_e = \theta_{eq}D^*$); and $C(z, t)$ is the interstitial O_2 concentration at time t (s) and position z (m). θ_a is often used instead of θ_{eq} (Yanful 1993; Aubertin et al. 1995); however, for a highly saturated material, θ_{eq} is needed to take into account oxygen fluxes in the air phase and water phase (dissolved oxygen). The equivalent porosity is defined as (Aubertin et al. 2000):

$$\theta_{eq} = \theta_a + H\theta \tag{3.54}$$

where H is the dimensionless Henry's equilibrium constant given by the ratio $H = C_w/C_a$ (for O_2, $H \cong 0.03$ at 20°C), and C_w and C_a are the oxygen concentrations in the water and air phases.

For a cover built using inert materials with a thickness L that is exposed to a fixed oxygen concentration from the atmosphere (C_0 = 20.9%-vol. ≈0.278 $kg.m^{-3}$ ≈8.7 $mol.m^{-3}$ at 20°C), if the oxygen concentration at the base is C_b at a given time t, a solution to Equation (3.53) can be written as:

$$F_{z=L} = \frac{(C_0 - C_b)D_e}{L} \tag{3.55}$$

In cases where there is rapid and complete consumption of O_2 just below the cover surface due to sulfide oxidation, C_b can be equal to zero.

The spatial distribution and evolution of the concentration $C(z,t)$ required for solving Equation (3.54) is obtained from a continuity equation that follows the same principles as in the case of water flow. For inert (non-reactive) materials that do not consume or produce oxygen, this continuity equation is given by Fick's second law:

$$\frac{\partial}{\partial t}(\theta_{eq}C) = \frac{\partial}{\partial z}\left(\theta_{eq}D^* \frac{\partial C}{\partial z}\right) = \frac{\partial}{\partial z}\left(D_e \frac{\partial C}{\partial z}\right) \tag{3.56}$$

Reactive tailings that are non-acid-generating, such as desulfurized tailings, can also be used as cover materials (Bussière et al. 2004; Demers et al. 2008, 2009; Rey et al. 2020). In these cases, oxygen is consumed over a given period before the eventual depletion of the residual sulfide minerals.

Oxygen consumption also occurs in sulfide tailings that are acid-generating. In these cases, Fick's second law can be modified as in Equation 3.57, assuming that the sulfide mineral oxidation follows an irreversible, first-order kinetic reaction (or exponential decay) (e.g., Elberling et al. 1994; Elberling and Nicholson 1996; Yanful et al. 1999; Mbonimpa et al. 2003, 2011).

$$\frac{\partial}{\partial t}\left(\theta_{eq}C\right)=\frac{\partial}{\partial z}\left(\theta_{eq}D^{*}\frac{\partial C}{\partial z}\right)-\theta_{eq}K_r^{*}C=\frac{\partial}{\partial z}\left(D_e\frac{\partial C}{\partial z}\right)-K_rC \tag{3.57}$$

where K_r^* and K_r are the bulk and effective reaction rate coefficients, respectively [T^{-1}]; $\left(K_r=\theta_{eq}K_r^*\right)$.

Equation (3.57) can be solved for both steady-state and transient conditions using numerical solutions for representative boundary conditions and using analytical solutions for relatively simple boundary conditions (Crank 1975; Nicholson et al. 1989; Elberling et al. 1994, Mbonimpa et al. 2003). One then obtains the concentration profile $C(z,t)$, which can provide the flux with Equation (3.53).

For sulfide tailings exposed to air (C_0 = atmospheric concentration), the oxygen concentration profile under steady-state conditions ($t \rightarrow \infty$) is then given by the following analytical solution when the following boundary and initial conditions apply before placing the cover: $C(z=0, t>0)=C_0$, $C(z=\infty, t>0)=C_\infty=0$, and $C(z>0, t=0)=0$ (e.g., Nicholson et al. 1989; Elberling et al. 1994):

$$C(z)=C_0\,exp\left(-z\sqrt{K_r^*/D^*}\right)=C_0\,exp\left(-z\sqrt{K_r/D_e}\right) \tag{3.58}$$

The corresponding steady-state flux, $F_{0,s}$, at any depth z becomes:

$$F_{0,s}(z)=\theta_{eq}\sqrt{D^*K_r^*}\,C(z)=\sqrt{D_eK_r}\,C(z) \tag{3.59}$$

The steady-state flux entering the surface of uncovered reactive tailings ($z=0$) can then be expressed as:

$$F_{0,s}(z=0)=\theta_{eq}C_0\sqrt{D^*K_r^*}=C_0\sqrt{D_eK_r} \tag{3.60}$$

However, in many situations, the boundary conditions described earlier for oxygen depletion at infinite depth ($C=0$ at $z\rightarrow\infty$) may be unrealistic, as was shown by Mbonimpa et al. (2003) who proposed an equation to estimate the depth of oxygen penetration. Mbonimpa et al. (2003) also presented numerical solutions that can be applied to assess the oxygen flux at the surface and bottom of a cover placed over highly reactive tailings so that oxygen is rapidly consumed at the tailings-cover interface where the concentration remains zero. However, these solutions only apply with the specific boundary conditions: $C(z=0, t>0)=C_a$, $C(z\geq L, t>0)=C_L=0$, and $C(z>0, t=0)=0$; that is, the initial concentration throughout the porous medium is zero.

3.5.2 Experimental Methods for Determining Diffusion Driving Coefficients

Various experimental approaches have been used to determine D_e, which is the unique diffusion driving coefficient for inert materials. In this chapter, emphasis is put on an easier test that can be conducted under controlled conditions in the laboratory; that is, the oxygen diffusion (OD) test (Figure 3.7). The OD setup was initially based on the work of Yanful (1993), but several modifications have been made in subsequent years. In brief, the setup consists of a closed cylindrical reservoir with a diameter D and height H. A sample is placed in the middle of the cylinder while ensuring empty source and receptor reservoirs on either side of the specimen. Calibrated

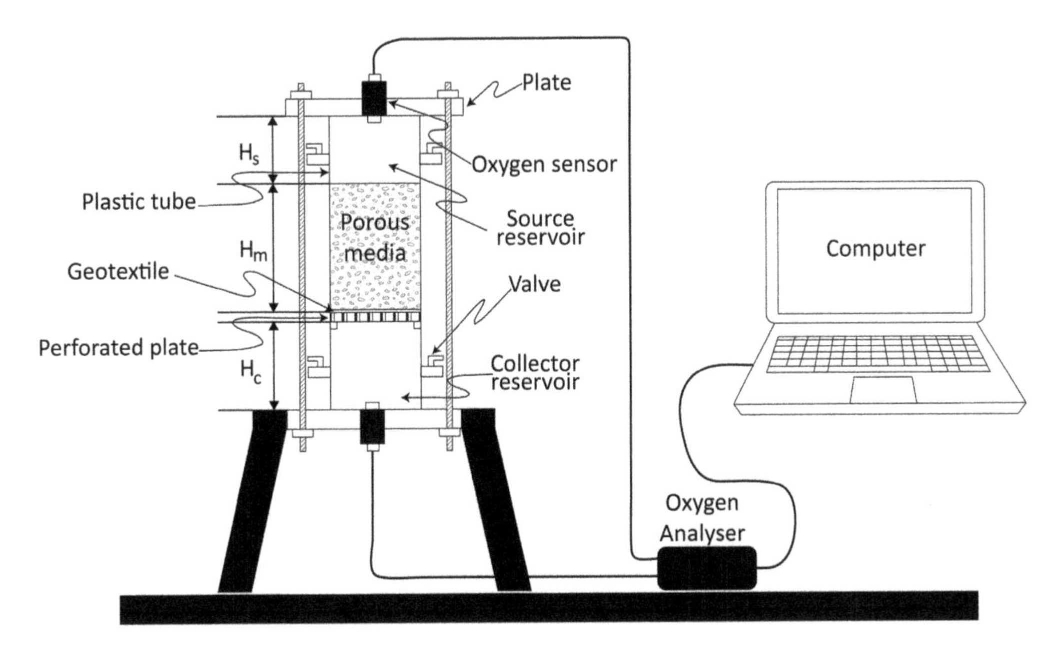

FIGURE 3.7 Schematic representation of the oxygen diffusion (OD) cell used to evaluate the effective oxygen diffusion coefficient (adapted from Aubertin et al. 1995).

oxygen probes are fixed to the reservoir plates to measure oxygen concentrations. Then, the following steps are performed (more details can be found in studies by Aubertin et al. 2000 and Aachib et al. 2004):

- Prepare the material layer to the targeted gravimetric water content.
- Place the material with a thickness L at the desired n and S_r.
- Purge the entire cell with humidified nitrogen until the oxygen concentration of the cell stabilizes to zero. The initial oxygen concentration within the sample must be zero for simplification of test interpretation.
- Briefly open the source reservoir to reach atmospheric conditions. Since in the receptor reservoir $C_0 = 0$, this creates a concentration gradient and induces oxygen migration by diffusion from the source to the receptor reservoir.
- Measure the temporal evolution of oxygen concentrations in the two reservoirs. These data are then used to interpret the results of the test.

Figure 3.8a presents typical results for an inert material with $n = 0.4$ and $S_r = 85\%$. The measured oxygen concentrations decrease in the source reservoir and increase in the collector reservoir. At a given time, equilibrium is reached such that the oxygen concentrations are equal and constant in both reservoirs. Equations for calculating oxygen concentrations were presented by Aachib et al. (2004). To obtain the actual value of D_e, the temporal evolution of measured oxygen concentrations is iteratively matched with concentrations calculated by solving Fick's laws for the initial and boundary conditions corresponding to the diffusion test. Here, the code POLLUTEv7 (Rowe and Booker 2005) was used, but other codes are available. Details can be found in a study by Aachib et al. (2004). For the results presented in Figure 3.8a, $D_e = 5.3 \times 10^{-5}$ m^2.h^{-1} or 1.5×10^{-8} m^2.s^{-1}.

For reactive materials, variations in oxygen concentrations with time are influenced by two coefficients, D_e and K_r. In this case, the more common laboratory tests are OD and consumption (ODC) tests, which also use a cell with two reservoirs (Figure 3.7). However, ODC tests can also be performed in a cell with one reservoir (Gosselin et al. 2007). The methodological approach presented

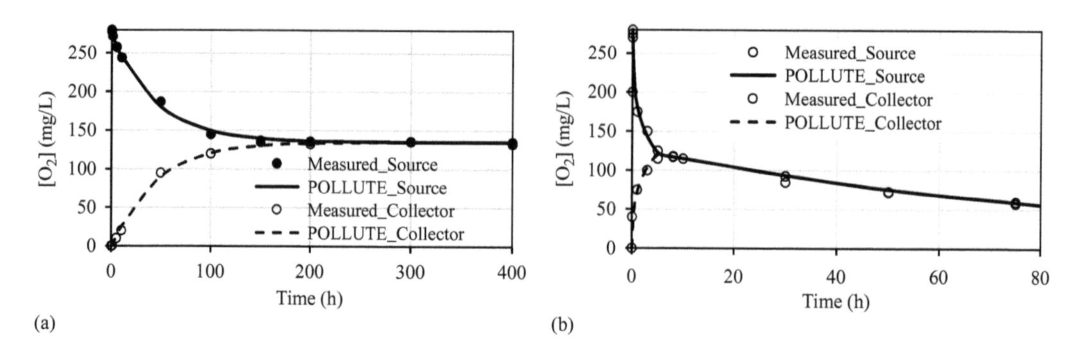

FIGURE 3.8 Typical results of (a) oxygen diffusion tests and (b) oxygen diffusion and consumption tests.

earlier for the OD tests applies similarly to ODC tests. Iterative numerical solutions allow for the determination of the actual values of D_e and K_r by matching measured and modeled oxygen concentrations as they evolve over time (Mbonimpa et al. 2003).

Figure 3.8b presents the results of ODC tests performed on reactive tailings, with $n = 0.46$ and $S_r = 60\%$. Measured oxygen concentrations decrease in the source reservoir and increase in the collector reservoir until an equilibrium is reached with equal oxygen concentrations in both reservoirs. However, oxygen concentrations continue to decrease due to oxygen consumption by the reactive tailings. To obtain the actual values of D_e and K_r, these variables are iteratively varied (here using the code POLLUTEv7, Rowe and Booker 2005) until measured and modeled oxygen concentrations match. In this case, $D_e = 1.2 \times 10^{-3}$ m^2.h^{-1} or $D_e = 3.3 \times 10^{-7}$ m^2.s^{-1} and $K_r = 3.80 \times 10^{-4}$ s^{-1}, or $K_r = 2.28 \times 10^{-2}$ h^{-1}, or $K_r = 200$ y^{-1}.

The oxygen consumption test (OCT), developed by Elberling et al. (1994) (see also Elberling and Nicholson 1996; Elberling and Langdahl 1998; Elberling 2001a, 2001b; Mbonimpa et al. 2011), can also be used in situ or in the laboratory to determine the D_e and K_r values of tailings exposed to air (with an oxygen concentration C_0). Figure 3.9a shows a schematic depiction of an OCT field test. This method is performed by first inserting an aluminum or steel tube into the tailings (which are assumed to be homogeneous) for field tests or by placing tailings in a cylindrical column for laboratory tests. In both cases, an empty space (headspace) is left at the top. After hermetically sealing the headspace with a cap equipped with an oxygen probe, a finite quantity of air is available in the headspace that acts as a source reservoir. For $t > 0$, the oxygen is consumed by oxidation, and the progressive decrease in oxygen concentrations in the headspace is monitored over time. Figure 3.9b shows a typical OCT setup and results (see Chapter 10 for a more detailed description of field OCTs).

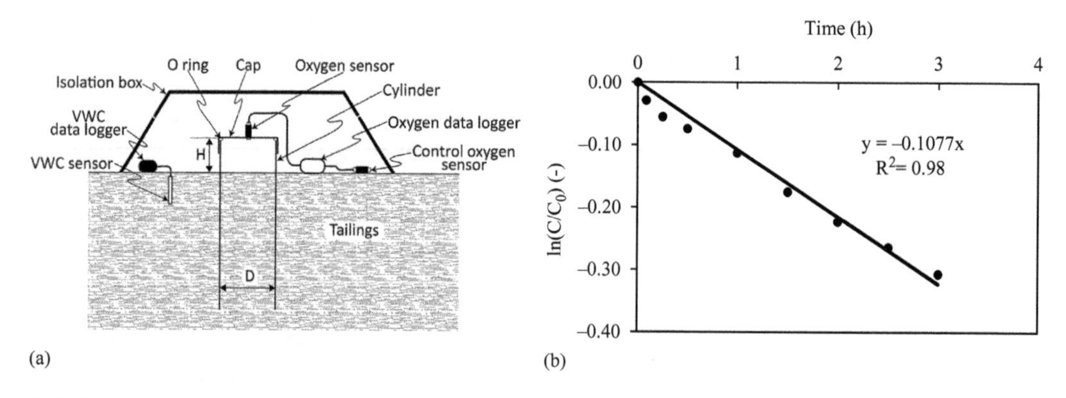

FIGURE 3.9 Schematic representation of (a) an OCT setup and (b) typical OCT results.

For short-duration OCTs, a simple analytical interpretation procedure was proposed by Elberling et al. (1994) to determine the value of (D_eK_r), which can be then used to calculate the steady-state oxygen flux, $F_{0,s}$ (with Equation 3.60) that is consumed by the tailings. Assuming that the oxygen flux at time t is represented by Equation (3.60), the continuity equation gives (Elberling et al. 1994):

$$V\frac{dC}{dt} = \theta_{eq}AC\left(K_r^*D^*\right)^{0.5} = -AC\left(K_rD_e\right)^{0.5} \tag{3.61}$$

where A is the cylinder area and V is the headspace volume. For the initial condition ($C = C_0$ at $t = 0$), integrating the above Equation (3.61) leads to:

$$\ln\left(\frac{C}{C_0}\right) = -t\left(K_rD_e\right)^{0.5}\frac{A}{V} \tag{3.62}$$

The slope of the linear plot of $ln(C/C_0)$ versus time, t, provides the value of $(D_eK_r)^{0.5}$ from a linear regression (see Equation 3.62). Figure 3.9b presents typical results obtained on reactive tailings with n = 0.44 and S_r = 60%. The A/V ratio, which corresponds to the height of the headspace, was 0.05 m. The slope of the linear regression (with $R^2 = 0.98$) was –0.1077 (when t is in h; the value of the slope changes depending on the unit used for t in Figure 3.9b). This slope corresponds to $-(K_rD_e)^{0.5}A/V$, which means that $(K_rD_e)^{0.5} = 5.4 \times 10^{-3}$ m.h^{-1}. D_e can be estimated using n and S_r (see Equation 3.63) as 3.3×10^{-7} m^2.s^{-1} or 1.2×10^{-3} m^2.h^{-1}. Knowing the value of D_e and $(K_rD_e)^{0.5}$, K_r can be calculated as 2.5×10^{-2} h^{-1} or 216 y^{-1}.

It should be recalled that Eq. (3.62) can only be valid for short-duration tests because a steady-state condition ($\delta C/\delta t = 0$) is assumed to exist before and during the tests. This condition is, however, no longer valid if the conditions during the test deviate significantly from this steady state, which can be expected because oxygen concentration changes with time. Mbonimpa et al. (2011) indicated that this interpretation approach underestimates oxygen fluxes. This underestimation becomes more significant with increasing test duration, decreasing height of the headspace, and decreasing S_r of the tailings. An alternative method for the interpretation of OCTs is to use numerical procedures that provide the individual values of D_e and K_r instead of their product.

For the analytical curve fit procedures mentioned for the OD test and the numerical solution mentioned for the OC, ODC, and OCTs, the initial values of D_e and K_r are generally estimated from semi-empirical expressions presented in the next section.

3.5.3 Predictive Methods of the Diffusion Driving Coefficients

The first semi-empirical models proposed in the literature use the air-filled porosity (θ_a) for estimating D_e (e.g., Reardon and Moddle 1985; Jin and Jury 1996), neglecting OD through the water phase. Consequently, these single-phase models lead to $D_e = 0$ for $S_r = 1$, and thus contradicting measured values. Millington and Quirke (1961) proposed the first dual-phase model that overcomes this limitation. This model has been modified and generalized over the years (Collin 1987, Millington and Shearer 1971; Aachib et al. 2004). According to Aachib et al. (2004), D_e can be estimated as follows:

$$D_e = \frac{1}{n^2}\left(D_a^0\theta_a^{p_a} + HD_w^0\theta^{p_w}\right) \tag{3.63}$$

where n is the porosity; D_a^0 is the free (undisturbed) OD coefficient in air ($\approx 1.8 \times 10^{-5}$ m^2.s^{-1} at 2°C); D_w^0 is the free OD coefficient in water ($\approx 2.2 \times 10^{-9}$ m^2.s^{-1} for oxygen at 20°C); θ_a and θ are the volumetric air and water contents, respectively; H is Henry's constant; and p_a and p_w are exponents

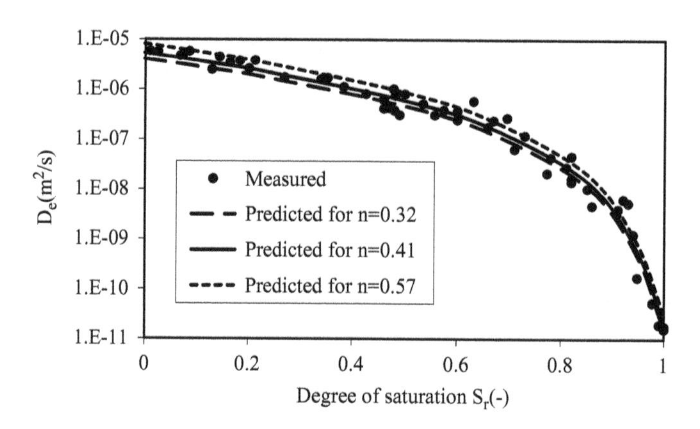

FIGURE 3.10 Effect of S_r on the measured D_e values obtained from tailings with $0.32 \leq n \leq 0.57$ (mean = 0.41) and on predicted D_e values for the minimum (n = 0.32), mean (n = 0.41), and maximum (n = 0.57) porosities.

associated with θ_a and θ for describing the tortuosity in the air and water phases, respectively, and are defined empirically with the following expressions:

$$p_a = 1.201\theta_a^3 - 1.515\theta_a^2 + 0.987\theta_a + 3.119 \quad (3.64)$$

$$p_w = 1.201\theta^3 - 1.515\theta^2 + 0.987\theta + 3.119 \quad (3.65)$$

Experimental values of D_e obtained from soils with porosities ranging between 0.39 and 0.59 (with a mean porosity of 0.45) are shown in Figure 3.10 with respect to the S_r. Values of D_e predicted with Equation 3.63 are also presented in Figure 3.10 for the minimum (n = 0.32), mean (n = 0.41), and maximum (n = 0.57) porosities. Predicted values are in good agreement with measured values. For a given S_r, the impact of porosity remains relatively negligible.

The other driving coefficient for OD in porous media is the reaction rate coefficient, K_r. Few predictive models have been developed to estimate K_r. For the case of pyrite, which is the Earth's most abundant sulfide mineral, Collin (1987, 1998) proposed a simple model based on surface kinetics, in which K_r varies linearly with the pyrite content of a material (Equation 3.66).

$$K_r = K' \frac{6}{D_H}(1-n)C_p \quad (3.66)$$

where k' is the reactivity of pyrite with oxygen ($\approx 5 \times 10^{-10}$ m^3 O_2.m^{-2}pyrite.s^{-1} $\approx 15.8 \times 10^{-3}$ m^3 O_2.m^{-2}pyrite.yr^{-1}), C_p is the pyrite content over mass of dry tailings (kg.kg^{-1}), n is the total porosity, and D_H is the equivalent grain-size diameter used to estimate the surface area of the tailings through a relationship with GSD parameters that were developed for hydraulic functions (Aubertin et al. 1998; Mbonimpa et al. 2003):

$$D_H = \left[1 + 1.17 log\left(C_U\right)\right] D_{10} \quad (3.67)$$

Measured values of K_r may differ from estimated ones as K_r also depends on a number of other factors that are not taken into account in the abovementioned equation, including the specific mineralogy and type of sulfides, temperature, oxidation state, and bacterial activities (e.g., Hollings et al. 2001). Attempts to assess the influence of the degree of saturation within a modified model (Gosselin et al. 2007) remain inconclusive.

3.5.4 Evaporation

Soil-atmosphere interactions influence the performance of cover systems for the reclamation of mine wastes. Evaporation rates at the surface of a cover system are particularly important for SR covers (see Chapter 5) and are a function of atmospheric forcing conditions (i.e., precipitation and potential evaporation) and soil properties. Fluxes within the soil/atmosphere continuum are fully coupled in terms of heat and mass transfer. Wilson et al. (1994) provide the following equation for the flow of liquid water (Darcy's law) and water vapor (Fick's law) within cover profiles considering potential volume change:

$$\frac{\partial h}{\partial t} = C_{vw}\left(k_z \frac{\partial h}{\partial z}\right) + C_v\left(D_v \frac{\partial P_v}{\partial z}\right) \tag{3.68}$$

where h is the hydraulic head (m), C_{vw} is the modulus of volume change with respect to the liquid water phase (see Equation 3.69), k_z is the saturated or unsaturated hydraulic conductivity in the vertical (z) direction, C_v is the modulus of volume change with respect to the water vapor phase (see Equation 3.69), D_v is the coefficient of diffusion for water vapor through soil (kg.m.kN^{-1}.s^{-1}), ρ_w is the density of liquid water (kg.m^{-3}), and P_v is the partial pressure in the soil due to water vapor (kPa).

$$C_{vw} = \frac{1}{g\rho_w m_{w2}} \quad \text{and} \quad C_v = \frac{P + P_v}{P\rho_w^2 g m_{w2}} \tag{3.69}$$

where g is the acceleration due to gravity (m.s^{-2}), m_{w2} is the coefficient of water volume change with respect to the change in matric suction or the slope of the WRC (kPa^{-1}), and P is the total atmospheric pressure (kPa).

The vapor pressure (P_v) required for solving this equation depends on the saturation vapor pressure (P_{vs}) of the soil-water (which increases with the temperature) and the relative humidity of the soil surface (H_r):

$$P_v = P_{vs} H_r \tag{3.70}$$

The following relationship for relative humidity and suction was given by Edlefsen and Anderson (1943):

$$H_r = e^{\frac{\psi W_v}{RT}} \tag{3.71}$$

The relative humidity of the soil surface is a function of total suction in the soil (kPa), the molecular weight of water (W_v = 0.018 kg/mole), the universal gas constant (R = 8.314 J.mol^{-1}.K^{-1}), and the absolute temperature T (K). A substantial decrease in H_r occurs only after the ψ > 2000–3000 kPa (i.e., for ψ = 3000 kPa, $H_r \approx 98\%$).

The temperature profile of the soil required for the calculation of the P_v is obtained from solving Equation (3.72), which governs the 1-D heat flow and conservation of thermal energy between soil, water, and water vapor (Wilson et al. 1994):

$$C_T \frac{\partial T}{\partial z} = \frac{\partial}{\partial z}\left(\lambda \frac{\partial T}{\partial z}\right) - L_v\left(\frac{P + P_v}{P}\right)\frac{\partial}{\partial z}\left(D_v \frac{\partial P_v}{\partial z}\right) \tag{3.72}$$

where C_T is the volumetric specific heat (J.m^{-3}.K^{-1}), T is the temperature (K), λ is the thermal conductivity (W.m^{-1}.K^{-1}), and L_v is the latent heat of vaporization (J.kg^{-1}).

The solution for Equations (3.68) and (3.72) during infiltration events is achieved by applying a flux equal to the rainfall intensity of the first term on the right side of Equation (3.68). The solution for evaporation events is equally complex because the rate of AE is determined by both the rate of potential evaporation that are established by climatic conditions and by the suction at the soil surface. The AE is equal to the potential rate of evaporation until the value of suction at the soil surface exceeds approximately 3000 kPa (Wilson et al. 1997). The value of AE progressively decreases with increasing suction when the soil suction exceeds 3000 kPa. The decline in evaporation occurs due to depression of vapor pressure within the voids of the soil with increasing suction. The AE from the surface of an unsaturated cover material may be computed using the modified form of the Penman (1948) method given by Wilson et al. (1994):

$$AE = \frac{\Delta Q_n + \gamma E_a}{\Delta + A} \tag{3.73}$$

where AE is the actual evaporative flux (mm.d^{-1}), Δ is the slope of the saturation vapor pressure versus the temperature curve at the mean temperature of the air (mm Hg.°C^{-1}), Q_n is the net radiant energy available at the surface (mm.d^{-1}), and γ is the psychrometric constant. In Equation 3.73, $E_a = e_a \times (B - A) \times f(u)$, where e_a is the water vapor pressure of the air above the soil surface (mm Hg), B is the inverse of the relative humidity in the air, A is the inverse of the relative humidity at the soil surface ($A = H_r^{-1}$, with $A = 1$ for saturated soil surfaces; in this case, AE becomes equal to the potential evaporation) and $f(u)$ is a turbulent exchange function that depends on the mixing characteristics of the air above the evaporating surface ($f(u) = 0.35 \times (1+0.146W_a)$, W_a is the wind speed (km.h^{-1})), and H_r is obtained by solving the coupled moisture and heat flow Equations (3.68) and (3.72).

When data for the net surface radiation (Q_n) are not available but the potential evaporation was measured, the AE can be estimated from the following equation (Wilson 1990):

$$AE = PE\left(\frac{H_r - \dfrac{P_{vs-air}}{P_{vs-soil}}H_a}{1 - \dfrac{P_{vs-air}}{P_{vs-soil}}H_a}\right) \tag{3.74}$$

where PE is the potential evaporation, H_r is the relative humidity at the soil surface, $P_{vs\text{-}air}$ is the saturated vapor pressure of air, $P_{vs\text{-}soil}$ is the saturated vapor pressure at the soil surface, and H_a is the relative humidity of air above the soil surface.

The solution for Equations 3.68 and 3.72 requires a surface boundary condition for temperature:

$$T_s = T_a + \frac{(Q_n - AE)}{\gamma f(u)} \tag{3.75}$$

where T_s is the temperature of the soil surface (°C), and T_a is the air temperature above the soil surface (°C). The general solutions for Equation 3.68 to 3.72 to compute infiltration, evaporation, and water flow conditions within soil cover profiles were developed in the numerical code SoilCover (Unsaturated Soil Group 1997).

The influence of root water uptake and loss at the leaf level due to plant transpiration must also be included in the calculation of ET when the soil cover is vegetated (see also Chapter 14). Tratch et al. (1995) showed how this can be implemented using a modified form of the method proposed by Ritchie (1972). The method provided by Tratch et al. (1995) relies on three principle variables to calculate actual plant transpiration: the leaf area index (LAI), plant limiting factor (PLF), and root depth. LAI is used to calculate potential transpiration (Fredlund et al. 2012). It corresponds to the total surface area of the leaves (one side) within a given canopy divided by the ground surface area beneath the canopy. In general, the rate of potential plant transpiration is equal to potential evaporation when the LAI is greater than 2.7. The potential rate of transpiration decreases according to empirical relationships given by Ritchie (1972) when the LAI is <2.7. The actual rate of root water uptake is subsequently computed as the product of the potential transpiration and PLF.

The PLF is calculated as a function of the matric suction profile in the root zone. Feddes et al. (1978) showed that, for most plants, the PLF is generally equal to 1 for matric suction values <100 kPa. Feddes et al. (1978) also suggest that PLF decreases on a log scale to zero as suction increases to 1500 kPa. The final step in applying the root water extraction rate to the cover system is to distribute the computed uptake over the specified root depth. Prasad (1988) showed that a linear distribution decreasing to zero with depth is a suitable assumption.

3.6 HEAT TRANSFER IN COVER MATERIALS

Heat transfers in cover materials can occur via conduction, convection, and/or radiation. Conduction is energy transmitted across and through the materials as a result of direct contact, whereas convection is the transfer of the energy associated with the movement of pore fluids. Thermal radiation occurs when electromagnetic energy is emitted by a mass and is transmitted through space to another mass. A summary of the main heat transfer mechanisms in porous media and their limits of predominance with respect to the equivalent particle diameter and degree of saturation is shown in Figure 3.11. Figure 3.11 shows that conduction (region 1) is the dominant heat transfer mechanism for most geomaterials. In fine-grained materials (i.e., silts and clays) at low S_r, heat transfer is mainly governed by the thermal redistribution of moisture (region 2; Figure 3.11) or temperature-dependent vapor diffusion (region 3; Figure 3.11). In coarse-grained materials (i.e., mostly gravels), heat transfer is associated with natural (or free) water convection (region 4; Figure 3.11) when the S_r is high or radiation (region 5; Figure 3.11) and natural (or free) air convection (region 6; Figure 3.11) when the S_r is low. In cover designs, silty to gravelly materials are commonly used as cover materials. Saturations ranging from about 15%–20% (e.g., for a drained sand in a capillary break layer) to 100% (e.g., for a silty material in a moisture-retaining layer or in tailings) are typically observed for cover materials, indicating that conduction is the main mechanism governing heat transfer. In some applications, coarser materials, such as rock-fill

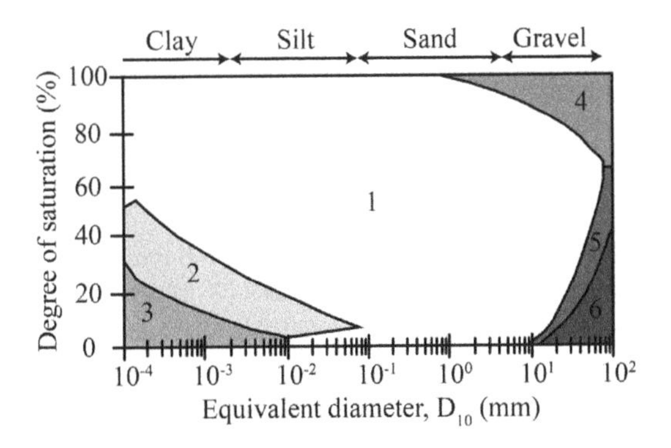

FIGURE 3.11 Heat transfer mechanisms in porous media and their predominant domain with respect to the degree of saturation and equivalent diameter (D_{10}) (redrawn and adapted after Johansen, 1975).

materials or waste rocks, are also used as covers materials. Since these materials have large particle sizes and typically show low S_r values, radiation and natural air convection are more likely to be important processes governing heat transfer.

3.6.1 Governing Equations

Equations describing heat transfer in porous media are extensively documented in the literature (e.g., Andersland and Ladanyi 2004; Fredlund et al. 2012; Nield and Bejan 2017). Thermal conduction can be visualized by an energy transfer due to interactions between particles from a region of high molecular activity to a region of lower molecular activity. Thus, in the presence of a thermal gradient, the movement of heat by conduction occurs from a zone of higher temperature toward a zone of lower temperature. Heat transfer by conduction is governed by mechanisms similar to that of water flow (i.e., Darcy's law) or OD (i.e., Fick's law). However, in this case, the driving force of heat transfer is a thermal gradient. Heat conduction occurs through a medium with a thermal conductivity (λ; $W.m^{-1}.K^{-1}$) in response to a thermal gradient. The thermal conductivity of a material represents its ability to conduct heat; that is, heat transfer occurs at a slower rate in materials with lower thermal conductivities. Thermal conduction is described by Fourier's law, which is expressed in the x, y, and z directions as:

$$q_{hx} = -\lambda_x \frac{\partial T}{\partial x} \tag{3.76}$$

$$q_{hy} = -\lambda_y \frac{\partial T}{\partial y} \tag{3.77}$$

$$q_{hz} = -\lambda_z \frac{\partial T}{\partial z} \tag{3.78}$$

where q_{hx}, q_{hy}, and q_{hz} (W/m^2) and λ_x, λ_y, and λ_z are the conductive heat fluxes and thermal conductivities in the x, y, and z directions, respectively; and T is temperature (K). In an isotropic porous media, $\lambda_x = \lambda_y = \lambda_z$, whereas in an anisotropic porous media, λ_x, λ_y, and λ_z do not have the same value.

Similar to the mass conservation equations developed in Section 3.4.1 for water flow, the thermal energy conservation principle requires that for steady-state heat flow analyses, the amount of heat flowing in and out of an elemental volume be equal. Considering Equations 3.76, 3.77, and 3.78 and the conservation of thermal energy, the 3-D heat flow equation can be expressed as:

$$\lambda_x \frac{\partial^2 T}{\partial_x^{\ 2}} + \frac{\partial \lambda_x}{\partial x}\frac{\partial T}{\partial x} + \lambda_y \frac{\partial^2 T}{\partial_y^{\ 2}} + \frac{\partial \lambda_y}{\partial y}\frac{\partial T}{\partial y} + \lambda_z \frac{\partial^2 T}{\partial_z^{\ 2}} + \frac{\partial \lambda_z}{\partial z}\frac{\partial T}{\partial z} = 0 \tag{3.79}$$

However, transient analyses are generally required for most engineering problems involving heat flow in cover designs and mine waste storage facilities. Transient heat flow analyses require that the net heat flux passing through an elemental volume be equal to its heat storage with respect to time. Accordingly, in the absence of energy production or loss by internal processes and considering the thermal energy conservation equation for transient heat flow, Equation 3.80 can be used to describe the 3-D conductive heat flow in unfrozen porous media and obtain $T_{(x,y,z,t)}$:

$$C_T \frac{\partial T}{\partial t} = \lambda_x \frac{\partial^2 T}{\partial_x^{\ 2}} + \frac{\partial \lambda_x}{\partial x}\frac{\partial T}{\partial x} + \lambda_y \frac{\partial^2 T}{\partial_y^{\ 2}} + \frac{\partial \lambda_y}{\partial y}\frac{\partial T}{\partial y} + \lambda_z \frac{\partial^2 T}{\partial_z^{\ 2}} + \frac{\partial \lambda_z}{\partial z}\frac{\partial T}{\partial z} \tag{3.80}$$

where C_T (J.m^{-3}.K^{-1}) is the volumetric heat capacity of the porous medium and t is time. C_T corresponds to the amount of heat that must be added to one unit of volume of the porous medium to increase its temperature by one unit, and $C_T \frac{\partial T}{\partial t}$ represents the term related to heat storage.

In the presence of sub-zero temperatures, phase change of interstitial water can occur, thus introducing the concept of latent heat. Latent heat is the heat absorbed by or released from an object (in this case, water) to change phase (melt, freeze, vaporize, condense, etc.). As temperatures are lowered to the freezing point, the latent heat of fusion ($L_f \cong 333.7$ kJ.kg^{-1}) contained in the interstitial water must be released before the material's temperature can be lowered further. During the freezing process, this release of latent heat acts as a heat source and contributes to maintaining the ground temperature close to the freezing point for a certain amount of time. Then, after most of the latent heat is extracted, ground temperatures can be lowered. As a result, latent heat often causes a discontinuity in the temporal evolution of the temperatures close to the freezing point. This discontinuity is referred to as the *zero-curtain effect* and happens at the freezing front of most saturated or partially saturated porous media (Harlan 1973; Outcalt et al. 1990). During the thawing process, the reverse behavior is observed. In order to change phase from ice to liquid water, latent heat must be absorbed by water, resulting in a heat sink–like behavior.

Because L_f is related to phase changes of water, the total energy involved depends on the amount of water contained in the volume of soil and the volumetric fraction of water that changes phases (Andersland and Ladanyi 2004). This means that materials with higher water contents contain greater amounts of latent heat than materials at lower water contents. As a result, the impact of latent heat on freezing and thawing processes is more significant for wetter materials, which usually take more time to freeze or thaw. Accordingly, Equation 3.80 can be modified to account for phase changes, which, considering the conservation of thermal energy, can be expressed as:

$$C_T \frac{\partial T}{\partial t} - L_f \rho_i \frac{\partial \theta_i}{\partial T}\frac{\partial T}{\partial t} = \lambda_x \frac{\partial^2 T}{\partial_x^{\ 2}} + \frac{\partial \lambda_x}{\partial x}\frac{\partial T}{\partial x} + \lambda_y \frac{\partial^2 T}{\partial_y^{\ 2}} + \frac{\partial \lambda_y}{\partial y}\frac{\partial T}{\partial y} + \lambda_z \frac{\partial^2 T}{\partial_z^{\ 2}} + \frac{\partial \lambda_z}{\partial z}\frac{\partial T}{\partial z} C_T \frac{\partial T}{\partial t} - L_f \rho_i \frac{\partial \theta_i}{\partial T}\frac{\partial T}{\partial t} \tag{3.81}$$

or as a vector equation:

$$\left(C_T - L_f \rho_i \frac{\partial \theta_i}{\partial T}\right)\frac{\partial T}{\partial t} = \vec{\nabla}\cdot\left[\lambda \cdot \vec{\nabla} T\right] \tag{3.82}$$

where ρ_i is the density of ice (kg.m^{-3}), θ_i is the volumetric ice content (m^3.m^{-3}); $\frac{\partial \theta_i}{\partial T}$ refers to the slope of the freezing curve, λ is the thermal conductivity tensor (W.m^{-1}.K^{-1}), ∇ (nabla) is a vector differential operator, and ∇T is the temperature gradient (K.m^{-1}).

Equation 3.82 considers that the continuity of the water phase in frozen soils is a function of θ_i and the unfrozen volumetric water content (θ_u; m^3.m^{-3}) as expressed by Equation 3.83 (Harlan 1973).

$$\theta = \theta_u + \frac{\rho_i}{\rho_u}\theta_i \tag{3.83}$$

where θ is the total volumetric water content (m^3.m^{-3}), and ρ_u is the density of unfrozen water (kg.m^{-3}).

In the case of freezing and thawing soils and geomaterials, the materials' thermal properties, such as thermal conductivity and volumetric heat capacity, can vary with the pore water state. The presence of unfrozen, frozen, or partially frozen pore water significantly influences the materials' thermal properties (as discussed below; also refer to Figure 3.13 for examples of unfrozen and frozen thermal conductivity and volumetric heat capacity functions). Therefore, materials' properties with respect to the state of pore waters must be considered during freezing/thawing thermal analyses. Additionally, it can be assumed that for most cover materials, pore water has a freezing point very close to that of pure water (i.e., 0°C). However, in some cases, the pore water can be concentrated with dissolved ions, thus affecting the freezing point. In the context of mine site reclamation, a freezing point depression is often observed for tailings and is usually attributed to the presence of oxidation products and process water in the tailings (MEND 2004). The chemistry of pore waters also has a significant impact on unfrozen water contents below the freezing point and on the progress of the freezing front (e.g., Yong et al. 1979; Watanabe and Mizoguchi 2002; Andersland and Ladanyi 2004; Arenson and Sego 2004).

The presence of unfrozen water below the freezing point is affected by several factors, including the specific surface area of soil particles (Anderson and Tice 1972; Dillon and Anderslan 1966) and the forces of adsorption exerted by certain soil particles (depending on the mineralogy) (e.g., Koopmans and Miller 1966; Miller 1980; Spaans and Baker 1996). The function representing θ_u in a porous medium as a function of temperature below the freezing temperature is the freezing curve (Figure 3.12). Because the continuity of the water phase in frozen soils is a function of θ_u and θ_i, the freezing curve can also be represented by the evolution of θ_i as a function of temperature (e.g., as required to solve Equation 3.82). The freezing curve is often described as being analogous to the WRC because the processes controlling wetting and drying are similar to those controlling freezing and thawing (e.g., Black and Tice 1989; Spaans and Baker 1996; Liu et al. 2012b; Williams 1964). This means that, for coarse-grained materials, most of the interstitial water freezes as the temperature drops below the freezing point. However, in finer-grained materials such as silts or clays, a more significant decrease in temperature is necessary to allow the water adsorbed around the particles or contained in the smallest pores to freeze.

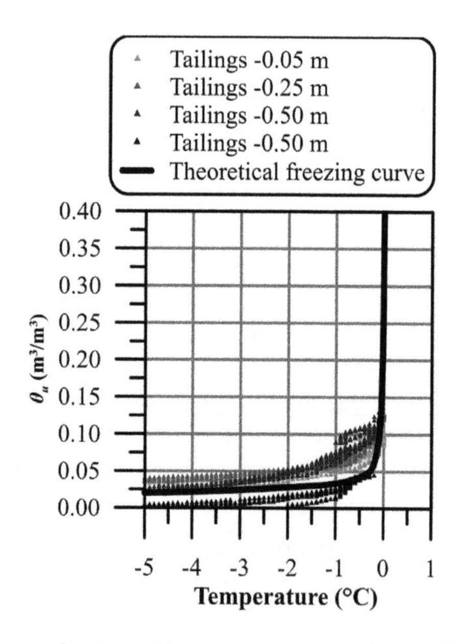

FIGURE 3.12 Examples of θ_u as a function of temperature measured in situ for nearly saturated mine tailings at n = 0.40, compared to a theoretical freezing curve for silty materials (data from Boulanger-Martel 2019).

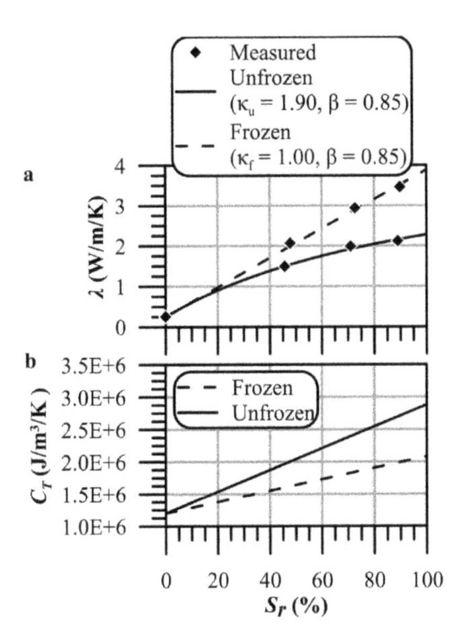

FIGURE 3.13 (a) Measured thermal conductivity (λ) as a function of the degree of saturation and $\lambda - S_r$ functions computed using the Côté and Konrad's (2005a, 2009) model for mine tailings with n = 0.40 and λ_s = 5.80 $W.m^{-1}.K^{-1}$ (data from Boulanger-Martel et al. 2018); (b) Volumetric heat capacity (CT) as a function of S_r estimated using Equation 3.92 for mine tailings with n = 0.40.

Some materials (especially clays) can even show significant unfrozen water contents far below 0°C (e.g., Nersesova and Tsytovich 1963; Williams 1964; Yong et al. 1979).

When the thermal gradient is sufficiently high and the properties of the cover materials are suitable for airflow in cover materials, natural convection can become an important mechanism governing heat transfer in cover materials (e.g., Arenson and Sego 2007; Pham 2013). In coarse geomaterials, natural convection is created by temperature gradients, which create air density gradients that induce movement in a mass of air under a gravity gradient. Natural convection occurs when the density gradient between the cold air at the surface and the warmer air at the base of a cover is high enough. The differences in density between the cold air at the top of the cover materials and the warm air at the bottom initiate a movement of the mass of air contained in the cover materials. The denser cold air tends to move downward, whereas the lighter warm air tends to rise up in the cover materials. For covers located in cold/permafrost environments, this mechanism can have significant effects on heat extraction and promote generally colder ground temperatures (e.g., Arenson and Sego 2007; Pham 2013). However, because the occurrence of natural convection essentially depends on the temperature/density gradient, natural convection can become dominant over conduction only when temperatures are low enough in the upper portion of the cover, that is, during winter. During the rest of the year, conduction is the governing heat transfer mechanism. The potential for convective cooling of a horizontal layer can be estimated using the Rayleigh-Darcy number (R_a), which is defined by Equation 3.84. R_a essentially represents the ratio of the factors that favor convection compared to those that favor conduction. Convection typically becomes significant at a Rayleigh number of $4\pi^2$, and its contribution to the overall heat transfer increases with increasing Rayleigh number (e.g., Nield and Bejan 2017).

$$R_a = \frac{\rho_a g \varphi C_{T,a} K_i H \nabla T}{\mu_a \lambda_e} \tag{3.84}$$

where ρ_a is the density of air (kg.m^{-3}), g is the gravitational acceleration constant (m.s^{-2}), φ is the coefficient of thermal expansion (K), $C_{T,a}$ is the volumetric heat capacity of air (J.m^{-3}.K^{-1}), K_i is the soil's intrinsic permeability (m^2; see Equation 3.12), H is the height of the layer (m), μ_a is the dynamic viscosity of air (kg.m^{-1}.s^{-1}), and λ_e is the effective thermal conductivity (W.m^{-1}.K^{-1}).

Overall, the heat transfer equation accounting for conduction, convection, phase changes, and energy conservation can be expressed as (Lebeau and Konrad 2009):

$$\left(C_T - L_f \rho_i \frac{\partial \theta_i}{\partial T}\right)\frac{\partial T}{\partial t} + C_{T,a}\vec{q}_a \cdot \vec{\nabla} T = \vec{\nabla} \cdot \left[\lambda \cdot \vec{\nabla} T\right] \tag{3.85}$$

where q_a is the volumetric flux vector of air (m.s^{-1}).

Finally, heat transfer by radiation is usually described by the Stephan-Boltzman law, which relates to the intensity of radiation as proportional to the surface temperature (Fillion et al. 2011):

$$q_r = \varepsilon \sigma T^4 \tag{3.86}$$

where q_r is the radiative heat flux vector (W.m^{-2}), ε is the emissivity of the surface (0–1 for an ideal black body), and σ is the Stephan-Boltzman constant (5.67 × 10^{-8} W.m^{-2}.K^{-4}). However, thermal radiation is often considered as a surface effect that is taken into account as a boundary condition (Lunardini 1991) or used in the calculation of evaporation and transpiration (Fredlund et al. 2012; Penman 1948; Wilson 1990).

3.6.2 Experimental Methods for Determining Thermal Properties

The governing equations presented earlier indicate that a number of different properties and parameters are required to solve heat transfer problems in engineered covers. Among these parameters are the thermal conductivity, volumetric heat capacity, soil-freezing curve, and intrinsic/air permeability, which are the most relevant properties to obtain through laboratory tests. Table 3.1 provides a summary of the most adapted test methods available for determining thermal conductivities, freezing curves, and intrinsic/air permeabilities of cover materials and mine wastes. Volumetric heat capacity can be measured in the laboratory using a calorimeter (Fredlund et al. 2012; Taylor and Jackson 1986) or a thermal needle probe (e.g., Kluitenberg et al. 1993; Bristow et al. 1994; Liu and Si 2011). However, for practical purposes, volumetric heat capacity is usually estimated and, thus, measurements will not be discussed further in this section.

Thermal conductivity is one of the most important parameters to obtain for assessing heat transfer processes in geomaterials, including reclamation covers (e.g., Côté and Konrad 2005a; Boulanger-Martel et al. 2018; Boulanger-Martel 2019). In general, the thermal conductivity of geomaterials is influenced by porosity, S_r, the state of the interstitial water, and the mineralogical composition of the solid particles (e.g., Kersten 1949; Johansen 1975; Côté and Konrad 2005a,b). Therefore, when designing a cover that requires analysis of heat transfer, it is important to obtain multiple measurements of the thermal conductivity at specific water contents/densities or measure the thermal conductivity as a function of S_r and under both frozen and unfrozen conditions. The latter usually requires calibrating a thermal conductivity model with experimental data. Among the test methods available, the thermal needle probe (Bristow et al. 1994; Putkonen 2003; ASTM 2014) and the thermal conductivity cell equipped with heat flux meters (Côté et al. 2013; Côté and Konrad 2005a,b) are recommended laboratory approaches for measuring the thermal conductivity of cover materials and other mining materials. These two apparatuses can also be used to assess the thermal conductivity of the solid particles (λ_s) by using the approaches described in Table 3.1. Measurements of λ_s can be used to calibrate thermal conductivity models (e.g., Côté and Konrad 2005b) or assess scale effects on thermal conductivity (Boulanger-Martel et al. 2018). In addition, thermal conductivity measurements can be made in coarse-grained materials, such as waste rocks, using large thermal conductivity probes that, similarly to the thermal needle probe, use the line heat source theory to deduce the thermal conductivity of the materials (e.g., Blackford and Harries 1985; Tan and Ritchie 1997; Pham 2013). Large heat transfer cells can also be used to assess the effective thermal conductivity (combined effects of thermal conductivity and radiation) of dry, coarse-grained materials, as well as assess the intrinsic permeability of coarse-grained materials (e.g., Goering et al. 2000; Côté et al. 2011; Fillion et al. 2011; Rieksts et al. 2019).

In addition to heat transfer cells, intrinsic permeability can also be deduced from the saturated hydraulic conductivity or the saturated air coefficient of permeability based on the density and dynamic viscosity of the fluid used for testing, as described by Equation 3.12. Air permeability can be measured using test methods similar to those used for saturated hydraulic conductivity, except that air is used as the fluid. ASTM Standard D6539-13 (ASTM 2013) describes two standard test methods to determine air permeability in the laboratory using a constant flux or a constant pressure gradient. Air permeability tests can also be performed in the field using air permeability probes (e.g., Bennett and Ritchie 1993; Smith et al. 2013).

The freezing curve can be determined in the laboratory or in the field based on principles similar to the WRC, for which measurements of a material's θ_u (m^3m^{-3}) are performed at several temperatures below the freezing temperature of interstitial water. In doing so, the reduction of θ_u with T is obtained. With recent improvements in techniques for measuring θ_u, such as time domain reflectometry¸ frequency domain, or nuclear magnetic resonance, it has become relatively simple to obtain a freezing curve with an acceptable level of precision (e.g., Azmatch et al. 2012a; Azmatch et al. 2012b; He and Dyck 2013; Watanabe and Wake 2009; Yoshikawa and Overduin 2005). For more details on the measurement of θ_u, refer to Chapter 10.

3.6.3 Methods for Estimating Thermal Properties

Several thermal conductivity models currently exist to estimate and model the thermal conductivity of partially saturated soils and granular materials (e.g., Kersten 1949; Woodside and Messmer 1961; Johansen 1975; Farouki 1982; Côté and Konrad 2005a,b, 2009). Among these models, the generalized Côté and Konrad's (2005a, 2009) model has been proven to be effective at computing the thermal conductivity of various cover materials, tailings, and waste rocks (e.g., Coulombe 2012; Boulanger-Martel et al. 2014, 2018; Lessard et al. 2018). The Côté and Konrad's (2005a, 2009) model computes the thermal conductivity (λ; W/m/K) of unfrozen and frozen granular materials as:

$$\lambda_{(u,f)} = \frac{\left(\kappa_{(u,f)}\lambda_{sat(u,f)} - \lambda_{dry}\right)S_{r(u,f)} + \lambda_{dry}}{1+\left(\kappa_{(u,f)}-1\right)S_{r(u,f)}} \tag{3.87}$$

where κ (-) is a soil type–based empirical parameter; λ_{sat} ($W.m^{-1}.K^{-1}$) and λ_{dry} ($W.m.^{-1}K^{-1}$) are the thermal conductivities of the material at the saturated and dry states, respectively; and S_r is the degree of saturation. In Equation 3.87, the u and f subscripts refer to the unfrozen and frozen states, respectively.

Typical values of κ at unfrozen and unfrozen states are shown in Table 3.2.

TABLE 3.2

Test Methods for Determining Thermal Conductivity, Freezing Curves, Intrinsic Permeability, and Air Permeability in Cover Materials and Other Mining Materials

Parameter	Apparatus	Type of Materials	Approach	Test Method	Examples for Cover or Mining Materials
Thermal conductivity of dry and moist soils, frozen and unfrozen	Thermal needle probe	Fine (<0.5 mm) intact or remolded	Measurement at a given S_r. Different temperatures can be tested.	ASTM D-5334 (ASTM, 2014)	Nixon (2000), Coulombe (2012)
	Thermal conductivity cell with heat flux meter	Fine to coarse (<20 mm) intact or remolded		Côté and Konrad (2005a,b), Côté et al. (2013)	Coulombe (2012), Boulanger-Martel et al. (2014, 2018)
	Thermal conductivity probe	Coarse-grained, in situ	Measurements at given time	Tan and Ritchie (1997)	Tan and Ritchie (1997), Smith et al. (2013), Pham et al. (2013), Amos et al. (2015)
Thermal conductivity of solid particles	Thermal needle probe	Fine (<0.5 mm) Remolded	Measurement at $S_r \approx 100$ % and indirect interpretation approach of Côté and Konrad (2007)	ASTM D-5334 (ASTM, 2014)	Coulombe (2012), Lessard et al. (2018), Boulanger-Martel et al. (2018)
	Thermal conductivity cell with heat flux meter	Intact rock samples	Direct measurement	Côté and Konrad (2005a,b), Côté et al. (2013)	Boulanger-Martel et al. (2018), Poirier (2019)
Effective thermal conductivity and intrinsic permeability	Heat transfer cells	Coarse (up to 250 mm) and dry	Steady state approach where heat flux and temperature are measured to establish λ_e and K	Côté et al. (2011), Fillion et al. (2011), Goering et al. (2000), Rieksts et al. (2019)	n.a.
Air permeability	Permeameter	Materials of intrinsic permeabilities from 10^{-10} to 10^{-15} m^2	Similar to k_{sat} tests Measurements at various S_r	ASTM D6539-13 (ASTM, 2013)	Godbout (2012)
	Air permeability probes	Coarse-grained, in situ	Measurements at a given time	Bennett and Ritchie (1993)	Bennett and Ritchie (1993), Smith et al. (2013), Amos et al. (2015)
Freezing curve	Experimental setup or field – instrumented to measure simultaneously T and θ_u	Limited by the volume of influence of the θ_u sensors	Measure θ_u at several T below the freezing point	Azmatch et al. (2012), Watanabe and Wake (2009)	Meldrum (1998), Elberling (2001), Coulombe (2012), Boulanger-Martel (2019)

n.a.: not available

The computation of $\lambda_{sat(u)}$ is performed using the geometric mean method (Johansen 1975) based on the thermal conductivity of the solid particles (λ_s; W.m^{-1}.K^{-1}) and of water (λ_w; W.m^{-1}.K^{-1}), as well as their respective volumetric fractions (Equation 3.88).

$$\lambda_{sat(u)} = \lambda_s^{1-n_u} \lambda_w^{n_u} \tag{3.88}$$

where n_u is the porosity of the unfrozen soil.

In saturated soils, a 9% void volume increase can be observed when all interstitial water is frozen (Côté and Konrad 2005b). Therefore, the porosity of a frozen saturated soil (n_f) can be estimated using the following equation (Côté and Konrad 2005b):

$$n_f = \frac{1,09n_u}{(1+0.09n_u)} \tag{3.89}$$

Accounting for phase changes and unfrozen volumetric water content, $\lambda_{sat(f)}$ is computed as:

$$\lambda_{sat(f)} = \lambda_s^{1-n_f} \lambda_i^{n_f-\theta_u} \lambda_w^{\theta_u} \tag{3.90}$$

where λ_i (W.m^{-1}.K^{-1}) is the thermal conductivity of ice.

Table 3.3 shows the densities and thermal conductivities for air, water, ice, and some selected rocks.

The volume change of freezing water also affects the degree of saturation in the frozen state ($S_{r(f)}$; Côté and Konrad 2005b), which represents the degree of saturation with respect to unfrozen water. $S_{r(f)}$ can be expressed in its generalized form as (Côté and Konrad 2005a):

$$S_{r(f)} = \frac{1,09S_{ru}n_u - 0,09\theta_u}{n + 0,09(S_{ru}n_u - \theta_u)} \tag{3.91}$$

Based on the work of Côté and Konrad (2005a), Côté and Konrad (2009) developed a dual-phase thermal conductivity model that takes into consideration the effect of matrix structure, including porosity, particle shape, and cementation. A formulation of the Côté and Konrad's (2009) model is presented in Equation 3.92 for the case of dry soils.

$$\lambda_{dry} = \frac{(\kappa_{2P}\lambda_s - \lambda_a)(1-\theta_a) + \lambda_a}{1 + (\kappa_{2P} - 1)(1-\theta_a)} \tag{3.92}$$

TABLE 3.3

Côté and Konrad (2005a)'s Suggested κ Values

Material	κ_u	κ_f
Gravel and coarse sand	4.60	1.70
Medium to fine sand	3.55	0.95
Silty and clayey materials	1.90	0.85
Peat	0.60	0.25

where θ_a is the volumetric air content ($m^3.m^{-3}$), λ_a ($W.m^{-1}.K^{-1}$) is the thermal conductivity of air, and κ_{2P} (-) is an empirical parameter accounting for structure.

For dry soils, κ_{2P} is expressed as a function of the λ_a/λ_s ratio and an empirical parameter, β, that accounts for the effect of structure on the thermal conductivity (Equation 3.92). For $\lambda_a/\lambda_s < 1/15$, the value of β is typically equal to 0.81, 0.54, and 0.34 for natural soils (i.e., rounded particles), crushed rock (i.e., angular particles), and cemented materials, respectively (Côté and Konrad 2009). For $\lambda_a/\lambda_s > 1/15$, β tends to be equal to 0.46 for all types of materials. For water-saturated soils, the simpler Equations 3.88 and 3.90 should be used to compute the unfrozen and frozen thermal conductivities.

$$\kappa_{2P} = 0.29\left(15\frac{\lambda_a}{\lambda_s}\right)^{\beta} \tag{3.92}$$

In most of the equations presented earlier, the value of λ_s is required to compute model parameters. Côté and Konrad (2005a) suggest estimating λ_s using a mineralogy-based geometric mean approach (Equation 3.93).

$$\lambda_s = \prod_{j=1}^{z} \lambda_{mj}^{x_j} \quad \text{with} \sum_{j=1}^{z} x_j = 1 \tag{3.93}$$

where λ_m ($W.m^{-1}.K^{-1}$) is the thermal conductivity of rock-forming mineral j, and x is the volumetric fraction of mineral j. Table 3.4 shows the thermal conductivities of some common

TABLE 3.4

Density and Thermal Conductivity of Air, Water, Ice, and Some Rocks

Constituents or Type of Rocks	ρ ($kg.m^{-3}$)	λ ($W.m^{-1}.K^{-1}$)
Air (10°C)	1.25	0.026
Water (0°C)	999.87	0.56
Water (10°C)	999.73	0.58
Ice (0°C)	900	2.21
Ice (–40°C)	900	2.66
Rocks		
Anorthosite	2730	1.8
Basalt	2900	1.7
Dolostone	2900	3.8
Gabbro	2920	2.2
Gneiss	2750	2.6
Granite	2750	2.5
Limestone	2700	2.5
Quartzite	2650	5.0
Sandstone	2800	3.0
Schist	2650	<1.5
Shale	2650	2.0

Values compiled by Andersland and Ladanyi (2004) and Côté and Konrad (2005a) from various sources.

TABLE 3.5
Thermal Conductivity of Selected Rock-Forming Minerals (From Côté and Konrad 2005a).

Mineral	λ_s (W.m^{-1}.K^{-1})
Actinolite	3.48
Albite	1.96
Anortite	1.68
Augite	3.82
Calcite	3.59
Chlorite	5.15
Chrysotile	5.30
Dolomite	5.51
Epidote	2.83
Fayalite	3.16
Forsterite	5.03
Hematite	11.28
Magnetite	5.10
Ilmenite	2.38
Muscovite	2.85
Orthoclase	2.32
Quartz	7.69
Pyrite	19.21
Siderite	3.01
Tremolite	4.78

rock-forming minerals. Additional thermal conductivities for rock-forming minerals have been presented by Horai (1971), Clauser and Huenges (1995), and Côté and Konrad (2005a, b).

Ultimately, it is desirable to calibrate the Côté and Konrad's (2005a, 2009) model on thermal conductivity measurements made in the laboratory to obtain representative unfrozen and frozen thermal conductivity functions. Figure 3.13a shows the results of Boulanger-Martel et al. (2018) who used the Côté and Konrad's (2005a, 2009) thermal conductivity model to match laboratory thermal conductivity measurements and describe the unfrozen and frozen thermal conductivity functions of tailings from Meadowbank mine. For this case, a λ_{dry} of 0.25 W.m^{-1}.K^{-1} was measured (and computed), whereas $\lambda_{sat(u)}$ and $\lambda_{sat(f)}$ of 2.28 and 3.89 W.m^{-1}.K^{-1}, respectively, were calculated using the Côté and Konrad's (2005a, 2009) model. These calculations were obtained for mine tailings at a porosity of 0.40 (i.e., the same as the tested tailings) in which no unfrozen volumetric water content was present (Boulanger-Martel et al. 2018).

The volumetric heat capacity (C_T) is estimated from the volumetric heat capacity of the constituents and their volume proportions as described by Equation 3.92 (Clauser 2011; de Vries 1963):

$$C_T = C_{T,s}\left(1-n\right)+C_{T,u}\theta_u + C_{T,a}\theta_a + C_{T,i}\theta_i \tag{3.94}$$

where n is the porosity; and $C_{T,s}$, $C_{T,u}$, and $C_{T,i}$ are the volumetric heat capacities of the solid, water, and ice phases (J.m^{-3}.K^{-1}), respectively.

Volumetric heat capacities of 2.0×10^6, 4.2×10^6, 1.2×10^3, and 2.2×10^6 J.m^{-3}.K^{-1} are typically observed for rocks (solids), water, air, and ice, respectively (Fredlund et al. 2012). Figure 3.13b shows an example of C_T estimated as a function of S_r using Equation 3.94 for mine tailings having a porosity of 0.40. The results indicate that C_T ranges from 1.2 MJ.m^{-3}.K^{-1} when the tailings are dry (unfrozen as well as frozen) to about 2.9 $MJ.m^{-3}.K^{-1}$ when the tailings are saturated and unfrozen and to about 2.1 $MJ.m^{-3}.K^{-1}$ when the tailings are saturated and frozen.

Because the permeability relative to one fluid is a function of its density and dynamic viscosity, as well as the intrinsic permeability of the porous medium, it is possible to estimate the intrinsic permeability based on basic geotechnical parameters using the equations given in Section 3.3.3 and the fluid properties (see Equation 3.12).

Some empirical models have been developed to estimate the amount of unfrozen water as a function of temperature (e.g., Dillon and Andersland 1966; Anderson and Tice 1972; Tice et al. 1976). However, these models are specific to fine (often plastic) materials. Therefore, they do not apply to most cover materials. Nonetheless, because the freezing curve is analogous to the WRC, a progressive reduction of θ_u in favor of an increasing θ_i produces an increase in soil suction as described by the Clausius-Clapeyron equation (e.g., Azmatch et al. 2012; Koopmans and Miller 1966; Kurylyk and Watanabe 2013; Ma et al. 2015; Shoop and Bigl 1997). The Clausius-Clapeyron equation can be expressed in many ways, but for a porous medium in which the air-liquid and water-ice phases exist and in which the ice pressure is considered atmospheric (freezing = drying), the relationship between suction in the soil and temperature can be given by:

$$\psi(T) = \frac{L_f}{g} ln\left(\frac{T_f - T}{T_f}\right) \tag{3.95}$$

where ψ is suction and T_f is the freezing temperature of the interstitial water.

For a T_f close to 0°C and under normal atmospheric pressure conditions, the increase in soil suction caused by freezing is around 1250 $kPa.°C^{-1}$ (Konrad 1994). Based on this principle, several authors have proposed predictive equations to develop a freezing curve derived from a WRC; the WRC is converted to the freezing curve using the soil-freezing suction as described by the Clausius-Clapeyron equation. Authors have used formulations of models such as those proposed by van Genuchten (1980) (e.g., Dall'Amico 2010; Dall'Amico et al. 2011), Durner (1994) (e.g., Watanabe et al. 2011), Brooks and Corey (1964) (e.g., Sheshukov and Nieber 2011), or Fredlund et al. (1994) (e.g., Azmatch et al. 2012) to describe (fit) soil-freezing curves.

3.7 RESEARCH NEEDS

Studies still need to be deepened on various issues related to water, gas, and heat flow in cover materials used for the reclamation of mine sites under arid, temperate, and cold climatic conditions. These issues include:

- advective and diffusive fluid flow in frozen or partially frozen materials that are either saturated or unsaturated;
- incorporation of non-Fickian diffusion of oxygen into frozen or partially frozen materials that are either saturated or unsaturated;
- understanding the effects of long-term climatic changes on the fluid flow properties of porous media;
- determining the influence of the degree of saturation S_r on reaction rate coefficients and developing of a predictive model for K_r that accounts for S_r;
- identifying parameters that affect the reaction order of sulfide mineral oxidation by oxygen;
- determining and quantifying the effect of organisms on OD and consumption;
- coupling thermal and hydrogeological mechanisms that govern the freezing and thawing of cover materials.

REFERENCES

Aachib, M., Mbonimpa, M., and Aubertin, M., 2004. Measurement and prediction of the oxygen diffusion coefficient in unsaturated media, with applications to soil covers. *Water, Air and Soil Pollution*, 156 (1–4): 163–193

Akindunni, F.F., Gillham, R.W. and Nicholson, R.V., 1991. Numerical simulations to investigate moisture-retention characteristics in the design of oxygen-limiting covers for reactive mine tailings. *Canadian Geotechnical Journal*, 28:446–451.

Almeida E.L., Teixeira A. S., Silva Filho C.S., Assis Júnior, R.N. and Oliveira Leão R. A., 2015. Filter paper method for the determination of the soil water retention curve. *Revista Brasileira de Ciência do Solo*, 39: 1344–1352.

Amos, R. T., Blowes D. W., Bailey B. L., Sego D. C., Smith L., Ritchie A. I. M., 2015. Waste-rock hydrogeology and geochemistry. *Applied Geochemistry*, 57:140–156.

Andersland, O. B., and B. Ladanyi, 2004. *Frozen Ground Engineering*. 2nd ed. Chichester: John Wiley & Sons.

Anderson, D.M., Tice, A.R., 1972. *Predicting Unfrozen Water Contents in Frozen Soils From Surface Area Measurements*. In *51st Annual Meeting of the Highway Research Board*, Highway Research Record, Washington, DC, USA. https://trid.trb.org/view/126420

Arenson, L., Sego D., 2007. Protection of mine waste tailing ponds using cold air convection. In *Assessment and Remediation of Contaminated Sites in the Arctic and Cold Climates (ARCSACC)*, ed. K. Biggar, G. Cotta, M. Nahir, et al., 256–264. Edmonton, AB.

Arenson, L.U., Sego, D.C., 2004. *Freezing Processes for a Coarse Sand With Varying Salinities, Proceedings of the Cold Regions* Engineering & Construction Conference, Edmonton, Alberta.

Arya, L.M., and Paris, J.F., 1981. A physico-empirical model to predict the soil moisture– characteristic from particle size distribution and bulk density data. *Soil Science Society of America Journal*, 45:1023–1030.

Arya, L.M., Leij, F.J., van Genuchten, M.T., and Shouse, P.J., 1999. Scaling parameter to predict the soil water characteristic from particle-size distribution data. *Soil Science Society of America Journal*, 63:510–519.

Askarinejad A, Beck, A, Casini F. and Springman S. M., 2012. Unsaturated hydraulic conductivity of a silty sand with the instantaneous profile method. In Claudio Mancuso, Cristina Jommi, and Francesca D'Onza (Eds.): *Unsaturated Soils: Research and Applications Volume 2*, p. 215–220.

ASTM 2012a. *Standard Test Methods for Laboratory Compaction Characteristics of Soil Using Standard Effort (12 400 ft-lbf/ft3 (600 kN-m/m3)) (D698-12e2)*. ASTM International, West Conshohocken, PA.

ASTM 2012b. *Standard Test Methods for Laboratory Compaction Characteristics of Soil Using Modified Effort (56,000 ft-lbf/ft3 (2,700 kN-m/m3)) (D1557-12e1)*. ASTM International, West Conshohocken, PA.

ASTM 2013. *Standard Test Method for Measurement of the Permeability of Unsaturated Porous Materials by Flowing Air (D6539-13)*, ASTM International, West Conshohocken, PA.

ASTM 2014. *Standard Test Method for Determination of Thermal Conductivity of Soil and Soft Rock by Thermal Needle Probe Procedure (D5334-14)*. ASTM International, West Conshohocken, PA.

ASTM 2015. *Standard Test Method for Measurement of Hydraulic Conductivity of Porous Material Using a Rigid-Wall, Compaction-Mold Permeameter (D5856-15)*. ASTM International, West Conshohocken, PA.

ASTM 2016a. *Standard Test Methods for Measurement of Hydraulic Conductivity of Saturated Porous Materials Using a Flexible Wall Permeameter (D5084-16a)*. ASTM International, West Conshohocken, PA.

ASTM 2016b. *Standard Test Methods for Freezing and Thawing Compacted Soil-Cement Mixtures (D560/D560M-16)*. ASTM International, West Conshohocken.

ASTM 2016c. *Standard Test Method for Measurement of Soil Potential (Suction) Using Filter Paper (D5298-16)*. ASTM International, West Conshohocken, PA.

ASTM 2017a. *Standard Test Methods for Liquid Limit, Plastic Limit, and Plasticity Index of Soils (D4318-17e1)*. ASTM International, West Conshohocken, PA.

ASTM 2017b. *Standard Practice for Classification of Soils for Engineering Purposes (Unified Soil Classification System) (D2487-17)*. ASTM International, West Conshohocken, PA.

ASTM 2019a. *Standard Test Method for Permeability of Granular Soils (Constant Head) (D2434-19)*. ASTM International, West Conshohocken, PA.

ASTM 2019b. *Standard Test Methods for Determining the Effect of Freeze-Thaw on Hydraulic Conductivity of Compacted or Intact Soil Specimens Using a Flexible Wall Permeameter (D6035/D6035M-19)*. ASTM International, West Conshohocken, PA.

Aubertin, M., Aachib M, and Authier, K., 2000. Evaluation of diffusive gas flux through covers with a GCL. *Geotextiles and Geomebranes*, 18: 1–19.

Aubertin, M., Chapuis, R.P., Aachib, M., Bussière, B., Ricard, J.F., and Tremblay, L., 1995. *Évaluation en laboratoire de barrières sèches construites à partir de résidus miniers.* École Polytechnique de Montréal, NEDEM/MEND Project 2.22.2a, Canada Center for Mineral and Energy Technology, Canada.

Aubertin, M., Mbonimpa, M., Bussière, B., Chapuis, R.P., 2003. A model to predict the water retention curve from basic geotechnical properties. *Canadian Geotechnical Journal*, 40(6): 1104–1122

Aubertin, M., Ricard, J.F., and Chapuis, R.P., 1998. A predictive model for water retention curve: application to tailings from hard-rock mines. *Canadian Geotechnical Journal*, 35: 55–69.

Aubertin, M., Bussière, B. and Chapuis, R.P., 1996. Hydraulic conductivity of homogenized tailings from hard rock mines, *Canadian Geotechnical Journal*, 33(3), 470–482.

Azmatch, T.F., Sego, D.C., Arenson, L.U., and Biggar, K.W., 2012a. New ice lens initiation condition for frost heave in fine-grained soils. *Cold Regions Science and Technology*, 82:8–13.

Azmatch, T.F., Sego, D.C., Arenson, L.U., and Biggar, K.W., 2012b. Using soil freezing characteristic curve to estimate the hydraulic conductivity function of partially frozen soils. *Cold Regions Science and Technology* 83–84, 103–109.

Baker, F. G., Veneman P. L. M., and Bouma J., 1974. Limitations of the instantaneous profile method for field measurement of unsaturated hydraulic conductivity. *Soil Science Society of America Journal*, 38 (6): 885–888

Bennett, J., and Ritchie, A.I.M., 1993. *Field Procedures Manual: Measurement of Gas Permeability.* MEND, project report 1.21.1a Mine Environment Neutral Drainage (MEND). Canada Center for Mineral and Energy Technology, Canada.

Black, P.B., and Tice, A.R., 1989. Comparison of soil freezing curve and soil water curve data for Windsor sandy loam. *Water Resources Research* 25(10), 2205–2210.

Blackford, M., and Harries, J., 1985. *A Heat Source Probe for Measuring Thermal Conductivity in Waste Rock Dumps.* Australian Atomic Energy Commission Research Establishment, Lucas Heights, p. 18.

Blight G., 2010. *Geotechnical Engineering for Mine Waste Storage Facilities.* CRC Press/Balkema, 634 p.

Boulanger-Martel, V., 2019. *Évaluation de la Performance de Recouvrements Miniers Pour Contrôler le Drainage Minier Acide en Climat Arctique*, Ph.D. Thesis, Polytechnique Montréal.

Boulanger-Martel, V., Bussière, B., and Côté, J., 2014. Laboratory evaluation of crushed rock-bentonite hydrogeotechnical properties, *67th Canadian Geotechnical Conference*, Regina, Saskatchewan, Canada.

Boulanger-Martel, V., Poirier, A., Côté, J., and Bussière, B., 2018. Thermal conductivity of Meadowbank's mine waste rocks and tailings, *71th Canadian Geotechnical Conference*, Edmonton, Alberta, Canada.

Bristow, K.L., White, R.D., and Kluitenberg, G.J., 1994. Comparison of single and dual probes for measuring soil thermal properties with transient heating. *Soil Research* 32(3), 447–464.

Brooks, R.H. and Corey, J.C., 1964. *Hydraulic properties of porous medium.* (Colorado State University Fort Collins), Hydrology Paper 3.

Burdine, N.T., 1953. Relative permeability calculations from pore-size distribution data. *Petroleum Transaction of the American Institute of Mining Engineering*, 198: 71–77.

Bussière, B., Aubertin, M., and Chapuis, R.P., 2003. The behaviour of inclined covers used as oxygen barriers. *Canadian Geotechnical Journal*, 40: 512–535.

Bussière, B., Benzaazoua, M., Aubertin, M., and Mbonimpa, M., 2004. A laboratory study of covers made of low sulphide tailings to prevent acid mine drainage. *Environmental Geology*, 45: 609–622

Chapuis, R.P., 2012. Predicting the saturated hydraulic conductivity of soils: a review. *Bulletin of Engineering Geology and The Environment*, 71, 401–434.

Chapuis, R.P. and Légaré, P.P., 1992. A simple method for determining the surface area of fine aggregates and fillers in bituminous mixtures, In *Effects of aggregates and mineral filler on asphalt mixture performance, American Society for Testing and Materials*, Special Technical Publication. 1147, 177–186.

Childs, E.C., and Collis-Georges, G.N., 1950. The permeability of porous materials. In *Proceedings of the Royal Society of London, Series A*, Vol. 2001: pp.392–405.

Clauser, C., 2011. *Thermal Storage and Transport Properties of Rocks, I: Heat Capacity and Latent Heat, Encyclopedia of Solid Earth Geophysics.* Springer, pp. 1423–1431.

Clauser, C., and Huenges, E., 1995. Thermal conductivity of rocks and minerals. *Rock Physics & Phase Relations: A Handbook of Physical Constants*, Published by the American Geophysical Union as part of the AGU Reference Shelf Series, John Wiley & Sons, Hoboken, NJ. Volume 3, 105–126.

Collin, M., 1987. *Mathematical modeling of water and oxygen transport in layered soil covers for deposits of pyritic mine tailings. Licenciate Treatise.* Royal Institute of Technology. Department of Chemical Engineering. S-10044 Stockholm, Sweden.

Collin, M., 1998. *The Bersbo Pilot Project. Numerical Simulation of Water and Oxygen Transport in the Soil Covers at Mine Waste Deposits*. Swedish Environmental Protection Agency. Report 4763, 46 pages, plus appendix.

Côté, J., Fillion, M.-H., and Konrad, J.-M., 2011. Intrinsic permeability of materials ranging from sand to rock-fill using natural air convection tests. *Canadian Geotechnical Journal* 48(5), 679–690.

Côté, J., Grosjean, V., and Konrad, J.-M., 2013. Thermal conductivity of bitumen concrete. *Canadian Journal of Civil Engineering* 40(2):172–181.

Côté, J., and Konrad, J.-M., 2005a. A generalized thermal conductivity model for soils and construction materials. *Canadian Geotechnical Journal* 42(2), 443–458.

Côté, J., and Konrad, J.-M., 2005b. Thermal conductivity of base-course materials. *Canadian Geotechnical Journal* 42(1), 61–78.

Côté, J., and Konrad, J.-M., 2007. Indirect methods to assess the solid particle thermal conductivity of Quebec marine clays. *Canadian Geotechnical Journal* 44(9), 1117–1127.

Côté, J., and Konrad, J.-M., 2009. Assessment of structure effects on the thermal conductivity of two-phase porous geomaterials. *International Journal of Heat and Mass Transfer* 52(3), 796–804.

Coulombe, V., 2012. *Performance de Recouvrements Isolants Partiels pour Contrôler L'oxydation de Résidus Miniers Sulfureux*. M.Sc.A. Thesis, Polytechnique de Montréal, Montreal, Canada.

Crank, J. 1975. *The Mathematics of Diffusion*. 2nd Ed. Clarendon Press, Oxford, UK.

Dall'Amico, M., 2010. *Coupled Water and Heat Transfer in Permafrost Modeling*. Ph.D. Thesis, University of Trento, Italy.

Dall'Amico, M., Endrizzi, S., Gruber, S., and Rigon, R., 2011. A robust and energy-conserving model of freezing variably-saturated soil. *The Cryosphere* 5, 469–484.

Dane J.H. , and Topp, G.C., 2002. *Methods of soil analysis. Part 4. Physical Methods*. SSSA Book Series 5. Soil Science Society of America, Madison, Wisconsin, USA.

Darcy, H., 1856. *Les Fontaines Publiques de la Ville de Dijon*, Victor Dalmont, Paris.

Davis DD, Horton R, Heitman JL, and Ren TS., 2009. Wettability and hysteresis effects on water sorption in relatively dry soil. *Soil Science Society of America* 73:1947–1951.

de Vries, D., 1963. Thermal properties of soils. *Physics of Plant Environment* 1, 57–109.

Demers, I., 2008. *Utilisation de la Désulfuration Environnementale dans la Restauration de Parcs a` Rejets Miniers*. Ph.D Thesis, UQAT, Rouyn-Noranda.

Demers, I., Bussiere, B., Mbonimpa, M. and Benzaazoua, M., 2009. Oxygen diffusion and consumption in low-sulphide tailings covers. *Canadian Geotechnical Journal*, 46(4): 454–469.

Demers, I., Bussiere, B., Benzaazoua, M., Mbonimpa, M., and Blier, A., 2008. Column test investigation on the performance of monolayer covers made of desulphurized tailings to prevent acid mine drainage. *Minerals Engineering*, 21(4): 317–329

Dillon, H.B., and Andersland, O.B., 1966. Predicting unfrozen water contents in frozen soils. *Canadian Geotechnical Journal* 3(2), 53–60.

Dirksen, C., 1999. *Soil Physics Measurements. GeoEcology Paperback*. Catena Verlag GMBH. Reiskirchen, Germany.

Durner, W., 1994. Hydraulic conductivity estimation for soils with heterogeneous pore structure. *Water Resources Research* 30(2): 211–223.

Edlefsen, N.E., and Anderson, B.C., 1943. Thermodynamics of soil moisture. *The Journal of Agricultural Science*, 15(2): 31–297.

Elberling, B., 2001a. Environmental controls of the seasonal variation in oxygen uptake in sulfidic tailings deposited in a permafrost-affected area, *Water Resources Research*, 37(1): 99–107.

Elberling, B. and Langdahl B.R., 1998. Natural heavy-metal release by sulphide oxidation in the High Arctic natural heavy-metal release by sulphide oxidation in the High Arctic. *Canadian Geotechnical Journal*, 35: 895–901.

Elberling, B. and Nicholson, R.V., 1996. Field determination of sulphide oxidation rates in mine tailings. *Water Resources Research*, 32(6): 1773–1784.

Elberling, B., 2001b. Environmental controls of the seasonal variation in oxygen uptake in sulfidic tailings deposited in a permafrost-affected area. *Water Resources Research* 37, 99–107.

Elberling, B., Nicholson, R.V., Reardon, E.J., and Tibble, P, 1994. Evaluation of sulphide oxidation rates: a laboratory study comparing oxygen fluxes and rates of oxidation product release. *Canadian Geotechnical Journal*, 13: 375–383.

Farouki, O.T., 1982. *Evaluation of Methods for Calculating Soil Thermal Conductivity*. US Army Corps of Engineers, Cold Regions Research and Engineering Laboratory, Hanover, N.H. CRREL Report 82-8.

Feddes, R., Kowalik, P., and Zaradny, H. 1978. *Simulation of Field Water Use and Crop Yield*, John Wiley and Sons, New York.

Fillion, M.-H., Côté, J., Konrad, J.-M., 2011. Thermal radiation and conduction properties of materials ranging from sand to rock-fill. *Canadian Geotechnical Journal* 48(4), 532–542.

Fredlund, D., Xing, A., Huang, S., 1994a. Predicting the permeability function for unsaturated soils using the soil-water characteristic curve. *Canadian Geotechnical Journal* 31(4), 533–546.

Fredlund, D.G. and Rahardjo, H., 1993. *Soil Mechanics for Unsaturated Soils*. John Wiley & Sons, New York.

Fredlund, D.G. and Xing, A., 1994. Equations for the soil-water characteristic curve. *Canadian Geotechnical Journal*, 31: 521–532.

Fredlund, D.G., Rahardjo, H., Fredlund M.D., 2012. *Unsaturated Soil Mechanics in Engineering Practice*. John Wiley & Sons, Hoboken, NJ.

Fredlund, D.G., Xing, A., and Huang, S., 1994b. Predicting the permeability function for unsaturated soils using the soil-water characteristic curve. *Canadian Geotechnical Journal*, 31: 533–546.

Fredlund, M.D., Wilson, G.W., and Fredlund, D. G., 2002. Use of the grain-size distribution for estimation of the soil-water characteristic curve. *Canadian Geotechnical Journal*, 39: 1103–1117.

Freeze, R.A. and Cherry, J.A., 1979 *Groundwater*. Prentice-Hall Inc., Englewood Cliffs, Vol. 7632, 604.

Gardner, W.R., 1958. Some steady state solutions of unsaturated moisture flow equations with application to evaporation from a water table. *Soil Science*, 85: 228–232.

Godbout, J., 2012. *Réactivité des remblais miniers cimentés contenants de la pyrrhotite et étude de paramètres d'influence d'importance telle la passivation des surfaces et le propriétés hydrogéologiques*, Ph.D. Thesis. Université du Québec en Abitibi-Témiscamingue, Rouyn-Noranda, Québec.

Goering, D., Instanes, A., Knudsen, S., 2000. Convective heat transfer in railway embankment ballast. *Ground Freezing* 2000, 31–36.

Gosselin, M., Mbonimpa, M., Aubertin, M., Martin, V., 2007. An investigation of the effect of the degree of saturation on the oxygen reaction rate coefficient of sulphidic tailings. *ERTEP 2007 - First International Conference on Environmental Research, Technology and Policy Building Tools and Capacity for Sustainable Production*. July 17–19, 2007, La Palm Royal Beach Hotel, Accra, Ghana.

Harlan, R., 1973. Analysis of coupled heat-fluid transport in partially frozen soil. *Water Resources Research* 9(5), 1314–1323.

Haverkamp, R., and Parlange, J.-Y., 1986. Predicting the water retention curve from particle size distribution: 1. Sandy soils without organic matter. *Soil Science*, 142: 325–339.

He, H., Dyck, M., 2013. Application of multiphase dielectric mixing models for understanding the effective dielectric permittivity of frozen soils. *Vadose Zone Journal* 12: 1-22.

Hillel, D., 1998. *Environmental Soil Physics Fundamentals, Applications, and Environmental Considerations*. Academic Press, Waltham.

Hollings, P., Hendry, M.J., Nicholson, R.V., and Kirkland, R.A., 2001. Quantification of oxygen consumption and sulphate release rates for waste rock piles using kinetic cells: Cluff Lake uranium mine, northern Saskatchewan, Canada. *Applied Geochemistry*, 16: 1215–1230.

Holtz, D.R. and Kovacs, D.W., 1981. *An Introduction to Geotechnical Engineering*. Prentice-Hall, New Jersey.

Horai, K.i., 1971. Thermal conductivity of rock-forming minerals. *Journal of Geophysical Research* 76(5), 1278–1308.

Jin, Y. and Jury, W.A., 1996 Characterising the dependence of gas diffusion coefficient on soil properties. *Soil Science Society of America Journal*, 60: 66–71.

Johansen, O., 1975. *Thermal Conductivity of Soils*. Ph.D. thesis, University of Trondheim, Trondheim, Norway. CRREL Draft English Translation 637, US Army Corps of Engineers, Cold Regions Research and Engineering Laboratory, Hanover, N.H.

Kersten, M.S., 1949. *Laboratory research for the determination of the thermal properties of soils*, Research Laboratory Investigations, Engineering Experiment Station. University of Minnesota, Minneapolis, MN.

Kluitenberg, G., Ham, J., Bristow, K.L., 1993. Error analysis of the heat pulse method for measuring soil volumetric heat capacity. *Soil Science Society of America Journal* 57(6), 1444–1451.

Klute, A. and Dirksen, C., 1986. Hydraulic conductivity and diffusivity: laboratory methods. *Methods of Soil Analysis, Part I*, 2nd Edition, A. Klute (ed.), Agron. Monogr. No. 9, ASA and SSSA, Madison, WI, 687–734.

Konrad, J. M. 1994. Sixteenth Canadian Geotechnical Colloquium: frost heave in soils: concepts and engineering. *Canadian Geotechnical Journal*, 31(2): 223–245.

Koopmans, R.W.R., Miller, R., 1966. Soil freezing and soil water characteristic curves. *Soil Science Society of America Journal* 30(6), 680–685.

Kovács, G., 1981. *Seepage Hydraulics*. Elsevier Science Publishers, Amsterdam.

Kurylyk, B.L., Watanabe, K., 2013. The mathematical representation of freezing and thawing processes in variably-saturated, non-deformable soils. *Advances in Water Resources* 60, 160 –177.

Kutzner, C., 1997. *Earth and Rockfill Dams. Principles of Design and Construction.* AA. Balkema, Rotterdam. 333p.

Lebeau, M. and J.-M. Konrad, 2009. Natural convection of compressible and incompressible gases in undeformable porous media under cold climate conditions. *Computers and Geotechnics* 36(3): 435 –445.

Lessard, F., Bussière, B., Côté, J., Benzaazoua, M., Boulanger-Martel, V., Marcoux, L., 2018. Integrated environmental management of pyrrhotite tailings at Raglan Mine: Part 2 desulphurized tailings as cover material. *Journal of Cleaner Production* 186, 883–893.

Liu, G., Si, B.C., 2011. Single- and dual-probe heat pulse probe for determining thermal properties of dry soils. *Soil Science Society of America Journal* 75(3), 787–794.

Liu Q., Yasufuku, N., Omine K., Hazarika H., 2012a. Automatic soil water retention test system with volume change measurement for sandy and silty soils. *Soils and Foundations*, 52(2): 368–380

Liu, Z., Zhang, B., Yu, X., Tao, J., 2012b. A new method for soil water characteristic curve measurement based on similarities between soil freezing and drying. *ASTM Geotechnical Testing Journal* 35(1), 2–10.

Lunardini, V.J., 1991. *Heat Transfer With Freezing and Thawing.* 1st Edition. (65). Elsevier Science, New York, NY.

Ma, W., Zhang, L., Yang, C., 2015. Discussion of the applicability of the generalized Clausius–Clapeyron equation and the frozen fringe process. *Earth-Science Reviews* 142, 47–59.

Maqsoud, A., Bussière, B., Aubertin M., and Mbonimpa, M., 2012. Predicting hysteresis of the water retention curve from basic properties of granular soils. *Geotechnical and Geological Engineering*, 30 (5):1147–1159.

Maqsoud, A., Bussière, B., Mbonimpa, M., Aubertin, M., Wilson, W.G., 2007. Instrumentation and monitoring techniques for oxygen barrier covers used to control acid mine drainage. *Energy and Mines*, April 29 - May 2, 2007, Montreal, Quebec, Canada, pp. 140–149. CIM.

Marinho F.A.M., Take W.A., Tarantino A., 2008. Measurement of matric suction using tensiometric and axis translation techniques. In: Tarantino A., Romero E., Cui YJ. (eds) *Laboratory and Field Testing of Unsaturated Soils.* Springer, Dordrech.

Mbonimpa M., Aubertin, M., Bussière, B., 2011. Oxygen consumption test to evaluate the diffusive flux into reactive tailings: interpretation and numerical assessment. *Canadian Geotechnical Journal*, 48(6): 878–890.

Mbonimpa, M., Aubertin, M, Bussière, B., and Maqsoud, A., 2006a. A predictive function for the water retention curve of compressible soils. *Journal of Geotechnical and Geoenvironmental Engineering, ASCE*, 132(9): 1121–1132.

Mbonimpa, M., Aubertin, M., and Bussière, B., 2006b. Predicting the unsaturated hydraulic conductivity of granular soils from basic geotechnical properties using the MK model and statistical models. *Canadian Geotechnical Journal*, 43(8): 773–787.

Mbonimpa, M., Aubertin, M., Aachib, M., and Bussière, B., 2003. Diffusion and consumption of oxygen in unsaturated cover materials. *Canadian Geotechnical Journal*, 40: 916–932.

Mbonimpa, M., Aubertin, M., Chapuis, R.P., Bussière, B., 2002. Practical pedotransfer functions for estimating the saturated hydraulic conductivity. *Geotechnical and Geological Engineering*, 20(3): 235–259.

McCarthy, D.F., 2007. *Essentials of Soil Mechanics and Foundations*, Prentice Hall, 7th Edition, Data Upper Saddle River, NJ, 850 p.

Meldrum, J.L., 1998. *Determination of the Sulfide Oxydation Potential of Mine Tailings fron Rankin Inlet, Nunavut, at Sub-zero Temperatures*, Department of Geological Sciences. Queen's University, Kingston, Ontario, Canada.

MEND. 2004. *Covers for Reactive Tailings Located in Permafrost Regions*, review project report 1.61.4. Mine Environment Neutral Drainage (MEND) Canada Center for Mineral and Energy Technology, Canada.

Miller, R.D., 1980. Freezing phenomena in soils. *Applications of Soil Physics* 278, p. 283.

Millington, R.J., and Quirk, J.P., 1961. Permeability of porous solids. *Transactions of the Faraday Society* 57: 1200–1206.

Millington, R.J., and Shearer, R.C., 1971. Diffusion in aggregated porous media. *Soil Science*, 111: 372–378.

Mitchell, J., D. Hooper, and Campanella, R., 1965. Permeability of compacted clay. *Journal of Soil Mechanics and foundation Division, ASCE* 91:41–65.

Morel-Seytoux, H.J., 1992. *The Capillary Barrier Effect at the Interface of Two Soil Layers With Some Contrast in Properties.* HYDROWAR Report 92.4, Hydrology Days Publications, Atherton, CA.

Mualem Y., Beriozkin A., 2009. General scaling rules of the hysteretic water retention function based on Mualem's domain theory. *European Journal of Soil Science*, 60:652–661.

Mualem, Y., 1976. A new model for predicting the hydraulic conductivity of unsaturated porous media. *Water Resources Research*, 12: 513 –522.

Nersesova, Z. A. and Tsytovich, N. A., 1963. *Unfrozen water in frozen soils. Proceedings of the Permafrost International Conference*, Washington, DC. pp. 230-233.

Nicholson, R. V., Gillham, R. W., Cherry, J. A., and Reardon, E. J., 1989. Reduction of acid generation in mine tailings through the use of moisture-retaining cover layers as oxygen barriers. *Canadian Geotechnical Journal*, 26: 1–8.

Nield, D.A., Bejan, A., 2017. *Convection in Porous Media*. 5th Edition. Springer International Publishing, New York.

Nixon, J., 2000. *Geothermal Analysis for Tailings Cover Design–Falconbridge Raglan Tailings Project*. Submitted to AGRA Earth and Environmental Ltd.

Noguchi, T., Mendes J., Toll D.G., 2012. Comparison of Soil Water Retention Curves Obtained by Filter Paper, High Capacity Suction Probe and Pressure Plate. *5th Asia-Pacific Conference on Unsaturated Soils 2012* – Pattaya, Thailand, 29 February to 2 March 2012.

O'Kane, M., Wilson, G.W., and Barbour, S.L., 1998. Instrumentation and monitoring of an engineered soil cover system for mine waste rock. *Canadian Geotechnical Journal*, 35: 828 –846.

Oleszczuk, R., Zając, E., Hewelke, E., Wawer, K., 2018. Determination of water retention characteristics of organic soils, using the indirect filter–paper method. *Acta Sci. Pol., Formatio Circumiectus*, 17(2), 13–21.

Outcalt, S.I., Nelson, F.E., Hinkel, K.M., 1990. The zero-curtain effect: Heat and mass transfer across an isothermal region in freezing soil. *Water Resources Research* 26(7), 1509–1516.

Penman, H.L. 1948. Natural evapotranspiration from open water, bare soil and grass. *Proceedings of the Royal Society of London, Series A*. 193: 120–146.

Pham, N.H., 2013. *Heat Transfer in Waste-Rock Piles Constructed in a Continuous Permafrost Region*. University of Alberta, Department of Civil and Environmental Engineering, Edmonton, Alberta.

Pham, N.H., Sego, D.C., Arenson, L.U., Blowes, D.W., Amos, R.T., Smith, L., 2013. The Diavik waste rock project: measurement of the thermal regime of a waste-rock test pile in a permafrost environment. *Applied Geochemistry* 36, 234–245.

Poirier, A., 2019. *Étude du Comportement Thermique d'une Halde à Stérile en milieu Nordique, Département des Génies Géologique, Civil et des Mines*. M.Sc.A. Thesis, École Polytechnique de Montréal, Montréal, Québec.

Poulovassilis A., 1962. Hysteresis of pore water, and application of concept of independent domains. *Soil Science* 93, 405–412.

Prasad, R. 1988. A linear root water uptake model. *Journal of Hydrology* 99, 297–306.

Putkonen, J., 2003. Determination of frozen soil thermal properties by heated needle probe. *Permafrost and Periglacial processes* 14(4), 343–347.

Reardon, E.J., and Moddle, P.M., 1985. Gas diffusion coefficient measurements on uranium mill tailings: implication to cover layer design. *Uranium*, 2: 111–131.

Rey, N.J., Demers, I., Bussière, B., Mbonimpa, M., 2020. Laboratory and field study of oxygen flux and hydrogeological behaviour of monolayer covers made of low-sulfide tailings combined with an elevated water table placed over acid-generating mine tailings. *Canadian Geotechnical Journal*. Published on the web 19 March 2020. https://doi.org/10.1139/cgj-2018-0875

Richards, L.A., 1931. Capillary conduction of liquids through porous medium. *Journal of Physics B* 1: 318–333.

Rieksts, K., Hoff, I., Scibilia, E., Côté, J., 2019. Laboratory investigations into convective heat transfer in road construction materials, *Canadian Geotechnical Journal* 57(7): 959–973.

Ritchie, J.T. 1972. Model for predicting evaporation from a row crop with incomplete cover. *Water Resources Research* 8:1204–1213.

Romero, E. 1999. *Characterization and Thermo-Hydro-Mechnical Behaviour of Unsaturated Boom Clay: An Experimental Study*. Ph.D. thesis, Universitat Politècnica de Catalunya, Barcelona, Spain

Rowe, R. K. and Booker, J.R., 2005. *POLLUTEv7 Pollutant migration through a nonhomogeneous soil*, ©1983-2005 Distributed by, GAEA Environmental Engineering, Ltd. Whitby, Ontario, Canada

Sass, J., Lachenbruch, A.H., Munroe, R.J., 1971. Thermal conductivity of rocks from measurements on fragments and its application to heat-flow determinations. *Journal of Geophysical Research* 76(14), 3391–3401.

Sheshukov, A.Y., Nieber, J.L., 2011. One-dimensional freezing of nonheaving unsaturated soils: model formulation and similarity solution. *Water Resources Research* 47(11), W11519.

Shoop, S.A., Bigl, S.R., 1997. Moisture migration during freeze and thaw of unsaturated soils: modeling and large scale experiments. *Cold Regions Science and Technology* 25(1), 33–45.

Smith, L.J., Moncur, M.C., Neuner, M., Gupton, M., Blowes, D.W., Smith, L., Sego, D.C., 2013. The Diavik waste rock project: design, construction, and instrumentation of field-scale experimental waste-rock piles. *Applied Geochemistry* 36, 187–199.
Spaans, E.J., Baker, J.M., 1996. The soil freezing characteristic: its measurement and similarity to the soil moisture characteristic. *Soil Science Society of America Journal* 60(1), 13–19.
Tan, Y., Ritchie, A., 1997. In situ determination of thermal conductivity of waste rock dump material. *Water, Air, and Soil Pollution* 98, 345 –359.
Tang, A. M. and Cui, Y. J., 2005. Controlling suction by vapour equilibrium technique at different temperatures, application to the determination of the water retention properties of MX80 clay. *Canadian Geotechnical Journal* 42(1):287–296.
Taylor, S.A., Jackson, R.D., 1986. Heat capacity and specific heat. *Methods of Soil Analysis: Part 1—Physical and Mineralogical Methods*, A. Klute, Ed., 1986, American Society of Agronomy and Soil Science Society of America, Inc, Madison, Wisconsin, 941–944.
Tice, A.R., Anderson, D.M., Banin, A., 1976. *The prediction of unfrozen water contents in frozen soils from liquid limit determinations*, CRREL Report 76(8). US Army Corps of Engineers, Cold Regions Research and Engineering Laboratory, Hanover, NH
Tratch, D., Fredlund, D., and Wilson, G. 1995. An Introduction to Analytical Modeling of Plant Transpiration for Geotechnical Engineers, *Proceedings 48th Canadian Geotechnical Conference*, Vancouver, BC, 771–780.
van Genuchten, M.Th., 1980. A closed-form equation for predicting the hydraulic conductivity of unsaturated soils. *Soil Science Society of America Journal*, 44: 892–898.
Watanabe, K., Kito, T., Wake, T., Sakai, M., 2011. Freezing experiments on unsaturated sand, loam and silt loam. *Annals of Glaciology* 52(58), 37–43.
Watanabe, K., Mizoguchi, M., 2002. Amount of unfrozen water in frozen porous media saturated with solution. *Cold Regions Science and Technology* 34(2), 103–110.
Watanabe, K., Wake, T., 2009. Measurement of unfrozen water content and relative permittivity of frozen unsaturated soil using NMR and TDR. *Cold Regions Science and Technology* 59(1), 34–41.
Watson K.K., 1966. An instantaneous profile method for determining the hydraulic conductivity of unsaturated porous materials. *Water Resources Research*, 2(4): 709–715.
Williams, P.J., 1964. Unfrozen water content of frozen soils and soil moisture suction. *Geotechnique* 14(3), 231–246.
Wilson, G.W., 1990. *Soil Evaporation Fluxes for Geotechnical Engineering Problems*. Ph.D. thesis, University of Saskatchewan, Saskatoon, Sask.
Wilson, G.W., Fredlund D.G., and Barbour S.L., 1997. The effect of soil suction on evaporative fluxes from soil surfaces. *Canadian Geotechnical*, 34: 145–155.
Wilson, G.W., Fredlund, D.G., and Barbour, S.L., 1994. Coupled soil-atmosphere modelling for soil evaporation. *Canadian Geotechnical Journal*, 31(2): 151–161. doi:10.1139/t94-021.
Woodside, W., Messmer, J., 1961. Thermal conductivity of porous media. I. Unconsolidated sands. *Journal of Applied Physics* 32(9), 1688–1699.
Yanful, E.K., 1993. Oxygen diffusion through soil covers on sulphidic mine tailings. *Journal of Geotechnical Engineering, ASCE*, 119 (8): 1207–1228.
Yanful, E.K., Simms, P.H., and Payant, S.C., 1999. Soil covers for controlling acid generation in mine tailings: a laboratory evaluation of the physics and geochemistry. *Water, Air, and Soil Pollution*, 114: 347–375.
Yong, R., Cheung, C., Sheeran, D., 1979. Prediction of salt influence on unfrozen water content in frozen soils. *Engineering Geology* 13(1), 137–155.
Yoshikawa, K., Overduin, P.P., 2005. Comparing unfrozen water content measurements of frozen soil using recently developed commercial sensors. *Cold Regions Science and Technology* 42(3), 250–256.

4 Low Saturated Hydraulic Conductivity Covers

Abdelkabir Maqsoud, Bruno Bussière, and Mamert Mbonimpa

4.1 INTRODUCTION: BACKGROUND AND HISTORICAL PERSPECTIVES

The storage of domestic waste in waste disposal facilities (i.e., landfills) remains the common method for waste management. Historically, landfills were often located in abandoned quarries, mines, or excavation holes where wastes were dumped. As early as the beginning of the twentieth century, regulations of waste management were developed to limit environmental degradation and protect human health (Michaud and Bjork 1995). These regulations were improved and, progressively, landfill sealing with covers was imposed. Hence, low saturated hydraulic conductivity covers (LSHCCs) were suggested to avoid exchange between waste disposal facilities and the natural environment. These covers can be made of natural soils (clay, sand, and gravel) and/or man-made materials (geotextile, geomembrane [GM], and geosynthetic clay liner [GCL]). These LSHCCs (also called water infiltration barriers or impermeable barriers) use low permeability physical barriers to limit water percolation infiltration into the waste and the generation of contaminated leachate.

Mine site reclamation, which began in the 1980s, is relatively recent, compared to the reclamation of landfills. However, most mining regulations in developed countries now require the creation of a mine reclamation plan where the control of contaminant generation from the mine wastes must be engineered before the start of the mine (e.g., Government of Western Australia 2014; Office of Surface Mining Reclamation and Enforcement 2016; MERN 2017). Different scenarios for mine site reclamation (see also Chapters 5–9) have been developed, including the use of LSHCCs. The main objective of LSHCCs is to control water infiltration to mine wastes so that the reactions that lead to the formation of acid mine drainage (AMD; see Chapter 1) can be controlled.

In this chapter, typical configurations of LSHCCs are first presented, followed by a more detailed description of low saturated hydraulic conductivity materials that can be used in LSHCCs to control water infiltration. Then, a cover design approach is presented, followed by some examples of LSHCCs used to reclaim AMD-generating mine sites. Finally, this chapter ends with the identification of the main advantages and limits and research needs associated with LSHCCs.

4.2 TYPICAL LSHCC CONFIGURATIONS

Typically, LSHCCs are made of a few layers of different materials, each playing a distinct role in the objective of controlling fluid movement (Aubertin et al. 1995, 2002; MEND 2004); the layers also play various other roles with respect to surface stability and future use (and esthetic) of the sites.

Figure 4.1 illustrates a typical LSHCC cover system that contains five layers. The number of layers can vary from two to more than five depending on a number of factors related to

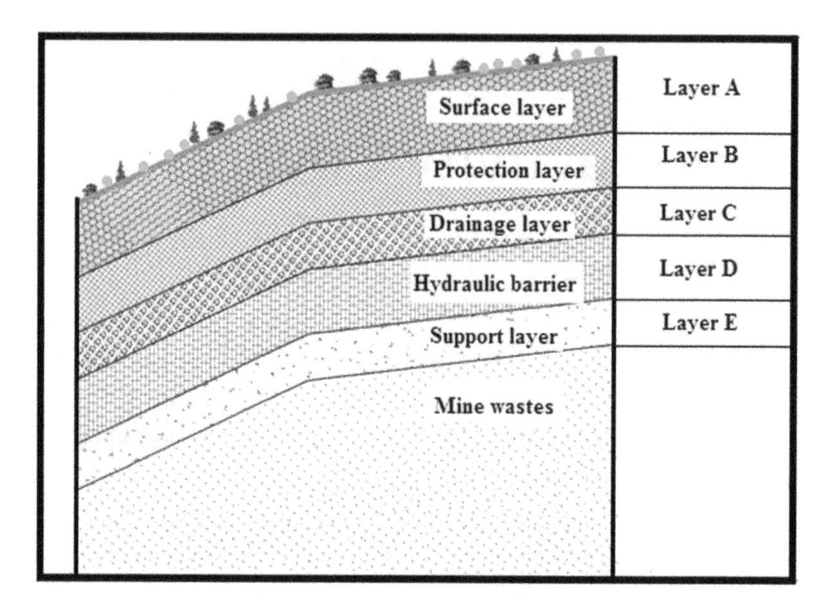

FIGURE 4.1 Typical configuration of LSHCCs (adapted from Aubertin et al. 1995).

site conditions and design objectives (Aubertin et al. 2016; Carson and Tolaymat 2017). In Figure 4.1, the top layer A, is usually made of organic soils that facilitate the establishment of vegetation. Layer B, made of coarse-grained materials with a high content of cobbles, is a protection layer that aims at physically stabilizing the site and preventing biointrusions. Layer C is a drainage layer made of a coarse-grained material (sand and/or gravel) that controls water inflow. The saturated hydraulic conductivity (k_{sat}) of this material is typically >10^{-5} m/s (Aubertin et al. 2016). The C layer may also play the role of a capillary break (see Chapter 3 for an explanation of this concept) that prevents upward moisture movement. Layers A, B, and C limit the impact of wet–dry cycles and freeze–thaw cycles on the underlying layers, and prevent ultraviolet (UV) degradation of GM materials when used as an impervious material. Layer D is the low saturated hydraulic conductivity layer (also called hydraulic barrier) that can be made of fine-grained soils (clay or fine silt), a GM, a GCL, soil-bentonite mix, or a combination of these. The typical k_{sat} value for layer D is < 10^{-9} m/s (Aubertin et al. 2016). Given the key role of layer D, characteristics of materials that can be used as this layer will be discussed in more detail in the next section. Layer E is the support layer placed on the reactive waste; the granular material used here may also act as a capillary break that prevents upward or downward moisture movement. As for layer C, the k_{sat} value for this layer is typically ≥ 10^{-5} m/s. The material used must also have a good bearing capacity because heavy equipment will circulate on it. This is particularly true when LSHCCs are built on tailing impoundments that contain tailings deposited in a loose state, at high water content (Ricard et al. 1997). Layer E may not be necessary when LSHCCs are built on coarse-grained materials such as waste rock (Zhan et al. 2000, 2001; Aubertin et al. 2009).

The thickness of the different layers may vary from a few mm (in the case of a GMs) to more than one meter (Aubertin et al. 1995, 2002, 2016). The different layers' thicknesses must be optimized as they significantly affect the efficiency and cost of the cover.

It is worth mentioning that initially only one low saturated hydraulic conductivity layer was used as a hydraulic barrier in landfills. After 1984, it was recommended to use double liners to control water infiltration in landfills efficiently (Das 2009). However, in the mining industry, there is no clear rule concerning the use of single or double protection; even today, most LSHCCs used to control water infiltration on mine waste disposal facilities only have a single protection.

4.3 TYPES OF LOW SATURATED HYDRAULIC CONDUCTIVITY MATERIALS

Different materials can be used as hydraulic barriers in LSHCCs; those proposed most often are (i) compacted clay, (ii) GM, and (iii) GCL. The following section provides a description of these materials and their properties. The main factors that influence their capacity to control water infiltration in the short to long term are also identified.

4.3.1 NATURAL CLAY MATERIAL

Historically, clay has been the most commonly used material as a hydraulic barrier in LSHCCs. In the literature, clay can refer both to a size of particle or to a class of minerals. In engineering classification, all particles < 2 μm are considered clay. The geotechnical classification of clays is based on the grain-size distribution and the Atterberg limits (liquid and plastic limits) (Holtz and Kovacs 1981). Mineralogically speaking, clay minerals are distinguished based on the following properties (Mitchell and Soga 2005): (1) small particle size; (2) a net negative electrical charge; (3) plasticity when mixed with water; and (4) high weathering resistance. Another particularity of clay minerals is their platy form. The mineralogical classification of clay minerals is carried out based on their crystal structure and stacking sequence of the layers. The main groups of clay minerals are (Mitchell and Soga 2005) kaolinite-serpentine, smectite (including montmorillonite), micalike (illite and vermicullite), and chlorite. The mineralogical composition of a given clay will play a significant role in its hydrogeotechnical behavior. In general, more clay in a soil results in a higher plasticity, a greater shrinking and swelling potential, a lower saturated hydraulic conductivity, a higher compressibility and cohesion, and a lower friction angle (Mitchell and Soga 2005). For LSHCCs, the reduction of k_{sat} is the main characteristic sought.

The clay barrier in LSHCCs is generally constructed in the form of layers with a thickness of 0.20–0.25 m before compaction, but after compaction, the thickness does not exceed 0.15 m, and the final thickness of the clay barrier is typically between 0.45 and 0.90 m. Different factors can affect the performance of LHCCs made of compacted clay such as clay properties, in situ water content and the compaction efforts, freeze–thaw cycles, wet–dry cycles, and the presence of roots from vegetation. Lower hydraulic conductivity values are reached when clays are compacted on the wet rather than the dry side of the compaction Proctor curve (Mitchell et al. 1965; Holtz and Kovacs 1981).

In general, clays with a high liquid limit and plasticity index (PI) have a greater quantity of clay particles or clay particles with higher surface activity (Mitchell 1976). It is well known that a relationship exists between saturated hydraulic conductivity and Atterberg limits (Terzaghi 1925); the higher the liquid limit and PI are, the lower the saturated hydraulic conductivity is (Benson et al. 1994). The clay material used as a hydraulic barrier should also have a proportion of coarse particles (> No. 200 sieve or 80 μm) lower than 30%. In general, if the criteria mentioned earlier are satisfied in terms of liquid limit, PI, percentage of fines (< No. 200 sieve or 80 μm), and compaction on the wet side of the optimum Proctor, the hydraulic barrier should satisfy, at least in the short term, the k_{sat} criterion of 10^{-9} m/s.

The following characteristics have been suggested for clay materials that constitute the hydraulic barrier in LSHCCs (Benson et al. 1994; Marcoen et al. 2000):

- Hydraulic conductivity is equal to or lower than 10^{-9} m/s.
- The percentage of fine particles must be greater than 30%.
- The percentage of clay fraction (<2 μm) should be more than 15%.
- The PI must be greater than 7% but less than 20% if freeze–thaw and wet–dry cycles are expected.
- Liquid limit must be higher than 20%.
- Gravel contents less than 50%.

The targeted in situ k_{sat} value of 10^{-9} m/s can be obtained if clods are eliminated during placement (Benson and Daniel 1990). Their presence can increase the k_{sat} value by typically two to three orders of magnitude (Mitchell et al. 1965). To avoid the influence of clods on the clay's k_{sat}, the control of the initial water content is important. It is usually recommended to compact clay on the wet side of the optimum Proctor to avoid the formation of clods (Mitchell and Soga 2005). The other way of eliminating clod formation is by increasing the compaction efforts; equipment that has higher compaction energy is then recommended (Benson et al. 1994).

Wet–dry cycles can affect significantly the properties of clays in LSHCCs. The phenomenon of desiccation due to wet–dry cycles is described further in Chapter 14. In summary, under the drying process, a decrease in water content in the clay induces an increase in matric suction. The suction that develops in the material leads to a decrease in the material volume by shrinkage (see Konrad and Ayad 1997; Saleh-Memba et al. 2016 for more details). At the large scale (no large horizontal displacement allowed), the stress concentration due to suction will initiate crack formation and propagation (Morris et al. 1992; Benson and Khire 1997; Kodikara et al. 2002; Wei et al. 2013). Due to the presence of cracks, preferential flow paths are created that favor water and gas movement. Figure 4.2 shows an increase in the percolation rates as a function of years due to the deterioration (mainly due to wet–dry cycles) of a clay layer. Similar results were obtained by many other researchers (e.g., Daniel and Wu 1993; Benson and Khire 1997; Albrecht and Benson 2001; see also Chapter 14 for more details).

Freeze–thaw cycles can also affect the performance of LSHCCs. Indeed, freeze–thaw cycles can modify the structure of fine-grained plastic soils and consequently modify their properties. The phenomenon is described in more detail in Chapter 14. In summary, when a LSHCC is exposed to cold temperatures, the freezing front in the clay layer generates a suction that attracts water molecules from the unfrozen zone to the freezing front. During this process, ice exerts pressure on the surrounding soil and causes it to move, rearrange, and consolidate (Benson and Othman 1993). The pore structure change can modify the soil structure and properties after thawing. In general, it is considered that low plasticity soils (with $PI < 10–20$; Othman et al. 1994; Eigenbrod 2003) and clayey soils composed mainly of swelling clay minerals (e.g., montmorillonite) can self-heal and are less affected by freeze–thaw cycles (Eigenbrod 2003). However, plastic soils with $20 < PI < 100$ can be destructured by freeze–thaw cycles and crack formation and subjected to a significant permanent increase in saturated hydraulic conductivity (k_{sat}) (Konrad and Samson 2000; Eigenbrod 2003).

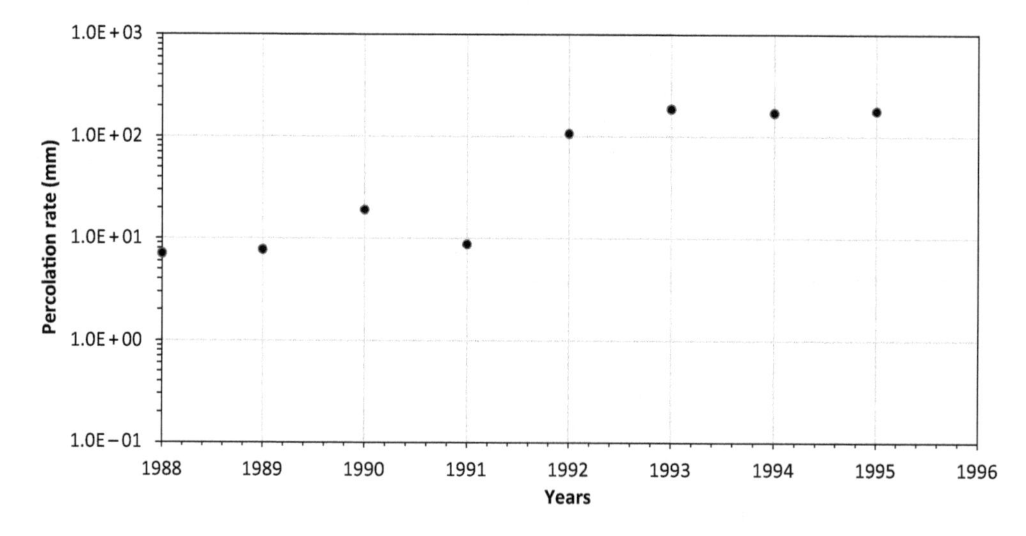

FIGURE 4.2 Percolation rate of a clay layer during time (data taken from Melchior 1997).

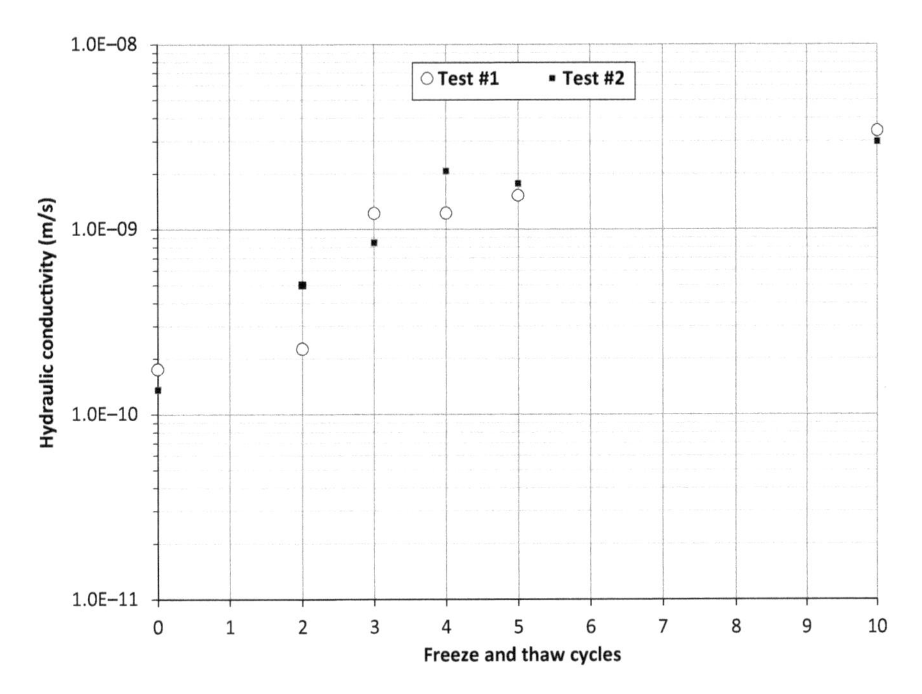

FIGURE 4.3 Hydraulic conductivity evolution of a clay material during freeze and thaw cycles.

As shown in Figure 4.3, the increase in k_{sat} can vary typically between one- and three-orders of magnitude after three to five freeze–thaw cycles (Daniel and Wu 1993; Albrecht and Benson 2001; Chapuis 2013).

Exhumation of existing LSHCCs with a hydraulic barrier made of clay showed that time clearly affects the capacity of the layer to maintain integrity. Albright et al. (2005) presented results from previous LSHCC exhumation studies (Montgomery and Parsons 1990; Benson and Wang 1996; Khire et al. 1997; Melchior 1997; Dwyer 2003) and their own results. In all cases, cracks were found in the clay layer, and the capacity of the layer to control water movement was reduced significantly. They conclude that pedogenic processes play an important role in the effectiveness of LSHCCs made of clay (Albright et al. 2005). The clay soil structure is significantly influenced by common surface processes (wetting and drying, propagation of roots, etc.) that can degrade covers relatively rapidly (within few years). They recommend adding an additional hydraulic barrier (e.g., GM or GCL) to ensure the long-term performance to control water infiltration (Albright et al. 2005).

Finally, it is worth mentioning that soil–bentonite mixtures (bentonite is mainly constituted of montmorillonite, a clay mineral) can be used as a cover material. However because, to the authors' knowledge, there is no real scale application in the mining industry, this low saturated hydraulic conductivity material will not be discussed further. More information on soil–bentonite mixtures can be found in Chapuis et al. (1992), Gueddouda et al. (2016), and Rosli et al. (2020).

4.3.2 GMs

GMs are thin polymer materials widely used as hydraulic barriers in civil engineering infrastructures (Rollin et al. 2002; Rowe et al. 2004; Müller 2007). The use of GMs in the mining industry is relatively recent but growing rapidly, mainly as components of heap leach pads (Rowe and Abdelaal 2016) and covers (Meiers et al. 2011; Maurice 2012; Power et al. 2017).

Different types of GMs are available, and their classification is based on the nature of the polymer resin used in their composition (Hsuan et al. 2008): polyethylene (PE), polypropylene (PP), polyvinyl chloride (PVC), and polyethylene terephthalate. For the polyethylene group (the most popular for LSHCCs built on mine waste disposal areas), the density is also considered in their own classification (see ASTM D 7700 2018):

- high-density polyethylene (HDPE), where the density is higher than 0.940 g/mL;
- medium-density polyethylene (MDPE), where the density is between 0.926 and 0.940 g/mL;
- low-density polyethylene (LDPE) with density included between 0.919 and 0.926 g/mL.

GMs are manufactured in the form of panels of variable dimensions. In the fabrication of GMs, polymer resins are combined with additives such as antioxidants, stabilizers, plasticizers, fillers, carbon black, and lubricants. These additives enhance the long-term performance of GMs by protecting the polyethylene from degradation (Ewais and Rowe 2014).

The different polymerization methods lead to different lengths and organization of carbon chains (Müller 2007). Two main zones can be identified in the GM structure: the crystalline zone and the amorphous zone (Scheirs 2000, 2009). The ratio between the crystalline and amorphous zones is called crystallinity. A higher crystallinity value corresponds to GMs with greater resistance against chemical attacks (Scheirs 2000, 2009; Müller 2007) and to lower saturated hydraulic conductivity (Sangam and Rowe 2001; Scheirs 2009). However, the greatest the crystallinity is, the more sensitive the GMs is against deformation and cracking. Typical properties of the main GMs used in the mining industry are summarized in Table 4.1.

The efficiency of LSHCCs made of GMs can be estimated in terms of percolation (the quantity of water that percolates through the cover for a given area). Water or fluids in general can flow through the LSHCC GMs by moving into the GMs (case of intact GMs) and through defects in the GMs. For the intact GMs, the watertightness can be evaluated in the laboratory by applying a water head difference of 100 kPa between both sides of a flat GM placed in a stainless steel cell (EN 14150, AFNOR 2006). The acceptable value of the flow rate corresponds to 10^{-4} m^3 m^{-2} d^{-1} (Rollin et al. 2002). Tests performed on PVC and HDPE GMs that were exposed for approximately 20 years in hydraulic application (without protection) showed that the flow rates were comparable to those of virgin GMs (10^{-6} m^3 m^{-2}·d^{-1}; Touze-Foltz 2015).

In the field, water volumes that percolate through LSHCCs made of GMs are significantly greater than those measured through intact GMs in the lab in idealized conditions due to defects appearing during and after cover's construction. Defects are mainly related to poor field seams, holes, tears, cuts, lack of welding, and burns. In their analysis of previous studies of GM defects, Rollin et al. (2002) found that 65% of defects were attributed to seaming and 35% were located in the sheet of the GMs itself. It is also reported by Forget et al. (2005) that most GM perforations occurred during their installation. To reduce these defects, it is crucial to implement a strict quality control program at the level of the manufacturer and the contractor. The climatic

TABLE 4.1
GM Properties

	Abbreviation	Resin (%)	Carbon Black	Additive	Crystallinity
High-density polyethylene	HDPE	95–98	2–3	0.25–1	55
Low-density polyethylene	LDPE	94–96	2–3	0.25–1	15
Polyvinyl chloride	PVC	30–40	1–2	2–5	0

Source: Adapted from Hsuan et al. (2008), US Department (2018), Scheirs (2009).

conditions during GM installation are also important. The creation of wrinkles due to GM dilation must be limited (GMs with a white instead of black side exposed to sun is being introduced to reduce this effect). Indeed, wrinkles can significantly increase percolation through defects (Brachman et al. 2011; Rowe et al. 2012). Furthermore, GMs are synthetic materials that can deteriorate over time. The degradation is caused by the oxidation of the polymer resin with the atmospheric oxygen (Scheirs 2000; Müller 2007). This degradation will affect the properties of GM. It is important to protect GMs from UV and high temperatures as much as possible to reduce the degradation rate. All these factors that can affect percolation through covers made of GMs have influenced regulators, who now often require a double-layer protection with the use of GMs as a hydraulic barrier; this practice, however, is rarely applied to LSHCCs built on mine sites (see Section 4.6).

Different approaches have been proposed to estimate the real in situ percolation rate through LSHCCs made of GMs (e.g., Giroud et al. 1998; Rowe and Booker 1998; Giroud and Touze-Foltz 2005). In these approaches, some of the proposed equations are empirical and others are based on experimental results. These equations are summarized and discussed by Cunningham (2018). Different parameters are considered in these equations such as the pinhole diameter, defects geometry, properties of the materials below, thickness of the GMs, hydraulic head, and GM slope. Different techniques, which can be grouped into two categories, can be used to detect and quantify these defects (Barroso 2005). The first ones are based on the electrical leak location and allow for detection and localization of leak paths, and the second ones are based on tracer tests, flood testing, and infrared thermography allowing only for detection of defects. More details about these techniques can be found in the study by Barroso (2005) and Forget et al. (2005).

The design of LSHCCs made of GMs also involves physical stability analysis. Designers must ensure that a sufficient factor of safety is reached against different types of failure from different types of solicitation (due to shear stress, tension, displacement, and dynamic loads; Koerner 2012; Aubertin et al. 2016). This aspect is particularly important when GMs are used in sloping covers; the friction angle between GMs and soils is relatively low. In these cases, textured instead of smooth GMs must be used. The effect of water and ice on the interface layer can also be a source of concern (Briançon et al. 2002). Many cases of failure along interfaces with GMs have been listed over the years (e.g., Seed et al. 1990; Stark et al. 2008; Datta 2010).

In the literature, LSHCCs with GMs are often referred to as water and gas barriers. Different authors measured some properties of GMs to control gas migration. Effective diffusion coefficient D_e for oxygen (see Chapter 3) between 1 and 4×10^{-11} m^2/s was measured by Epacher et al. (2000) and Patterson et al. (2006). Kim and Benson (1999) evaluated the oxygen flux through GMs, and the obtained values ranged between 1.7 and 1.9 mol-O_2 m^{-2} y^{-1}. These values are comparable to those obtained by Power et al. (2017) at the Sydney Coalfield from 2012 to 2016 with the total mean flux equal to 2.5 mol-O_2 m^{-2} y^{-1}.

4.3.3 GCLs

GCLs consist of a clay material (bentonite) layer placed between two geotextiles or/and between a geofilm (Rowe et al. 2004). Bentonite (that contains mainly montmorillonite from the smectite group) is a natural clay that appears largely as a product of weathering, through a chemical transformation from volcanic ash in the presence of water. The bentonite may also contain other minerals such as kaolinite, interbedded minerals, and zeolites. The montmorillonite content in the bentonite ranges typically between 60 and 90%; however, for high-quality bentonite, its content in montmorillonite ranges between 75 and 90% (Egloffstein 2001; Bouazza et al. 2002; Lee et al. 2005).

The GCL thickness ranges between 4.9 and 12.9 mm (Dickinson and Brachman 2008). Different types of GCL are available, and their differentiation is based on the montmorillonite content (Rowe et al. 2004), its chemical nature (sodium or potassium montmorillonite), and the mass of bentonite

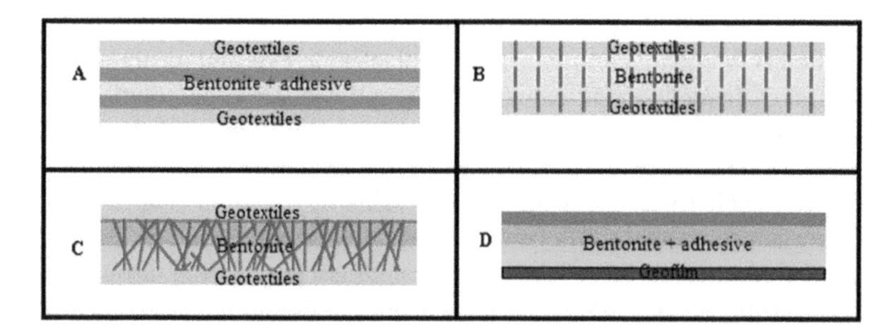

FIGURE 4.4 Different GCLs (modified from Koerner 1994).

per unit area (kg/m^2). There are four different types of methods to fix the bentonite between the two geotextiles (Koerner 1994; see Figure 4.4):

- adhesive-bound bentonite geotextile (A)
- stitch-bonded bentonite geotextile (B)
- needle-punched geotextile (C)
- adhesive-bound geofilm (D).

GCLs are characterized by a low saturated hydraulic conductivity typically between 5×10^{-12} and 2×10^{-10} m/s (Daniel et al. 1997; Bouazza 2002; Silvestre et al. 2003; Rowe et al. 2004; Shackelford et al. 2010). These values are valid for permeability tests performed on the fully hydrated GCL with non-cationic water. It is important to recall that during GCL permeation, a decrease of the hydraulic conductivity is observed during the first days (Xue et al. 2012; Figure 4.5). In fact, when montmorillonite (sodium Na-Mnt or calcium Ca-Mnt type) is exposed to water, layers of water molecules bound to the clay sheets and sheets move away from each other as the number of layers increases. The pathway for free water to circulate through the structure is then more tortuous (Jo et al. 2001; Mitchell and Soga 2005). This clearly demonstrates the importance of GCL hydration in obtaining a good performance in the field.

The hydration and permeant liquids can also have a significant impact on GCL's k_{sat} (Petrov and Rowe 1997; Xue et al. 2012; Yesiller et al. 2019). The weakly bonded Na^+ monovalent cations of the sodium montmorillonite (Na-Mnt type is taken here as an example but the concept

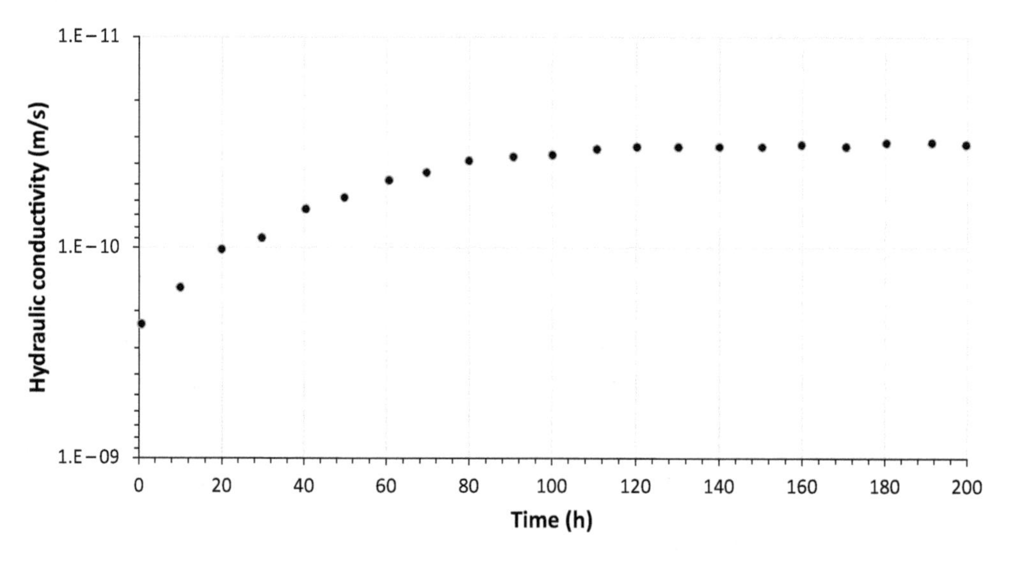

FIGURE 4.5 GCL hydraulic conductivity variation during time and permeated with water (data from Xue et al. 2012).

TABLE 4.2
Saturated Hydraulic Conductivity for GCL Samples Hydrated and Permeated with Multivalent Inorganic Cation Water

Study	Moisturizing Liquid	Permeant Liquids	k_{sat} (m/s)
Petrov and Rowe (1997)	Deionized water	0.6 M NaCl	8.0×10^{-10}
	0.6 M NaCl	0.6 M NaCl	4.0×10^{-9}
	Deionized water	2.0 M NaCl	4.7×10^{-10}
	2.0 M NaCl	2.0 M NaCl	1.2×10^{-8}
Shackelford et al. (2000)	Deionized water	6.7 M CaCl2	10^{-10}
	6.7 M $CaCl_2$	Deionized water	10^{-8}
	Deionized water	pH 8.7 et 330 mg/L	7×10^{-11}
	pH 8.7 et 330 mg/L	pH 8.7 et 330 mg/L	9×10^{-8}
Shan and Lai (2002)	Hard water	Hard water	4.4×10^{-11}
	Hard water	Sea water	2.3×10^{-10}
	Sea water	Sea water	1.7×10^{-7}
Shackelford et al. (2010)	Groundwater	Groundwater	1.7×10^{-11}
	Groundwater	Synthetic AMD	7.9×10^{-11}
	Synthetic AMD	Synthetic AMD	3.9×10^{-8}

Source: Translated from Chevé (2019).

can also be applied to Ca-Mnt type) are likely to be replaced by multivalent cations with greater strength of attraction within the montmorillonite sheets. As Na^+ ions are replaced, strong attractive forces between the multivalent cations and the montmorillonite sheets prevent the bonding of a large number of water layers on the clay sheets. This results in less swelling and a higher hydraulic conductivity (Mesri and Olson 1971; Shackelford 1994; Shackelford et al. 2000, 2010; Egloffstein 2001). Different authors demonstrate how salt solutions of various concentrations affect the GCL's swelling and saturated hydraulic conductivity; typical values are presented in Table 4.2. One can see that the increase of k_{sat} can be as high as a factor of 1000 compared to values obtained for the GCL hydrated with a solution that has a low multivalent inorganic cation concentration.

Another factor that can affect the properties of GCLs is the confining stress. Figure 4.6 shows an example of the influence of confining stress on k_{sat}. As the confining stress increases from 3 to 200

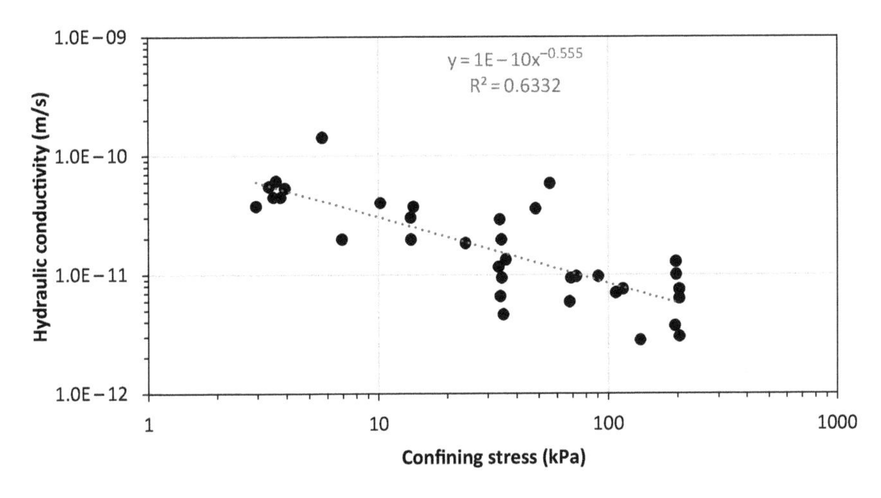

FIGURE 4.6 GCL hydraulic conductivity vs. confining stress (data from Estomell and Daniel (1992), Ruhl and Daniel (1997), Rad et al. (1994), Petrov et al. (1997), and Barroso (2005).

kPa, the GCL's saturated hydraulic conductivity decreases by a factor of about 50 (Estomell and Daniel 1992; Rad et al. 1994; Petrov et al. 1997; Ryhl and Daniel 1997; Shackelford et al. 2000; Barroso 2005). However, in the context of the use of the GCL as a cover material, it is probable that the GCL will support a low confining pressure.

As mentioned in the previous section on clay covers, swelling clays with high plasticity are not significantly affected by freeze–thaw and wet–dry cycles. Permeability tests performed on GCLs (made of swelling clays) hydrated with water that have a low multivalent inorganic cation concentration showed that k_{sat} is not significantly affected by freeze–thaw (Kraus et al. 1997; Podgorney and Bennett 2006) and wet–dry (Shan and Daniel 1991; Boardman and Daniel 1996) cycles, confirming their self-healing capacity. However, the self-healing potential of the GCL is significantly reduced when they are hydrated and permeated with multivalent inorganic cationic water, and a decrease of k_{sat} is observed. For example, Chevé et al. (2019) showed that the influence of freeze–thaw cycles can be significant for GCL samples hydrated and permeated with multivalent inorganic cationic water.

Most of the studies in the literature on the GCL focused on the saturated hydraulic conductivity. In the context of LSHCCs built on reactive mine waste, it could be also interesting to know other hydrogeological properties such as the water retention curve and the effective diffusion coefficient for oxygen D_e (see Chapter 3). Chevé et al. (2019) obtained air entry values (AEVs) between 0.25 and 0.4 MPa (see Figure 4.7) while Southen and Rowe (2007) obtained AEVs between 0.5 and 0.6 MPa for samples hydrated with non-cationic water. Yesiller et al. (2014) and Chevé et al. (2019) suspect an influence of the water quality on the WRC, but more work is needed to clarify this impact. Finally, the work of Aubertin et al. (2000) and Chevé et al. (2019) showed that, as for soils, the D_e of GCLs is mainly a function of the degree of saturation and that D_e values can be estimated by predictive models such as the one proposed by Aachib et al. (2004) (see Chapter 3).

As for GMs, defects can also appear during construction with GCLs. The most well-known problems are: openings between panels due to piles movement or GCL retreat (Thiel and Richardson 2005); bentonite erosion in sloping areas that can create an heterogeneity in the thickness of bentonite in the GCL (Brachman et al. 2015); and drying of bentonite during installation. Many other factors have been identified in the literature; the interested reader is referred to Rowe (2020) for

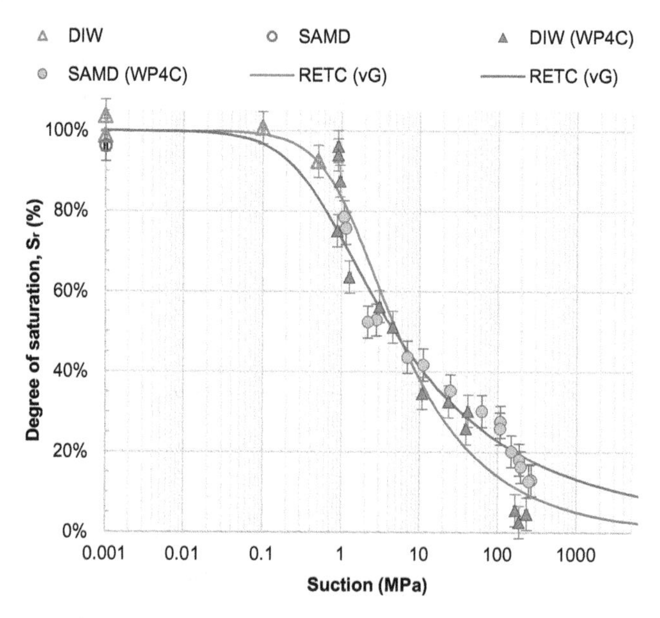

FIGURE 4.7 Water retention curves of GCL specimens permeated with deionized water (data from Chevé 2019, and figure modified from Chevé et al. 2019).

more details. Finally, slope stability can also be an issue for LSHCCs with the GCL as a hydraulic barrier (Daniel et al. 1998). Different factors are involved when determining short-term and long-term stability of LSHCCs made of GCLs; more information can be found in the study by Fox and Stark (2004).

4.4 COVER DESIGN

When designing LSHCCs, various questions need to be answered in order to select the best materials for the cover. The most important questions are as follows:

- What is the availability and volume of adequate natural materials around the mine site for the construction of the hydraulic barrier and the other LSHCC layers?
- What are the typical climatic conditions on site?
- What is the desired performance targeted in terms of water percolation and oxygen flux?
- Will the hydraulic barrier be subjected to wet–dry cycles, freeze–thaw cycles, and biointrusions?
- Are significant movements of the waste storage facility expected due to consolidation, creep, or other phenomenon?
- What is the expected life of the hydraulic barrier?
- Is the hydraulic barrier in contact with AMD?
- What is the geometry of the site and what is the inclination of the sloping areas?

Answers to these questions should help identify the best material (or combination) to be used as a hydraulic barrier in the LSHCCs. In general, GCLs would not be recommended if hydration could be altered due to the permeation of multivalent cationic solution. If low percolation is targeted, the use of double protection is recommended. Slopes greater than 3:1 are not recommended in the long term for GCLs and GMs. GCLs and GMs are sensitive to differential movements. Indeed, differential movements can create significant stresses in GMs that could accelerate their degradation and lead to failure. GCLs can also be affected by soil movement by creating openings between GCL panels. When significant movements are expected, LSHCCs made of geological materials should be favored. For dry climates, hydraulic barriers made of clay or a GCL should be avoided because of their sensitivity to wet–dry cycles. Based on the aforementioned questions and recommendations described in the previous paragraphs, the following steps are proposed to design a LSHCC (see Figure 4.8):

- Step 1: On the basis of the regulations concerning the mine site reclamation, it is necessary to define the objectives to be achieved. These objectives can be identified in terms of water percolation and oxygen flux.
- Step 2: The site parameters must be identified such as climate conditions, topography, materials availability, and vegetation.
- Step 3: From results of Steps 1 and 2, the type of hydraulic barrier (single or double protection) is selected.
- Step 4: The detailed LSHCC configuration is then established with the calculation of the volume of materials necessary and the costs associated.
- Step 5: The detailed design can then be performed by integrating notions such as landscaping, hydrology, hydrogeology, and slope stability and erosion, for actual and future conditions that take into account climate changes (see Chapter 14).
- Step 6: Construction and monitoring is the last stage of the process (see Chapter 10). It is important to validate that the system is performing as expected using equipment to monitor piezometric levels, surface flow, water quality evolution (surface and underground), and movement of the storage area. Exhumation, sampling, and testing of the hydraulic barrier should also be integrated in the long-term monitoring plan.

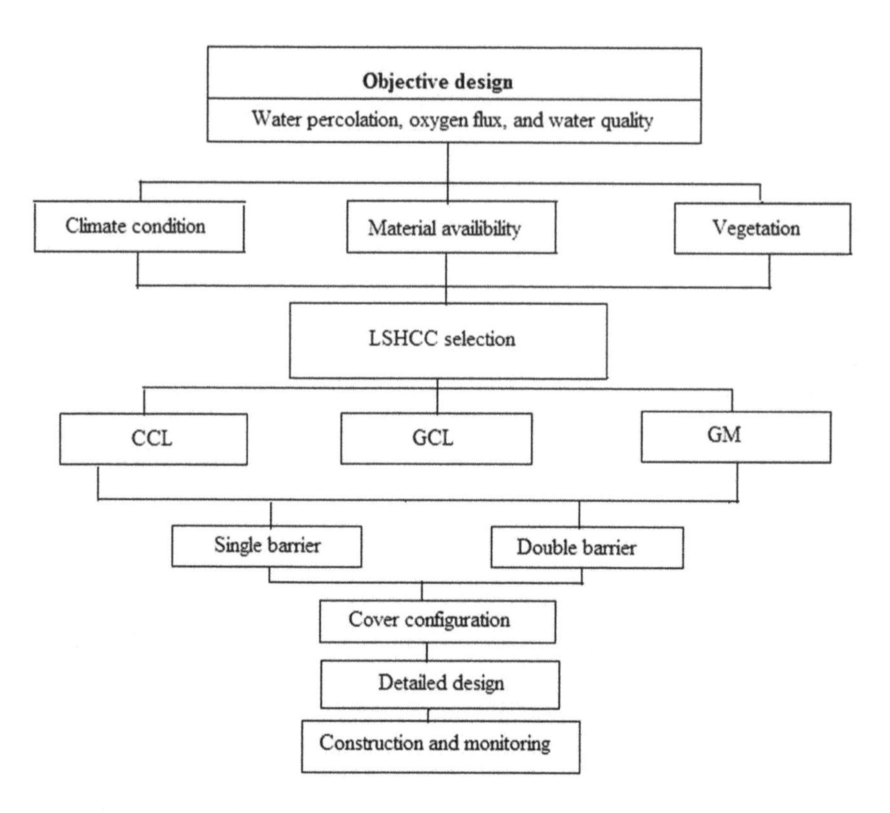

FIGURE 4.8 Steps of the LSHCC design.

4.5 CASE STUDIES OF HYDRAULIC BARRIERS USED IN MINE SITE RECLAMATIONS

Different mine sites across the world were reclaimed with LSHCCs. Table 4.3 presents some examples from Canada. This table shows the different configurations and highlights the variability of the number of layers and their thickness (see Figure 4.9). Furthermore, one can observe that some LSHCCs use double protection where clay is used in conjunction with GMs. However, many case studies only have one protection (GM or GCL).

As shown in Table 4.3, for many cases, no information is available concerning LSHCC performance. Sometimes, some information related to water percolation, piezometric level evolution, and groundwater quality is presented. In most cases, quantifiable environmental objectives are not stated, and the capacity of the cover system to reach it is usually absent.

4.6 ADVANTAGES AND LIMITS

LSHCCs have advantages and disadvantages that are summarized in Table 4.4.

For those made of natural compacted clay, the main advantage is that the hydrogeotechnical behavior of the material is well known and many examples can be found in the literature. Clay can also retain some contaminants by sorption, but in the case of covers, this aspect is not necessarily relevant (a significant advantage for liners). Finally, the experience gained over the years with clay as an impervious material can help avoiding some design fatal flaws. However, clay is not an easy material to work with, and reaching the adequate in situ properties can be challenging. If the material does not have the right properties (see Section 4.3.1), the material can

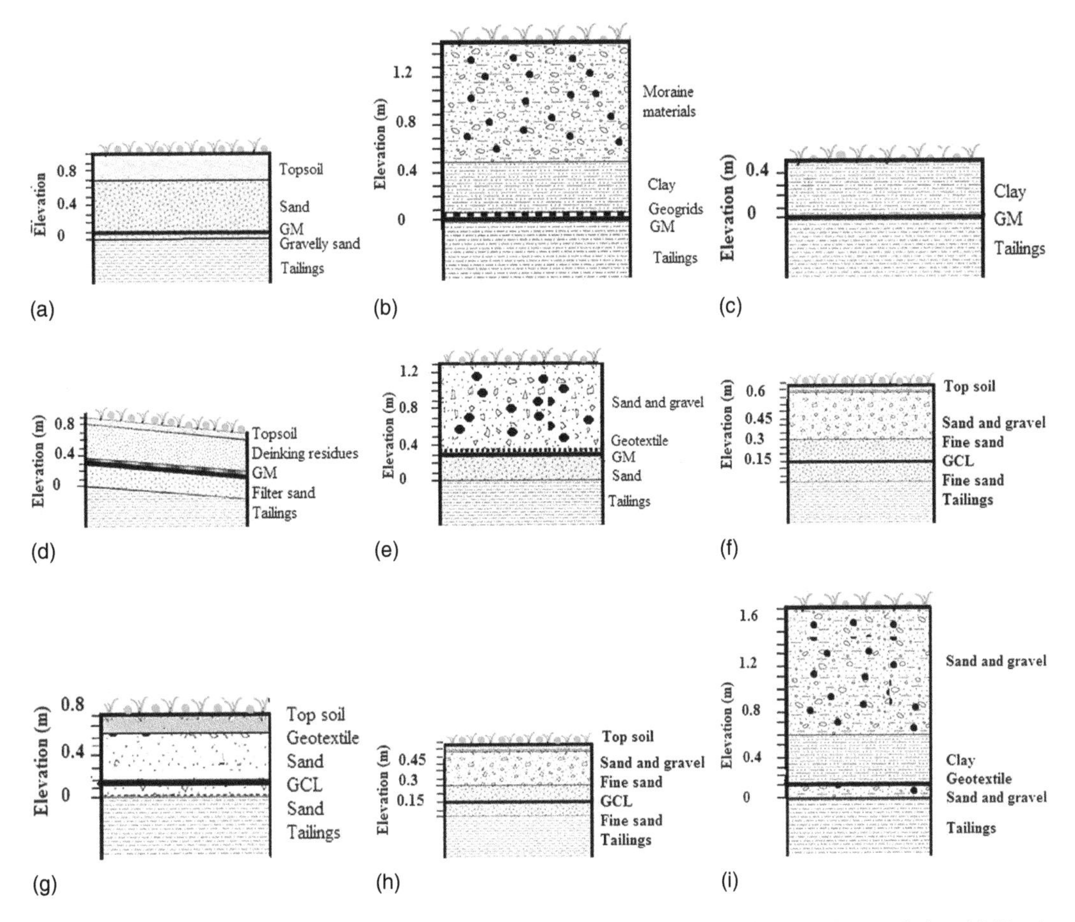

FIGURE 4.9 Mine site cover configurations: (a) Weedon site, (b) Poirier site, (c) Normetal site, (d) Eustis site, (e) Aldermac site, (f) Barvue site, (g) New Calumet site, (h) Premier Gold site and (i) Whistle site. ([a–f] adapted from Rarison 2017).

be affected by freeze–thaw and wet–dry cycles (see also Chapter 14). Roots from vegetation can also alter the properties of the clay and reduce its capacity to control percolation (see Chapter 14). In terms of construction costs, the most important parameter is the distance between the borrow pit and the mine site to reclaim. Typically, the construction cost increases significantly when the distance between the two sites (borrow pit and mine site) is greater than 10 km (Bussière et al. 1999).

GMs are more and more proposed as a hydraulic barrier in LSHCCs built on mine waste storage facilities mainly because of their low k_{sat} and D_e values that help control fluid (liquid and gas) migration. Other advantages mentioned in the literature include chemical resistance, high puncture and tensile strength, and high modulus of elasticity (Rowe et al. 2008). With the years, the installation technique has improved, and the costs associated with GMs have become competitive compared to other low saturated hydraulic conductivity materials. However, construction with GMs automatically involves a detailed quality control program that must be carefully followed to obtain final results close to those expected at the design stage. Even with good quality control, a certain number of defects per surface area have to be considered in the percolation calculations, which are usually much greater than those estimated based on GM's k_{sat}. But the main objection regarding the use of a GM in LSHCCs for acid-generating mine sites is their limited lifetimes, which is usually

TABLE 4.3
LSHCCs Used as Mine Site Covers

	Cover Configuration	Cover Performance	References
a) Weedon mine site, QC, Canada	• 0.3 m of top soil • 0.6 m of sand • 1.5 mm of HDPE • 0.3 m of sand and gravel	• Information not available	Landry and Senay (2010)
b) Poirier mine site, QC, Canada	• 1 m of tills • 0.5 m of clay • 1.5 mm of HDPE *	• Piezometric level decrease (1.5–2.4 m) • Reduction of effluent flow and in acidity, iron and zinc concentrations	Maurice (2002)
c) Normetal mine site, QC, Canada	• 0.5 m of clay • 1.5 mm of HDPE *	• Piezometric level decrease (≈3 m) • pH increase and decrease in dissolved element	Hofton and Schwenger (2009)
d) Eustis Mine site, QC, Canada	• 0.50 m of deinking residues • 1.5 mm of HDPE0.3 m of sand	• Information not available	Cyr (2011)
e) Aldermac mine site, QC, Canada	• 1 m of sand and gravel • 1.5 mm of HDPE • 0.3 m of sand	• Reduction of effluent flow but AMD is still flowing from the pile • The equivalent hydraulic permeability evaluated on the exhumed GM after 13 years is similar to a pristine GM • Significant movement are observed with zone of tension in the GM	Daudet (2013); Maqsoud et al. (2015); Rarison et al. (2020)
f) Barvue mine site, QC, Canada	• 0.7 m of protective layer • 0.3 m of sand • 1.5 mm of HDPE • 0.3 m of sand	• Information not available	Zetchi and Fouquet (2017)
g) New Calumet, QC, Canada	• 0.15 m of top soil • geotextile • 0.45 m of sand • GCL • 0.15 of sand	• Decrease in dissolved element	WSP (2018)
h) Premier Gold project, BC, Canada	• 0.05–0.10 m top soil • 0.25–0.35 m of coarse sand and gravel • 0.1–0.15 m fine sand • GCL • 0.1–0.15 m fine sand	• Higher oxygen concentrations were observed in two out of three soil profiles in the tailings • Some GCL specimens, exhumed in 2004, had air entry values of less than 10 kPa	Renken et al. (2005)
i) Whistle Mine, ON, Canada	• 1.2 m of sand and gravel • 0.45 m of compacted clay • Geotextile • 0.1 m of sand and gravel	• cover system is performing as expected based on field datacollected over the past six years • Based on field data collected over the past six years, influx of atmospheric oxygen and meteoric water to the waste rock backfill has been substantially reduced	Ayres et al. (2012)

Source: * Adapted from Rarison (2017).

TABLE 4.4
Advantages and Disadvantages of the Main Low Saturated Hydraulic Conductivity Materials Used in LSHCCs

	Main Advantages	Main Disadvantages
Compacted Clay material	• Clay can mitigate pollutants • Long life and self-healing capability for appropriate PI • Well known geological material • Inexpensive, if materials are available locally	• Construction is difficult and requires heavy equipment • Can be affected by freeze-thaw cycles and desiccation cracks • Vegetation can alter the properties of the clay layer • Low shear strength • Relatively high cost, if materials are not available locally • In general, performance is not at the level expected at the design stage
GM	• Low saturated hydraulic conductivity and effective O_2 diffusion coefficient • Easy installation • High shear strength • Good short-term performance if installed properly	• Degrades with time due to physico-chemical effects • Need to develop a strict quality control during construction (to avoid punctures, wrinkles, etc.) • Prone to instability in sloping areas • High leakage in cases of perforation or poor welding • Durability dependent to in situ conditions • Only few documented cases in the mining industry
Geosynthetic clay liner	• Easy installation • No welding required • Self-healing capability • Low k_{sat} when hydrated and permeated with non-multivalent cationic solutions	• Low thickness, making it vulnerable to perforation or roots intrusion • Instability on slopes • Subject to shrinkage • Not a good O_2 diffusion barrier • Not many applications on mine sites • Vulnerable to cation exchange that affects their properties

expressed in decades (or centuries in the best conditions) depending on various physicochemical effects (swelling, UV degradation, degradation by extraction, biological degradation, and oxidative degradation; Rowe et al. 2004) that are themselves a function of temperature, oxygen availability, mechanical stress applied on the GM, and chemical quality of the water. Such fairly short lifetimes are hardly compatible with the very long-term requirements for covers that aim to control AMD for centuries (to not saying perpetuity). The effect of settlements and risk of sliding along the slopes are other major concerns (Aubertin et al. 2016).

The main advantages of the GCL are related to the homogeneity of the properties when hydrated with non-multivalent cationic solutions (k_{sat} can be as low as $< 10^{-10}$ m/s), their self-healing capability, and the relatively low construction costs. GCLs are also relatively simple to install and, with an appropriate overlap, the joints are not as problematic as for GMs. GCLs have disadvantages such as being sensitive to chemical quality of hydration and permeant liquids; with cation exchange, the GCLs can lose their advantageous hydrogeological properties, particularly after freeze–thaw and wet–dry cycles. The low friction angle (internal and external) reduces their potential use on steep slopes. The relatively low thickness (<10 mm) can favor gas migration, such as oxygen, through a diffusion mechanism (Aubertin et al. 2000; Shackelford 2014; Chevé et al. 2019). Downslope erosion of bentonite has also been observed between the moment of GCL placement and the completion of cover's installation. This can induce a performance heterogeneity with zones with more or less the thickness of bentonite.

Because of uncertainties related to the capacity of the three low saturated hydraulic conductivity materials presented in this chapter to maintain their properties in the long term, it is recommended to combine them to create a double protection against percolation, as is imposed by many regulators for landfill covers.

4.7 RESEARCH NEEDS

LSHCCs can be used to reclaim mine sites. However, their performance can vary greatly depending on many factors, as seen previously. For the three main hydraulic barriers presented in this chapter, some aspects need to be better understood for the development of more efficient LSHCC design:

- Compacted clay: The influence of time on clay properties still needs more investigation, particularly the influence of roots and vegetation on the water budget of reclaimed sites and the cover material properties (see Chapter 14). Modification of clays with amendments could also be an option when available clays don't have the appropriate properties. The influence of the other layers on top of the compacted clay on LSHCCs hydrogeological behavior is another aspect that should be investigated further.
- GCL: The synergy between chemical quality of the permeant and hydration water, freeze–thaw and wet–dry cycles (see Chapter 14), and low confining pressure should be further studied for an optimal application of GCLs in LSHCCs. The long-term physical stability in sloping areas reclaimed with GCLs is another topic of interest. The capacity of in situ hydrated GCLs to control water and gas movement is still not well understood. Finally, vegetation will eventually invade most reclaimed mine sites, but its influence on the long-term capacity to control fluid movement needs further investigation (see Chapters 12 and 14).
- GMs: A lot of research efforts are actually dedicated to GMs. However, the particularity of the mining context should involve specific research efforts, particularly those related to the long-term performance. One can think of the influence of mine storage facilities' movement on the stress in the GM and its capacity to maintain the performance. Surprisingly, there are only few in situ performance assessments of LSHCCs made of GMs built on mine waste disposal facilities. This in situ performance and its evolution with time must be better understood. The use of GMs in Arctic climates should also be investigated further (e.g., influence of freeze–thaw cycles on physical and hydrogeological properties).
- Climate changes and single vs. double protection: The way climate changes can affect long-term performance of LSHCCs, no matter the type of hydraulic barrier used, should be investigated further. It is important to integrate future extreme conditions in the design of LSHCCs (see Chapter 14). Double protection is very rarely used in the mining industry, while it is a common practice for covers used to reclaim landfills. There is a need for a comparison between the performance of single and double protection LSHCCs to control water infiltration in the mining context.

REFERENCES

Aachib, M., M. Mbonimpa, and M. Aubertin (2004). Measurement and prediction of the oxygen diffusion coefficient in unsaturated media, with applications to silt covers. *Water, Air and Soil Pollution*, 156: 163–193.

AFNOR-(Association Française de Normalisation) (2006). *Geosynthetic barriers — Determination of permeability to liquids*. La Plaine Saint-Denis, France: AFNOR.

Albrecht, B.A., and C.H. Benson (2001). Effect of desiccation on compacted natural clays. *Journal of Geotechnical and Geoenviromental Engineering*, 127(1): 67–75.

Albright, W.H., C.H. Benson, and G.W. Gee (2005) Field performance of a compacted clay landfill final cover at humid site. *Journal of Geotechnical and Geoenvironmental Engineering*, 132(11): 1393–1403.

ASTM D 7700 (2018). *Standard Guide for Selecting Test Methods for Geomembrane Seams*. West Conshohocken, PA: ASTM International, www.astm.org.

Aubertin, M., B. Bussière, T. Pabst, M. James, and M. Mbonimpa (2016). Review of the reclamation techniques for acid-generating mine wastes upon closure of disposal sites. In *Geo-Chicago 2016*, pp. 343–358.

Aubertin, M., B. Bussière, and L. Bernier (2002). Environnement et gestion des rejets miniers. *Manuel sur CD-ROMom*, Presses Internationales Polytechniques.

Aubertin, M., M. Aachib, and K. Authier (2000). Evaluation of diffusive gas flux through co vers with a GCL. *Geotexti les and Geomembranes*, 18(2–4): 215–233.

Aubertin, M., R.P. Chapuis, M. Aachib, B. Bussière, J.F. Ricrad, and L. Tremblay (1995). *Évaluation en laboratoire de barrières sèches construites à partir de résidus miniers.* École Polytechnique de Montréal, NEDEM/MEND Projet 2.22.2a. École Polytechnique de Montréal.

Aubertin, M., S. Cifuentes, E. Apitthy, B. Bussière, J. Moslson, and R.P. Chapuis (2009). Analyses of water diversion along inclined covers with capillary barrier effects. *Canadian Geotechnical Journal*, 46: 1146–1164.

Ayres, B.K., L. Lanteigne, M. O'Kane, and G. Meiers (2012). Whistle mine backefilled pit dry cover case study – performanece based on six years of field monitoring. *Proceeding of the 9th International on acid rock drainage*, Ottawa, ON, Canada.

Barroso, M. (2005). Fluid migration through geomembrane seams and through the interface between geomembrane and geosynthetic clay liner. PhD Thesis diss., Joseph-Fourier University, Grenoble, France.

Benson, C., and D. Daniel (1990). Influence of clods on hydraulic conductivity of compacted clay. *Journal of Geotechnical Engineering, ASCE*, 116: 1231–1248.

Benson, C., and M. Khire (1997). Earthen materials in surface barriers. In *Barrier technologies for environmental management: Summary of a workshop*, pp. 79–89. Washington, DC: National Academic Press.

Benson, C., and X. Wang (1996). *Field hydraulic conductivity assessment of the NCASI final cover test plots.* Environmental geotechnics report 96-9, Department of Civil and Environmental Engineering, University of Wisconsin-Madison.

Benson, C., H. Zhai, and X. Wan (1994). Estimating hydraulic conductivity of compacted clay liner. *Journal of Geotechnical Engineering*, 120(2), Paper No. 4052.

Benson, C.H., and M.A. Othman (1993). Hydraulic conductivity of clay frozen and thawed in situ. *Journal of Geotechnical Engineering*, 192(2): 276–294.

Boardman, B. T., and D. E. Daniel. (1996). Hydraulic conductivity of desiccated geosynthetic clay liners. *Journal of Geotechnical Engineering*, 122(3): 204–208.

Bouazza, A. (2002). Geosynthetic clay liners. *Geotextiles and Geomembranes* 20(1): 3–17.

Bouazza, A., J. G. Zornberg, and D. Adam (2002). Geosynthetics in waste containment facilities: Recent advances. *Keynote paper in Proceedings of Seventh International Conference on Geosynthetics*, Nice, France, vol. 2, 445–511.

Brachman, R. W. I., A. Rentz, and R. K. Rowe (2015). Classification and quantification of downslope erosion from a geosynthetic clay liner (GCL) when covered only by a black geomembrane. *Canadian Geotechnical Journal* 52, no. 4: 395–412.

Brachman, R. W. I., P. Joshi, R.K. Rowe, and S. Gudina (2011). Physical response of geomembrane wrinkles near GCL overlaps. In *Advances in Geotechnical Engineering, Proceedings of GeoFrontiers, Dallas, Texas, 13–16 March 2011. Geotechnical Special Publication (GSP) 211.* Edited by J. Han and D.A. Alzamora. New York: American Society of Civil Engineers, pp. 1152–1161.

Briançon, L., H. Girard, and D. Poulain (2002). Slope stability of lining systems? Experimental modeling of friction at geosynthetic interfaces. *Geotextiles and Geomembranes* 20(3): 147–172.

Bussière, B., M. Aubertin, M. Benzaazoua, and D. Gagnon (1999). Modèle d'estimation des coûts de restauration de sites miniers générateurs de DMA. Séminaire Mines écologiques présentés dans le cadre du congrès APGGQ 1999: Les sciences de la terre: Une vision planétaire, Rouyn-Noranda.

Carson, D., and T. Tolaymat (2017). Post-Closure Performance of liner Systems at RCRA Subtitle C Landfills. U.S. Environmental Protection Agency, Washington, DC, EPA/600/R-17/205, 2017.

Chapuis, R. P. (2013). Full-scale evaluation of the performance of three compacted clay liners. *Geotechnical Testing Journal*, 36(4), 575–583.

Chapuis, R.P., Lavoie, J. and Girard, D. (1992), Design, construction, and repair of the soil-bentonite liners of two lagoons. *Canadian Geotechnical Journal*, 29: 638–649

Chevé, N. (2019). Évaluation de la performance des géocomposites benthoniques comme barrière aux fluides dans un contexte de recouvrement minier. Mémoire de maîtrise Master's thesis, Université du =Québec en Abitibi-Témiscamingue.

Chevé, N., and Bussière, B. (2019). Hydrogeological properties of geosynthetic clay liners permeated with synthetic acid mine drainage. *Proceedings of the 72th Canadian Geotechnical Conference*, St John's, Canada.

Cunningham, H. L. (2018). Evaluating the rate of leakage through defects in a geomembrane. Master's thesis, University of Saskatchewan, Canada.

Cyr J. (2011). Présentation - Restauration des sites miniers. In *Resources Naturelles Québec. Proceedig of the Envioromine Symposium Rouyn-Noranda*, Québec, Canada.

Daniel, D. E., and Y. K. Wu (1993). Compacted clay liners and covers for arid sites. *Journal of Geotechnical Engineering* 119(2): 223–237.

Daniel, D. E., J. J. Bowders, Jr., and R. B. J. J. Gilbert (1997). Laboratory hydraulic conductivity testing of GCLs in flexible-wall permeameters. In L.W. Well (Ed.), *Testing and Acceptance Criteria for Geosynthetic Clay Liners*, pp. 208–226.

Daniel, D. E., R. Koerner, R. Bonaparte, R. Landerth, D. Carson, and H. Scranton (1998). Slop stability of geosynthetic clay liner test plots. *Journal of Geotechnical and Geoenvironmental Engineering* 124(7): 628–637.

Das, B. M. (2009). *Principles of Geotechnical Engineering, - SI Version*. Stanford, CT: Stanford, CT: Category: Cengage Learning Publisher, p. 544.

Datta, M. (2010). *Factors Affecting Slope Stability of Landfill Covers*. Berlin: Springer. https://doi.org/10.1007/978-3-642-04460-1_65.

Daudet, A. (2013). *Restauration de site minier Abandonné: cas du site Aldermac*. Rapport de stage URSTM-UQAT.

Dickinson, S., and R. W. I. Brachman (2008). Assessment of alternative protection layers for a geomembrane – geosynthetic clay liner (GM–GCL) composite liner. *Canadian Geotechnical Journal*, 45(11): 1594–1611.

Dwyer, S. F. (2003). *Water balance measurements and computer simulations of landfill covers*. PhD Diss University of New Mexico Albuquerque, p. 250.

Egloffstein, T. A. (2001). Natural bentonites influence of the ion exchange and partial desiccation on permeability and self-healing capacity of bentonites used in GCLs. *Geotextiles and Geomembranes* 19(7): 427–444.

Eigenbrod, K. D. (2003). Self-healing in fractured fine-grained soils. *Canadian Geotechnical Journal*, 40(2): 435–449.

Epacher, E., J. Tolyéth, C. Krönke, and B. Pukansky (2000). Processing stability of high density polyethylene: Effect of adsorbed and dissolved oxygen. *Polymer* 41(23): 8401–8408.

Estomell, P., and D. E. Daniel (1992). Hydraulic conductivity of three geosynthetic clay liners. *Journal of Geotechnical Engineering* 18(10). https://doi.org/10.1061/(ASCE)0733-9410(1992)118:10(1592).

Ewais, A. M. R., and R. K. Rowe (2014). Effects of blown film process on initial properties of HPDE geomembranes of different thicknesses. *Geosynthetics International*, 21(1): 62–82.

Forget, B., A. L. Rollin, and T. Jacquelin (2005). Lessons learned from 10 years of leak detection surveys on geomembranes. *Proceeding of the tenth International Waste Management and Landfill Symposium*, Italy

Fox, J. P., and T. D. Stark (2004). State of the art: GCL shear strength and its measurement. *Geosynthetics International*, 11(3): 141–175.

Giroud, J. P., and N. Touze-Foltz (2005). Equations for calculating the rate of liquid flow through geomembrane defects of uniform width and finite or infinite length. *Geosynthetics International*, 12(4): 191–204.

Giroud, J. P., K. L. Soderman, M. V. Khire, and K. Badu-Tweneboah (1998). *New developments in landfill liner leakage evaluation*. In *Proceedings of Sixth International Conference on Geosynthetics*, IFAI, Atlanta, USA, Vol. 1, pp. 261–268.

Government of Western Australia, Department of Mines and Petroleum. 2014. Mining Rehabilitation Fund fact sheet. Department of Mines and Petroleum. http://www.dmp.wa.gov.au/Documents/Environment/ENV-MEB-381.pdf.

Gueddouda, M. K., I. Goual, and A. Demdoum (2016). Experimental tests on the Physical and hydro-mechanical properties of dune sand-Bentonite mixtures for use in engineered barriers. *Journal of Rock Mechanics and Geotechnical Engineering*, 8(4): 541–550.

Hofton, T., and R. Schwenger (2009). Geomembrane Cover on Normétal Reactive Tailings: A Case Study. http://bc-mlard.ca/files/presentations/2009-9-HOFTON-SCHWENGER-geomembrane-cover-normetal-reactive-tailings.pdf

Holtz, D. R., and D. W. Kovacs (1981). *An introduction to geotechnical engineering*. Upper Saddle River, NJ: Prentice-Hall.

Hsuan, Y. G., H. F. Schroeder, R. K. Rowe, W. Müler, J. Greenwood, D. Cazzuffi, and R. M. Koerner (2008). Long-term performance and lifetime prediction of geosynthetics. Keynote paper at *EuroGeo4, the Fourth European Geosynthetics Conference*, Edinburgh, UK, September 7–10.

Jo, H. Y., T. Katsumi, C. H. Benson, and T. B. Edil (2001). Hydraulic conductivity and swelling of nonprehydrated GCLs permeated with single-species salt solutions. *Journal of Geotechnical and Geoenvironmental Engineering*, 127(7): 557–567.

Khire, M. V., C. H. Benson, and P. J. Bosscher (1997). Water balance modeling of earthen landfill covers. *Journal of Geotechnical and Geoenvironmental Engineering*, 123(8): 744–754.

Kim, H., and C. Benson (1999). Oxygen transport through multilayer composite caps over mine waste. In *Proceedings of. Sudbury"99—Mining and the Environment II, Centre in Mining and Mining Environment Research, Laurentian University*, Sudbury, Ontario, ON, pp. 183–192.

Kodikara, J., S. L. Barbour, and D. G. Fredlund (2002). Structure development in surficial heavy clay soils: A synthesis of mechanisms. *Australian Geomechanics: Journal and News of the Australian Geomechanics Society*, 37(3): 25–40.

Koerner, R. M. (1994). *Designing with Geosynthetics*. 3rd edition. Prentice Hall; (1709), ASIN: B011MEWBHM.

Koerner, R. M. (2012). *Designing with Geosynthetics*. 6th edition. New York: Xlibris Publishing Co.

Konrad, J. M., and R. Ayad (1997). Desiccation of a sensitive clay: Field experimental observations. *Canadian Geotechnical Journal*, 34(6): 929–942.

Konrad, J. M., and M. Samson (2000). Hydraulic conductivity of kaolinite-silt mixtures subjected to close-system freezing and thaw consolidation. *Canadian Geotechnical Journal*, 37(4): 857–869.

Kraus, J. F., C. H. Benson, A. E. Erikson, and E. J. Chamberlain (1997). Freeze-thaw cycling and hydraulic conductivity of bentonitic barriers. *Journal of Geotechnical and Geological Engineering*, 123(3).

Landry, B., and D. Senay (2010). *Portrait et enjeux miniers de l'Estrie*. Commission régionale sur les ressources naturelles et le territoire- https://www.yumpu.com/fr/document/view/17208160/commission-regionale-sur-les-ressources-naturelles-et-le-territoire

Lee, J. M., C. D. Shackelford, C. H. Benson, and H. Y. Jo (2005). Correlating index properties and hydraulic conductivity of geosynthetic clay liners. *Journal of Geotechnical and Geoenvironmental Engineering*, 131(11): 1319–1329.

Maqsoud, A., M. Mbonimpa, B. Bussière, and M. Benzaazoua (2015). The hydrochemical behaviour of the Aldermac abandoned mine site after its rehabilitation. *Proceeding of the Canadian Geotechnical Conference*, Québec.

Marcoen, J.-M., J. Thorez, A. Manjoie, and C. Schroeder (2000). Manuel relatif aux matières naturelles pour barrières argileuses ouvragées pour C.E.T. http://hdl.handle.net/2268/76216.

Maurice, R. (2002). Restauration du site minier Poirier (Joutel) – expériences acquises et suivi des travaux. *Défis & Perspectives: Symposium 2002 sur l'Environnement et les Mines*, Rouyn-Noranda, 3-5 novembre 2002, Comptes-rendus sur CD-ROM. Développement Économique Canada/Ministère des Ressources Naturelles du Québec/CIM, papier s32 a1021 p545.

Maurice, R. (2012). Normétal mine tailings storage facility HDPE cover: Design considerations and performance monitoring. In *Proceedings of 9th International Conference on Acid Rock Drainage*, Ottawa, ON

Meiers, G. P., S. L. Barbour, C. Qualizza, and B. S. Dobchuk (2011). Evolution of hydraulic conductivity of reclamation covers over sodic/saline mine overburden. *Journal of Geotechnical and Geoenvironmental Engineering*, 137(10): 9. doi: 10.1061/(ASCE)GT.1943-5606.0000523.

Melchior, S. (1997). In-situ studies on the performance of landfill caps. In *Proceedings of the International. Containment Technology Conference*, St. Petersburg, FL, pp. 365–373. https://www.osti.gov/servlets/purl/576553.

MEND (2004). Mine environment neutral drainage: Design construction and performance monitoring of cover systems for waste rock and tailings. Report 2.21.4.

MERN (2017). *Plan de réaménagement et de restauration des sites miniers au Québec*. Gouvernement du Québec Ministère de l'Énergie et des Ressources naturelles, M08-03-1710, ISBN: .

Mesri, G., and R. E. Olson (1971). Mechanisms controlling the permeability of clays. *Clay and Clay Minerals*, 19(3): 151–158.

Michaud, L. H., and D. Bjork (1995). The feasibility of constructing solid waste landfills as a reclamation method for abandoned mine lands. Paper presented at *Sudbury '`95: Mining and the Environment*, Sudbury, ON, May 23–June 1, 1995, (Canada), http://pdf.library.laurentian.ca/medb/conf/Sudbury95/MiningSociety/MS1.PDF.

Mitchell, J. K. (1976). *Fundamentals of Soil Behaviour*. Toronto: John Wiley & Sons.

Mitchell, J. K., and K. Soga (2005). *Fundamentals of soil behavior*. 3rd ed. Hoboken, NJ: Wiley.

Mitchell, J., D. Hooper, and R. Campanella (1965). Permeability of compacted clay. *Journal of Soil Mechanics and Foundation Division, ASCE*, 91(4): 41–65.

Montgomery, R., and L. Parsons (1990). The Omega hills cover test plot study: Fourth year data summary. In *Proceedings of the 22nd Mid-Atlantic Industrial Waste Conference*. Lancaster, PA: Technomic Publishing Company.

Morris, P.-H., J. Graham, and D.-J. Williams (1992). Cracking in drying soils. *Canadian Geotechnical Journal*, 29(2): 263–277.

Müller, W. (2007). *HDPE geomembranes in geotechnics*. Berlin: Springer-Verlag.

OSMRE - Office of Surface Mining Reclamation and Enforcement (OSMRE) (2016) *Reclamation performance bonds*. https://www.osmre.gov/resources/bonds.shtm.

Othman, M., C. Benson, C. Chamberlain, and T. Zimmie (1994) Laboratory testing to evaluate changes in hydraulic conductivity caused by freeze-thaw: State-of-the-art. In *Hydraulic conductivity and waste containment transport in soils*, ed. D. Daniel and S. Trautwein, 227–254. West Conshohocken, PA: ASTM International. STP 1142, ASTM, S. Trautwein and D. Daniel, Eds.

Patterson, B. M., B. S. Robertson, R. J. Woodbury, B. Talbot, and G. Davis (2006). Long-term evaluation of a composite cover overlaying a sulfidic tailings facility. *Mine Water and the Environment*, 25: 137–145.

Petrov, R. J., and R. K. Rowe (1997). Geosynthetic clay liner (GCL) – Chemical compatibility by hydraulic conductivity testing and factors impacting its performance. *Canadian Geotechnical Journal*, 34(6): 863–885.

Petrov, R. J., R. K. Rowe, and R. M. Quigley (1997). Selected factors influencing GCL hydraulic conductivity. *Journal of Geotechnical and Geoenvironmental Engineering*, 123(8): 683–695.

Podgorney, R. K., and J. E. Bennett (2006). Evaluating the long-term performance of geosynthetic clay liners exposed to freeze-thaw. *Geotechnical and Geo Environmental Engineering*, 132(2). https://doi.org/10.1061/(ASCE)1090-0241(2006)132:2(265).

Power, C., M. Ramasamy, D. MacAskill, J. Shea, J. MacPhee, D. Mayich, F. Baechler and M. Mkandawire (2017). Five-year performance monitoring of a high-density polyethylene (HDPE) cover system at a reclaimed mine waste rock pile in the Sydney Coalfield (Nova Scotia, Canada). *Environmental Science and Pollution Research*, 24(34): 1–19.

Rad, N. S., B. D. Jacobson, and R. C. Bachus (1994). Compatibility of geosynthetic clay liners with organic and inorganic permeants. In *Proceedings of the 5th International Conference on Geotextiles, Geomembranes and Related Products*, Singapore, pp. 1165–1168.

Rarison, F., M. Mbonimpa, B. Bussière, and A. Maqsoud (2020). *Properties of an HPDE geomembrane used for mine site reclamation: preliminary results 13 years after installation. Proceeding of the 7th European Geosynthetic Conference (EuroGeo7).*

Rarison, F. (2017). Utilisation des Géoembranes dans les recouvrements miniers. Seminaire Géomatériaux, Unpublished report, Polytechnique Montréal.

Renken, K., E. K. Yanful, and D. M. Mhina (2005). Field performance evaluation of soil-based cover systems to mitigate ARD for the closure of a potentially acid generating tailings storage facility. *Proceeding of the British Columbia Symposium.*

Ricard, J. F., M. Aubertin, F. W. Firlotte, R. Knapp, and J. McMullen (1997). Design and construction of a dry cover made of tailings for the closure of Les Terrains Aurifères site, Malalrtic, Québec, Canada. In *Proceedings of the 4th International Conference on Acid Rock Drainage*, Vancouver, 4: 1515–1530.

Rollin, A., S. Lambert, and P. Pierson (2002). *Geomembranes: G guide de choix sous l'angle des matériaux.* Montreal: Presses Internationales Polytechnique.

Rosli, R. N., M. R. Selamat, and H. Ramli (2020). The permeability and strength of compacted laterite soil-bentonite mixtures for landfill cover application. In: Mohamed Nazri F. (eds) *Proceedings of AICCE'19: Transforming the nation for a sustainable tomorrow*, AICCE 2019. Lecture notes in civil engineering, vol. 53. Cham: Springer.

Rowe, R. K. (2020). Geosynthetic clay liners: Perceptions and misconceptions. *Geotextiles and Geomembranes*, 48(2): 137–156. https://doi.org/10.1016/j.geotexmem.2019.11.012.

Rowe, R. K., and F. B. Abdelaal (2016). Antioxidant depletion in high-density polyethylene (HDPE) geomembrane with hindered amine light stabilizers (HALS) in low-pH heap leach environment. *Canadian Geotechnical Journal*, 53(10): 1612–1627.

Rowe, R. K., and J. R. Booker (1998). Modelling impacts due to multiple landfill cells and clogging of leachate collection systems. *Canadian Geotechnical Journal*, 35: 1-14.

Rowe, R. K., M. J. Chappel, R. W. Brachman, and W. A. Take (2012). Field monitoring of geomembrane winkles at a composite liner test site. *Canadian Geotechnical Journal*, 49(10): 1196–1211.

Rowe, R. K., R. M. Quigley, R. W. Brachman, and J. R. Booker (2004). *Barrier systems for waste disposal facilities.* Boca Raton, FL: Spon CRC Press.

Rowe, R. K., M. Z. Islam, and Y. G. Hsuan (2008). Leachate chemical composition effects on OIT depletion in HDPE geomembranes. *Geosynthetics International*, 15(2): 136–151.

Ruhl, J. L., and D. E. Daniel (1997). Geosynthetic clay liners permeated with chemical solutions and leachates. *Journal of Geotechnical and Geoenvironmental Engineering*, 123(4): 369–381.

Saleh-Mbemba, F., M. Aubertin, M. Mbonimpa, and L. Li (2016) Experimental characterization of the shrinkage and water retention behaviour of tailings from hard rock mines. *Geotechnical Geological Engineering*, 34(1): 251–266.

Sangam, H. P., and R. K. Rowe (2001). Migration of dilute aqueous organic pollutants through HDPE geomembranes. *Geotextiles and Geomembranes*, 19(6): 329–357.

Scheirs, J. (2000). *Compositional and failure analysis of polymers: A practical approach.* New York, USA: John Wiley & Sons.

Scheirs, J. (2009). *A guide to polymeric geomembranes: A practical approach.* Chichester, UK: John Wiley & Sons.

Seed, R. B., J. K. Mitchell and H. B. Seed (1990). Kettleman Hills waste landfill slope failure. II: Stability analysis. *Journal of Geotechnical Engineering, ASCE,* Vol. 116, (4): 669–689.

Shackelford, C. D. (1994). Waste-soil interactions that alter hydraulic conductivity. In *Hydraulic conductivity and waste contaminant transport in soil,* ed. D. Daniel and S. Trautwein, 111–168. West Conshohocken, PA: ASTM International. https://doi.org/10.1520/STP23887S.

Shackelford, C. D., C. H. Benson, T. Katsumi, B.T. Edil, and L. Lin (2000). Evaluating the hydraulic conductivity of GCLs permeated with non-standard liquids. *Geotextiles and Geomembranes,* 18(2–4): 133–161.

Shackelford, C. D., G. W. Sevick, and G. R. Eykholt (2010). Hydraulic conductivity of geosynthetic clay liners to tailings impoundment solutions. *Geotextiles and Geomembranes* 28(2): 149–162.

Shackelford, J. F. (2014). *Introduction to materials science for engineers.* 8th edition. Pearson: Prentice Hall.

Shan, H.-Y., & Y.L. Lai (2002). Effect of hydrating liquid on the hydraulic properties of geosynthetic clay liners. *Geotextiles and Geomembranes,* 20(1): 19–38. http://www.sciencedirect.com/science/article/pii/S026611440 1000231

Shan, H.-Y., and D. E. Daniel (1991). Results of laboratory tests on a geotextile/bentonite liner material. In *Proceedings, Geosynthetics "91,* vol. 2, 517–535. St. Paul, MN: Industrial Fabrics Association.

Silvestre, P., V. Norotte, and O. Oberti (2003). Les géosynthétiques en couverture. *Rencontres géosynthétiques francophones,* 97–113.

Southen, J. M., and K. Rowe (2007). Evaluation of the water retention curve for geosynthetic clay liners. *Geotextiles and Geomembranes,* 25(1): 2–9.

Stark, T. D., T. R. Boerman, and J. Connor (2008). Puncture resistance of PVC geomembranes using the truncated cone test. *Geosynthetic International,* 15(6): 480–486.

Terzaghi, K. (1925). *Erdbaumechanik auf Bodenphysikalisher Grundlage.* Venna, Austria: Deuticke.

Thiel, R., and G. N. Richardson (2005). Concern for GCL shrinkage when installed on slopes. In: *Geosynthetics Research and Development in Progress.* Austin, United States, pp. 1–7. https://doi.org/10.1061/40782(161)64

Touze-Foltz, N. (2015). Quantification of flow rates through virgin and exposed geomembranes and multicomponent geosynthetic clay liners. In *The 2nd International GSI-Asia Geosynthetics Conference (GSI-Asia 2015),* Seéoul, South Korea, 4.

US Department (2018). Design Standards No. 13 Embankment Dams Chapter 20: Geomembranes Phase 4 (Final). U.S. Department of the Interior Bureau of Reclamation. https://www.usbr.gov/tsc/techreferences/designstandards-datacollectionguides/finalds-pdfs/DS13-20.pdf

Wei, X., M. Hattab, J.M. Fleureau, R. Hue (2013). Macro-micro experimental study of two clayey materials on drying paths. *Bulletin of Engineering Geology and the Environment,* 72: 495–508.

WSP (2018). *Suivi environnemental de l'ancien site minier New Calumet,* Québec. MERN report.

Xue, Q., Q. Zhang, and L.Q. Liu (2012). Impact of high concentration solutions on hydraulic properties of geosynthetic clay liner materials. *Materials,* 5(11): 2326–2341.

Yesiller, N., J. L. Hanson, L. Risken, C. H. Benson, T. Abichou, B. Jenner, and J. Darius (2019). Hydration fluid and field exposure effects on moisture-suction response of geosynthetic clay liners. *Journal of Geotechnical and Geoenvironmental Engineering,* 145(4): 04019010.

Yesiller, N., J. Risken, J. Hanson, and J. Darius (2014). Effects of hydration fluid on moisture-suction relationships for geosynthetic clay liners. In *Unsaturated Soils: Research & Applications,* pp. 1023–1029.

Zhan, G. S., A. Mayer, J. McMullen, and M. Aubertin (2000). Capillary cover design for a spent leach pad. In *Proceeding of the International Symposium on Hydrogeology and the Environment,* Wuhan, China, ed. Y.X. Wang, 144–150. Beijing: China Environmental Science Press.

Zhan, G., M. Aubertin, A. Mayer, K. Burke, and J. McMmullen (2001). Capillary cover design for leach pad closure. *SME Transaction,* 1: 104–110.

5 Store-and-Release Covers

Bruno Bussière and G. Ward Wilson

5.1 INTRODUCTION: BACKGROUND AND HISTORICAL PERSPECTIVES

One way to limit acid mine drainage (AMD) generation is to control the infiltration of water by constructing cover systems on mining waste disposal sites. Various materials and configurations are used to design and construct covers (e.g., Koerner and Daniel 1997). In this regard, an interesting solution for sites located in arid and semi-arid climates is to make use of capillary barrier (CB) effects (see Chapter 3 for more details on this phenomenon) and evapotranspiration (the combined process of evaporation and plant transpiration) to control water infiltration at the surface of waste disposal sites (Benson et al. 2001; Albright et al. 2010). In the literature, studies on such types of covers, referred to herein as store-and-release (SR) covers, are mainly dedicated to the comparison between the performance of more traditional covers (with low saturated hydraulic conductivity) and SR covers in intermediate-scale field experimental setups (e.g., Nyhan et al. 1990, 1997; Benson et al. 1994; Hakonson et al. 1994; Melchior 1997; Rock et al. 2012; Apiwantragoon et al. 2015). In some cases, results from these *in situ* experiments are compared with numerical simulations (Khire et al. 1999; Scanlon et al. 2005; Ogorzalek et al. 2008; Bohnhoff et al. 2009; Albright et al. 2010). The capacity of inclined CBs to divert water has also been investigated (Ross 1990; Steenhuis et al. 1991; Stormont and Morris 1998).

It is important to note that all studies cited above were performed mainly in the context of controlling water percolation through landfill covers. Fewer studies can be found on the use of SR covers to control AMD generation from mine waste disposal areas by reducing deep water percolation (see Section 5.5).

5.2 CONCEPTS AND MAIN FACTORS OF INFLUENCE FOR UNIDIMENSIONAL CONDITIONS (1D)

A typical SR cover is composed of a SR layer (usually a fine-grained soil) placed over a capillary break (CB) layer (usually a coarse-grained material; Figure 5.1). Monolithic SR covers (without a CB layer) can also be found in the literature but will not be discussed here due to their lower efficiency for a similar cover thickness (see Albright et al. 2010 for more details). When precipitation is applied to an SR cover using capillary barrier effects (initially dry), the water accumulates in the SR material until the pressure at the interface reaches the water entry value (ψ_w) of the coarse-grained material (Steenhuis et al. 1991; Morel-Seytoux 1992; see Figure 5.1). Usually, this critical condition is reached when the fine-grained soil is nearly saturated (Morel-Seytoux 1992). It is also important to mention that even if the critical suction at the interface is reached (ψ_w), the flux of water can be low for a significant period of time due to the low unsaturated hydraulic conductivity (k_u) of the coarse-grained material at ψ_w. More details on capillary barrier effects are provided in Chapter 3.

An efficient SR cover should have a maximum storage capacity (MSC) that is able to accept precipitation until the water is removed by evapotranspiration. Stormont and Morris (1998) suggested a simple approach based on the water retention curve (WRC) of the fine-grained soil, the thickness of the cover (*b*), and the suction at the interface with the coarse-grained soil to estimate MSC. This approach assumes that a hydrostatic equilibrium exists in the cover. The area below the WRC, above

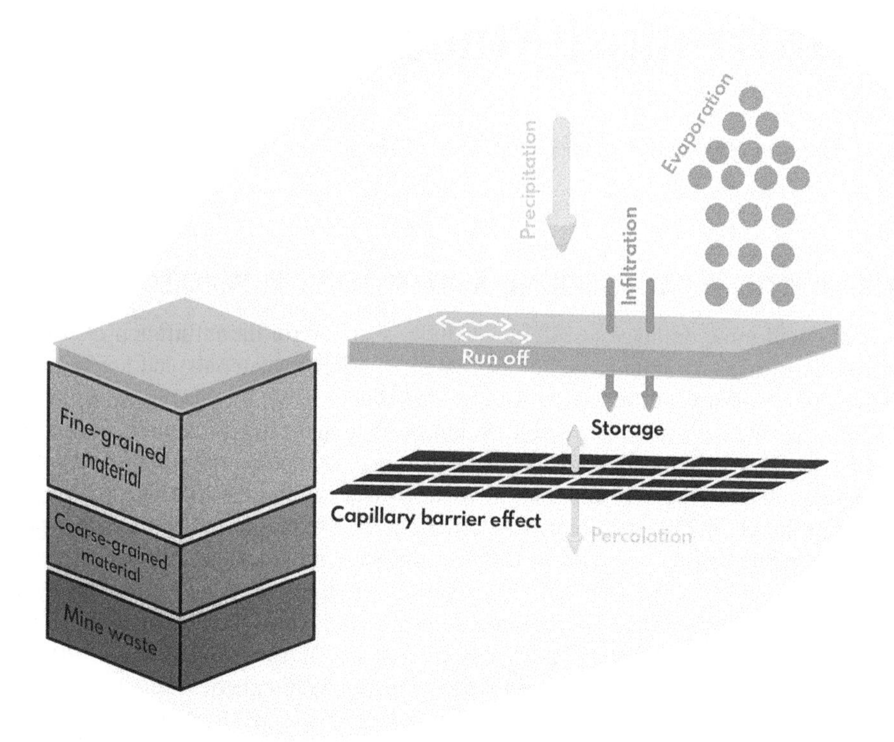

FIGURE 5.1 Schematic representation of the SR cover functioning.

the residual water content, and between the suction at the base and at the top of the fine-grained layer corresponds to the MSC. Equation 5.1 is the mathematical expression of the MSC parameter:

$$MSC = \int_0^b \theta\left(z + \psi_r\right)dz \tag{5.1}$$

Their study showed that Equation 5.1 could be used as a preliminary tool to assess the capacity of a cover material (or to compare different available materials) to be used in a given climatic context (Stormont and Morris 1998). However, the design of an SR cover cannot be limited to this simple method, and more sophisticated numerical modeling should be performed.

The main parameters that control the performance of SR covers are the materials' unsaturated properties (the water retention curve and the permeability function), the thickness of the layers, the climatic conditions (particularly, the extreme precipitation events), and the inclination of the cover (Khire et al. 2000; Aubertin et al. 2009). The impact of the first three parameters is presented in the following section, while the impact of the slope is presented in a separate section (Section 5.3). Other factors such as erosion, vegetation, biota intrusion, and long-term evolution of the materials' properties could also affect the performance of an SR cover; these aspects will be addressed in Section 5.7.

5.2.1 Influence of SR Cover Material Properties on Controlling Water Percolation

Many authors have investigated the influence of the unsaturated properties of the cover materials on the SR covers' capacity to store and release water and to control water percolation (e.g., Stormont and Morris 1998; Khire et al. 2000; Ogorzalek et al. 2008; Bohnhoff et al. 2009; Bossé et al. 2015b).

Using UNSAT-H software, Khire et al. (2000) performed a numerical parametric study on the hydrogeological behavior of SR covers made of different types of materials. This study is summarized below to assess the main factor of influence on a 1D SR cover. Four fine-grained materials (classified as clay CL, silty sand SM, between a silty sand and a non-plastic silt SM-ML, and a non-plastic silt ML by the USCS classification system) and two coarse-grained (poorly graded sand SP and poorly graded gravel GP) materials were tested as SR layers and CB layers, respectively (see Table 5.1 for the unsaturated properties). The stratigraphy of the cover simulated was 0.3 m of a fine-grained SR layer overlying a 0.75-m thick coarse-grained CB layer. Climatic conditions applied at the top were from different locations in the United States. In the following summary, the emphasis is on the Wenatchee climate (1992–1993; see Figure 5.2). Cheatgrass was assumed to be the vegetation on the covers. In this study, the potential evapotranspiration (upper limit on actual evapotranspiration) is estimated from the daily minimum and maximum air temperatures, net solar radiation, relative humidity, and daily wind speed using a modified form of Penman's equation as given by Doorenbos and Pruitt (1977). If precipitation during a time step exceeds the infiltration capacity of the SR material, the extra water is shed as runoff, which prevents ponding from occurring.

Results from Khire et al.'s (2000) simulations showed that the percolation through the SR cover decreases from 0.014 to 0.0015 m when the ML or SM-ML is used instead of SM for the topmost SR layer. Soil water storage is more stable (and lower) for the finer soils (SM-ML and ML) than for the SM material (see Figure 5.3), which can be explained mainly by the higher cumulative runoff for the SM-ML and ML soils (no runoff is predicted for the SM material). Evapotranspiration is also smaller for the finer soils (ML and CL) because less water is available. These results clearly show the importance of the unsaturated properties of the SR material on the performance of an SR cover in terms of limiting water infiltration. Ideally, the soil must retain water that could later be released

TABLE 5.1

Parameters for Water Retention Curves and Permeability Functions

USCS Classification	Particle Size	k_{sat} (m/s)	θ_{s1}	θ_{r2}	α_{v3} (1/m)	n_{v3}	Source
SM	Finer	2.7×10^{-6}	0.42	0.02	0.5	1.48	Khire et al. (1994)
SM-ML	Finer	9.0×10^{-8}	0.35	0.02	1.2	1.123	CEC (1997)
ML	Finer	3.2×10^{-8}	0.52	0.08	3.5	1.25	Khire et al. (1994)
CL	Finer	1.0×10^{-11}	0.38	0.22	0.124	1.34	Tinjum et al. (1997)
SP	Coarser	2.9×10^{-5}	0.40	0.01	3.8	2.69	Khire et al. (1994)
GP	Coarser	1.0×10^{-2}	0.30	0.01	57.4	2.44	CEC (1997)

Source: Khire et al. (2000).

[1] θ_s = volumetric water content at saturation; [2] θ_r = residual volumetric water content; α_v, n_v = van Genuchten (1980) equation parameters (see Equation 3.39).

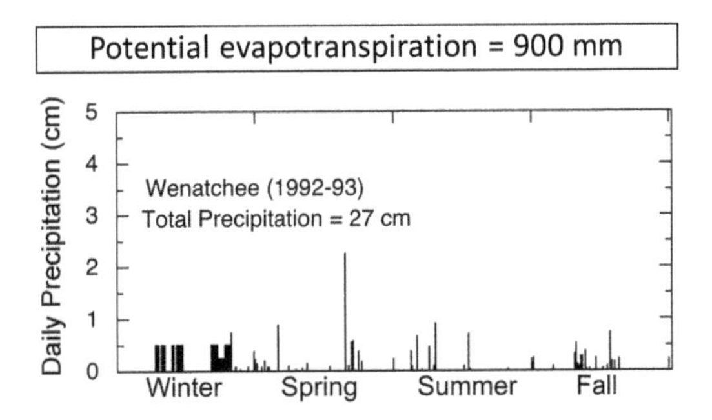

FIGURE 5.2 Wenatchee climatic data used by Khire et al. (2000).

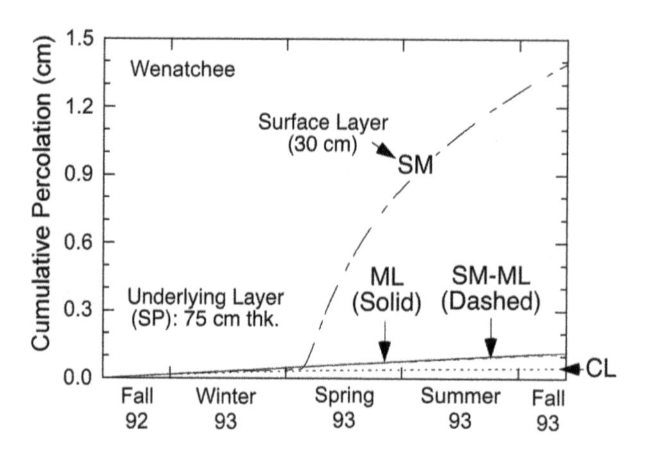

FIGURE 5.3 Influence of unsaturated SR cover material properties on water percolation through SR covers (Khire et al. 2000).

into the atmosphere and must also favor runoff. SM, SM-ML, and ML soils are usually preferred to clayey soils to constitute the SR layer. Clayey materials are usually susceptible to desiccation, which produces cracks and significantly increases water flow through the material (e.g., Benson and Othman 1993; Chapuis 2002; Eigenbrod 2003).

Khire et al. (2000) also investigated the impact of the unsaturated properties of the underlying coarse-grained material (the CB layer). They modeled two different materials (a sand (SP) and a gravel (GP)) for the 0.75-m CB layer placed below an SR layer made of 0.3 m of SM material exposed to Wenatchee climatic conditions (see Table 5.1 for the unsaturated properties of the two soils). Figure 5.4 depicts some of the results. The results show that the storage capacity is slightly higher when the GP soil is used. However, because of the higher unsaturated hydraulic conductivity, the cumulative percolation at the base is higher. The results also demonstrate that the hydraulic contrast between the soils is significant (when the CB effects are well developed), and the impact of the coarse-grained CB layer properties is not as significant as the impact of the fine-grained SR layer. Bossé et al. (2013, 2015a) obtained similar results concerning the storage behavior of SR covers in an in situ experiment in Morocco.

5.2.2 Influence of Climatic Conditions on the Performance of SR Covers

Khire et al. (2000) also numerically investigated the impact of climatic conditions on an SR cover made of a 0.3-m-thick fine-grained (moisture-retaining) layer (SM material in Table 5.1) placed over a coarse-grained layer (SP in Table 5.1) with a thickness of 0.75 m. Four different climatic conditions typical of wet years at Wenatchee, Reno, Phoenix, and Denver were tested (see Khire et al. 2000 for details).

Figure 5.5 shows that the percolation is not directly related to the annual amount of precipitation. For example, the Denver climate has the greatest annual precipitation (0.492 m), but the SR cover transmits the least percolation (0.001 m). Conversely, the Wenatchee SR cover received an intermediate amount of annual precipitation (0.272 m) but transmitted the most percolation (0.014 m). The differences in cumulative percolation between the four scenarios can be explained by examining soil water storage in the moisture-retaining layer. In three of the studied cases (Wenatchee, Phoenix, and Reno), precipitation is more frequent during fall and early winter, when the evapotranspiration is usually low due to lower solar radiation, lower air temperatures, and dormant vegetation. The water in this period accumulates in the moisture-retaining layer (the soil water storage increases), and occasionally, the soil water storage exceeds the total water storage capacity of the cover (estimated at about 0.11 m in this SR cover), which causes percolation through the bottom coarse-grained layer.

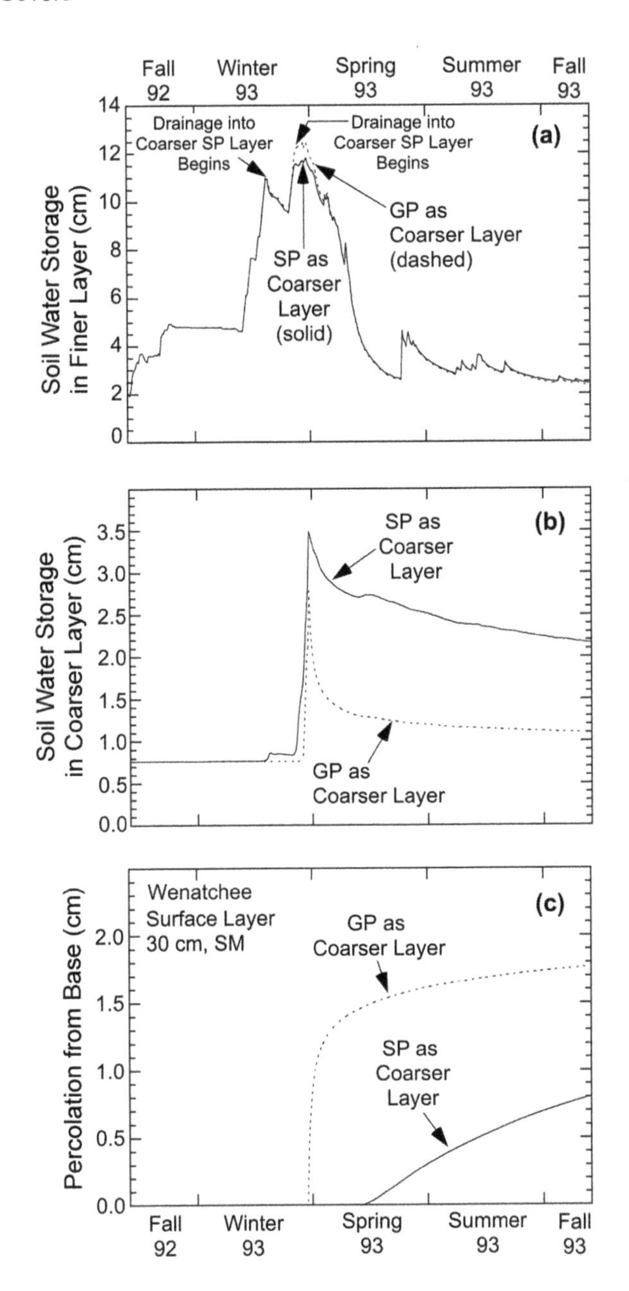

FIGURE 5.4 Impacts of coarse-grained material unsaturated properties on water percolation through SR covers (Khire et al. 2000).

In contrast, the Denver climatic conditions are characterized by frequent precipitation during periods where solar radiation and air temperature are relatively high and vegetation is active. Consequently, most of the water is rapidly evacuated by runoff and evapotranspiration, which limits percolation through the SR cover.

These results indicate that the critical meteorological conditions for an SR cover are site-dependent and are not only a function of the annual precipitation but also of their repartition in time and intensity. The most critical conditions are when precipitations are frequent and when evapotranspiration is low (more information on evapotranspiration can be found in Chapter 3). Snowfall that might accumulate on the cover and later melt can provoke critical conditions. Finally, extreme

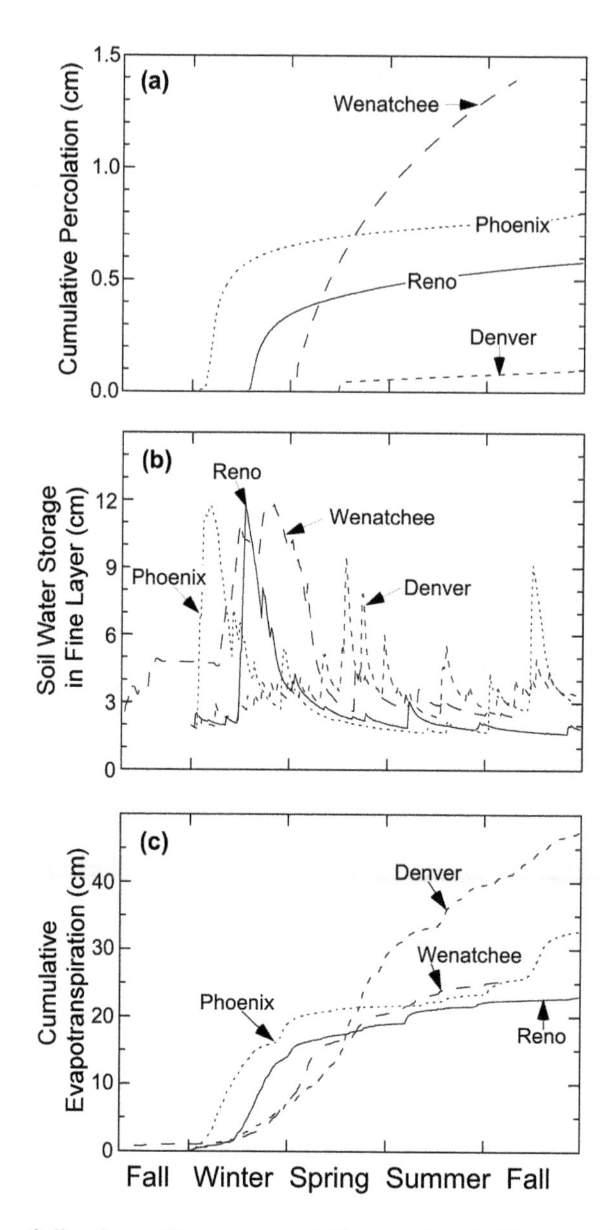

FIGURE 5.5 Impacts of climatic conditions on water infiltration through SR covers (Khire et al. 2000).

conditions that can happen in a short period of time must be identified, and the capacity of the cover to deal with these conditions must be evaluated.

5.2.3 Influence of SR Cover Materials' Properties on Releasing Capacity

In addition to water storage, the ideal SR material must rapidly release water to the atmosphere, allowing the system to regain its full capacity to manage the next precipitation event (without significant water accumulation remaining in the SR layer). Nevertheless, due to the difficulty in understanding the main mechanisms that control the release process by evaporation (or evapotranspiration), only a few studies have taken into account the natural release capacity of fine-grained materials to design SR cover systems. The emphasis in the design process is usually on the storage capacity (e.g., Stormont and Morris 1998; Khire et al. 2000; Ogorzalek et al. 2008; Bohnhoff et al. 2009).

TABLE 5.2
Hydrogeological Properties and van Genuchten (1980) Parameters for the SR Materials Used to Assess the Impact of Extreme Rainfall Events on the Releasing Capacity of SR Covers

Material	Symbol	θ_r	θ_s	α_v (kPa^{-1})	n_v	m_v^4	k_{sat} (m/s)
SR material (1, 2, 3)	SM-ML	0.01	0.43	0.122	1.43	0.301	$5 \times 10^{-7} - 5 \times 10^{-8}$
SR material (4)	CL	0.03	0.41	0.151	1.55	0.356	1×10^{-8}
Capillary Break	GP	0.01	0.38	0.73	4.94	0.797	1×10^{2}

Source: Modified from Bossé et al. (2015a).
m_v = van Genuchten (1980) equation parameters (see Equation 3.39).

Bossé et al. (2015b) simulated the transient unsaturated hydrogeological behavior (using the HYDRUS-1D code) of four SR cover systems made with different fine-grained materials as an SR layer (see Table 5.2 for the materials' properties) under natural and extreme rainfall conditions typical of a semi-arid climate. The four tested SR covers have a total height of 1.30 m and consist of 1 m of a fine-grained material as an SR layer overlying 0.3 m of a coarse-grained material used as a CB layer. The top boundary conditions applied to the model were consecutive extreme rainfall events for the arid conditions tested (155 mm of water is considered an extreme event close to Marrakech, Morocco; see Bossé 2014 for details) on the first day of the year over 10 years (see Figure 5.6a).

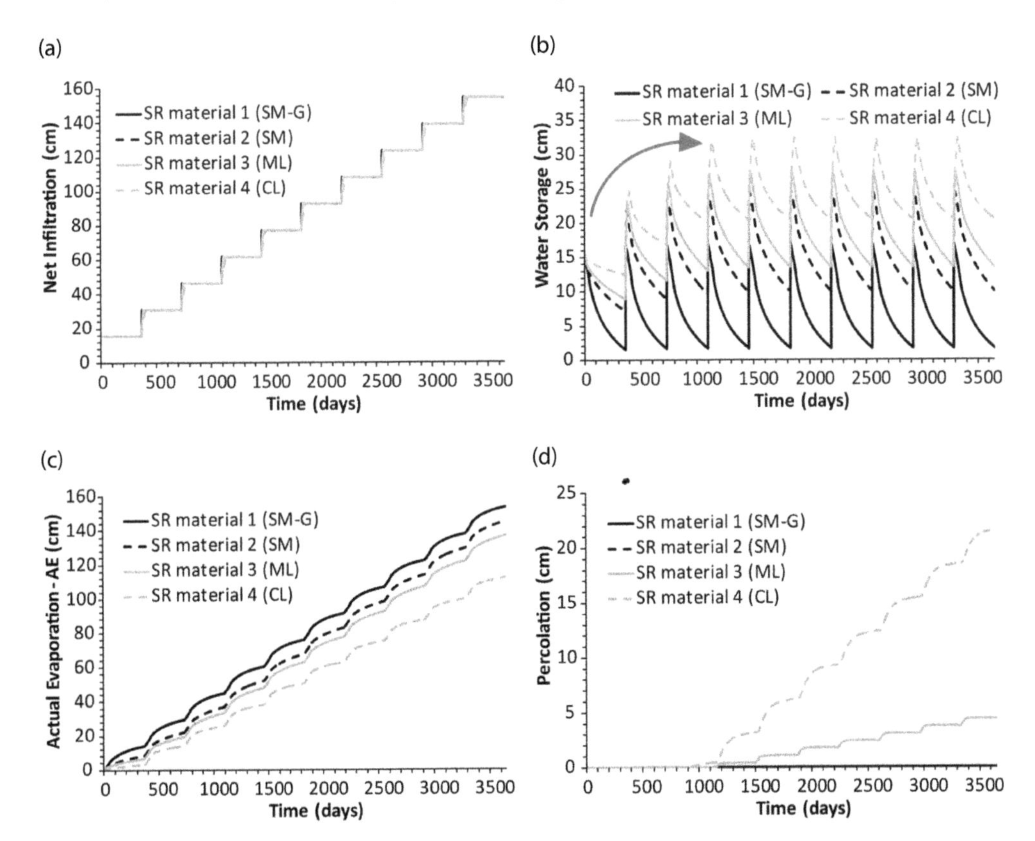

FIGURE 5.6 (a) Net infiltration, (b) water storage time trends during consecutive extreme rainfall events for the four SR materials, (c) actual evaporation, and (d) percolation time trends during consecutive extreme rainfall events for the four SR materials; the arrow represents the increased water storage during the first consecutive extreme rainfall events (Bossé et al. 2015b).

Climatic conditions representing a typical year for the same site in Morocco were applied for the next 364 days. The main objective of the study was to compare the ability of the different SR cover scenarios to regain their storage capacity after an extreme precipitation event that is repeated every year. More details of the numerical model are presented in Bossé et al. (2015b).

The numerical predictions indicated higher release rates for the coarser SR material (SM-G) (0.3 mm/d), whereas a finer SR material (CL) (0.03 mm/d) accumulated water above the interface between the SR and CB layers and did not prevent water percolation into the reactive mine wastes under consecutive extreme rainfall events. Coarser SR materials (i.e., silty sands SM and SM-G), which recover their full storage capacity more rapidly, proved more suitable to ensure the long-term performance of these systems under natural and extreme climatic conditions. Figure 5.6b shows that the short-term storage capacities of the SR materials SM, ML, and CL are increasingly affected during the first five years when the extreme event is simulated (red arrow). For the coarser SR material (SM-G), the water storage was only slightly affected. Indeed, after each simulation of the extreme rainfall event, the material recovered its full storage capacity, whereas for the SR materials CL, SM, and ML, water accumulation above the interface of the cover systems increased progressively. In other words, these three cover systems are not able to recover their entire storage capacity between two extreme events (when applied every year). Due to their lower release capacity (i.e., actual evaporation; see Figure 5.6c), the SR materials ML and CL accumulated water above the interface of the cover systems and generated water percolation approximately three years after the first simulation (Figure 5.6d). At the end of the period, the predicted water percolation for the SR materials ML and CL was 0.004 and 0.022 m, respectively. Note that although the SR material SM accumulated water at the interface of the cover, no percolation was predicted after 5 years.

5.3 CONCEPTS AND MAIN FACTORS OF INFLUENCE FOR INCLINED SR COVERS (2D)

The previous sections show the influence of factors such as unsaturated cover material properties and climatic conditions on the capacity of 1D SR covers in controlling water percolation into mine wastes. However, many SR covers are built on inclined surfaces that provoke a slope-induced effect on water flow, which engineers often overlook, although this effect can limit the capacity of the cover system to control water percolation (Ross 1990; Steenhuis et al. 1991; Stormont and Morris 1998; Aubertin et al. 2009).

5.3.1 Diversion Capacity of Inclined SR Covers

Among the main factors that affect water flow in an inclined SR cover is the angle of the slope. Indeed, moisture is not evenly distributed along the slope length; a higher water content is usually observed in the lower elevations of the cover system. When there is a significant inflow of water, the water accumulates above the interface until the local negative pressure reaches the water entry value (ψ_w, or water entry pressure) of the coarse material (Steenhuis et al. 1991). The ψ_w value corresponds, on the WRC, to the suction (ψ) at which water starts to penetrate the material on a wetting path (causing a rise of the water content). At this location (where the pressure at the interface is equal to the ψ_w of the coarse-grained soil), water infiltration into the coarse material becomes significant and the capillary barrier effect progressively disappears. The location of this "point" along the slope (which is actually a zone) is called the downdip limit or DDL point (Ross 1990). The amount of water flowing laterally up to the DDL point (see the breakthrough point in Figure 5.7) is called the diversion capacity (Q_{max}) of the capillary barrier (Ross 1990; Steenhuis et al. 1991; Stormont 1995). The horizontal distance between the top of the slope and the DDL point is referred to as the effective length of the capillary barrier (L_{eff}).

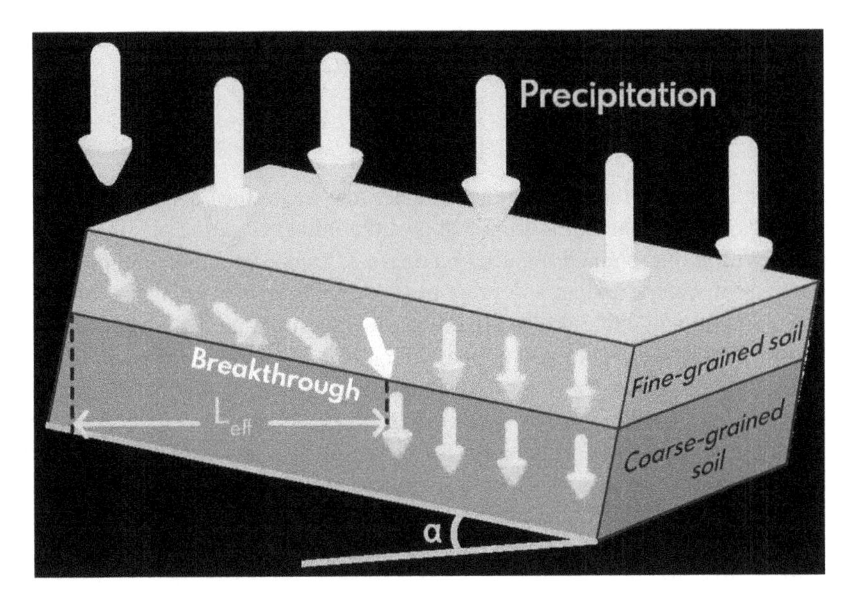

FIGURE 5.7 Schematic representation of a layered cover, showing water movement in an inclined SR cover following a precipitation event; the water infiltrates the coarse-grained soil when suction reaches its water entry value (adapted from Bussière 1999).

5.3.2 Parameters of Influence on the Diversion Capacity of Inclined SR Covers

The main factors that affect the capacity of an inclined SR cover to divert water are related to the cover material properties, slope angle, and climatic conditions (more specifically, precipitation rate). Analytical solutions can be used to characterize the capacity of an inclined cover to divert water; the most well-known equations are the Ross (1990) equation, the modified version of the Ross (1990) equation proposed by Steenhuis et al. (1991), and the Morel-Seytoux equation (1994). These solutions have typically been developed for steady-state conditions, with a pressure (or suction) that does not depend on the location along the slope (except for the Morel-Seytoux equation). Furthermore, they usually consider the coarse-grained material layer to be infinitely thick and the phreatic surface to be far from the interface. Some of these analytical solutions have been reviewed and compared (e.g., Bussière et al. 1998, 2007). Equation 2 presents the Ross equation modified by Steenhuis et al. (1991; indices 1 and 2 in the equations refer to the coarse and fine material, respectively):

$$Q_{sat2} tan\alpha \left[a_{s2}^{-1} \left(1 - \frac{P}{k_{sat2}} \right) + \psi_{a2} - \psi_{w1} \right]_{max} \tag{5.2}$$

where α is the slope angle; k_{sat} is the saturated hydraulic conductivity (m/s); a_s is the sorptive number related to the pore size distribution (Philip 1969); ψ_a is the air entry value (m of water); ψ_{w1} is the water entry pressure of the coarse material (m of water); and P is the vertical water flux (m/s). The value of L_{eff} corresponding to a given Q_{max} can be obtained mathematically with the following expression:

$$L_{eff} = \frac{Q_{max}}{P} \tag{5.3}$$

$$L_{eff} = \frac{Q_{max}}{P} = L_D \times \cos\alpha \tag{5.4}$$

To illustrate the influence of the different parameters, the L_{eff} values of inclined SR covers with different properties are calculated. Figure 5.8 shows the influence of the precipitation rate on L_{eff} for an SR cover with a fine-grained material having a k_{sat} value of 5.1×10^{-6} m/s, a ψ_a value of 7.5 kPa (or 0.75 m of water), and an a_s^{-1} value of 0.45; a CB layer with a ψ_{w1} value of 2 kPa (or 0.2 m of water); a slope angle of 3:1 (18.3°); and a precipitation rate that varies between 5×10^{-6} and 5×10^{-8} m/s. These values correspond to a wide range of precipitation rates from 26 to 2600 mm/month. Results showed that for $P > 5 \times 10^{-7}$ m/s, the L_{eff} value would be less than 3 m, but the value increases rapidly between 1×10^{-7} (≈20 m) and 1×10^{-8} m/s (≈170 m). Figure 5.9 demonstrates the influence of the slope for the same SR cover when P is fixed at 1×10^{-7} m/s. One can see that reducing the slope angle from 18° to 10° reduces the L_{eff} value by half. Finally, Figure 5.10 shows the influence of the fine-grained material k_{sat}. Based on the equation proposed by Steenhuis et al. (1991), this parameter is clearly the one that has the greatest influence on L_{eff}; increasing the k_{sat} value of the fine-grained SR material by one order of magnitude increases the L_{eff} value by the same factor. Hence, materials with too low of a k_{sat} value are not the best material for diverting water along the slope. However, materials with too high of a k_{sat} value might not create the necessary CB effects with the coarse-grained material below and also would not be efficient for diverting water.

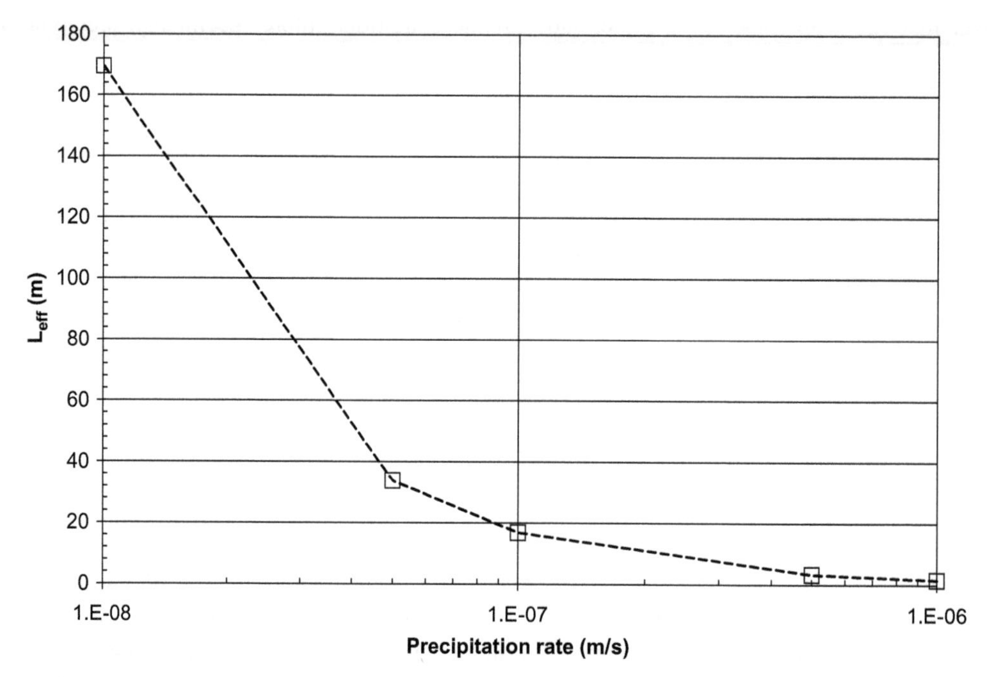

FIGURE 5.8 Influence of the precipitation rate on L_{eff} of an inclined (3:1) CB based on the Ross equation modified by Steenhuis et al. (1991); fine-grained SR layer properties: $k_{sat} = 5.1 \times 10^{-6}$ m/s, $\psi_a = 7.5$ kPa, and $a_s^{-1} = 0.45$; CB layer with $\psi_w = 2$ kPa.

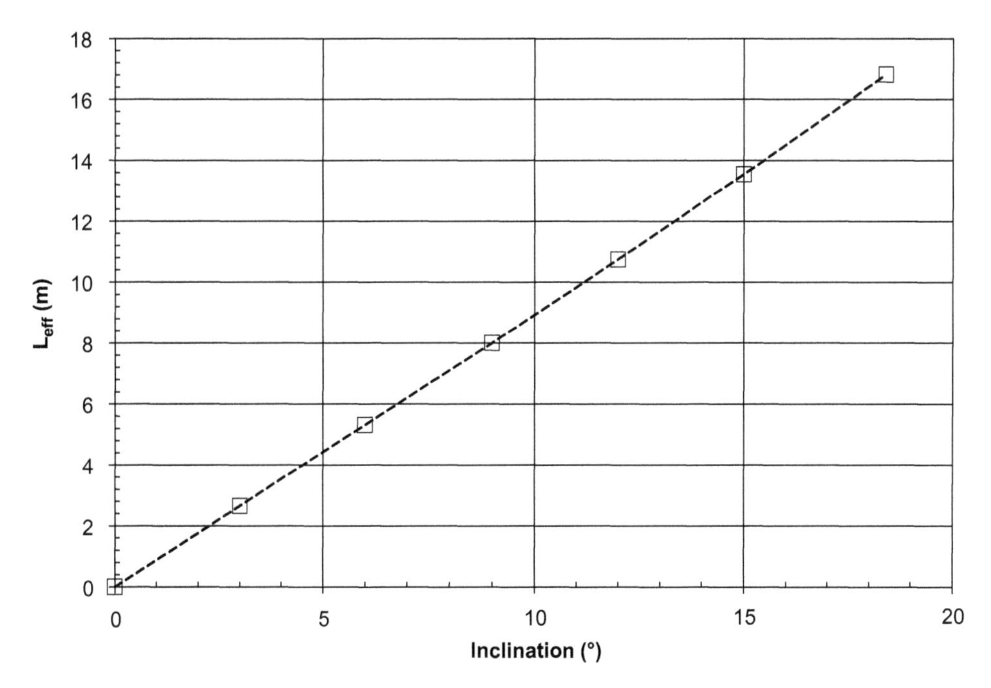

FIGURE 5.9 Influence of slope angle on L_{eff} of an inclined (3:1) CB exposed to a precipitation rate of 1×10^{-7} m/s based on the Ross equation modified by Steenhuis et al. (1991); fine-grained SR layer properties: $k_{sat} = 5.1 \times 10^{-6}$ m/s, $\psi_a = 7.5$ kPa, $a_s^{-1} = 0.45$; CB layer with $\psi_w = 2$ kPa.

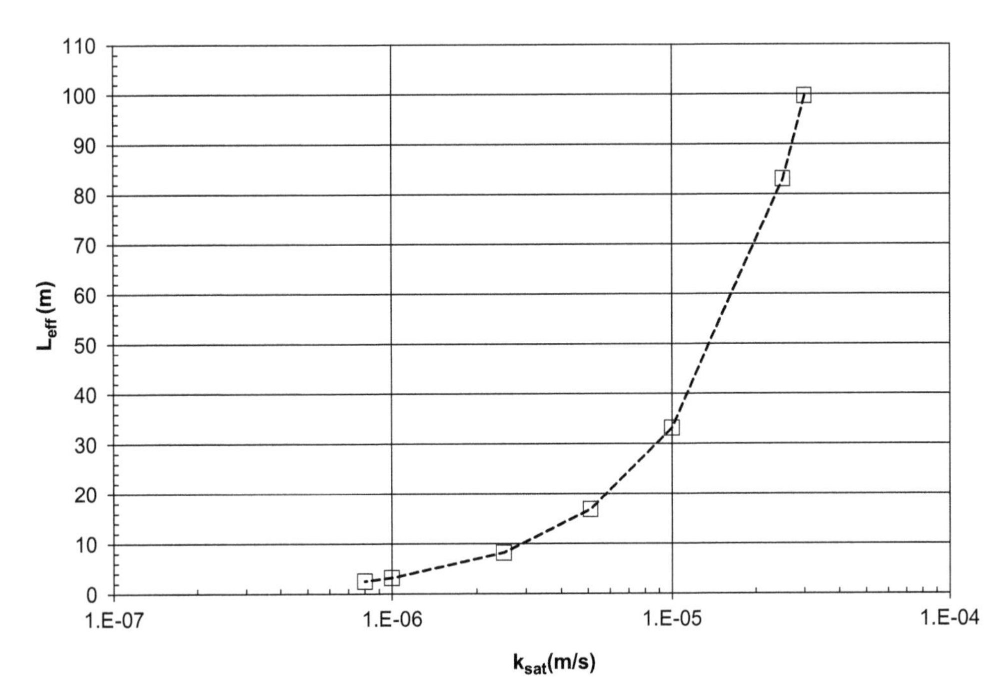

FIGURE 5.10 Influence of SR cover material saturated hydraulic conductivity on L_{eff} of an inclined (3:1) CB exposed to a precipitation rate of 1×10^{-7} m/s based on the Ross equation modified by Steenhuis et al. (1991); other SR cover material properties: $\psi_a = 7.5$ kPa, $a_s^{-1} = 0.45$; CB layer with $\psi_w = 2$ kPa.

As a final remark, it is worth mentioning that runoff was not considered in the above as a factor that could influence the overall water budget of the inclined SR cover. In a previous study on an inclined (14.5°) SR cover, Knidiri et al. (2017) showed that the proportion of runoff on the water budget was small (3–4%) during extreme precipitation conditions when a 10^{-4} m/s cover material was used in an arid climate. However, with a finer material, this proportion could be greater.

5.4 METHODOLOGY FOR THE DESIGN

The results presented in the previous section identified the most important parameters for designing an efficient SR cover to limit water infiltration. Ideally, the soil must retain water that could later be released into the atmosphere and must also favor runoff. Capillary barrier effects must be well developed between cover layers to maximize the performance. The fine-grained material must retain water during wet periods but must also be able to efficiently and rapidly release water during dry periods; the SR cover must be ready to rapidly absorb the next major precipitation event. Moreover, the soil must be resistant to desiccation cracking to avoid preferential water flow through the cover. Hence, SM, SM-ML, and ML soils are usually preferred to clayey soils to constitute the moisture-retaining layer because they have the qualities listed above. Finally, when an SR cover is built on an inclined surface, it is important to consider the accumulation of water along the interface and the possible disappearance of capillary barrier effects. The value of L_{eff} must be equal to or greater than the length of the slope for normal and extreme conditions.

Figure 5.11 presents a schematic diagram of how to design an SR cover on a mine waste disposal facility. Steps 1 and 2 involve collecting data related to the material properties and site conditions. The properties that must be evaluated for each material are grain size distribution, specific gravity, Atterberg limits (if applicable), compaction properties, saturated hydraulic conductivity (k_{sat}), and unsaturated properties (water retention curve and unsaturated hydraulic conductivity). At the design stage, it is also essential to collect typical climatic data for the site. The climatic information needed is temperature, wind speed and direction, solar radiation, air humidity, and amount of precipitation. These data can be used later to estimate the evapotranspiration rate using indirect approaches and as inputs in the numerical models. For long-term performance evaluation, a typical wet year must be identified. The short-term response of the cover following major storms is also important to evaluate (see Chapter 14). Water uptake from vegetation roots and transpiration are important components of the cover's water balance. The evapotranspiration estimated by numerical software is usually a function of different parameters such as root density and depth, percentage of bare area, growing season length, and the leaf area index (e.g., Fayer and Jones 1990; Khire et al. 1999, 2000; Dwyer 2003). These parameters must be estimated or measured before the numerical analysis (if there is vegetation on the cover; see Chapter 14).

The next step (Step 3) is to assess the cover's hydrogeological behavior using numerical modeling. The unsaturated soil properties obtained in the laboratory (for porosity values similar to those expected in the field) are integrated into the numerical models along with the critical climatic conditions (short- and long-term conditions) and the vegetation data. A parametric numerical study should be performed to optimize the layers' thicknesses. The variability of soil properties is assessed and the worst-case properties should be used at the design stage. Similar (independent) numerical analyses, but in 2D, must be performed when the cover system is inclined. It is crucial to distinctly analyze the flat and inclined portions of a site when designing an SR cover, given that its hydrogeological behavior can vary due to geometric effects.

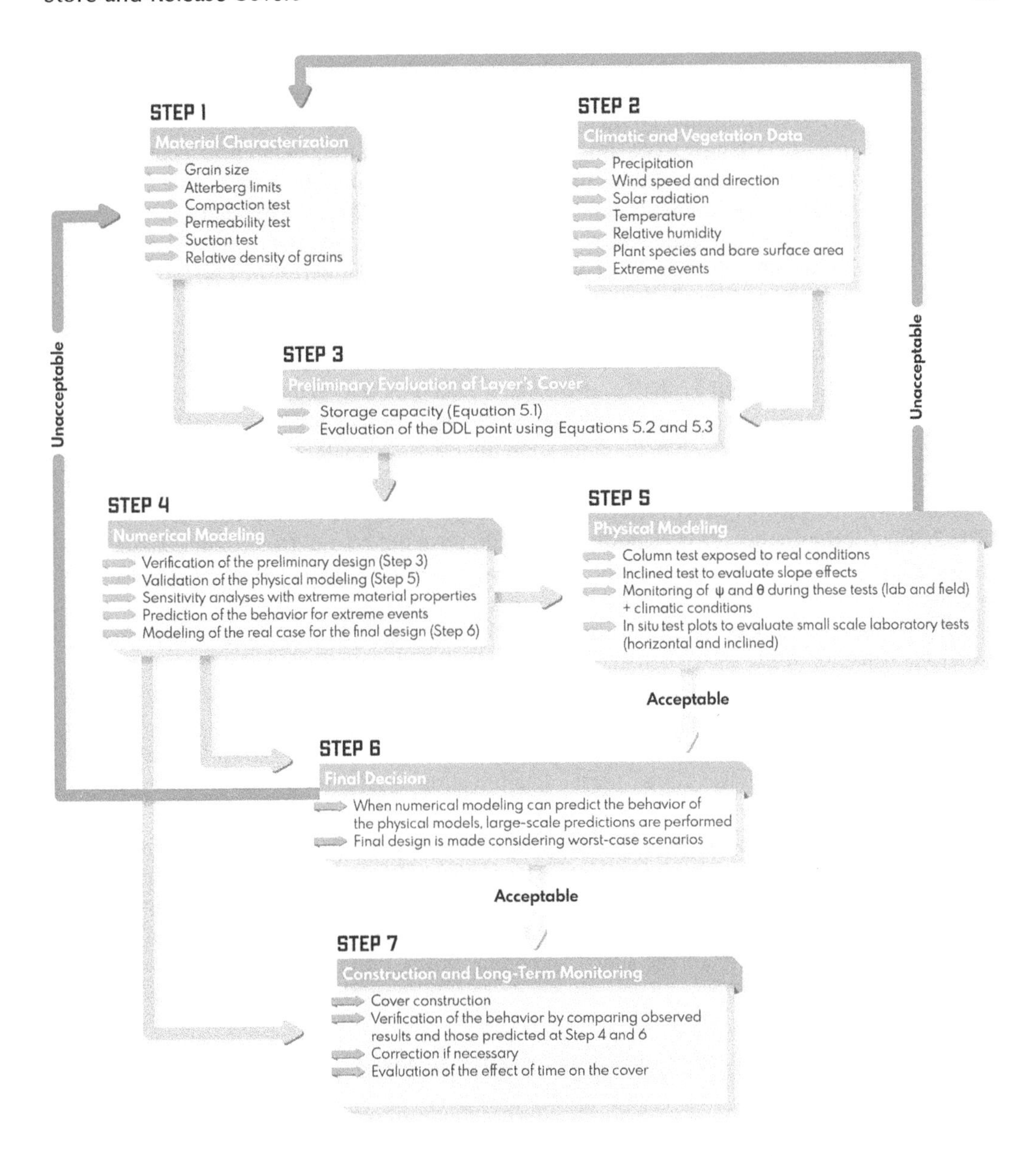

FIGURE 5.11 Schematic diagram representing the different steps to design an SR cover.

The results of the numerical analyses are used to identify the materials and configuration that will reach the targeted performance in terms of water infiltration. The maximum acceptable water percolation depends on the environmental objectives and regulations applicable in the country. Typically, an infiltration rate between 1 and 10 mm/year is considered acceptable for hazardous and solid waste landfill sites (Khire et al. 2000). For an inclined SR cover, an additional objective is to avoid the appearance of the DDL point in the sloping cover for extreme conditions.

Following the first series of numerical analyses, it is important to validate the numerical results using controlled conditions (Step 4). For cover systems installed on a flat surface, physical models such as instrumented columns can be used for this validation. Different variables can be studied in these columns, such as the amount of precipitation and moisture and pressure distributions in the layers. The columns can be placed in the laboratory or outside in order to expose the columns to real climatic conditions (Bossé et al. 2013). The data obtained from these experiments should correlate well with unsaturated flow modeling results. If not, a comprehensive analysis of the physical and numerical results must be performed to understand the discrepancies between the two. A laboratory physical model can also be used to evaluate the diversion length of an inclined SR cover system, with different slope angles and precipitation rates (e.g., Bussière 1999; Bussière et al. 2002; Kämpf et al. 2003; Tami et al. 2004). Different variables are usually studied in the layered system, such as the amount of precipitation and moisture, pressure distributions, and water flow at the base of the model. Physical modeling of SR covers can also be performed in the field at an intermediate scale. Indeed, experimental test plots can be constructed (on a flat or inclined surface) to compare various cover scenarios and to identify how the various cover systems would behave under real conditions (e.g., Zhan et al. 2001, 2014; Albright et al. 2004; Knidiri et al. 2017). Again, results from these experimental test plots can be used to validate the numerical tools and to refine the final configuration of the cover.

Using a numerical model that represents the real case, the final design steps (Steps 5 and 6) predict the behavior of the selected cover for typical and extreme conditions. When the design is acceptable, the cover can be constructed. After the construction stage, the cover must be instrumented and monitored, and a comparison between predicted and measured results should be performed. The numerical model used at the design stage is assessed, and the model is recalibrated if necessary. The monitoring also allows for the identification of any significant long-term changes in the efficiency of the cover to limit water percolation. If the performance of the cover is significantly affected by time, actions must be taken at the mine site. Among the factors that can affect the performance of an SR cover, biointrusion, freeze-thaw effects, geometrical effects on actual evaporation rate, desiccation cracking, and water and wind erosion are the most important (see Chapter 14). Since these factors can have an impact on the capacity of the cover to limit water percolation, a safety factor should be added to the final cover design. However, no clear rules are proposed in the literature.

5.5 CASE STUDIES OF SR COVERS TO RECLAIM MINE WASTES

Table 5.3 shows some mines sites where SR covers (with different configurations) were used or tested to reclaim acid-generating mine wastes; Albright et al. (2010) presented other case studies of SR covers built on landfills. In most cases, the thickness of the SR covers was less than 1.5 m, and the cover materials used were usually natural soils, except for the Ketarra study (phosphate mine waste was used as an SR cover material) and Richmond Hill (a bentonite–ore mixture was integrated into the design). In general, the efficiency of these covers in controlling water percolation was high (< 1% of *P*) when the ratio of potential evaporation/precipitation was high (e.g., Goldstrike, Kettara, and Kidston) but much less at a lower ratio (e.g., Richmond Hill). It is also worth mentioning that, in most cases, the design approach proposed in Figure 5.11 was not strictly followed.

TABLE 5.3
Mine Sites with SR Covers Located in Semi-Arid and Arid Climates

Mine Site	Location	Climate	Reclamation Work: Intermediate Field Scale	Reclamation Work: Full Scale	Type of Mine Waste	Sources
Bisbee	AZ - USA	P = 452 mm/yr PE > 1700 mm/yr	Cover of 0.6 m	-	Tailings	Milczarek et al. (2011b)
Golden Sunlight	Whitehall MT - USA	P = 243 mm/yr PE = 1048 mm/yr	Cover of 0.8 to 1.8 m	-	Tailings	Junqueira et al. (2006)
Goldstrike	Elko NV - USA	P = 260 mm/yr PE = 1270 mm/yr	Inclined cover of 0.6 m (+ CB)	Cover of 1.2 to 1.5 m (+ CB)	Waste rock	Zhan et al. (2001, 2006, 2014)
Richmond Hill	SD - USA	P = 730 mm PE = 900 mm	-	Inclined cover of 1.65 m	Waste rock	Zhan et al. (2019)
Kettara	Morocco	P = 150 mm PE = 1980 mm	6 1D experimental cells with cover of 0.5–1 m (+ CB) 1 inclined experimental cell of 1m (+ CB)		Tailings	Bossé et al. (2015b) Knidiri et al. (2017)
Kidston	QLD - Australia	P = 700 mm/yr PE = 2800 mm/yr	-	Cover of 1.5 m	Waste rock	Williams et al. (2006)
Monticello	UT - USA	P = 390 mm/yr PE = 1000 mm/yr	-	Cover of 1.6 m (+ CB)	Tailings	Waugh et al. (2008)
Morenci	AZ - USA	P = 328 mm/yr PE = 2300 mm/yr	Cover of 0.3 to 0.6 m	-	Tailings	Milczarek et al. (2003, 2011a, 2011b)
Peak Gold	Cobar NSW-Australia	P = 415 mm/yr PE = 2550 mm/yr	Cover of 1.5 to 2 m	-	Tailings	Meiers et al. (2005)
Phoenix	Battle Mountain, NV - USA	P = 207 mm/yr PE = 1787 mm/yr	-	Cover of 2 m	Waste rock	Keller et al. (2011)
Questa	NM - USA	P = 310 mm/yr PE = 1715 mm/yr	Cover of 0.2 to 0.6 m	-	Tailings	Wels et al. (2001, 2002)
Sullivan	Kimberley BC - Canada	P = 402 mm/yr PE = 700 mm/yr	Cover of 0.45 m (+ CB)	Cover of 1 m (+ CB)	Tailings	Gardiner et al. (1997) O'Kane et al. (1999)
Mt. Whaleback	Newman WA - Australia	P = 300 mm/yr PE = 3000 mm/yr	Cover of 2 to 4 m (1D and 2D)	-	Tailings	O'Kane et al. (2000), O'Kane and Waters (2003) Wilson et al. (2003)

P = precipitation, *PE* = potential evaporation
Source: Modified from Bossé (2014).

5.6 ADVANTAGES AND LIMITS

Given their relative simplicity, long-term stability, and potentially lower construction costs (Morris and Stormont 1997; Ward and Gee 1997; Scanlon et al. 2005), SR covers are considered an advantageous alternative to more traditional covers that rely on materials with low saturated hydraulic conductivity (see Chapter 4). Using numerical modeling, Morris and Stormont (1997) showed that SR covers can be as efficient as traditional low saturated hydraulic conductivity covers (called Subtitle D covers) for a variety of climates. Because they are made of geological materials with low plasticity, SR covers are more durable than covers made of low saturated hydraulic conductivity materials that can be affected by wet–dry cycles. Indeed, fine-grained materials with high plasticity tend to crack when exposed to high suction conditions (see Chapter 14), and geomembranes degrade with time, which can significantly affect the long-term performance or integrity (i.e., permeability) of low saturated hydraulic conductivity cover systems (e.g., Albrecht and Benson 2001; Benson et al. 2002; Aubertin et al. 2016). Recent research showed that mining materials could be reused or valorized as components of SR covers; tailings have adequate properties to be used as material for the SR layer, while waste rock has good potential for use as a material for the CB layer (Bossé et al. 2015b; Knidiri et al. 2017).

In addition to climatic conditions, the main limitation in the application of SR covers is the availability of materials with appropriate hydrogeological properties. As previously shown, the performance of SR covers can change drastically with a change of these properties. Furthermore, the best materials for an efficient SR cover on a given site can be different for flat or inclined surfaces, which increases the difficulty in finding one unique material for the cover. Finally, the cover's ability to undergo extreme precipitation conditions must be carefully identified, and the eventual evolution of such extreme conditions due to climate change should be integrated into the design. However, there are no clear guidelines at the moment on how to do so.

5.7 RESEARCH NEEDS

SR covers are an attractive option to reclaim acid-generating mine waste disposal areas. They can be built at relatively low costs and can efficiently control deep water percolation and AMD generation. With better waste management at a mine, it is possible to build the cover partly or entirely with mining materials such as clean tailings or waste rock. For a more efficient application of this technology, further research is needed on the following.

- Recently, attention has been dedicated to the role of vegetation in SR cover performance (Albright et al. 2010; Apiwantragoon et al. 2015; Schnabel et al. 2012a, 2012b, 2015). Plants can return water to the atmosphere and intercept rainfall before it hits the soil. They can also stabilize the soils and reduce erosion. On the other hand, runoff can be reduced by vegetation, and plant roots can modify the properties of the soil cover in the long term (see Chapter 14). Arnold et al. (2015) showed that in arid and semi-arid climates, the vegetation impact on water balance is low. However, the interaction between plants and SR covers and their influence on soil water storage, evapotranspiration, runoff, and erosion, should be more thoroughly investigated, given that only few studies have looked at this cover component (Arnold et al. 2015).
- Evapotranspiration is a key parameter to control water percolation. However, this parameter is usually deduced from the other components of the water balance and not directly measured. There is a need for direct measurements of evapotranspiration on existing SR covers.
- Most numerical studies used single WRCs for their analysis of the hydrogeological behavior of SR covers. However, these covers are subjected to many cycles of precipitation and drainage during a typical year, which means that hysteresis effects could be present in the fine-grained material (Bossé et al. 2016). Depending on the amplitude of the phenomenon, it

might be appropriate to use numerical codes that address hysteresis effects to better predict the performance of SR covers.

- An SR cover must be able to handle the most critical climatic conditions. Considering climate change, it would be important to define future extreme conditions and not only the design based on actual extreme conditions (see Chapter 14).
- Runoff could contribute to water diversion via the inclined SR cover. More work is thus necessary to assess this component of the water budget and to identify the main parameters of influence.

REFERENCES

Albrecht, B. A., and C. H. Benson. 2001. Effect of desiccation on compacted natural clays. *Journal of Geotechnical and Geoenvironmental Engineering* 127, no. 1: 67–76.

Albright, W. H., C. H. Benson, and W. J. Waugh. 2010. *Water balance covers for waste containment – Principles and practice*. Reston, VA: American Society of Civil Engineers (ASCE) Press.

Albright, W. H., C. H. Benson, G. W. Gee, et al. 2004. Field water balance of landfill final covers. *Journal of Environmental Quality* 33: 2317–2332.

Apiwantragoon, P., C. H. Benson, and W. H. Albright. 2015. Field hydrology of water balance covers for waste containment. *Journal of Geotechnical and Geoenvironmental Engineering* 141, no. 2: 04014101-1-04014101-20.

Arnold, S., A. Schneider, D. Doley, and T. Baumgartl. 2015. The limited impact of vegetation on the water balance of mine waste cover systems in semi-arid Australia. *Ecohydrology* 8, no. 3: 355–367.

Aubertin, M., B. Bussière, T. Pabst, M. James, and M. Mbonimpa. 2016. Review of reclamation techniques for acid generating mine wastes upon closure of disposal sites. In *Proceedings of Geo-Chicago: Sustainability, Energy and the Geoenvironment*, Chicago, August 14–18

Aubertin, M., E. Cifuentes, S. A. Apithy, B. Bussière, J. Molson, and R. P. Chapuis. 2009. Analyses of water diversion along inclined covers with capillary barrier effects. *Canadian Geotechnical Journal* 46, no. 10: 1146–1164.

Benson, C. H., T. Abichou, W. H. Albright, G. W. Gee, and A. C. Roesler. 2001. Field evaluation of alternative earthen final covers. *International Journal of Phytoremediation* 3, no. 1: 1–21.

Benson, C. H., W. H. Albright, A. C. Roesler, and T. Abichou. 2002. Evaluation of final cover performance: Field data from the alternative cover assessment program (ACAP). In *Proceedings of Waste Management'02 Conference*, Tucson, AZ.

Benson, C., P. Bosscher, D. Lane, and R. Pliska. 1994. Monitoring system for hydrologic evaluation of landfill final covers. *Geotechenical Testing Journal* 17, no. 2: 138–149.

Benson, C. H., and M. A. Othman. 1993. Hydraulic conductivity of compacted clay frozen and thawed in-situ. *Journal of Geotechnical Engineering* 119, no. 2: 276–294.

Bohnhoff, G. L., A. S. Ogorzalek, C. H. Benson, C. D. Shackelford, and P. Apiwantragoon. 2009. Field data and water-balance predictions for a monolithic cover in a semiarid climate. *Journal of Geotechnical and Geoenvironmental Engineering* 135, no. 3: 333–348.

Bossé, B. 2014. Évaluation du comportement hydrogéologique d'un recouvrement alternatif constitué de rejets calcaires phosphatés en climat semi-aride à aride. PhD diss., Université du Québec en Abitibi-Témiscamingue.

Bossé, B., B. Bussière, R. Hakkou, A. Maqsoud, and M. Benzaazoua. 2013. Assessment of phosphate limestone wastes as a component of a store-and-release cover in semiarid climate. *Mine Water and the Environment* 32, no. 2: 152–167.

Bossé, B., B. Bussière, R. Hakkou, A. Maqsoud, and M. Benzaazoua. 2015a. Field experimental cells to assess the hydrogeological behaviour of store-and-release covers made with phosphate mine waste. *Canadian Geotechnical Journal* 52, no. 9: 1255–1269.

Bossé, B., B. Bussière, A. Maqsoud, M. Benzaazoua, and R. Hakkou. 2015b. Influence of extreme events and hydrogeological properties on the release capacity of store-and-release covers in a semiarid climate. In *Proceedings of the Canadian Geotechnical Conference, GéoQuébec 2015, Des défis du Nord au Sud*, Québec, Canada, September 20–23.

Bossé, B., B. Bussière, A. Maqsoud, R. Hakkou, and M. Benzaazoua. 2016. Hydrogeological behavior of a store-and-release cover: A comparison between a field column test and numerical predictions with or without hysteresis Effects. *Mine Water and the Environment* 35, no. 2: 221–234.

Bussière, B. 1999. Étude du comportement hydrique de couvertures avec effets de barrière capillaire inclinées à l'aide de modélisations physiques et numériques. PhD diss. École Polytechnique de Montréal.

Bussière, B., M. Aubertin, and R. P. Chapuis. 2002. A laboratory set up to evaluate the hydraulic behavior of inclined capillary barriers. *Physical Modelling in Geotechnics: Proceedings of the International Conference on Physical Modelling in Geotechnics, ICPMG '02*, St. John's, Newfoundland, Canada, July 10–12, ed. R. Phillips, pp. 391–396. Lisse: Balkema.

Bussière, B., M. Aubertin, H. J. Morel-Seytoux, and R. P. Chapuis. 1998. A laboratory investigation of slope influence on the behavior of capillary barriers. In *Proceedings of the 51th Canadian Geotechnical Conference*, Edmonton, Alberta, October 4–7, vol. 2, pp. 831–836.

Bussière, B., M. Aubertin, and G. Zhan. 2007. Design of inclined covers with capillary barrier effect by S.-E. Parent and A. Cabral. *Geotechnical and Geological Engineering* 25: 673–678.

Civil Engineering Consultants (CEC). 1997. *Hydraulic characteristics of barrier soils for the alternative landfill cover demonstration, City of Glendale*, AZ. Report prepared for City of Glendale, AZ, Verona, WI.

Chapuis, R. P. 2002. The 2000 R.M. Hardy Lecture: Full-scale hydraulic performance of soil-bentonite and compacted clay liners. *Canadian Geotechnical Journal* 39, no. 2: 417–439.

Dwyer, S. F. 2003. Water balance measurements and computer simulations of landfill covers. PhD diss., University of New Mexico, Albuquerque.

Eigenbrod, K. D. 2003. Self-healing in fractured fine-grained soils. *Canadian Geotechnical Journal* 40, no. *3*: 435–449.

Doorenbos, J., and W. Pruitt. 1977. Guidelines for predicting crop water requirements. FAO Irrigation Paper No. 24, 2nd ed., Food and Agricultural Organization of the United Nations, Rome, 1–107.

Fayer, M., and T. Jones. 1990. UNSAT-H version 2: Unsaturated Soil-Water and Heat Flow Model, ver. 2.0, Pacific Northwest Laboratory, Richland, WA, Report prepared for the US Department of Energy, under contract DE-AC06-76RLO 1830.

Gardiner, R. T., D. B. Dawson, and G. G. Gray. 1997. Application of ARD abatement technology in reclamation of tailings ponds at Cominco Ltd., Sullivan Mine. In *Proceedings of the 4th International Conference on Acid Rock Drainage (ICARD)*, Vancouver, Canada, vol. 1, pp. 47–63.

Hakonson, T. E., K. V. Bostic, G. Trujillo, et al. 1994. Hydrologic evaluation of four landfill cover designs at Hill Air Force Base, Utah. LA-UR-93-4469. Los Alamos National Laboratory, Los Alamos, NM.

Junqueira, F. J., G. W. Wilson, C. Nichol, and S. Dunlap. 2006. The influence of climate, vegetation, layer thickness, and material properties for performance of the cover systems at the golden Sunlight mine. In *Proceedings of the 7th International Conference on Acid Rock Drainage (ICARD), and 23rd Annual Meetings of the American Society of Mining and Reclamation*, St. Louis, Missouri, March 26–30, ed. R. I. Barnhisel, pp. 849–885. Lexington, KY: American Society of Mining and Reclamation.

Kämpf, M., T. Holfelder, and H. Montenegro. 2003. Identification and parameterization of flow process in artificial capillary barriers. *Water Resources Research* 39, no. 10: SBH 2-1-SBH 2-9.

Keller, J. M., L. T. Busker, M. A. Milczarek, R. C. Rice, and M. A. Williamson. 2011. Monitoring of the geochemical evolution of waste rock facilities at Newmont's Phoenix Mine. In *Proceedings of the Sixth International Conference on Mine Closure*, Lake Louise, Canada, Australian Centre for Geomechanics, ed. A. B. Fourie, M. Tibbett and A. Beersing, pp. 251–260.

Khire, M. V., C. H. Benson, and P. J. Bosscher. 1999. Field data from capillary barrier and model predictions with UNSAT-H. *Journal of Geotechnical and Geoenvironmental Engineering* 125, no. 6: 518–527.

Khire, M. V., C. H. Benson, and P. J. Bosscher. 2000. Capillary barriers: Design variables and water balance. *Journal of Geotechnical and Geoenvironmental Engineering* 126, no. 8: 695–708.

Khire, M., C. H. Benson, and P. J. Bosscher. 1994. Final cover hydrologic evaluation—phase III. Envir. Geotechnics Report 94-4, Department of Civil and Environmental Engineering, University of Wisconsin, Madison, WI.

Knidiri, J., B. Bussière, R. Hakkou, B. Bossé, A. Maqsoud, and M. Benzaazouaa. 2017. Hydrogeological behaviour of an inclined store-and-release cover experimental cell made with phosphate mine wastes. *Canadian Geotechnical Journal* 54, no. 1: 102–116.

Koerner, R. M., and D. E. Daniel. 1997. *Final covers for solid waste landfills and abandoned dumps.* Reston, VA: American Society of Civil Engineers (ASCE) Press

Meiers, P. G., S. Shaw, N. Overdevest, and M. O'Kane. 2005. Linking tailings seepage geochemistry to the performance of cover field trials for a tailings storage facility at peak gold mine operations. In *Proceedings of the 9th International Mine Water Congress, Mine Water 2005 – Mine Closure*, Ovido, Spain, ed. J. Loredo and F, pp. 317–329. Pendás, .

Melchior, S. 1997. In-situ studies of the performance of landfill caps (compacted soil liners, geomembranes, geosynthetic clay liners, capillary barriers). *Land Contamination Reclamation* 5: 209–216.

Milczarek, M., F. M. Steward, W. B. Word, M. M. Buchanan, and J. M. Keller. 2011a. Final results for the Morenci tailings experimental reclamation plots. In *Mine Closure 2011: Proceedings of the 6th International Conference on Mine Closure*, Lake Louise, Alberta, Canada/ACG, September 18–21, Australian Centre for Geomechanics, vol. 1, ed. A. Fourie and M. Tibbett, pp. 281–293.
Milczarek, M., F. M. Steward, W. B. Word, M. M. Buchanan, and J. M. Keller. 2011b. Salinity/pH interactions and rooting morphology in monolayer soil covers above copper tailings. In *Tailings and Mine Waste Conference*, Vancouver, BC, Canada, November 6–9, pp. 677–686. Norman B. Keevil Institute of Mining Engineering.
Milczarek, M., T. Yao, J. Vinson, J. Word, S. Kiessling, B. Musser, and R. Mohr. 2003. Performance of monolayer evapotranspiration covers in response to high precipitation and extended drought periods in the southwestern United States. In *Proceedings of the 6th International Conference on Acid Rock Drainage & Society for Mining, Metallurgy, and Exploration (U.S.)*, Cairns, Qld., AusIMM, Carlton South, Vic, 14–17.
Morel-Seytoux, H. J. 1992. The capillary barrier effect at the interface of two soil layers with some contrast in properties. HYDROWAR Report 92.4, Hydrology Days Publications, Atherton, CA.
Morel-Seytoux, H. J. 1994. Steady-state effectiveness of a capillary barrier on a sloping interface. *14th Hydrology Days*, ed. H. J. Morel-Seytoux, pp. 335–346. Atherton, CA: Hydrology Days Publications.
Morris, C. E., and Stormont, J. C. (1997). Capillary barriers and subtitle D covers: Estimating equivalency *Journal of Environmental Engineering., ASCE*, 23(1), 3–10.
Nyhan, J. W., T. Hakonson, and B. Drennon. 1990. A water balance study of two landfill cover designs for semiarid regions. *Journal of Environmental Quality* 19: 281–288.
Nyhan, J. W., T. G. Schofield, and R. H. Starmer. 1997. A water balance study of four landfill cover designs varying in slope for semiarid regions. *Journal of Environmental Quality* 26, no. 5: 1385–1392.
Ogorzalek, A. S., G. L. Bohnhoff, C. D. Shackelford, C. H. Benson, and P. Apiwantragroon. 2008. Comparison of field data and water-balance predictions for a capillary barrier cover. *Journal of Geotechnical and Geoenvironmental Engineering* 134, no. 4: 470–486.
O'Kane, M., and P. Waters. 2003. Dry cover trials at Mt Whaleback: A summary of overburden storage area cover system performance. In *Proceedings of the 6th International Conference on Acid Rock Drainage & Society for Mining, Metallurgy, and Exploration (U.S.)*, Cairns, Qld., AusIMM, Carlton South, Vic, pp. 147–154.
O'Kane, M., L. Ryland, and R. T. Gardiner. 1999. Field performance monitoring of the Kimberley operations siliceous tailings test plots. In *Proceedings of the Sixth International Conference on Tailings and Mine Waste '99*, Fort Collins, CO, January 24–27, pp. 23–33.
O'Kane, M., D. Porterfield, A. Weir, and A. L. Watkins. 2000. Cover system performance in a semi-arid climate on horizontal and sloped waste rock surfaces. In *Proceedings of the 5th International Conference on Acid Rock Drainage (ICARD)*, Denver, CO, vol. 2, pp. 1309–1318. Society for Mining, Metallurgy and Exploration.
Philip, J. R. 1969. Theory of infiltration. *Advances in Hydroscience* 5: 215–296.
Rock, S., B. Myers, and L. Fiedler. 2012. Evapotranspiration (ET) covers. *International Journal of Phytoremediation* 14, no. S1: 1–25.
Ross, B. 1990. The diversion capacity of capillary barriers. *Water Resources Research* 26: 2625–2629.
Scanlon, B. R., R. C. Reedy, K. E. Keese, and S. F. Dwyer. 2005. Evaluation of evapotranspirative covers for waste containment in arid and semiarid regions in the Southwestern USA. *Vadose Zone Journal* 4: 55–71.
Schnabel, W., J. Munk, and A. Byrd. 2015. Field note: Comparative efficacy of a woody evapotranspiration landfill cover following the removal of aboveground biomass. *International Journal of Phytoremediation* 17, no. 2: 159–164.
Schnabel, W. E., J. Munk, T. Abichou, D. Barnes, W. Lee, and B. Pape. 2012a. Assessing the performance of a cold region evapotranspiration landfill cover using lysimetry and electrical resistivity tomography. *International Journal of Phytoremediation* 14, no. S1: 61–75.
Schnabel, W. E., J. Munk, W. J. Lee, and D. L. Barnes. 2012b. Four-year performance evaluation of a pilot-scale evapotranspiration landfill cover in Southcentral Alaska. *Cold Region Science and Technology* 82: 1–7.
Steenhuis, T. S., J.-Y. Parlange, and K.-J. S. Kung. 1991. Comment on "The diversion capacity of capillary barriers" by Benjamin Ross. *Water Resources Research* 27, no. 8: 2155–2156.
Stormont, J. C. 1995. The effectiveness of two capillary barrier on a 10% grade. *Geotechnical and Geological Engineering* 14, no. 4: 243–267.
Stormont, J. C., and C. Morris. 1998. Method to estimate water storage capacity of capillary barriers. *Journal of Geotechnical and Geoenvironmental Engineering* 124, no. 4: 297–302.

Tami, D., H. Rahardjo, E.-C. Leon, and D. G. Fredlund. 2004. Design and laboratory verification of a physical model of sloping capillary barrier. *Canadian Geotechnical Journal* 41, no. 5: 814–830.

Tinjum, J., Benson, C., and Blotz, L. 1997. Soil-water characteristic curves for compacted clays. *Journal of Geotechnical and Geoenvironmental Engineering* 123, no. 11: 1060–1069.

van Genuchten, M. T. 1980. A closed form equation for predicting the hydraulic conductivity of unsaturated soils. *Soil Science Society of America Journal* 44: 892–898.

Ward, A. L., and G. W. Gee. 1997. Performance evaluation of a field-scale surface barrier. *Journal of Environmental Quality* 26, no. 3: 694–705.

Waugh, W. J., M. K. Kastens, L. R. L. Sheader, C. H. Benson, W. H. Albright, and P. S. Mushovic. 2008. Monitoring the performance of an alternative landfill cover at the Monticello, Utah, uranium mill tailings disposal site. In *Proceedings of the Waste Management Conference*, Phoenix, AZ, February 24–28.

Wels, C., M. O'Kane, and S. Fortin. 2001. Assessment of water storage cover for Questa tailings facility, New Mexico. In *Proceedings of the National Meeting of the American Society for Surface Mining and Reclamation*, Albuquerque, New Mexico, June 3–7, pp. 1–14. Lexington, KY: ASSMR.

Wels, C., S. Fortin, and S. Loudon. 2002. Assessment of store-and-release cover for Questa tailings facility, New Mexico. In *Proceedings of the 9th International Conference on Tailings and Mine Waste '02*, Fort Collins, CO, January 27–30, vol. 2, 459–468.

Williams, D. J., D. J. Stolberg, and N. A. Currey. 2006. Long-term monitoring of Kidston's "Store/Release" cover system over potentially acid forming waste rock piles. In *Proceedings of the 7th International Conference on Acid Rock Drainage (ICARD), and 23rd Annual Meetings of the American Society of Mining and Reclamation*, St. Louis, MI, March 26–30, ed. R. I. Barnhisel, pp. 463–472. Lexington, KY: American Society of Mining and Reclamation.

Wilson, G. W., D. J. Williams, and E. M. Rykaart. 2003. The integrity of cover systems: An update. In *Proceedings of the 6th International Conference on Acid Rock Drainage & Society for Mining, Metallurgy, and Exploration (U.S.)*, Cairns, Qld., AusIMM, Carlton South, Vic, vol. 3, 14–17.

Zhan, G., M. Aubertin, A. Mayer, K. Burke, and J. McMullen. 2001. Capillary cover design for leach pad closure *SME Transaction* 1: 104–110.

Zhan, G., J. Keller, M. Milczarek, and J. Giraudo. 2014. 11 years of evapotranspiration cover performance at the AA leach pad at Barrick Goldstrike mines. *Mine Water and the Environment* 33, no. 3: 195–205.

Zhan, G., D. Lattin, J. Keller, and M. Milczarek. 2019. 20 years of evapotranspiration cover performance of the leach pads at Richmond Hill Mine. *Mine Water and the Environment* 38: 402–409. https://doi.org/10.1007/s10230-019-00592-7.

Zhan, G., W. Schafer, M. Milczarek, K. Myers, J. Giraudo, and R. Espell. 2006. The evolution of evapotranspiration cover systems at Barrick Goldstrike Mines. In *Proceedings of the 7th International Conference on Acid Rock Drainage (ICARD), and 23rd Annual Meetings of the American Society of Mining and Reclamation*, St. Louis, Missouri, March 26–30, ed. R. I. Barnhisel, pp. 2585–2603. Lexington, KY: American Society of Mining and Reclamation.

6 Water Covers

Akué Sylvette Awoh, Mamert Mbonimpa, and Bruno Bussière

6.1 INTRODUCTION AND BACKGROUND

Oxygen barriers are known as the most effective approach to control acid mine drainage (AMD) in humid climates (SRK 1989; Aubertin et al. 2016). One of the techniques used to create oxygen barriers involves placing a water cover over potentially acid-generating (PAG) mine wastes (i.e., waste rocks and tailings) (MEND 2001a). The water cover significantly reduces oxygen availability to the mine wastes (as explained below), thus inhibiting acid generation.

The water cover technique was first presented in 1971 as submarine disposal (including shore deposition, shallow, and deep-sea disposal) of mine wastes (MEND 1997; Koski 2012). However, submarine disposal of mine wastes has been banned in most countries (Canada banned the practice in 1977; Coumans 2002) because its effects on the marine environment cannot be easily addressed, and the environmental impacts on submarine flora and fauna can be significant, particularly in waters frequented by fish. For example, turbidity caused by tailings disposal in lakes can drive away mobile species, smother native species, destroy rare organisms, and reduce biodiversity (Coumans 2002). Nonetheless, some researchers continue to defend submarine deposition of tailings under specific conditions (Dold 2014). For example, the deposition of inert inorganic geological material into the sea can be permitted (Dold 2014).

Natural lakes were also used for underwater deposition of PAG sulfide wastes (MEND 1989; Table 6.1). Under current legislation, this practice is approved only in very special cases where all other options are inapplicable (Aubertin et al. 2002). In Canada, several approvals defined by the Canadian Navigable Waters Act and Metal and Diamond Mining Effluent Regulations are required for this kind of disposal.

Nowadays, the mining industry is mostly focusing on artificial water covers. In these cases, mine wastes are deposited

- underwater (subaqueous) in constructed (or engineered) reservoirs surrounded by watertight dikes and filled with water;
- underwater (subaqueous) in pits flooded after shutting off the sump pump;
- in pits that are flooded at reclamation and;
- in surface waste ponds where mine waste are flooded at reclamation.

Plenty of studies evaluate the efficiency of these types of water covers in controlling environmental issues. This chapter begins by describing the concept of artificial water covers and discusses the hydrogeological, hydrodynamic, and geochemical processes that influence the efficiency of these water covers. A design methodology is then provided. Finally, three case studies are presented, and the advantages and limitations of the techniques are discussed. It is worth mentioning that additional information on in-pit disposal of mine wastes can be found in Chapter 13.

6.2 CONCEPT

The water cover technique consists of maintaining a water layer above PAG mine wastes in order to limit the oxygen supply to the underlying tailings. The water cover acts as oxygen barrier because the concentration of oxygen and diffusion coefficient are much lower in water than in air, respectively.

TABLE 6.1
Sites Using the Water Cover Technique: Submarine and in-Lake Disposal

	Mine site	Location	Climate	Intermediate Field Scale	Full scale (hwc)	Type of Mine Waste	Sources
Submarine disposal	Atlas Mine	Philippines	Annual precipitation = 0–550 mm Average wind speed = 10–30 km/h	(——)	Shallow sea	Tailings	Morello et al. (2016), Kwong et al. (2019)
	Black Angel Mine	Greenland	Annual precipitation = 0–100 mm Average wind speed = 21.6 km/h	(——)	Shallow fjord	Tailings	Berkun (2005), Kwong et al. (2019)
	Boulby Potash Mine	England	Annual precipitation = 0–200 mm Average wind speed = 10–40 km/h	(——)	North sea	Tailings	Morello et al. (2016)
	Cayeli Bakir copper Mine	Turkey	Annual precipitation = 0–300 mm Average wind speed = 5–25 km/h	(——)	Deep black sea	Tailings	Berkun (2005), Kwong et al. (2019)
	Copper Mine	Barents Sea, Norway	Precipitation = 625 mm/year mm Average wind speed = 23.6 km/h	(——)	Fjord, depth=230 m	Tailings	Vogt (2013), Kwong et al. (2019)
	Huasco Iron Pelletising Plant	Chile	Annual precipitation = 0–100 mm Average wind speed = 5–25 km/h	(——)	Shallow sea	Tailings	Morello et al. (2016), Kwong et al. (2019)
	Iron Mine	Norway–Russia border, Norway	Annual precipitation = 625 mm/year mm Wind speed = 23.6km/h	(——)	Fjord	Tailings	Vogt (2013), Kwong et al. (2019)
	Island Copper Mine	British Columbia, Canada	Annual precipitation = 0–150 mm Average wind speed = 5–20 km/h	(——)	Fjord	Tailings	Ellis and Poling (1995), Kwong et al. (2019)
	Kitsault Molybdenum Mine	British Columbia, Canada	Annual precipitation = 0–500 mm Average wind speed = 10–40 km/h	(——)	Fjord	Tailings	Ellis and Poling (1995), Kwong et al. (2019)
	Lihir Mine	Papua New Guinea	Annual precipitation = 0–750 mm Average wind speed = 5–35 km/h	(——)	Shallow sea	Tailings	Morello et al. (2016), Kwong et al. (2019)
	Minahasa Raya Mine	Indonesia	Annual precipitation = 0–400 mm Average wind speed = 5–15 km/h	(——)	Shallow sea	Tailings	Morello et al. (2016), Kwong et al. (2019)
	Misima Mine	Papua New Guinea	Annual precipitation = 0–850 mm Average wind speed = 5–25 km/h	(——)	Shallow sea	Tailings	Morello et al. (2016), Kwong et al. (2019)
	Norway's west coast Site	Norway	Average Precipitation = 2250/year mm Average wind speed = 8–18 km/h	(——)	Deep sea	Tailings	Vogt (2013), Kwong et al. (2019)

In-lake disposal	Anderson Lake	Manitoba, Canada	Precipitation = +200 mm Average wind speed = 5–25 km/h	(——)	Max depth = 8 m, min depth = 3-3.6 m	Tailings	MEND (1989), Fraser and Robertson (1994)
	Babine Lake	British Columbia, Canada	Precipitation = 307.5 mm Average wind speed = 20 km/h	(——)	Lake elevation = 711 m	Waste rock and Tailings	MEND (1989)
	Benson Lake	British Columbia, Canada	Average precipitation = 116 mm Average wind speed =11–74 km/h	(——)	Max depth = 54 m, Mean depth = 25.5 m	Tailings	MEND (1989), Fraser and Robertson (1994), MEND (2001a)
	Brucejack Lake	British Columbia, Canada	Average precipitation = 3 mm Average wind speed = 6–52 km/h	(——)	Mean depth = 88 m	Tailings	MEND (1989), Rescan (2013)
	Buttle Lake	British Columbia, Canada	Average precipitation= 2000–2500 mm Wind speed = 10–56 km/h	(——)	Min depth = 61.5 m, mean depth = 61.3 m, Max depth = 87 m	Tailings	MEND (1989), Fraser and Robertson (1994), Rodenhuis et al. (2009)
	Fox Lake	Manitoba, Canada	Average precipitation = 499.4 mm Average wind speed = 14–17.8 km/h	(——)	Max depth = 7 m	Tailings	MEND (1989), MHEC (2020)
	Garrow Lake	Nunavut, Canada	Average precipitation = 250 mm Average wind speed = 20 km/h	(——)	Mean depth = 24.5 m	Tailings	MEND (1989), Donald (2005)
	Kootenay Lake	British Columbia, Canada	Mean precipitation = 365 mm Average wind speed= 8–39 km/h	(——)	Min depth = 94 km, Max depth = 154 m	Tailings	MEND (1989), Rodenhuis et al. (2009)
	Mandy	Manitoba, Canada	Average precipitation = 343 mm Average wind speed = 14–17.8 km/h	(——)	Max depth = 5.5 m, Mean depth = 3.6 m	Tailings	MEND (1989), Petersen et al. (1993), Fraser and Robertson (1994), MEND (2001a, 2006)
	Silver Bay Lake	Minnesota, USA	Average precipitation = 813 mm Average wind speed = 10.6–13.3 km/h	(——)	Depth = 200–300 m	Tailings	MEND (1989)

Indeed, the dissolved oxygen (DO) concentration and diffusion coefficient in water are 8.6 mg/L and 2×10^{-9} m^2/s at 25°C, compared to 285 mg/L and 1.8×10^{-5} m^2/s at 25°C in air, respectively (Yanful and Verma 1999; Romano et al. 2003; Davé et al. 2003; Awoh et al. 2013b). Only the concepts of artificial water covers (subaqueous disposal in constructed reservoirs, in open pits, and flooding of existing waste disposal ponds) are presented below due to the low social acceptability and low probability of implementation of submarine disposal.

6.2.1 Subaqueous Disposal in Constructed Water Reservoirs

This subaqueous or underwater disposal consists of discharging tailings (no case is known for such disposal for waste rocks) directly under water contained in an engineered impoundment or reservoir surrounded by impervious dikes (MEND 1989, 2001a). The physical (geotechnical) stability of the dikes must be ensured for the entire life of the structure, as a failure could have catastrophic consequences (Aubertin et al. 1997). Monitoring of structures and appropriate corrective actions are then required in the very long term (Aubertin et al. 2002). Figure 6.1 illustrates the concept of this disposal/reclamation method. This technique was implemented on several mine sites (see Section 6.5). The minimum height of the water cover ranges generally between 1 and 2 m as described in Section 6.4.

6.2.2 Subaqueous Disposal in Pit Lakes

Open pits or quarries and underground openings (mine shafts) can also be used as a reservoir for underwater mine waste storage (once the mining is completed in and below the pit). A pit lake usually forms after mining stops, and the groundwater pumping is stopped. Subaqueous or underwater disposal in a pit lake consists of discharging tailings or waste rocks into flooded pits (MEND 1995, 2015). Figure 6.2 illustrates the concept of this disposal/reclamation method. This technique has been implemented on several mine sites (see Section 6.5). Particular attention must be paid to the risks of contaminant migration and its effects on groundwater quality and surrounding surface water (Molson et al. 2012; Awoh et al. 2013a; Abdelghani et al. 2015). To avoid any discharge of impacted cover water to groundwater, a long-term passive hydraulic sink or trap must be maintained. The disposal can be done without any barrier, with a surface barrier, with a groundwater barrier, or with surface and groundwater barriers (MEND 1995).

Subaqueous in-pit disposal without any barrier is the simplest, most common approach, applicable in an ideal pit where convective groundwater transport is expected to be minimal and contaminant release would occur mainly through diffusion from the reactive mine wastes into the pit water. If this option is inadequate, additional engineered barriers become necessary.

Underwater disposal in pits can be done with a surface barrier made of clean soil (sand, till, clay, etc.) placed over the surface of the submerged reactive wastes to reduce upward contaminant transport into the water. Subaqueous in-pit disposal with a groundwater barrier is applied when a substantial convective groundwater flow through the reactive mine waste is expected. Three types of potential barriers were proposed in MEND (1995): (1) a barrier that blocks the groundwater flow through the tailings and acts as a barrier to contaminant diffusion (liners, clay, dense till), (2) a barrier that provides a low resistive flow path (i.e., the pervious shell) to groundwater, creates a

FIGURE 6.1 Schematic illustration of the concept of subaqueous disposal (constructed reservoirs).

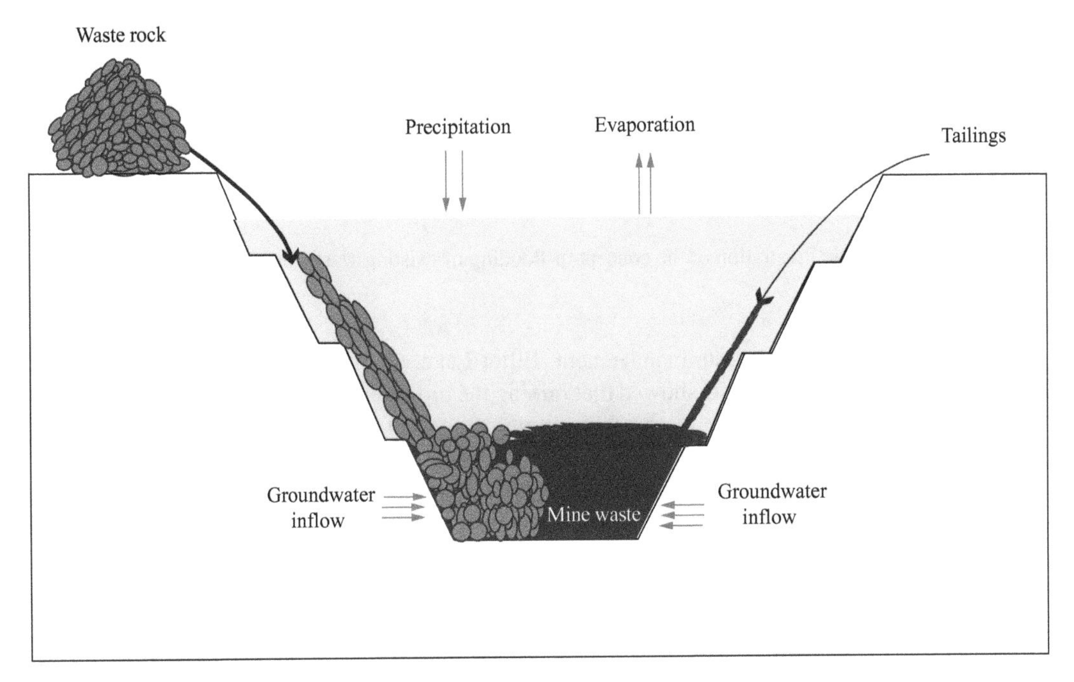

FIGURE 6.2 Schematic illustration of the concept of subaqueous disposal in pit lake.

pervious envelope around the waste to create a preferential flow path for groundwater, and equalizes the gradient across the pit so that groundwater cannot pass through the tailings (Thériault 2004), and (3) a barrier that remediates contamination entering the groundwater (a redox barrier or an acid-consuming barrier).

Underwater disposal in pits with surface and groundwater barriers consists of placing a complete envelope around the wastes. The best example of this engineered pit disposal concept was applied at the Rabbit Lake pit of the Collins "B" site, Saskatchewan, Canada (MEND 1995), where a bottom rock drain, an engineered pervious envelope, and a soil/sand diffusion barrier were used. The tailings were placed as a dry filter cake.

6.2.3 Flooding of Existing Waste Disposal Ponds

The flooding consists of submerging existing surface or in-pit impoundment where the tailings or waste rocks were discharged with a water cover (MEND 2001a). Figure 6.3 illustrates the concept of this disposal/reclamation method. This technique has been used on several mine sites to limit the generation of AMD of pre-oxidized mine wastes (see Section 6.5). The main objectives of the flooding technique are to inhibit further oxidation of the mine wastes and improve the cover water quality to allow the discharge of excess water without treatment after a transition period. The transition period refers to between saturation by flooding and the moment when the water quality is sufficiently improved for it to be released into the environment (MEND 2001a). Before applying the flooding technique to mine wastes stored in valley or in impoundment, it is necessary to build additional dikes to create the reservoir. Also, flooding can involve resurfacing the tailings to ensure a uniform thickness of the water cover (MEND 1997, 2001a).

Sometimes lime was added to the oxidized mine waste to increase pH and reduce metals' mobility in water cover and interstitial water. Lime can be mixed to the top layer of the tailings (mixing the upper 15 cm of tailings with lime in Solbec-Cupra, QC, Canada) (MEND 1997, 2001a). Lime can also be added simply to the surface in the tailings (extend limestone to the surface of tailing before

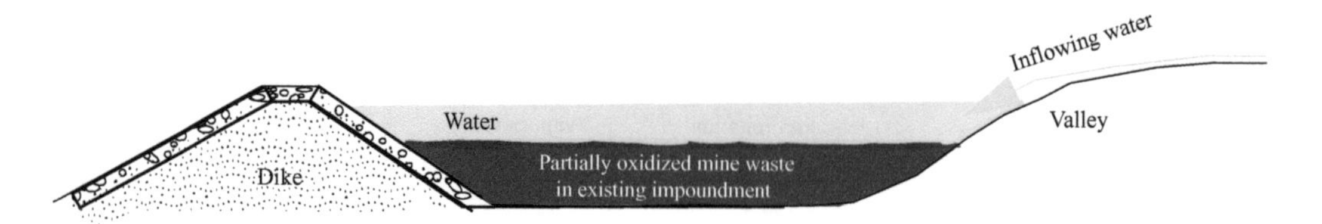

FIGURE 6.3 Schematic illustration of the concept of flooding of existing waste disposal areas.

flooding in Cell 14, Quirke waste management, Elliot Lake, ON, Canada) (MEND 1997, 2001a; Kam et al. 1997). Laboratory tests showed that mixing the top layer of the pre-oxidized tailings with lime prior to flooding was more effective than simply applying lime to the tailings surface (MEND 1997). Lime can also be added as hydrated lime to the tailings pond in areas where the tailings were already submerged (MEND 2001a). The flooding of pre-oxidized tailings has been used on several mine sites to limit the generation of AMD, including the following Canadian sites: Solbec-Cupra in Quebec (Canada) (Karam and Guay 1994; Vézina and Amyot 1999), Mattabi in Ontario (Canada) (Simms et al. 2000), and Quirke near Elliot Lake in Ontario (Canada) (Peacey et al. 2002; Martin et al. 2003).

6.3 EFFECTIVENESS OF WATER COVERS: PARAMETERS OF INFLUENCE

A water cover is a complex and dynamic system subjected to several phenomena that may affect the efficiency of the cover to control the quality of the shallow cover water and groundwater in the surrounding environment. Coupled hydrogeological, hydrodynamic, and geochemical processes can be distinguished (Li et al. 1997; Aubertin et al. 2002) (Figure 6.4). Hydrogeological processes relate to exchanges with the atmosphere (precipitations, runoff, and evaporation) and with the surrounding environment (water inflow and outflow and infiltration within the tailings and to the groundwater). Hydrodynamic processes include wind-driven waves and return currents that can induce mixing of the cover water as well as tailings erosion and resuspension. Geochemical processes relate to oxygen transport through the water column to the tailings and to metal release. These processes are briefly addressed below.

6.3.1 Hydrogeological Processes

The performance of water covers is closely related to the water budget components of the site and their interrelationships. The following hydrogeological parameters help to define the water budget given in equation 6.1 (which is an extension of the water budget presented in Chapter 3) for any water cover.

$$P + R_{in} + G_{in} = E + T + G_{out} + \Delta S \tag{6.1}$$

where P represents precipitation falling on the surface of the water cover, R_{in} represents surface watershed runoff entering the water cover, G_{in} represents groundwater entering the water cover (G_{in} = 0 for engineered reservoirs), E represents evaporation at the surface of the water cover, T represents transpiration (applicable only when aquatic plants and trees have grown), G_{out} represents groundwater leaving the water cover, and ΔS represents water stored in the facility.

In engineered water reservoirs, the water storage ΔS may be controlled to keep the maximal height of the water cover to avoid overflowing (see Section 6.4.1). In the case of in-pit disposal, ΔS may also be controlled to keep a long-term passive hydraulic sink or trap (i.e., no discharge of

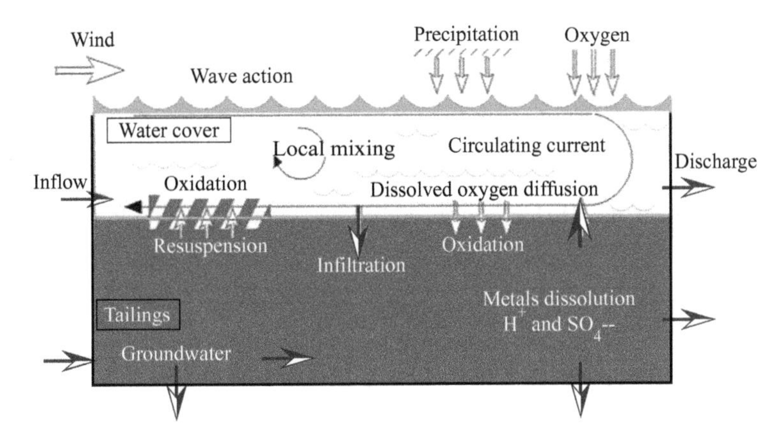

FIGURE 6.4 Factors affecting the performance of water covers (Modified from Aubertin et al. 2002).

impacted water to groundwater) taking into account the groundwater barriers mentioned in Section 6.2.2. An adequate system for regulating the water level in the basin (e.g., using a weir) must be installed to control the maximal water cover height. For the hydrological design, the probable maximum precipitation corresponding to a return period prescribed by regulatory requirements must be used. The design must also take into account the predicted future extreme precipitations with climatic changes (see Chapter 14). In Equation (6.1), ΔS includes the water passing through the system for regulating the water level. In cold climates, where ice can accumulate in the outlet during winter and minimize outlet flows, the risk for overtopping must not be overlooked.

6.3.2 Hydrodynamic Processes: Erosion and Resuspension

In a water cover system, the water column is exposed to the atmosphere, and it undergoes wind effects. Wind-induced waves and return currents apply shear stresses τ_{wa} and τ_{cu} at the waste-water interface, respectively. In other words, the cover water is not any time stagnant and unmixed. When the combined stress τ_{tot} (= $\tau_{wa} + \tau_{cu}$) becomes greater than a threshold shear stress that is specific to the mine wastes in place, called the critical shear stress (τ_{cr}) (i.e., $\tau_{tot} > \tau_{cr}$), erosion and resuspension of waste particles occur (Adu-Wusu et al. 2001; Catalan and Yanful 2002; Mian and Yanful 2004). This phenomenon may particularly affect fine particles such as tailings (the authors are not aware for a study on such phenomena in the case of submerged waste rocks, although fine particles in waste rocks can be affected). Resuspended particles can be transported to the final effluent, which can negatively affect water quality and exceed contaminant limits allowed by regulations. For example, in Quebec, Directive 019 on the mining industry recommends monthly and maximum average suspended solids concentrations of 15 mg/L and 30 mg/L for the final effluent, respectively. Furthermore, resuspended sulfide particles are more exposed to DO in the cover water (see Section 6.3.3) and thereby can more easily oxidize, potentially increasing the release of dissolved metals. This phenomenon of erosion and resuspension is much more accentuated in very shallow water covers (Catalan and Yanful 2002). A minimum water depth h_{min} is then required to avoid this phenomenon (see Section 6.4).

Many studies have been conducted on the erosion and resuspension of tailings submerged under a shallow water cover. Some studies have determined the concentration of suspended particles by filtration of sample collected at different depths in the water cover (Awoh et al. 2013a) using sediment traps (allowing for sampling water at different elevations and analyzing the samples for particle size and mass; Adu-Wusu et al. 2001; Catalan and Yanful 2002) and also using suspended particles measurement probes (Kachhwal et al. 2010; Awoh et al. 2013a). Based on work performed by the Coastal Engineering Research Center of the United States Army (CERC 1984, 2002), complementary studies have focused on methods for determining the shear stresses induced by winds effect (Yanful and Mian 2003; Mian et al. 2007) and the critical shear stress (Krishnappan 1993; Krishnappan and Marsalek 2002; Yanful and Catalan 2002; Davé et al. 2003; Mian et al. 2007). These latter studies were carried out considering the coherent or non-coherent nature of the tailings under the water cover. Suspended sediment concentrations measured at several sites in Canada and in laboratory tests have shown that resuspension cannot be eliminated completely (Adu-Wusu et al. 2001; Yanful and Catalan 2002; Awoh et al. 2014). However, adding a layer of granular non-reactive material can help to control the resuspension of the reactive particles and thus reduce their oxidation.

6.3.2.1 Critical Shear Stress (τ_{cr})

The critical shear stress (τ_{cr}) represents the tailings-specific threshold stress above which erosion and resuspension of tailings take place. A critical velocity V_{cr} at the bed of the submerged sediment can be associated with τ_{cr} (see Section 6.4.1). For preliminary investigations, τ_{cr} can be estimated using semi-empirical equations (e.g., Chien and Wan 1998). The critical shear stress should be determined by experimental models in the laboratory using, among others methods, a rotating circular flume (RCF), a wave flume (WF), and a laboratory column calibrated using laser Doppler anemometry.

The RCF such as the one installed at the National Water Research Institute, Burlington, Ontario, Canada (Krishnappan 1993; Krishnappan and Marsalek 2002; Davé et al. 2003) consists of a circular channel (or flume) welded to a rotating bottom platform. The circular channel is associated with an annular cover plate (ring) attached to an upper turntable. The sampling ports placed on the wall of the flume allow sampling of suspended sediments by applying suction at timed intervals. To operate the RCF, a sediment bed layer is placed in the circular channel underwater. The flume and the cover plate are then rotated in the same direction at maximum speeds of up to 2.5 RPM for about 20 minutes in order to allow the sediment to be uniformly distributed in the flume. The speed is then decreased progressively up to zero, and the system is left for sedimentation. After that, the flume and the cover plate are again rotated in opposite directions this time to produce a uniform tangential stress on the sediment bed. The rotational speed of the flume is intensified at a uniform time step. For each speed of rotation, which corresponds with a tangential stress at the surface of the sediment bed, the concentration of suspended particle above the sediment bed is measured through the sampling ports. A particle size analyzer and a Doppler laser velocimeter are used to monitor the particle size distribution of suspended particles and the tangential component of the applied speed (that can be linked to the shear stress $\tau = \rho \nu \frac{du}{dz}$, where ρ is the fluid density, ν is the kinematic viscosity of the fluid, and u is the horizontal velocity), respectively. The representation of the concentration of suspended sediment, as a function of the stress applied to the sediment bed, is used to obtain the critical shear stress that corresponds to the minimal stress inducing the suspension of solid particles (solid concentrations > 0) in the water cover.

The WF, such as the one used at the Canadian Hydraulics Centre at the National Research Council of Canada, Ottawa, Ontario (Davé et al. 2003), is a linear channel at the base of which the material intended for the test is placed under the water layer. The channel is equipped with wave generator and an active wave absorption system. To determine the critical shear stress in the WF, the sediment is subjected to a series of regular and irregular waves with different heights and periods, which are monitored using wave gauges. An acoustic Doppler current probe and sediment suction samplers are then used to measure the wave orbital velocity near the sediment bed and the concentration of suspended particles through the height of the water column, respectively. The wave orbital velocity can be linked to the shear stress for any level of wave heights and periods. The plot of the wave orbital-velocity-dependent shear stress versus the suspended particles concentrations determines the critical shear stress as the minimal stress that induces the suspension of solid particles.

The laboratory column calibrated using laser Doppler anemometry was developed by Mian et al. (2007). The sediment is deposited at the base of a Plexiglas column. A stationary two-blade

Teflon stirrer placed about 60 cm above the surface of the sediment is used to induce resuspension. Sampling ports are also installed along the column. The critical shear stress is assessed in two steps. The first step is conducted with the column filled with 85 cm of water (without sediment) and aims at calibration (i.e., determining the bottom shear stress induced by the stirrer for each speed of rotation). For that purpose, a Doppler laser probe allows to establish vertical profiles of the azimuthal velocity for different radial positions (from the center to the wall) and for each stirrer speed. For a given stirrer speed, a normalization process merges velocity profiles at the different radial positions in a single profile. The slope of vertical profile was converted into shear stress exerted at the bottom, and a linear relationship was established between the stirrer speed and the induced bottom shear. In the second step, the sediment to be tested is placed in the column filled with 85 cm of water so that a final sediment thickness of 2 to 3 cm is obtained after sedimentation/settlement. Once the suspended solid concentration is zero, the stirrer is rotated at a constant speed and induces the resuspension of the sediment. Once the steady state is reached, samples are taken from the sampling ports installed along the column, and the sediment concentration is determined by filtration. This procedure is repeated for different stirrer speeds. For each applied stirrer speed, the erosion rate can be determined and plotted against the wall shear stresses determined in step 1. A trend line is determined and used to assess the critical shear stress critical at zero sediment concentration.

Table 6.2 presents typical values of the critical shear stress measured on tailings. The type of tailings, available physical properties (grain-size characteristics and specific gravity G_S), and the source of the data are also given. The data presented confirm that, in general, $\tau_{cr} < 0.5$ Pa, which represents the critical shear stress for dense cohesive sediment bed (Adu-Wusu et al. 2001).

6.3.2.2 Bottom Shear Stress Due to Waves (τ_{wa})

The shear stresses generated at the bottom of the water cover by wind-induced waves can be determined using Equations 6.2 to 6.4 presented in Table 6.3 (Adu-Wusu et al. 2001; Samad and Yanful 2005), based on wave characteristics (i.e., significant wave height H_s, significant wave period T_s, and wave length L), among other factors. A significant wave characteristic (H_s or T_s) is defined as the average highest one-third of all waves (CERC 1984, 2002). Equations for T_s and H_s were developed for two waves conditions distinguished according to the ratio h/L between the water depth h and the calculated wave length L: shallow-water waves or long waves when $h < 0.5$ L and deep-water waves or short waves when $h/L > 0.5$ (Lawrence et al. 1991, Adu-Wusu et al. 2001; Yanful and Catalan 2002; Peacey and Yanful 2003; Samad and Yanful 2005).

For water cover shallower than one-half the wave length ($h < 0.5L$), H_s, T_s, and L are given by Equations 6.5, 6.6, and 6.7, respectively (see Table 6.3). For water cover deeper than one-half the wave length ($h > 0.5$ L), H_s, T_s, and L are given by Equations 6.7, 6.8, and 6.9, respectively (see Table 6.3). Equations 6.5 to 6.9 involve the wind stress factor U_A (defined in Equation 6.10 using the wind speed U_{10} measured at 10 m above the ground surface), the wind fetch F (unobstructed distance of water over which wind can blow in a single direction), and the water cover height (thickness) h.

In the case of in-pit disposal, the downward transfer of wind energy can be ineffective to induce waves and the ensuing vertical mixing due to the small fetch and large depth. However, end uses that can induce waves and water mixing such as active recreation and fishery must be avoided to reduce the potential resuspension of the mine wastes.

6.3.2.3 Bottom Shear Stress Due to the Return Currents (τ_{cu})

The shear stress due to the return currents τ_{cu} (Pa) at the water-tailings interface is often estimated using the density of air (temperature dependent, 1.24 kg/m^3 at 10°C) and the wind stress factor and U_a (m/s) as presented in Equation 6.11 in Table 6.3 (Lawrence et al. 1991; Yanful and Catalan 2002; Mian and Yanful 2004). The shear stress induced by the return currents (τ_{cu}) is generally negligible compared with the wave-induced shear stress τ_{wa} (Adu-Wusu et al. 2001). Consequently, waves are the process that contribute the most to resuspension of submerged waste bed.

TABLE 6.2
Typical Values of Critical Shear Stress Measured on Tailings and Similar Materials

Site	Type of Waste	D_{10} (μm)	D_{50} (μm)	D_{60} (μm)	C_u	G_s	τ_{cr} (N/m²)	Experimental Models	References
INCO Limited, Copper Cliff, Ontario, Canada	Total milling tailings	(——)	150	(——)	(——)		0.16	RCF	Davé et al. (2003)
	Coarse tailings	(——)	0.230	(——)	(——)	2.61	0.21–0.25	WF	
	Fine tailings	(——)	0.083	(——)	(——)	2.37	0.20–0.22	WF	
	Fine tailings covered with 0.5 m layer of fine silica sand	(——)	0.12	(——)	(——)	2.40	0.16	WF	
Kingston, Ontario, Canada	Bottom fine sediment (Kingston pond)	(——)	(——)	(——)	(——)	(——)	0.12	RCF	Krishnappan and Marsalek (2002)
Heath Steele, New Brunswick, Canada	Tailings (upper cell)	3	12	16	5.3	(——)	0.102	Laboratory column calibrated using laser Doppler anemometry	Mian et al. (2007)
	South London silt	≈1.5	14	16	10.7	(——)	0.114	Laboratory column calibrated using laser Doppler anemometry	Mian et al. (2007)
Heath Steele, New Brunswick, Canada	Tailings (upper cell)	≈3.4	12	≈15	4.4	4	0.12	RCF	Mian and Yanful (2007), Mian et al. (2007)
Northwestern, British Columbia	Tailings (Premier Gold Project)	≈2.2	10	≈16	7.3		0.10	RCF	Mian and Yanful (2007), Mian et al. (2007)

TABLE 6.3
Equations Used for Describing the Bottom Shear Tress Generated by Wind-Induced Waves and Return Currents

Equations	Numbers	Description of Parameters
$\tau_{wa}=\rho_w\sqrt{\frac{v_w u_{bm}^3}{a_m}}$	(6.2)	τ_{wa}: bottom shear stress due to waves ρ_w: fluid (water) density (kg/m^3) ν_w: kinematic viscosity of the fluid (m^2/s) ($\nu_w = \mu_w/\rho_w$ where is the dynamic viscosity in Pa.s.) u_{bm}: maximum horizontal bottom fluid velocity near the tailings bed (m/s) (given by Equation 6.4): a_m: maximum displacement of the individual fluid particles from their mean position (given by Equation 6.5)
$u_{bm}=\frac{\pi H_s}{T_s}\left[\frac{1}{sinh\left(\frac{2\pi h}{L}\right)}\right]$	(6.3)	u_{bm}: maximum horizontal bottom fluid velocity near the tailings bed (m/s) H_s: significant wave height (m) T_s: significant wave period (s) L: Significant wave length (m)
$a_m=\frac{H_s}{2\,sinh\left(\frac{2\pi h}{L}\right)}$	(6.4)	
$H_s=0.283\left(\frac{U_A^2}{g}\right)tan\,h\left[0.53\left(\frac{gh}{U_A^2}\right)^{0.75}\right]$ $tan\,h\left\{\frac{0.00565\left(\frac{gF}{U_A^2}\right)^{0.5}}{tan\,h\left[0.53\left(\frac{gh}{U_A^2}\right)^{0.75}\right]}\right\}$	(6.5)	g: gravitational acceleration (m/s^2) U_A: wind stress factor (defined in Equation 6.10 using the wind speed U_{10} measured at 10 m above the ground surface) F: wind fetch F (unobstructed distance of water over which wind can blows in a single direction) h: water cover height (m)
$T_s=7.54\left(\frac{U_A}{g}\right)tan\,h\left[0.833\left(\frac{gh}{U_A^2}\right)^{0.375}\right]$ $tan\,h\left\{\frac{0.0379\left(\frac{gF}{U_A^2}\right)^{0.25}}{tan\,h\left[0.833\left(\frac{gh}{U_A^2}\right)^{0.375}\right]}\right\}$	(6.6)	
$L=\frac{gT_s^2}{2\pi}tanh\left(\frac{2\pi h}{L}\right)$	(6.7)	
$H_s=0.0016\sqrt{\frac{FU_A^2}{g}}$	(6.8)	
$T_s=0.2714\sqrt[3]{\frac{FU_A}{g^2}}$	(6.9)	
$U_A=0.71U_{10}^{1.23}$	(6.10)	
$\tau_{cu}=10^{-4}\left(0.75+0.067U_A\right)\rho_a U_A^2$	(6.11)	ρ_a: density of air (1.24 kg/m^3 at a temperature of 10°C)

6.3.3 Geochemical Processes: Oxygen Transport

The performance of the water cover to limit the availability of DO to underlying tailings depends on the molecular diffusion of DO in the cover water and in the water-filled voids in the tailings. Oxygen is first dissolved from the air into the water through the water-air interface. The oxygen transfer from air to water cover is a complex process studied by several authors (Liss 1973; Li et al. 1997; Lee 2002; Schladow et al. 2002; Mölder et al. 2005). This transfer depends on the oxygen concentration gradient between the air and water phases, temperature (that controls the solubility), and turbulence. It should be mentioned that the maximal oxygen concentration that can be dissolved in water increases with decreasing temperature (≈12.8 mg/L at 5°C and mg/L ≈8.3 at 25°C; Otwinowski 1995).

The DO then diffuses through the water column until it reaches underwater tailings. When the cover water is stagnant (which represents an idealized case never encountered in nature), the DO concentration should be linearly decreasing from the top of the water column to the water-waste interface, whereas when the water is mixed over a given water column, the DO concentration is constant over the entire thickness of the water column (Li et al. 1997). Figure 6.5 illustrates this statement for the cases described above. These results were obtained via column test performed in the laboratory under two hydrodynamic conditions for 1 year: (1) 0.80 m stagnant water cover, and (2) 0.80 m mixed of water cover with stirrer at 87 RPM (Awoh et al. 2014).

The vertical transport of DO through the tailings relies on the gradient of oxygen concentration at the water-tailings interface, the reaction rate coefficient (K_r) of submerged sulfide tailings, the effective DO diffusion coefficient (D_e) in the pore water voids, and the rate of infiltration of water (if applicable) as described in Equation 6.12 (Li et al. 1997; Li and St-Arnaud 2000; Elberling and Damgaard 2001; Awoh et al. 2013b).

$$\frac{\partial C}{\partial t} = D_e \frac{\partial^2 C}{\partial z^2} \pm v_z \frac{\partial C}{\partial z} - K_r C \tag{6.12}$$

where C represents oxygen concentration in the mine wastes, D_e represents oxygen diffusion coefficient of the submerged wastes, K_r is the oxygen reaction rate coefficient of the wastes, v_z is the velocity of water infiltration into mine tailings (v_z is positive) or of the upward water movement

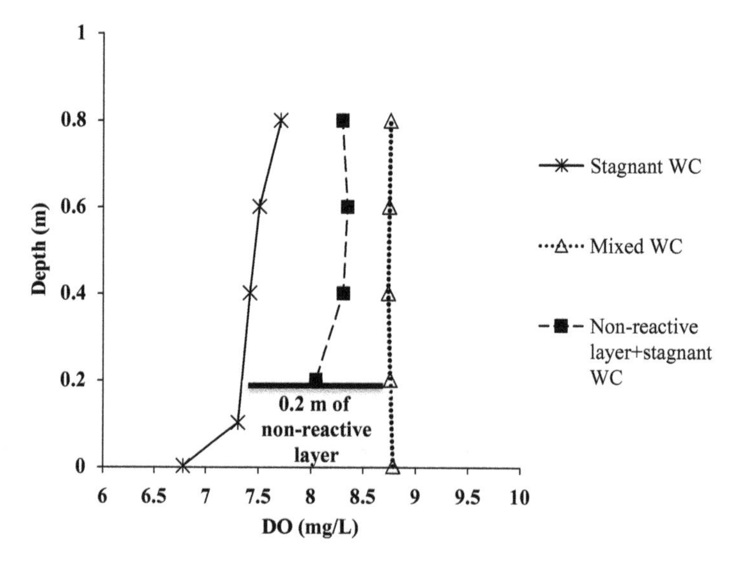

FIGURE 6.5 DO profile for stagnant water cover, mixed water cover, and stagnant water cover with protective layer placed on the mine wastes (Modified from Awoh et al. 2014).

from the wastes (v_z is negative), and z is the depth (z = 0 at the waste-water interface, positive in the downward direction).

D_e can be derived from the equation proposed by Aachib et al. (2004) (see Chapter 3, Section 3.5.3), considering that $\theta_a = 0$ and $\theta_w = n$. Awoh et al. (2013b) present an experimental method to determine K_r of submerged sulfide tailings. Placing a layer of non-reactive material on the tailings can contribute to reduce the oxygen flux toward the reactive tailings (Li et al. 1997; Yanful and Simms 1997; Awoh et al. 2014). Figure 6.5 illustrate the theoretical statement for this case, and we can see that the DO diffusion is similar to that of stagnant cover water.

Li et al. (1997) determined analytically the oxygen flux consumed by tailings placed in experimental cells with a water cover with a height of 30 cm. The average oxidation rate was 0.065 nmol of FeS_2 per m^2 pyrite surface per second. The annual oxygen flux were 3.5, 36, 46, and 76 $g/m^2/y$ in the case where the cover is (1) stagnant, (2) mixed, and fully oxygenated (constant DO concentration) without infiltration, (3) fully oxygenated with a downward infiltration at a velocity v = 1 m/y, and (4) fully oxygenated with a downward infiltration at a velocity v = 1 m/y and resuspension of tailings (of 2 mm layer), respectively. In the case of stagnant water cover, DO concentrations of 11.3 mg/L at the surface and 0 mg/L were considered; the penetration of DO in the tailings was about 5 mm. For the remaining cases 2 to 4, the OD concentrations at the surface and interface were similar (11.3 mg/L). In these cases, the penetration of DO increased to about 8 mm.

6.3.4 Geochemical Processes: Metal Release

The performance of a water cover is evaluated in terms of the quality of the cover water resulting from geochemical processes. This quality, controlled by regulatory requirements for final effluents, can be expressed by concentrations of dissolved metals and sulfates, and pH. The oxidation of thiosalts in the tailings that are disposed underwater can also contribute to lowering of the pH and hence acidification (Yanful and Samad 2004).

Dissolved metals and sulfates are released in the pore water and in the water cover by oxidation of sulfide minerals contained in settled tailings and suspended tailings, respectively (see Chapter 1). The pH can also be reduced depending on the intensity of sulfide oxidation and the neutralization capacity of tailings. Exchanges of dissolved metals between the water cover and the pore water occur according to three mechanisms: molecular diffusion, ascending advection, and tailings resuspension (Li et al. 1997). Molecular diffusion is the phenomenon by which metals move from higher to lower or nonexistent metal concentrations (due to a concentration gradient). Advection is induced by the upward infiltration of pore water transporting dissolved metals. Pore water can also be lifted together with the tailings due to resuspension.

Many studies on cover and interstitial water geochemistry focused on metal release. These studies were conducted in the laboratory (Karam and Guay 1994; Aubé et al. 1995; Catalan and Kumari 2005; Awoh et al. 2014), in the field (Holmström and Öhlander 1999; Vézina and Amyot 1999; Peacey et al. 2002; Awoh et al. 2013a), and by geochemical modeling (Bain et al. 2001; Peacey et al. 2002; Romano et al. 2003; Molson et al. 2004). Results indicated that the metal release phenomenon is less critical for underwater disposal of fresh tailing than flooding of AMD mine wastes that can be pre-oxidized and pre-contaminated (MEND 2001a). However, even for flooding, the impact of this metal release on the quality of the cover water may be negligible in the long term because of dilution from precipitation and fresh water inflows and flushing of the oxidation products down into the pore water (MEND 2001a; Peacey et al. 2002).

Laboratory tests and numerical modeling have shown that installing a protection or attenuation layer made of “inert” material (Li et al. 1997; Yanful and Simms 1997; Awoh et al. 2014) or peat (Yanful and Simms 1997) could eliminate the tailings resuspension, reduce metals and sulfate release, and improve the efficiency of the water cover in terms of the water quality (Li et al. 1997; Yanful and Simms 1997; Awoh et al. 2014).

6.3.5 Thermal and Chemical Stratification

Changes during the annual temperature cycle can produce thermal stratification in deep temperate pit lakes (Gammons and Duaime 2006; Geller et al. 2013) that distinguish holomictic and meromictic lakes. Stratification processes in the water column induce seasonal thermal and chemical turnover events and related DO profiles.

6.3.5.1 Holomictic Lakes

Holomictic lakes undergo a complete top-to-bottom turnover during the spring and fall as explained below (Gammons and Duaime 2006; España et al. 2008; Geller et al. 2013). In spring, the temperature is uniform throughout the water column. The heat exchange between the atmosphere and the water increases the temperature of the surface water. With the arrival of summer, the low efficiency of the diffusion/conduction processes of the sun energy in water induces an increasingly significant thermal gradient between water in direct contact with the atmosphere and water in depth. Thus, a temperature gradient is formed and the turbulent action due to the wind develops stratification, that is, several layers with different physical properties due to the density-temperature relationship of water. The upper warm layer, called epilimnion, is of low density with an almost homogeneous temperature and can be easily stirred by the wind. It should be recalled that the density of freshwater water reaches its maximum (1000 kg/m^3) at the temperature of around 4°C and decreases on either side of this maximum to about 997 kg/m^3 at 25°C and to about 999 kg/m^3 for liquid water at 0°C and 915 kg/m^3 for solid water (ice). The presence of salt can increase the water density. Furthermore, the density change per degree of temperature change is higher in warm water than in cold. The deep cold layer, called hypolimnion, is characterized by almost homogeneous temperature and high density. A transition layer of rapid temperature and density change (the metalimnion) separates the epilimnion and hypolimnion. The depth with the maximum of the thermal gradient is called a thermocline. Figure 6.6a illustrates this summer stratification.

In autumn, the cooling of the epilimnion induces the appearance of convection currents (colder temperature and higher density), which descend in the water column to the layers of water of equivalent density. The more the temperature decreases, the more these currents deepen the stratification until its complete destruction by a complete mixing of the water column; this is the autumnal turnover. During turnover periods, the thermal gradient disappears and allows the convection of oxygen and other chemical compounds throughout the body of water.

In winter, a reverse thermal stratification develops where the coldest waters with the highest density are found above waters with warmer temperatures. A spring turnover establishes a uniform temperature throughout the water column and the cycle continues.

6.3.5.2 Meromictic Lakes

Meromictic lakes undergo only a partial annual mixing (Gammons and Duaime 2006; España et al. 2008; Geller et al. 2013). This is the case for lakes that contain a bottom layer of highly saline water that has a density too high to mix with overlying waters, regardless of the temperature profile. In general, the effect of changes in salinity on the density of water is greater than the effect of changes in temperature. A shallow epilimnion and a mid-level hypolimnion (together called mixolimnion) are subjected to mixing, while the lower layer (monimolimnion) is dense and salty and doesn't overturn seasonally. The interface between the mixolimnion and monimolimnion, which corresponds to an interface in chemical composition or salinity is often referred to as a chemocline. This interface separates oxygenated mixolimnion with dominantly oxidizing conditions from an anaerobic and dominantly reducing monimolimnion (España et al. 2008). That is, there is a chemical stratification. Figure 6.6b illustrates this stratification in meromictic lakes. The relative depth Z_{rel} (ratio of maximum depth of the pit Z_{max} and the average diameter of the surface area A, in percent [Z_{rel} (%) $= 100 \times Z_{max} \times \pi^{0.5})/(2 \times A^{0.5})$] is the main factor in the development of meromixis (Gammons and Duaime 2006). Meromixis occurs in deep pit lakes with high relative depth more commonly than in

a) Holomictic pit lake

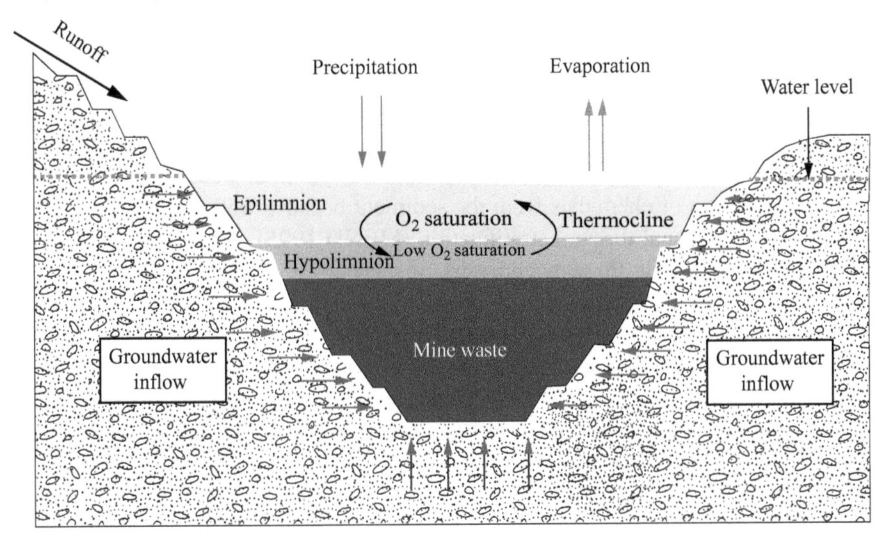

b) Meromictic pit lake

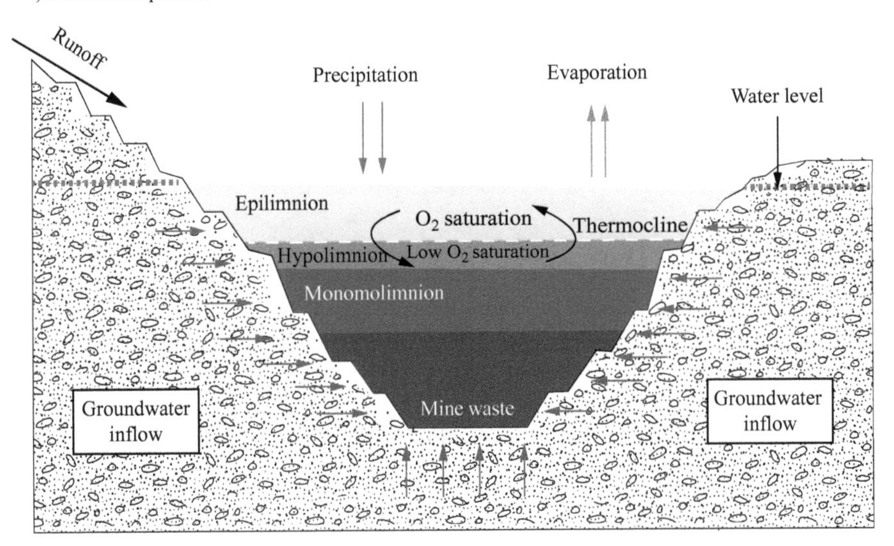

FIGURE 6.6 Schematic illustration of the concept of (a) holomictic and (b) meromictic lakes.

natural lakes because of the small relative depth in natural lakes. Metal-mine pit lakes have much higher values with 15% ≤ Z_{rel} ≤ 45%, while well-mixed natural lakes usually have Z_{rel} ≈2% or less (Geller et al. 2013). Pit lakes are meromictic when Z_{rel} ≥ 10% to 40% and holomictic for Z_{rel} < 10% (Gammons and Duaime 2006; España et al. 2008). The wide range of values Z_{rel} leading to meromictic conditions is imputable to the influence of the vertical gradients in water temperature and salinity and the velocity, duration, and direction of surface winds on the chemical stratification (Gammons and Duaime 2006). Hence, the relative depth Z_{rel} alone cannot constitute a meromixis criterion.

6.3.5.3 Practical Significance

Meromixis is an option to prevent and manage environmental impacts of hazardous mine wastes and mine water. Meromixis is helpful in the successful long-term storage of waste in pit lakes by establishing the stability of the conditions inside the waste and the water column (MEND 1995; Schultze et al. 2011). Indeed, meromixis is accompanied by strong anoxia, occurrence of hydrogen sulfide, and precipitation of metal sulfides in the monimolimnion. Meromixis condition represents an advantage due to the very limited exchange with the rest of the lake (the stratification forms an additional barrier to contaminant migration) (Boehrer and Schultze 2006). When the lake undergoes a major turnover event, chemical or biological reactions occur that can transform the water quality of the mixolimnion. The shaping of the final mine void and the final water level have considerable influence on the recycling of chemicals from the sediment because water depth is a decisive factor for the occurrence of seasonal thermal stratification (MEND 1995). For deep water pits, the mass transfer equation must include these phenomena of meromixis (chemical stratification), thermal stratification, or anaerobic conditions.

MEND (2015) presents cases of in-pit disposal of reactive mine wastes where meromixis occurs: the Owl Creek pit lake located 18 km northeast of Timmins, Ontario, Canada. The Owl Creek pit lake is a meromictic lake that is roughly 23 m deep after being backfilled with waste rock, possibly due to a higher concentration of dissolved salts and metals (including Cd, Co, Fe, Pb, and Ag) and sulfate in the lower levels. The upper 8 m of waters (chemocline) are affected by climatic conditions and maintain 60% to 100% oxygen saturation depending on the season. Waters below the chemocline are anoxic (<20% saturation) and stagnant for many years.

6.4 DESIGN METHODOLOGY

The design of water cover technology is site specific as it depends on the waste properties, site topography, impoundment volume, exposed surface, and climatic conditions. However, there are necessary steps for designing the right water cover for a given site and type of water cover (subaqueous disposal in engineered reservoir or pit lake).

6.4.1 Direct Underwater Disposal in Constructed Reservoirs

For constructed reservoirs surrounded by impervious dikes, Aubertin et al. (2002) proposed a logical sequence of steps to follow for designing water covers that submerge tailings. The main steps are summarized in Figure 6.7:

1. Collect available climatic and hydrological conditions on the selected site and data on the sedimentation characteristics of the tailings.
2. Collect and consult current and previous studies that have focused, directly or indirectly, on the technology of water cover.
3. Determine regulatory requirements, focusing mainly on the quality criteria of the final effluent and on the risk levels considered acceptable (particularly about recurrence periods for the hydrogeological and hydrodynamic design).
4. Obtain data from milling/processing (production rate) in cooperation with those responsible for the mineralurgical process of the mine, from the physical (relative density and grain-size distribution), chemical, and mineralogical characterization tests of the tailings and from the tailings reactivity tests, etc.
5. Design the reservoir from the available information: Calculate the initial impoundment volume, which can be adjusted over the years depending on expansion strategies, considering wind direction and intensity to properly select the location (e.g., avoiding placing the largest side of the tailings pond in the direction of the prevailing winds to limit the fetch).
6. Integrate recurrence periods into the design, considering critical values of winds and waves characteristics, precipitation, and runoff.

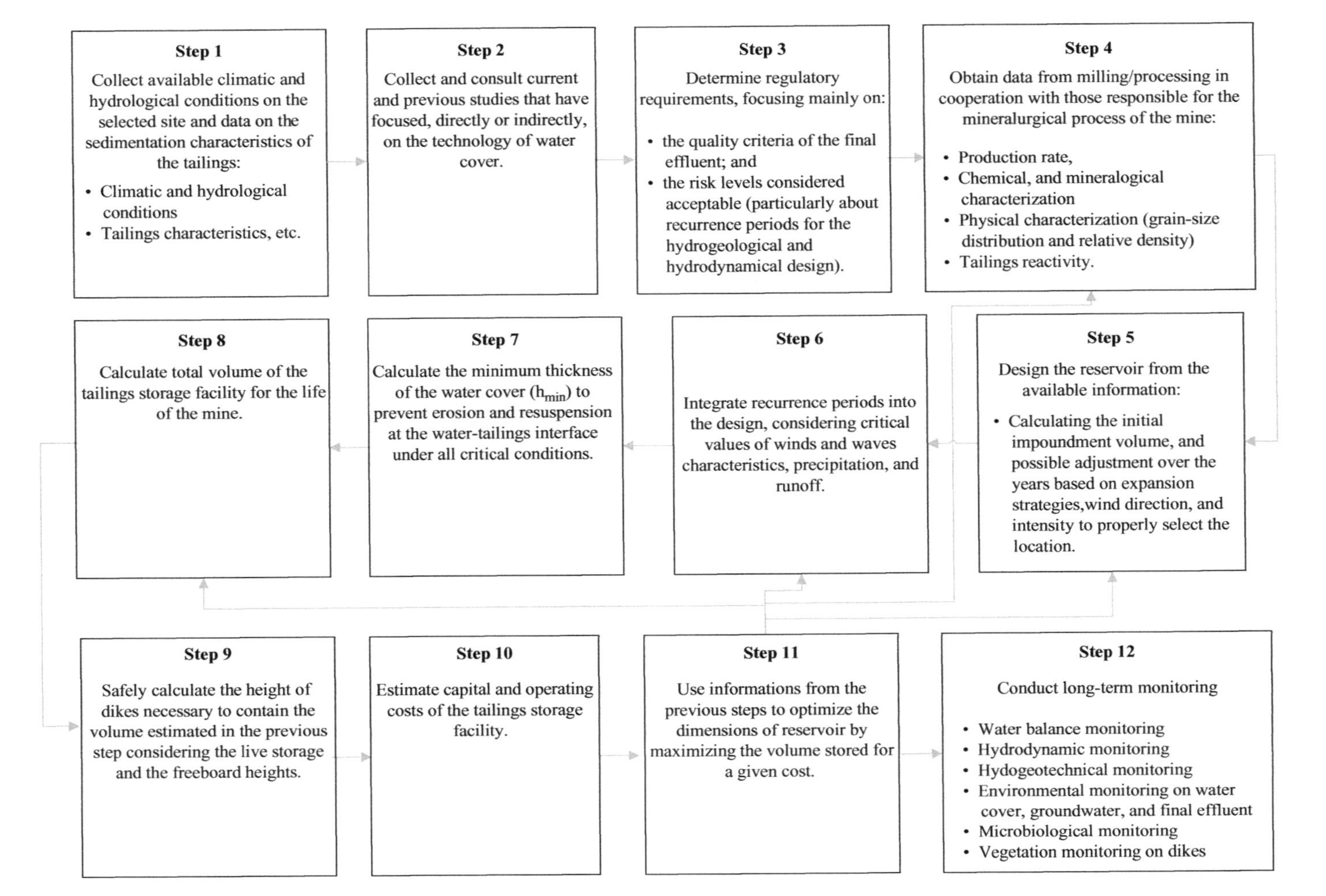

FIGURE 6.7 Flow diagram showing the steps for designing water cover in constructed reservoir (inspired and translated from Aubertin et al. 2002).

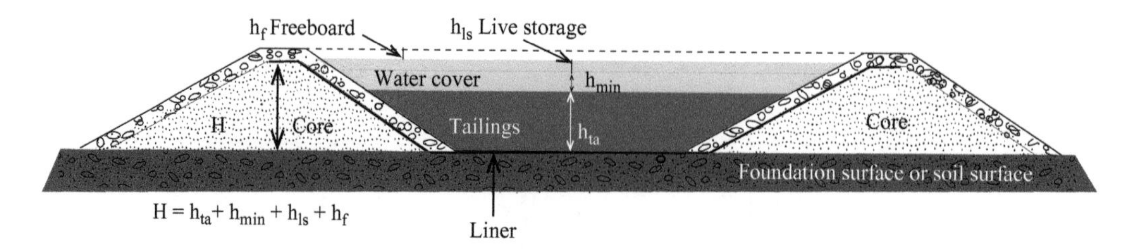

FIGURE 6.8 Determination of the containment dikes height (Modified from Atkins et al. 1997).

7. Calculate the minimum thickness of the water cover (height h_{min}, see Figure 6.8) to prevent erosion and resuspension at the water-tailings interface under all critical conditions.
8. Calculate total volume of the tailings storage facility (height h_{ta}; see Figure 6.8) for the life of the mine.
9. Safely calculate the height of dikes necessary to contain the volume estimated in the previous step considering the live storage (h_{ls}) and the freeboard (h_f) heights (see Figure 6.8).
10. Estimate capital and operating costs of the tailings storage facility.
11. The information from the previous steps is used to optimize the dimensions of the reservoir by maximizing the volume stored for a given cost. This can be done in collaboration with the production conditions (operation), especially if it is possible to adapt the production rate or other characteristics of the tailings.
12. Conduct long-term monitoring (see Section 6.4.3)

Step 7, determining the minimum thickness of the water cover (h_{min}) to prevent the erosion and resuspension of submerged tailings (see Section 6.3.1), is crucial for a state-of-the-art design of a shallow water cover. The minimum water depth h_{min} is the value that satisfies the condition τ_{tot} ($= \tau_{wa} + \tau_{cu}$) $< \tau_{cr}$ in general, and particularly $\tau_{wa} < \tau_{cr}$ when τ_{cu} is negligible so that $\tau_{tot} \approx \tau_{wa}$. As the water cover depth (h) is involved in the simple Equations 6.7 to 6.9 used to calculate the shear stress τ_{wa} for deep-water waves ($h/L_s > 0.5$), a simple analytic relationship was developed to estimate h_{min} (Lawrence et al. 1991; Atkins et al. 1997; Yanful and Simms 1997). The minimum water depth h_{min} is expressed as a function of the wind stress factor U_A (Equation 6.10), the fetch (F), and the critical velocity at the tailings bed (V_{cr}) (Lawrence et al. 1991; Atkins et al. 1997) using Equation 6.13.

$$h_{\min} = 1.58\times10^{-3}\, g\left(U_a F\right)^{2/3} ln\left[5.14\times10^{-2}\frac{\left(FU_a^4\right)^{1/6}}{V_{cr}}\right] \tag{6.13}$$

The critical velocity at the bed V_{cr} used in Equation 6.13 can be obtained with the following Equation 6.14:

$$V_{cr} = \left[0.21g\left(\frac{\rho_s - \rho_w}{\rho_w}\right)\left(\frac{D_{50}T_s}{\pi}\right)^{1/2}\right]^{2/3} \tag{6.14}$$

where g is the gravitational acceleration (m/s^2); ρ_s and ρ_w are the density of the sediment and water (kg/m^3), respectively; D_{50} is the mean tailings diameter (mm); and T_s is the significant wave period (s) (given by Equation 6.9 for deep-water waves).

For shallow-water waves ($h/L_s < 0.5$), the Equations 6.5 and 6.6 describing H_s and T_s, respectively, are complex so that an analytical solution of h_{min} cannot be easily derived. In this case, h_{min} can be calculated numerically or determined graphically.

In cold climates, the minimum thickness of the water cover must avoid freezing of the water cover at the bottom, which can cause resuspension in the same way as for waves. This has to be considered for each specific case using numerical simulations.

Equation 6.14 shows that h_{min} is directly proportional to the fetch F. Dividing the tailings impoundment in cells separated by wave breaker reduces the fetch F and consequently the thickness of the water cover h_{min} required to eliminate tailings resuspension.

A hydrological and hydrogeological design is required to maintain the minimum water depth h_{min} any time considering the water balance (see Section 6.3.1). Figure 6.8 (inspired from the study by Atkins et al. 1997) presents a typical water cover confined within constructed dikes where an impervious core and bottom liner are used to minimize water outflow. If the height h_{ta} (and volume) of the tailings to be stored and the minimum water depth h_{min} is known (from steps 7 and 8, respectively), the height of the containment dike H (step 9) can be determined as $H = h_{ta} + h_{min} + h_{ls} + h_f$. h_{ls} is the live storage height required to compensate water loss by potential evaporation, infiltration, and percolations through dikes. h_f is the freeboard height required to avoid dam overflowing even if an adequate system for regulating the water level in the basin must be installed to keep the maximal water cover depth to $h_{min} + h_{ls}$. The freeboard height is linked to the level of uncertainty attached to the determination of design events. The dimensions of the basin can be optimized for maximizing the volume of tailings stored, considering all costs (step 10).

6.4.2 Direct Underwater Disposal in Pit Lake

In the cases of direct in-pit disposal of mine wastes, steps 4, 5, and 9 described earlier (see also Figure 6.7) are not required for the design methodology. However, information concerning the pit (that may not be needed for mining) is necessary for pit disposal design and engineering, including the following:

- Pit hydrology and hydrogeology: to determine where the final location of the water table and the rate at which it will be reestablished considering mining-induced fracture zones, open drill holes and zones of high hydraulic conductivity (e.g., fault zones);
- Physical stability of the pit walls: to avoid the walls becoming unstable due to an increased hydraulic pressure as a pit lake develops;
- Wall rock characteristics: to predict the potential for acid generation and metal leaching by broken and exposed wall rock and determine the hydraulic conductivity of undisturbed and blast-fractured wall rock.

Furthermore, the physical, mineralogical, and chemical characteristics of wastes must be determined at the time of potential placement in the open pit (these characteristics can be different to those of wastes when they were produced).

As mentioned above, the water cover must act as a long-term passive hydraulic sink or trap. Special attention must be taken when the water table is too deep so that an adequate system for regulating the water level in the basin (e.g., using a weir) cannot be installed. When groundwater barriers are used to control the discharge of impacted water to groundwater, these barriers must be designed according to the state of the art (MEND 1995; Thériault 2004).

6.4.3 Flooding of Mine Wastes Stored in Existing Structures

In the case of flooding of mine wastes already stored in existing surface and in-pit ponds, the design methodology can be limited to steps 2, 3, 7, 9, 10, and 11 described above (see also Figure 6.7).

For surface tailings impoundments, flooding can result in raising of the dikes to create a sufficient height above the tailings that corresponds to the sum of the minimum thickness of the water cover h_{min}, the live storage height (h_{ls}), and the freeboard height (h_f) (see Figure 6.8).

The physical, mineralogical, and chemical characteristics of wastes (eventually pre-oxidized) must be determined during flooding to help understand the hydrogeochemical and environmental behavior of the water cover as monitored in step 11.

6.4.4 Monitoring Program

The last and important step for the design of all types of water cover involves implementing a long-term monitoring program of the installations. Such monitoring is required to evaluate the performance of reclamation techniques (see more details in Chapter 10). Mitigative measurements must be quickly implemented to address any anomaly. The monitoring program for water covers includes several components (monitoring equipment is not provided here):

- Water balance monitoring to assess the post-closure water balance: measuring hydrological parameters (liquid and solid precipitations, evaporation, and water inflows and outflows) and meteorological data (temperature, relative humidity, and solar radiation).
- Hydrodynamic monitoring to check if the calculated cover depth is appropriate: measuring additional meteorological data (wind speed and direction, and temperature), concentration of suspended solids in water.
- Hydrogeotechnical monitoring to assess the physical stability of dams and dikes in the case of engineered reservoirs: the most critical structure when a water cover is used is the physical stability of the dikes. The development of pore pressures and hydraulic gradients can affect the stability of dikes and can cause internal erosion. Understanding the evolution over time of water pressures and hydraulic gradients is therefore an important aspect that should not be overlooked. Then, special measures must be dedicated to the maintenance of dikes, spillways, and trenches. Regular inspections, annual geotechnical inspections, and dam safety revisions are also required. Special attention should also be paid to the spillways. Indeed, beaver activities can block the spillways and increasing the water cover depth. The water level in the water cover and underground must be monitored to verify if the hydraulic trap is maintained at all times.
- Environmental monitoring to assess the risks of contaminant migration from water cover to the final effluent (which can cause damage to the neighboring ecosystems, in particular the fauna and flora) and also from the water cover to the groundwater if the principle of hydraulic sink or track is compromised (which can affect groundwater quality used as drinking water). The quality of the cover water, interstitial water in the submerged wastes, final effluent, and groundwater should be monitored in terms of pH, conductivity, redox potential, sulfate, dissolved metals, toxicity, and suspended particles concentration in the final effluent.
- Microbiological monitoring to assess the evolution of the microbiological populations dominant in the acid generation process and the formation of biofilm over submerged wastes that can affect sulfide oxidation and metal mobilization. Methods to characterize the biofilm are described in MEND (2010).
- Surface vegetation monitoring to limit the growth of trees that could affect the physical integrity of the dike. Vegetation on dikes can be characterized to identify species to be maintained and removed.

6.5 CASE STUDIES

The water cover reclamation techniques described in this chapter have been used in plenty of mine sites. Table 6.4 presents sites where direct underwater disposal and flooding in constructed reservoirs were used. Table 6.5 lists sites where underwater in-pit disposal was applied. These tables are limited to sites for which studies have been conducted, and therefore, information can be found in the literature. The tables are undoubtedly incomplete (e.g., they lack information for several countries with an industrial mining culture). Location, the annual precipitation and average wind speed, type of submerged waste (tailings and/or waste rocks), and the source of data are provided.

TABLE 6.4
Mine Sites Using the Water Cover Technique: Direct Underwater Disposal and Flooding Constructed Reservoirs

	Mine Site	Location	Intermediate Field Scale	Full Scale (hwc)	Type of Mine Waste	Sources
Direct underwater disposal	Hjerkinn	Norway	(——)	Area = 100 ha, Depth = 1.5 m, design depth = 1.5 m	Tailings	MEND (1997), Haakenstad and Haugen (2013)
	Lokken	Norway	(——)	Area = 20 ha, Depth = 1–4 m	Tailings	MEND (1997)
	Louvicourt	Quebec, Canada	Experimental cells 21 m × 21 m and 0.3 m of water cover	Area = 150 ha, Depth = 0.3–1 m	Tailings	MEND (2001b), Julien et al. (2003), Vigneault et al. (2007)
Flooding	Equity Silver	Houston, British Columbia, Canada	(——)	Area = 120 ha, Min depth =1.5–1.7 m, Max depth = 7.5 m, Median depth = 4.3 m, Design depth = 1.4 m	Tailings	MEND (1997, 1998, 2001), Price and Aziz (2019)
	Heath Steele	Miramichi, north-central New Brunswick, Canada	(——)	Area = 215 ha, Min depth = 0.3 m, and Max depth = 6 m, Fetch = 1 km	Oxidized and non-oxidized Tailings	Mian et al. (2007), MEND (2010)
	Mattabi	Ontario, Canada	Experimental cells 70 m × 70 m and 1-1.7 m of water cover	Area = 125 ha, Max depth = 10 m,	Tailings	MEND (1997, 2000)
	Quirke cell 14, Elliot Lake	Northern, Ontario, Canada	(——)	Area = 192 ha, Min depth = 0.35 m, and Mean depth =1 m, Fetch =1.5 km	Tailings	Adu-Wusu et al. (2001), Mian and Yanful (2003)
	Shebandowan	Thunder Bay, northwestern Ontario, Canada	(——)	Area = 115 ha, Min depth ≈ 1 m, and Max depth > 2 m, Fetch = 1.65 km	Tailings and waste rock	Kachhwal et al. (2011)
	Stekenjokk	Sweden	(——)	Area = 110 ha, Min depth = 0.2, Max depth = 9 m and Mean depth = 2.2 m, Fetch = 0.05–1.1 km, adding 0.5 m of sand layer	Fresh and partially oxidized tailings and dissolution of gypsum in the oxidized tailings	MEND (1997), Holmström and Öhlander (1999), MEND (2006)

TABLE 6.5
Mine sites using the water cover technique: In-pit disposal

	Mine site	Location	Intermediate Field Scale	Full Scale	Type of Mine Waste	Sources
In-pit disposal	Berkeley Lake	Montana, USA	(——)	Area = 128 ha, Max depth = 175 m	Lime sludge treatment	Davis and Ashenberg (1989), MEND (1995), Doyle and Runnells (1997), Golder (2017), Castendyk and Eary (2009)
	B-Zone pit lake	Northern, Saskatchewan, Canada	(——)	Area = 24 ha, Max depth = 57 m	Tailings	Rowson and Tremblay (2005), Pollock (2008), Golder (2017)
	City Resources Limited (Consolidated Cinola Mines Ltd.) site	Queen Charlotte Islands, British Columbia, Canada	(——)	(-)	Waste rock	MEND (1989)
	Deilmann TMF site	Key lake, Saskatchewan, Canada	(——)	Max depth = 40 m	Tailings	MEND (1995), Rowson and Tremblay (2005), Pollock (2008), Golder (2017)
	DJX pit lake	Cluff lake, Saskatchewan, Canada	(——)	Area = 8 ha; Max depth = 90 m	Tailings	Rowson and Tremblay (2005), Pollock (2008), Golder (2017)
	Don-Rouyn pit lake	Rouyn-Noranda, Quebec, Canada	(——)	Area = 7.7 ha, Max depth = 2.81m, Average depth = 1.23 m, median depth = 1.16 m, and Min depth = 0.81 m, Design depth = 1 m Fetch = 392 m	Tailings	GEOCON (1997), Mbonimpa et al. (2008), Awoh et al. (2013a)
	Endako Mines Division site	Fraser lake, British Columbia, Canada	(——)	(-)	Tailings	MEND (1989)
	Gunnar pit lake	Uranium city, Saskatchewan, Canada	(——)	Area = 7 ha, Max depth = 110 m	Waste rock	Tones (1982), MEND (1995), SRK (2008), Golder (2017)
	JEB pit	Northern, Saskatchewan, Canada	(——)	Area = 14 ha, Max depth = 118 m	Tailings	Rowson and Tremblay (2005), Pollock (2008), MEND (2015), Golder (2017)
	Jundee pit	Little Sandy desert, Australia	(——)	(——)	Tailings and Waste rock	Shaw et al. (2006), MEND (2015), Golder (2017)
	Geierswald pit lake	Germany	(——)	Area = 620 ha, Max depth = 35 m	Lime Waste	Golder (2017), Schultze et al. (2011)
	Kepwari pit lake	Collie, Western Australia	(——)	Area = 98 ha, Max depth = 65 m	Waste rock	McCullough et al. (2012b), Golder (2017)
	Main zone pit lake, Equity silver mine	Houston, British Columbia, Canada	(——)	Area = 20.5 ha, Max depth = 120 m	Waste rock	Crusius et al. (2003), MEND (2015), Golder (2017)

In-pit disposal	Nifty Copper Pit Lake	Western Australia	(——)	Area = N/A, Max depth = 60 m	Waste rock	McCullough et al. (2012a), Golder (2017)
	No. 3 Pit Lake	Stratford, Quebec, Canada	(——)	Area = 1.4 ha, Max depth = 25 m	Waste rock	Golder (2017)
	Owl Creek Pit Lake	Timmins, Ontario, Canada	(——)	Area = 15 ha, Max depth = 100 m	Waste rock	MEND (1995), MEND (2015), Golder (2017), Martin et al. (2016)
	Phoenix Copper Mine	Phoenix, British Columbia, Canada	(——)	(-)	Tailings	MEND (1989)
	Rabbit Lake In-Pit TMF	Rabbit lake, Saskatchewan, Canada	(——)	Area = 17 ha, Max depth = 122 m	Tailings	Rowson and Tremblay (2005), Pollock (2008), MEND (1995), MEND (2015), Golder (2017)
	Runstädter See	Germany	(——)	Area = 622 ha, Max depth = 28 m	Industrial waste ash and nitrate production	Fritz et al. (2001), Golder (2017)
	Solbec-Cupra Mine	Montreal, Quebec, Canada	Experimental cells 3 m × 3 m and 1.34 m of water cover without sand and 0.74m of water cover with sand	Area = 66 ha, Design depth = 1m, Fetch = 1.5 m	Oxidized tailings and calcite dust	MEND (1993), Karam and Guay (1994), MEND (1995, 2002, 2015), Vézina and Amyot (1999),
	South Mine Pit Lake	Tennesse, USA	(——)	Area = 8 ha, Max depth = 61 m	Water sludge treatment	Wyatt et al. (2006), Golder (2017)
	South Pit	Campbell River, British Columbia, Canada	(——)	Area = 2.5 ha, Max depth = 2 m	Coarse coal waste	Golder (2011, 2017)
	Sphinx Lake	Cadomin, Alberta, Canada	(——)	Area = 6.4 ha, Max depth = 50 m	Waste rock	EPLWG (End Pit Lake Working Group) (2004), Golder (2017)
	Springer Pit Lake	Mount Polley Mine, British Columbia, Canada	(——)	Area = 65.2 ha, Max depth = 50 m	Tailings	McCullough et al. (2020), Golder (2017)
	Trojan Pond	Logan lake, British Columbia, Canada	(——)	Area = 26 ha, Max depth = 10 m	Tailings	Otchere et al. (2002), Golder (2017)
	Waterline Pit Lake, Equity Silver Mine	Houston, British Columbia, Canada	(——)	Area = 2.6 ha, Max depth = 40 m	Waste rock	Crusius et al. (2003), MEND (2015), Golder (2017)

The three cases studies in this section involve a subaqueous or underwater disposal of tailings in a constructed reservoir (Louvicourt site), in an open pit (Don Rouyn site), and on a flooding (Heath Steele site). Emphasis will be placed on results gathered from studies conducted on these sites to assess the performance of each water cover.

6.5.1 Louvicourt (Quebec, Canada)

Louvicourt Mine was an underground Cu/Zn mine located about 15 km east of Val d'Or, Quebec (Figure 6.9). It was in operation between July 1994 and July 2005. When the mine closed in July 2005, 6.6 million tons of PGA tailings containing 30% to 50% pyrite, 5% to 24% carbonates, and 0.6% sphalerite (MEND 2000) were disposed directly underwater in a 1.5×10^6 m^2 constructed reservoir (MEND 2007; Ouellet et al. 2011). The shallow water cover design depth defined after a hydrogeological evaluation was 1 m (Julien et al. 2003). The average wind speed was 3.6 m/s or 13 km/h (Li et al. 1997). The average precipitation, evaporation, and wind speed were typically 920 mm, 510 mm, and 7 to 11 km/h, respectively (MEND 2001b; Julien et al. 2003; Vigneault et al. 2007).

The monitoring program performed since the construction of the tailings management facility consisted mainly of geotechnical monitoring (monthly routine inspection, annual geotechnical inspection, and dams inspection every five years), environmental monitoring at the tailings pond and groundwater and vegetation monitoring on dikes (every three years, vegetation on dikes is removed) (Ouellet et al. 2011). Since 2007, a decreasing trend in Cu and Zn concentrations (initially 0.7 mg/L for Zn and ≈0.12 mg/L for Cu) has been observed. Results obtained up to 2010 showed that the final effluent and seepage water quality complied with Canadian and Quebec (Directive 019) regulations (acceptable average monthly concentrations of 0.3 and 0.5 mg/L for Cu and Zn, respectively). Groundwater showed no signs of contamination. According to Ouellet et al. (2011), the challenge with the Louvicourt water cover remains the consistent maintenance to ensure the physical stability of the dams. A monitoring campaign initiated in 2009 (MEND 2010) revealed patches of iron oxyhydroxide of less than 5 mm thickness scattered over the tailings. These patches were colonized by a microbial population. The depth of the water cover was uniform, and the water cover was relatively free of suspended tailings (under the monitoring conditions). DO measurements indicated that the water column was well oxygenated with concentrations ranging from 9.3 to 9.5 mg/L at temperature around 16°C, which almost correspond to the maximal oxygen concentration that can be dissolved in water at this temperature (Otwinowski 1995).

FIGURE 6.9 View of water cover at the Louvicourt site (Courtesy of Louvicourt Mine, Industrial Chair Polytechnique-UQAT on Environment and Mine Waste Management).

6.5.2 Don Rouyn (Rouyn-Noranda, Quebec, Canada)

The Don Rouyn site is an old porphyry copper deposit located 4 km west of Rouyn-Noranda (QC, Canada). This deposit was exploited intermittently in former open-pit mine from 1959 to 1980. When the operation was completed, the pit covering an area of 7.7 ha was flooded with natural water (precipitation, infiltration etc.). Then, this pit was used to deposit highly sulfide-rich and PAG tailings from the Gallen Mine between 1999 and 2000, thus forming the Don Rouyn water cover (Figure 6.10). The average annual precipitation was about 868.7 mm and the wind speed ranged between 10 and 20 km/h.

The minimal water cover thickness defined at the design stage was 1 m (GEOCON 1997). Water level measurements made on this site have shown that the maximum, average, median, and minimum depths of the water cover above tailings surface are 2.81, 1.23, 1.16, and 0.81 m, respectively (Mbonimpa et al. 2008). Also, to verify the hydraulic connection between the water cover and the groundwater, monitoring wells were installed around the Don Rouyn pit in 1997. The water elevation at the Don Rouyn site relative to mean sea level is maintained at 308.5 m by a weir built in a channel to discharge the water into the environment. This water lever control maintains the hydraulic trap.

A monitoring program was performed by Awoh et al. (2013a) during the summer and autumn of 2008, 2009, and 2010. For that purpose, various in situ measurements and laboratory tests were conducted: (1) characterizing the tailings samples; (2) monitoring the water quality in the final effluent, shallow water cover, and groundwater around the pit. In situ measurements included the vertical profile and spatial distribution of pH, temperature, DO, electric conductivity, and redox potential. Suspended tailings and wind speed and direction were also continuously monitored. Groundwater, cover water, and final effluent water samples were chemically analyzed in the laboratory while suspended particles were determined only on water cover. Physical, mineralogical, and chemical tailings properties were also determined. Results obtained showed that Don Rouyn tailings contained a high pyrite content between 82% and 84%, which gives a high specific gravity (4.6) that is close to that of pure pyrite (5.01; Landry et al. 1995). The geochemical quality of final effluent and groundwater complied with Canadian and Quebec regulations. The concentrations of As, Pb, and Ni were generally below the detection limit. Fe and Zn concentration ranged from 0.006 to 0.057 mg/L and from 0.03 to 0.12 mg/L, respectively. Sulfate concentrations varied

FIGURE 6.10 View of the water cover at the Don Rouyn site.

between 91 and 128 mg/L. The suspended particles measured were ranged from 0.91 to 18.5 mg/L. According to the results of the water levels in the water cover and in the piezometers (set up around the pit), the water cover acts as a hydraulic trap, which captures the groundwater toward the pit. In addition, the hydrochemical facies of groundwater (bicarbonate facies) determined with Piper diagram is different from that of cover water (calcium sulfate water). These results suggested that for the monitoring period from 2008 to 2010, the cover water didn't affect the surrounding groundwater.

6.5.3 Heath Steele (New Brunswick, Canada)

The Heath Steele Mine site is located 50 km northeast of Miramichi, north-central New Brunswick, Canada. Tailings are mainly composed of 50% to 65% pyrite and pyrrhotite, sphalerite, and galena in small amount, as well as quartz, smectite, calcite, and mica (Mian and Yanful 2003, 2004).

About 21 million tons of tailings were flooded in 1996 (St-Germain et al. 1997; Peacey and Yanful 2003). In the northeast corner of the tailings pond, the tailings were covered with high-density sludge (MEND 2004). The water level in 2001 varied between 0.3 and 6 m. To reduce the fetch F m, this tailings pond with a total area of 215 ha is divided into three cells (north, upper, and lower cells), which are all hydraulically connected (Mian and Yanful 2003, 2004; MEND 2004). The mean annual precipitation, evaporation, and wind speed were typically 1118 mm, 714 mm, and 8 to 20 km/h, respectively.

The performance of water cover at the Heath Steele Mine was monitored in terms of surface water quality at various locations in the three cells (Peacey and Yanful 2003; MEND 2004). Zn was the most significant metal detected downstream (Peacey and Yanful 2003; MEND 2004). The values of pH and Zn in the water covers (north, upper, and lower cells) right after flooding (1996) varied between 3.7 and 6.3 and between 0.02 to 10.7 mg/L, respectively. The Zn concentration at the final effluent was 0.12 mg/L. To respect the environmental criteria, it was necessary to add lime regularly in the water cover (from the internal dam spillway). Thus, the results obtained from investigations conducted in 2002 showed that pH values ranged from 7 to 11, while Zn concentrations ranged from ≈ 0.01 to 0.9 mg/L in water covers and 0.09 mg/L in the final effluent (MEND 2004). Mian and Yanful (2003) also investigated hydrodynamic conditions for erosion and tailings resuspension at the water-tailing interface. The results showed that resuspension likely occurred. Indeed, the shear stresses exceeded the critical shear stresses required for erosion and resuspension when wind speeds were greater than 9 m/s; the average hourly wind speed on the site exceeded 9 m/s frequently (Mian and Yanful 2003). The results of the resuspension measurements using sediment traps showed that resuspension depended on water cover depth and wind-generated stresses, but also on the cohesion and self-consolidation of the wastes as well as the surface topography of the area that determines whether more or less wind obstruction occurs. Monitoring conducted between 1996 and 1999 showed that Zn concentrations increase with suspended particles (Peacy and Yanful 2003). This indicates that the oxidation of suspended particles contributes significantly to Zn dissolution.

6.6 MAIN ADVANTAGES AND LIMITATIONS OF THE TECHNIQUE

6.6.1 Advantages

Well-designed water covers have several advantages compared to other reclamation techniques:

- They immediately limit the reactivity of the mine waste and prevent AMD generation.
- They are an economical approach for countries in which the climate is favorable (precipitation significantly higher than evaporation) such as Canada, even if the regulations allow less and less the application of water cover (Kachhwal 2011).

- They eliminate wind erosion and dust emission from potentially PGA mine wastes and contribute to improving the aesthetic and quality of the environment surrounding the mining sites.
- When pit lakes are used for in-pit mine wastes disposal, the following advantages can be mentioned (Golder 2017):
  - reduce the need for aboveground mine wastes storage facilities as the pit allows the storage of a large amount of mine wastes;
  - offer geotechnical stability because the pit lake is completely enclosed by rock walls compared to the constructed reservoir surrounded by built dikes;
  - act as a hydraulic trap and avoid contamination of surrounding groundwater; and
  - ensure a long-term physical and geochemical stability of mine waste.

6.6.2 Limits

The water cover technology has also major limits:

- There is a necessity to ensure a regular long-term structural integrity and safety of the impoundments (dams and dikes) for constructed reservoirs because a failure could cause significant and dramatic environmental disasters.
- The long-term quality of the final effluent must be ensured to meet environmental criteria. If the environmental criteria are not respected, which can likely occur in the case of flooding, this could lead to the establishment of a water treatment plant before returning final effluents to the environment.
- The installation of a barrier to discourage human access is required to ensure public and wildlife safety measures around the water cover, particularly for in-pit disposal.
- In the context of climate change, a temporary decrease in the thickness of the cover water (generally 1 m), due to an increase in temperatures in dry periods, would favor the underwater reactive tailings resuspension without a protective layer and, therefore, the generation of contaminated drainage (Bussière et al. 2017; see also Chapter 14). On the other hand, an increase in frequency of extreme events (e.g., precipitation, storm and strong winds) could allow this resuspension, especially under the action of waves and could also cause overflows (Bussière et al. 2017).
- The social acceptability aspect is becoming more and more difficult because of fear of fatal accidents and significant environmental consequences in the case of dam failure for constructed reservoirs. Therefore, the long-term maintenance of dike (due to physical and chemical instability) is a risk that mining companies want less and less to assume.
- Pit lakes used as mine wastes disposal show some major limits.
  - The main challenge is the deposition of mine waste in the pit (how to return the mine waste in the pit mainly in the case of waste rock), particularly when the underwater disposal is performed at the end of the operations (wastes initially stored in the surface).
  - All mine waste generated at a given mine site cannot be returned to the pit considering the volume expansion due to the rock fragmentation at blasting and the grinding of the ore and the maximal storage capacity of the pit that is controlled by the level of the groundwater table to ensure the hydraulic trap condition. Consequently, part of the mine wastes is left to be managed on the surface.
  - The concept of hydraulic trap can change over time due to modifications of water balance caused by climate change. Furthermore, if the pit does not act as a hydraulic trap, the quality of the surrounding groundwater can be affected by the water quality.

6.7 PERSPECTIVES AND RESEARCH NEEDS

Even though the water cover technique has been applied since the 1970s, and the parameters of influence of the performance are well known due to extensive research, the fact remains that certain aspects require further study.

- Given the efficiency of the attenuation layer of inert material observed in the laboratory and predicted by numerical modeling, its installation method (constructability) at large-scale needs more investigation, especially for subaqueous disposal of mine wastes.
- Coupled hydrogeological and geochemical modeling of the quality of water cover would be interesting for an evaluation of the performance of water covers in the long term, taking into consideration climate change.
- Considering that the finest particles of submerged tailings are more prone to resuspension than coarse particles, an adaptation of the predictive equations for the critical shear stress and the minimum depth of the water cover could be revisited to better consider the contribution of fines particles. Currently, the average diameter D_{50} of the initial tailings is used.
- There is a need to develop robust empirical equations for estimating the critical shear stress for erosion and resuspension of submerged sediments that can be applied for the preliminary design of water covers.
- Given the possible development of organisms on the water-tailings interface, their impacts on the tailings resuspension, oxygen consumption, and metal release need to be deeply studied.
- The influence of biological processes (bacteria, etc.) on the oxidation of submerged tailings need to be studied to better explain the geochemical behavior of water cover.
- For underwater in-pit disposal, further work is still needed on an engineered design of the groundwater barriers that aim to avoid any discharge of impacted cover water to groundwater.
- Generalized criteria should be developed for creating meromictic conditions in pit lakes containing mine wastes.

REFERENCES

Aachib, M., Mbonimpa, M., Aubertin, M. 2004. Measurement and prediction of the oxygen diffusion coefficient in unsaturated media, with applications to soil cover. *Water, Air, and Soil Pollution*, 156: 163–193.

Abdelghani, F.B., Aubertin, M., Simon R., Therrien R. 2015. Numerical simulations of water flow and contaminants transport near mining wastes disposed in a fractured rock mass. *International Journal of Mining Science and Technology*, 25: 37–45.

Adu-Wusu, C., Yanful, E.K., Mian, M.H. 2001. Field evidence of resuspension in a mine tailings pond. *Canadian Geotechnical Journal*, 38: 796–808.

Atkins, R.J., Hay, D., Robertson, J. 1997. Shallow water cover design methodology and field verification. *4th International Conference on Acid Mine Drainage*, May 31–June 6, Vancouver, Canada, 211–228.

Aubé, B.C., St-Arnaud, L.C., Payant S.C., Yanful, E.K. 1995. *Laboratory evaluation of the effectiveness of water covers for preventing acid generation from pyritic rock. Mining and the Environment Conference*, May 28–June 1, Sudbury, Ontario, Canada, 495–504.

Aubertin, M., Aachib, M., Monzon, M., Joanes, A.-M., Bussière, B., and Chapuis, R.P. 1997. *Étude de laboratoire sur l'efficacité des barrières de recouvrement construites à partir de résidus miniers. Mine Environment Neutral Drainage (MEND) Report 2.22.2b*. CANMET, Ottawa, Canada, 141p.

Aubertin, M., Bussière, B., Bernier, L. 2002. *Environnement et gestion des rejets miniers*. CD-ROM, Les Presses Internationales Polytechnique, École Polytechnique de Montréal, Québec, Canada.

Aubertin, M., Bussière, B., Pabst, T., James, M., Mbonimpa, M. 2016. *Review of reclamation techniques for acid generating mine wastes upon closure of disposal sites. Geo-Chicago: Sustainability, Energy and the Geoenvironment*, August 14–18. ASCE, Chicago.

Awoh, A.S., Mbonimpa, M., Bussière, B. 2013a. Field study of the chemical and physical stability of highly sulphide-rich tailings stored under a shallow water cover. *Mine Water and the Environment*, 32: 42–55.

Awoh, A. S., Mbonimpa, M., Bussière, B. 2013b. Determination of the reaction rate coefficient of sulphide mine tailings deposited under water. *Journal of Environmental Management*, 128: 1023–1032. doi: 10.1016/j.jenvman.2013.06.037.

Awoh, A.S., Mbonimpa, M., Bussière, B., Plante, B., Bouzahzah, H. 2014. Laboratory study of highly pyritic tailings submerged beneath a water cover under various hydrodynamic conditions. *Mine Water and the Environment*, 33: 241–255

Bain, J.G., Mayer, K.U., Blowes, D.W., Frind, E.O., Molson, J.W., Kahnt, R., Jenk, U. 2001. Modeling the closure-related geochemical evolution of groundwater at a former uranium mine. *Journal of Contaminant Hydrology*, 52: 109–135.

Berkun, M. 2005. Submarine tailings placement by a copper mine in the deep anoxic zone of the Black Sea. *Water Research*, 39: 5005–5016.

Boehrer, B., Schultze, M. 2006. *On the relevance of memorixis in mine pit lakes. Proceedings of the 7th ICARD*. ASMR, Lexington, KY, pp. 200–213.

Bussière, B., Demers, I., Charron, P., Bossé, B. 2017. *Analyse de risques et de vulnérabilités liés aux changements climatiques pour le secteur minier québécois*. Disponible sur https://mern.gouv.qc.ca/wpcontent/uploads/analyse-changements-climatiques-secteur-minier.pdf consulté le 20 juin 2018. 331p.

Castendyk, D.N., Eary, L.E. 2009. The nature and global distribution of pit lakes. In: D.N. Castendyk, L.E. Eary (Eds.) *Mine pit lakes: characteristics, predictive modeling, and sustainability*. Littleton, Colorado, Society for Mining, Metallurgy, and Exploration, 1–11.

Catalan, L.J.J., Kumari, A. 2005. Efficacy of lime mud residues from kraft mills to amend oxidized mine tailings before permanent flooding. *Journal of Environmental Engineering and Science*, 4: 241–256.

Catalan, L.J.J., Yanful, E.K. 2002. Sediment-trap measurements of suspended mine tailings in shallow water cover. *Journal of Environmental Engineering*, 128: 19–30.

CERC. 1984. *Shore protection manual*. Army Corps of Engineers, Coastal Engineering Research Center, U.S. Government Printing Office, Washington, DC.

CERC. 2002. *Coastal engineering manual*. Washington, DC: U.S. Army Corps of Engineers, Coastal Engineering Research Center, U.S. Government Printing Office.

Chien, N., Wan, Z. 1998. *Mechanics of sediment transport*. Reston, VA: ASCE Press.

Coumans, C. 2002. *Submarine tailings disposal - STD toolkit*. https://miningwatch.ca/sites/default/files/01.STDtoolkit.intr_.pdf

Crusius, J., Pieters, R., Leung, A., Whittle, P., Pedersen, T., Lawrence, G., McNee, J.J. 2003. Tale of two pit lakes: Initial results of a three-year study of the Main Zone and Waterline pit lakes near Houston, British Columbia, Canada. *Mining Engineering*, 552: 25–30.

Davé, N.K., Krishnappan, B.G., Davies, M., Reid, I. 2003. *Erosion characteristics of underwater deposited mine tailings. Mining and the Environment Conference*, May 25–28, Sudbury, Canada, 11p.

Davis, A., Ashenberg, D. 1989. The aqueous geochemistry of the Berkeley pit, Butte, Montana, USA. *Applied Geochemistry*, 41: 23–36.

Dold, B. 2014. Submarine tailings disposal (STD)—A review. *Minerals*, 4: 642–666.

Donald, B.J. 2005. Polaris Mine: a case study of reclamation in the high Arctic. *British Columbia Mine Reclamation Symposium*, University of British Columbia. Norman B. Keevil Institute of Mining Engineering. Available at: http://dx.doi.org/10.14288/1.0042473.

Doyle, G.A., Runnells, D.D. 1997. Physical limnology or existing mine pit lakes. *Mining Engineering*, 49: 76–80.

Elberling, B., Damgaard, L.R. 2001. Microscale measurements of oxygen diffusion and consumption in subaqueous sulphide tailings. *Geochemica et Cosmochimica Acta*, 65: 1897–1905.

Ellis, D.V., Poling, G.W. 1995. Special issue: submarine tailings disposal. *Marine Georesources and Geotechnology*, 131/2: 1–233.

EPLWG (End Pit Lake Working Group). 2004. *Guidelines for lake development at coal mine operations in mountain foothills of the northern east slopes*. Report # ESD/L/M/00-1. Alberta Environnent, Environnent Service.

España, J.S., Pamo, E.L., Pastor, E.S., Ercilla, M.D. 2008. The acidic mine pit lakes of the Iberian Pyrite Belt: An approach to their physical limnology and hydrogeochemistry. *Applied Geochemistry*, 23: 1260–1287.

Fraser, W.W.S., Robertson, J.O. 1994. *Subaqueous disposal or reactive mine waste: An overview and update of case studies, MENO/Canada, Proceedings of the Third International Conference on the Abatement or Acidiç Drainage*, Pittsburgh, PA.

Fritz, W., Tropp, P., Meltzer, A. 2001. A Remediation and reclamation strategy for disused brown coal mines in the Geiseltal area, Surface Mining. *Braunkohle and Other Minerals*, 53: 155–166.

Gammons, C.H., Duaime, T.E. 2006. Long term changes in the limnology and geochemistry of the Berkeley Pit Lake, Butte, Montana. *Mine Water and the Environment*, 25: 76–85.

Geller, W., Schultze, M., Kleinmann, R., Wolkersdorfer, C. 2013. *Acid Pit Lakes: The legacy of coal and metal surface mines*. Environnemental Science of Engineering, Berlin: Spring-Verlag.

GEOCON. 1997. *Projet de déposition de résidus miniers, ancienne carrière Don Rouyn: Étude hydrogéologique*. Rapport préparé pour Métallurgie Noranda Inc., Fonderie Horne, Rouyn-Noranda, Québec, Canada, dossier M-6148 600996.

Golder. 2017. *Global review of pit lake case studies*. Report number 1777450. 91p.

Haakenstad, H., Haugen, J.E. 2013. *A 15-year high resolution meteorological dataset for risk assessment in southern Norway*. MET report, 47p.

Holmström, H., Öhlander, B. 1999. Oxygen penetration and subsequent reactions in flooded sulphidic mine tailings: a study at Stekenjokk, northern Sweden. *Applied Geochemistry*, 14: 747–759.

Julien, M., Lemieux, M., Cayouette, J., Talbot, D. 2003. *Performance and monitoring of the Louvicourt mine tailings disposal area. 21st ICOLD Congress*, Montréal Québec, Canada.

Kachhwal, L.K. 2011. *Evaluation of wind-induced resuspension on the performance of a mine tailings storage facility*. Electronic Thesis and Dissertation Repository. 300p.

Kachhwal, L.K., Yanful, E.K., Rennie, C.D. 2010. *Field measurement of re-suspension in a tailings pond by acoustic and optical backscatter instruments. 63rd Canadian Geotechnical Conference & 6th Canadian Permafrost Conference*, September 12–16, Calgary, Canada.

Kam, S.N., Knapp, R.A., Balins, J.K., Payne, R.A. 1997. *Interim assessment of flooded tailings performance, Quirke mine waste management area*. In *Proceedings of the 4th International Conference on Acid Rock Drainage*, Vancouver, Canada, 853–870.

Karam, A., Guay, R. 1994. *Inondation artificielle du parc à résidus Solbec-Cupra: Études microbiologiques et chimique*. MEND Report, 2.13.2c.

Koski, R.A. 2012. Metal dispersion resulting from mining activities in coastal environments: A pathways approach. *Oceanography*, 25:170–183.

Krishnappan, B.G. 1993. Rotating circular flume. *ASCE Journal of Hydraulic Engineering*, 119: 658–767.

Krishnappan, B.G., Marsalek, J. 2002. Transport characteristics of fine sediments from an on-stream stormwater management pond. *Urban Water*, 4: 3–11.

Kwong, Y.T.J., Apte, S.C., Asmund, G., Haywood, M.D.E., Morello, E.B. 2019. Comparison of environmental impacts of deep-sea tailings placement versus on-land disposal. *Water, Air, & Soil Pollution*, 230: 287

Landry, B., Pageau, J.G., Gauthier, M., Bernard, J., Beaudin, J., Duplessis, D. 1995. *Prospection minière*. Modulo Éditeur, Mont-Royal, Québec, 240p.

Lawrence, G.A., Ward, G.A., MacKinnon, P.R.B. 1991. Wind wave-induced suspension of mine tailings in disposal ponds. A case study. *Canadian Journal of Civil Engineering*, 18: 1047–1053.

Lee, M. 2002. Visualization of oxygen transfer across the air–water interface using fluorescence oxygen visualization method. *Water Research*, 36: 2140–2146.

Li, M., Aubé, B., St-Arnaud, L. 1997. *Consideration in the use of shallow water covers for decommissioning reactive tailings. 4th International Conference on Acid Mine Drainage*, May 31–June 6, Vancouver, Canada, 115–130.

Li, M., St-Arnaud, L. 2000. *Reactivity assessment and subaqueous oxidation rate modelling for Louvicourt tailings - Final Report. MEND report No. 2.12.1d*, Natural Resources Canada, CANMET, Ottawa, Canada, 204p.

Liss, P.S. 1973. Processes of gas exchange across an air–water interface. *Deep-Sea Research*, 20: 221–238.

MHEC (Manitoba Hydro's Environmental Commitment). 2020. *Keeyask transmission project-KRI tap decommissioning environmental protection plan*. Available at: https://www.gov.mb.ca/sd/eal/registries/5614keeyask_transmission/kr1_tap_decommissioning_plan.pdf

Martin, J., Sulatyky, T., Fraser, B., Parker, R., Rodgers, B., Nicholson, R. 2016. *The Owl Creek Pit Part 1: Relocating mine rock from surface stockpiles to the pit to mitigate acid drainage. Mine Environment Neutral Drainage MEND BC Conference*. Vancouver, Canada.

Martin, A.J., Pedersen, T.F., Crusius, J., McNee, J.J., Yanful, E.K. 2003. *Mechanisms of metal release from subaqueous mine waste at circum-neutral pH: Examples from four case studies. 6th International Conference on Acid Mine Drainage*, July 12–18, Cairns, QLD, 10p.

Mbonimpa, M., Awoh, A.S., Beaud, V., Bussière, B., Leclerc, J. 2008. *Spatial water quality distribution in the water cover used to limit acid mine drainage generation at the Don Rouyn site QC, Canada. 61th Canadian Geotechnical Conference and the 9th Joint CGS/IAH-CNC Groundwater Conference*, September 21–24, Edmonton, Canada, 855–862.

McCullough, C.D., Marchand, G., Unseld, J., Robinson, M., O'Grady, B. 2012a. *Pit lakes as evaporative 'terminal' sinks: an approach to best available practice mine closure. Proceedings of the International Mine Water Association IMWA Congress*. Bunbury, Australia. International Mine Water Association IMWA, 167–174.

McCullough, C.D., Kumar, N.R., Lund, M.A., Newport, M., Ballot, E., Short, D. 2012b. *Riverine breach and subsequent decant of an acidic pit lake: evaluating the effects of riverine flow-through on lake stratification and chemistry. Proceedings of the International Mine Water Association IMWA Congress*. Bunbury, Australia. 533–540.

McCullough, C.D., Schultze, M., Vandenberg, J. 2020. Realizing beneficial end uses from abandoned pit lakes. *Minerals*, 10: 133. Available at: http://dx.doi.org/10.3390/min10020133

MEND. 1989. *Subaqueous disposal of reactive mine wastes: An overview. MEND report 2.11.1a.* CANMET, Ottawa, Canada, 207p.

MEND. 1993. *Expérimentation de l'inondation du parc à résidus miniers Solbec Cupra phase IV. MEND report 2.13.2a.* CANMET, Ottawa, Canada, 233p.

MEND. 1995. *Review of in-pit disposal practices for the prevention of acid drainage-case studies. MEND report 2.36.1.* CANMET, Ottawa, Canada, 323p.

MEND. 1997. *Review of water cover sites and research projects. MEND report 2.18.1.* CANMET, Ottawa, Canada, 136p.

MEND. 2000. *Flooding of Pre-Oxidized Mine Tailings: Mattabi Case Study. MEND Project 2.15.1a.* CANMET, Ottawa, Canada, 231p.

MEND. 2001a. *MEND Manual: Volume 4 —— prevention and control. MEND report 5.4.2d,* CANMET, Ottawa, Canada, 385p.

MEND. 2001b. *Evaluation of man-made subaqueous disposal option as a method of controlling oxidation of sulfide minerals: Column studies. MEND report 2.12.1e.* CANMET, Ottawa, Canada, 112p.

MEND. 2004. *Case Study Assessment-Heath Steele Tailings Area, Miramichi, New Brunswick. MEND report 9.2c.* CANMET, Ottawa, Canada, 25p.

MEND. 2006. *Update on cold temperature effects on geochemical weathering. MEND report 1.61.6.* CANMET, Ottawa, Canada, 103p.

MEND. 2007. *Assessing the long term performance of a shallow water cover to limit oxidation of reactive tailings at Louvicourt Mine. MEND report 2.12.2.* CANMET, Ottawa, Canada, 63p.

MEND. 2010. *Field assessment of the occurrence of algal biofilm on submerged tailings. MEND report 2.12.2b.* CANMET, Ottawa, Canada, 52p.

MEND. 2015. *In-pit disposal of reactive mine wastes: Approaches update and case study results. MEND report 2.36.1b.* CANMET, Ottawa, Canada, 250p.

Mian, M.H., Yanful, E.K. 2003. Tailings erosion and resuspension in two mine tailings ponds due to wind waves. *Advances in Environmental Research*, 7: 745–765.

Mian, M.H., Yanful, E.K. 2004. Analysis of wind-driven resuspension of metal mine sludge in a tailings pond. *Journal of Environmental Engineering and Science*, 3: 119–135.

Mian, M.H., Yanful, E.K. 2007. Erosion characteristics and resuspension of sub-aqueous mine tailings. *Journal of Environmental Engineering and Science*, 6: 175–190.

Mian, M.H., Yanful, E.K., Martinuzzi, R. 2007. Measuring the onset of mine tailings erosion. *Canadian Geotechnical Journal*, 44: 473–489.

Mölder, E., Mashirin, A., Tenno, T. 2005. Measurement of the oxygen mass transfer through the air–water interface. *Environmental Science and Pollution Research*, 12: 66–70.

Molson, J., Aubertin, M., Bussière, B. 2012. Reactive transport modelling of acid mine drainage within discretely fractured porous media: Plume evolution from a surface source zone. *Environmental Modelling & Software*, 38: 259–270.

Molson, J.W., Frind, E.O., Aubertin, M., Blowes, D. 2004. *POLYMIN: A reactive mass transport and sulphide oxidation model.* École Polytechnique, Montréal: User Guide.

Morello, E.B., Haywood, M.D.E., Brewer, D.T., Apte, S.C., Asmund, G., Kwong, Y.T.J., Dennis, D. 2016. The ecological impacts of submarine tailings placement. In: R.N. Hughes, J. Hughes, I.P. Smith, A.C. Dale (Eds.) *Oceanography and marine biology: an annual review*, vol. 54, pp. 315–366.

Otchere, F.A., Veiga, M.M., Hinton, J.J., Farias, R.A., Hamaguchi, R. 2002. Transforming open mining pits into fish farms: Moving towards sustainability. *Natural Resources Forum*, 28: 216–223.

Otwinowski, M. 1995. *Scaling analysis of acid rock drainage. MEND Project 1.19.2.* CANMET, Ottawa, Canada, 71p.

Ouellet, R., Filion, M.P., Edwards, M., Davies, G.W. 2011. *Louvicourt Mine – a recent mine closure case study. Proceedings of the 6th International Conference on Mine Closure*, Lake Louise, vol. 2, pp. 441–450.

Peacey, V., Yanful, E.K. 2003. Metal mine tailings and sludge co-deposition in a tailings pond. *Water, Air, and Soil Pollution*, 145 (1): 307–339.

Peacey, V., Yanful, E.K., Payne, R. 2002. Field study of geochemistry and solute fluxes in flooded uranium mine tailings. *Canadian Geotechnical Journal*, 39: 357–376, DOI: 10.1023/A:1023624827766.

Petersen, T.F., Mueller, B., McNee, J.J., Pelletier, C.A. 1993. The early diagenesis of submerged sulphide-rich mine tailings in Anderson Lake, Manitoba. *Canadian Journal of Earth Science*, 30: 1099–1109.

Pollock, R. 2008. *Tailings and waste rock management best practice at AREVA's McClean Lake operation.* Presentation at *IAEA-WNA Technical Meeting on the implementation of Sustainable Global Best Practices in Uranium Mining and Processing*, October 15–17, Vienna, Austria.

Price, W.A., Aziz, M. 2019. *The flooded tailings impoundment at the Equity Silver mine.*
Rescan. 2013. *Brucejack Gold Mine Project: 2012 Meteorology Baseline Report.* Prepared for Pretium Resources Inc. by Rescan Environmental Services Ltd.: Vancouver, Canada.
Rodenhuis, D.R., Bennett, K.E., Werner, A.T., Murdock, T.Q., Bronaugh, D. 2009. *Hydro-climatology and future climate impacts in British Columbia. Pacific Climate Impacts Consortium*, University of Victoria, Victoria, Canada, 132pp.
Romano, C.G., Mayer, K.U., Jones, D.R., Ellerbroek, D.A., Blowes, D.W. 2003. Effectiveness of various cover scenarios on the rate of sulfide oxidation of mine tailings. *Journal of Hydrology*, 271: 171–187.
Rowson, J., Tremblay, M.A.J. 2005. Tailings management best practice: case study of the McClean Lake JEB Tailings Management Facility. *Canadian Nuclear Society Bulletin*, 262: 17–23.
Samad, M.A., Yanful, E.K. 2005. A design approach for selecting the optimum water cover depth for subaqueous disposal of sulfide mine tailings. *Canadian Geotechnical Journal*, 42: 207–228.
Schladow, S.G., Lee, M., Hürzeler, B.E., Kelly, P.B. 2002. Oxygen transfer across the air–water interface by natural convection in lakes. *Limnology and Oceanography*, 47: 1394–1404.
Schultze, M., Boehrer, B., Friese, K., Koschorreck, M., Stasik, S., Wendt-Potthoff, K. 2011. Disposal of waste materials at the bottom of pit lakes. In: A.B. Fourie, M. Tibbett, A. Beersing (Eds.) *Mine Closure 2011: Proceedings of the Sixth International Conference on Mine Closure, Lake Louise, Canada.* Australian Centre for Geomechamics ACG, Perth, Australia, 555–564.
Shaw, S., Martin, J., Meiers, G., O'Kane, M., Wefs, C. 2006. *Characterization of the waste rock and pit walls at the Jundee gold mine site in Western Australia and implications for long-term issues. 7th International conference on acid drainage ICARD*, St. Louis: American Society of Mining and Reclamation.
Simms, P.H., Yanful, E.K. St-Arnaud, L., Aubé, B. 2000. A laboratory evaluation of metal release and transport in flooded pre-oxidized mine tailings. *Applied Geochemistry*, 15: 1245–1263.
SRK (Steffen, Robertson and Kirsten). 1989. *Draft Acid Rock Technical Guide.* BC AMD Task Force, Vol. 1.
SRK Consulting, Inc. 2008. *Environmental assessment: Gap analysis: Gunnar mine site rehabilitation project. Report Prepared for Saskatchewan Research Council*, 71pp.
St-Germain, P., Larratt, H., Prairie, R. 1997. *Field studies of biologically supported water covers at two Noranda tailings ponds. Proceedings of the Fourth International Conference on Acid Rock Drainage*, vol. 1, May 31–June 6, Vancouver, Canada, pp. 133–145.
Thériault, V. 2004. *Étude de l'écoulement autour d'une fosse remblayée par une approche de fracturation discrète.* MSc thesis, École Polytechnique de Montréal, Québec, Canada.
Tones, P.I. 1982. *Limnological and fisheries investigation of the flooded open pit at the Gunnar uranium mine.* Saskatchewan Research Council Publication No. C-805-10-E-82, Saskatoon, Canada.
Vézina, S., Amyot, G. 1999. *Le suivi de la restauration par inondation au parc à résidus miniers - Solbec. MEND report 9-3a-F.* CANMET, Ottawa, Canada, 13p.
Vigneault, B., Kwong, Y.T.J., Warren, L. 2007. *Assessing the long term performance of a shallow water cover to limit oxidation of reactive tailings at Louvicourt Mine. MEND report 2.12.2*, CANMET, Ottawa, Canada, 63p.
Vogt, C. 2013. *International assessment of marine and riverine disposal of mine tailings.* Full report, November 2013. CANMET, Ottawa, Canada, 138pp.
Wyatt, G., Miller, F. Chermak, J. 2006. Innovative water treatment plant utilizing the South Mine Pit at the Copper Bas in mining site in Tennessee. USA. In *Proceedings 7th ICARD. R.I. Barnhisel ed, 7Ih International Conference on Acid Rock Drainage ICARD*, March 26–30, St. Louis, MI, American Society of Mining and Reclamation ASMR, Lexington, 2529–2539.
Yanful, E.K., Samad, M. 2004. Shallow water cover technology for reactive sulphide tailings management. *Geotechnical News*, September 2004, 42–51.
Yanful, E.K., Mian, M.H. 2003. *The nature and implications of resuspension in subaqueous sulfide tailings. 6th International Conference on Acid Mine Drainage*, July 12–18, Cairns, QLG, CD-ROM, 1177–1183.
Yanful, E.K., Catalan, L.J.J. 2002. Predicted and field measured resuspension of flooded mine tailings. *Journal of Environmental Engineering*, 128: 341–351.
Yanful, E.K., Simms, P.H. 1997. *Review of water cover sites and research projects. MEND report 2.18.1*, CANMET, Ottawa, Canada, 136p.
Yanful, E.K., Verma, A. 1999. Oxidation of flooded mine tailings due to resuspension. *Canadian Geotechnical Journal*, 36: 826–845.

7 Covers with Capillary Barrier Effects

Isabelle Demers and Thomas Pabst

7.1 INTRODUCTION AND BACKGROUND

Covers to prevent acid mine drainage (AMD) generation from mine tailings have evolved from a single layer of organic topsoil to complex multilayered systems. In cases where the water table is well below the surface of the tailings, the incentive to develop and apply more complex covers arose from the low performance of single-layer covers, which were intended to favor revegetation of the site but not to prevent AMD formation (INAP 2009). Engineered covers that include several layers of different materials became an interesting reclamation method in the 1990s as researchers investigated the unsaturated properties of materials and transferred that knowledge to cover systems (Yanful 1993; Barbour and Yanful 1994; Aubertin et al. 1995). The main purpose of multilayered covers in a humid climate is to significantly reduce the oxygen transport toward the sulfidic tailings and thus prevent AMD generation. As discussed in this chapter, covers with capillary barrier effects (CCBEs) are considered an efficient reclamation method for unsaturated sulfidic tailings (Nicholson et al. 1989) and are used on both fresh and oxidized tailings.

7.2 CONCEPT

7.2.1 Configuration of CCBE Systems

A phenomenon called the capillary barrier effect, as described in Chapter 3, is used in covers to maintain one of the layers near full water saturation even if the water table is low. A high water content is required to significantly reduce the oxygen diffusion coefficient through the cover, thus preventing contact between acid-generating tailings and oxygen. Capillary barrier effects are observed when a fine-grained material is placed over a coarser one in unsaturated conditions. Therefore, the CCBE design uses the hydrogeological contrast between coarse- and fine-grained soils to create the oxygen barrier.

Typically, a CCBE contains three to five layers, made of different materials. Each layer has one (or more) function(s), as described in the next section. A schematic illustration of a five-layer CCBE placed on acid-generating mine waste is presented in Figure 7.1. The core of the CCBE is composed of three layers, which are essential to creating the oxygen barrier. The bottom layer is made of a fairly coarse-grained material, which acts as both a mechanical support and a capillary break layer (CBL). The fine-grained material, used as the moisture-retaining layer (MRL), is placed over the first layer to create the oxygen barrier with the capillary barrier effect. Another coarse-grained material is placed over the MRL to prevent water loss by evaporation and to promote lateral drainage. Usually, the properties of the material used in this layer are similar to those of the bottom layer. The top two layers do not take an active role in capillary barrier effects but are intended to provide protection against the natural elements. The number of layers and their materials can vary depending on the situation; some layers can be combined and have more than one role.

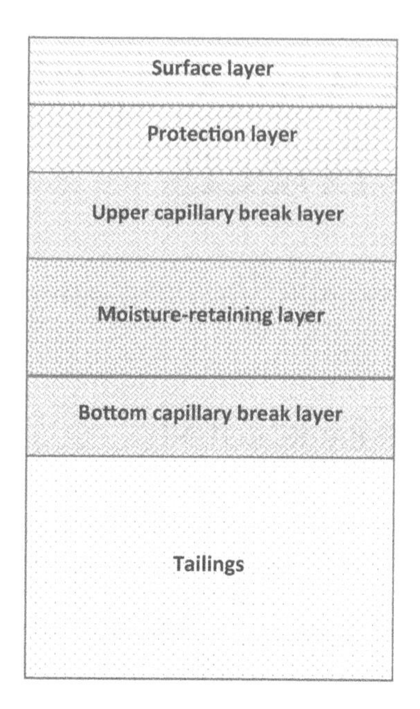

FIGURE 7.1 Schematic representation of a CCBE.

7.2.2 Role and Materials for Each Cover Layer

Each cover layer has a specific function in the cover systems (Aubertin et al. 1999). Table 7.1 summarizes the role and materials for each cover layer; each is discussed further in this section. It is important to keep in mind that each site is different, and, therefore, each cover system will be different to accommodate the specificity of the site. This section provides generic information about the cover layers.

7.2.2.1 Surface Layer

The surface layer, the topmost layer, is exposed to natural elements. As such, it acts as a protective layer against water and wind erosion and provides a growing medium for vegetation (see Chapter 12). It is not involved directly in the reduction of oxygen flux. Typical organic soils are often used as relatively thin layers, 15–20 cm.

7.2.2.2 Protection Layer

The protection layer aims to protect the core of the cover system against erosion and bio-intrusion (by plant roots and/or burrowing animals) and to reduce the effects of wet–dry and freeze–thaw cycles on the deeper layers. Thus, a coarse material, such as sand and gravel with cobbles, is often used. The thickness of the protection layer varies; it is often combined with the surface layer to make one protection/surface layer or with the upper CBL.

7.2.2.3 Upper CBL

The upper CBL is the first essential component of the oxygen barrier created by the CCBE. It serves as a drainage layer to evacuate excess water accumulation and reduce pore pressure. Indeed, it is generally built with a gentle slope for efficient drainage. Coarse materials, such as sand and gravel or crushed inert waste rock, may be used because they have a high saturated hydraulic conductivity that ensures quick drainage. Furthermore, when drained, this layer limits water loss by evaporation from the MRL. The thickness of this layer should be designed to provide sufficient drainage of excess water so that the layer reaches its residual water content, typically between 30 and 100 cm.

TABLE 7.1
Main Characteristics of the CCBE Layers (Inspired by Aubertin et al. 1995).

Layer (see Figure 7.1)	Objectives	Typical Materials	Typical k_{sat}	Typical Thickness
Surface	• Transition zone between the cover and nature • Reduces fluctuation of T° and humidity • Creates adequate conditions for vegetation • Minimizes erosion	• Organic soils • Can also contain geosynthetic components (e.g., Geogrid)	N/A	15–20 cm
Protection	• Bio-intrusion barrier • Protect the underlying layers against climatic effects (freeze–thaw cycles, desiccation) • Temporary water storage	• Sand and gravel with high percentage of cobbles (when limiting bio-intrusion is critical) • Crushed rock, crushed waste rock	$10^0 < k_{sat}$ $<10^{-4}$ cm/s	<1 m
Upper capillary break	• Creates capillary barrier effects • Reduces water accumulation • Reduces interstitial pore pressure in the cover • Protects the MRL from water losses through evaporation	• Sand • Sand and gravel	$10^{-1} < k_{sat} <$ 10^{-3} cm/s	30–50 cm
Moisture-retaining	• Oxygen barrier • Reduction of water infiltration	• Silty and clayey soils (low plasticity) • Inert mine tailings	$k_{sat} < 10^{-5}$ cm/s	50–100 cm
Bottom capillary break	• Create capillary barrier effects • Avoid AMD capillary rise from the mine waste • Support layer for heavy equipment	• Sand • Sand and gravel	$10^{-1} < k_{sat} <$ 10^{-3} cm/s	30–50 cm

7.2.2.4 MRL

The MRL is the actual oxygen barrier of the CCBE. By keeping a high degree of saturation, the MRL allows the oxygen diffusion coefficient to remain very low and allows very little oxygen to penetrate the moisture-retaining layer. The material for this layer should have a relatively low saturated hydraulic conductivity (k_{sat}) compared to the CBL (i.e., k_{sat} two–three orders of magnitude lower than the material below it), an air entry value (AEV) higher than the water entry value (WEV) of the CBL, and a significant moisture-retaining capacity. Natural materials such as silt and clay can be used, as well as fined-grained materials that have similar hydrogeological properties. Inert mine tailings were successfully used as a moisture-retaining layer at the Les Terrains Aurifères (LTA) site (Quebec, Canada) (Bussière et al. 2006). High-plasticity materials should be avoided because of the risk of cover efficiency reduction after several climatic cycles. The layer thickness should be optimized so that matric suction remains below the AEV of the material. Typically, a thickness of 50–100 cm is used.

7.2.2.5 Bottom CBL

The bottom CBL is made from a coarse material to provide the required contrast to create a capillary barrier effect and prevent desaturation of the moisture-retaining layer. This layer usually drains well, and thus, its unsaturated hydraulic conductivity becomes very low, which also prevents the rise by capillarity of contaminated water from the mine waste placed underneath. A thickness of 30–50 cm is often suggested to favor drainage. This layer allows construction of the other layers using heavy machinery, which would not be able to circulate directly on the tailings because of their low bearing capacity and high liquefaction potential.

For all layers, materials should be compatible to minimize fine particle movement from one material to another, which could modify the hydrogeological behavior of the entire system. It is important to recall that materials are subject to climatic conditions such as freeze–thaw cycles, and some types of materials may be more susceptible to alteration of their properties after several freeze–thaw cycles. For example, high-plasticity materials such as clay can be significantly affected by freeze–thaw cycles (Hillel 1998).

7.3 FACTORS THAT INFLUENCE COVER PERFORMANCE

Given that CCBEs are engineered systems that are integrated into their environment, their performance, that is, their ability to limit AMD generation, is influenced by the CCBE design, site characteristics, and regional environment. Evaluation of cover performance is often described in terms of oxygen flux or a target degree of saturation that will provide a low enough oxygen flux to limit AMD generation. The literature suggests that a degree of saturation between 85% and 90% should be the design target (e.g., Aubertin et al. 1998; Bussière et al. 2007).

7.3.1 Material Properties

Because the CCBE relies on capillary barrier effects, and such effects are created by a contrast in hydrogeological properties, these material properties have a major influence on the cover performance. Numerical modeling in one dimension was performed by several authors to illustrate the impact of particle size distribution variations on the hydrogeological behavior of the cover system (Akindunni et al. 1991; Woyshner and Yanful 1995; Aubertin et al. 1996). An example of a typical CCBE configuration is represented in the following numerical simulations performed with SEEP/W (Geo-Slope 2016). A schematic of the one-dimensional model is presented in Figure 7.2, where the CCBE is illustrated as a bottom CBL made of 30 cm of sand, a 50-cm silt moisture-retaining layer, covered by a 30-cm sand upper CBL. The chosen materials were typical sample materials provided by the software; their main hydraulic properties are presented in Table 7.2. The model was initially

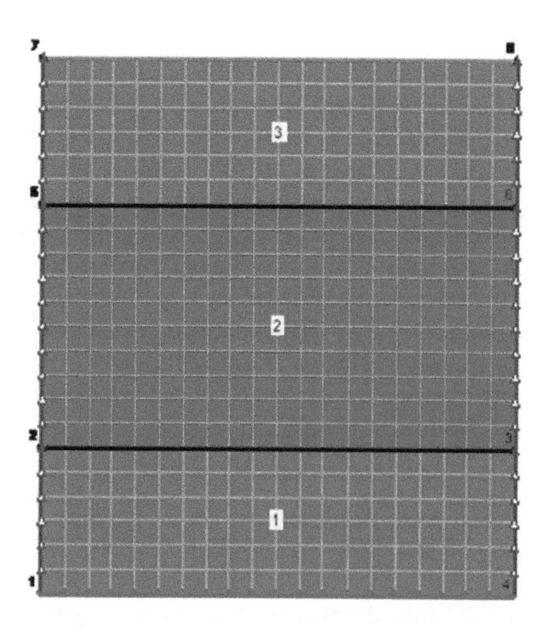

FIGURE 7.2 Numerical model representation, using SEEP/W version 2016.

TABLE 7.2
Properties of Materials Used in the Numerical Model

Material	Porosity	AEV (kPa)	Saturated Hydraulic Conductivity (m/s)
Coarse sand	0.4	0.9	1×10^{-2}
Fine sand	0.4	3	1×10^{-3}
Silt	0.4	7	1×10^{-7}
Fine silt	0.4	11	1×10^{-9}

saturated for all simulations, and water flow was vertical. Evaporation was not included in the simulation because it was assumed that the upper CBL prevents evaporation, as observed in laboratory and field experiments (Bussière et al. 2007).

The first simulation involved coarse sand as CBLs and fine silt as the MRL, and a water table level 2 m below the bottom sand layer. Figure 7.3 shows the distribution of volumetric water content across the CCBE with time. Both sand layers desaturate rapidly and reach a low volumetric water content, around 0.1 for the upper layer and around 0.05 for the bottom layer. The moisture-retaining layer remains at a high volumetric water content, slightly below 0.4, which is the porosity. Therefore, the degree of saturation in the silt layer remains close to 100%. The capillary break effect occurs because the contrast between the materials' hydrogeological properties is adequate; that is, the difference in k_{sat} is of seven orders of magnitude, and the difference in the AEV is of one order of magnitude, and the AEV of silt is reached before the WEV of CBL (10 kPa).

A second simulation was performed using fine sand as CBLs and silt as the MRL. The k_{sat} difference is four orders of magnitude; however, the AEV is close, 3 kPa for fine sand and 7 kPa for silt. The simulation result, presented in Figure 7.4, shows that both sand layers desaturate, but not as much as in the previous simulation and that the MRL cannot maintain a high volumetric water content with time. Indeed, the capillary break is weak at the interface between silt and bottom sand layers, and water can flow from the silt toward the sand. After 10 days, volumetric water content values around 0.3 can be observed in the low section of the silt layer, corresponding to

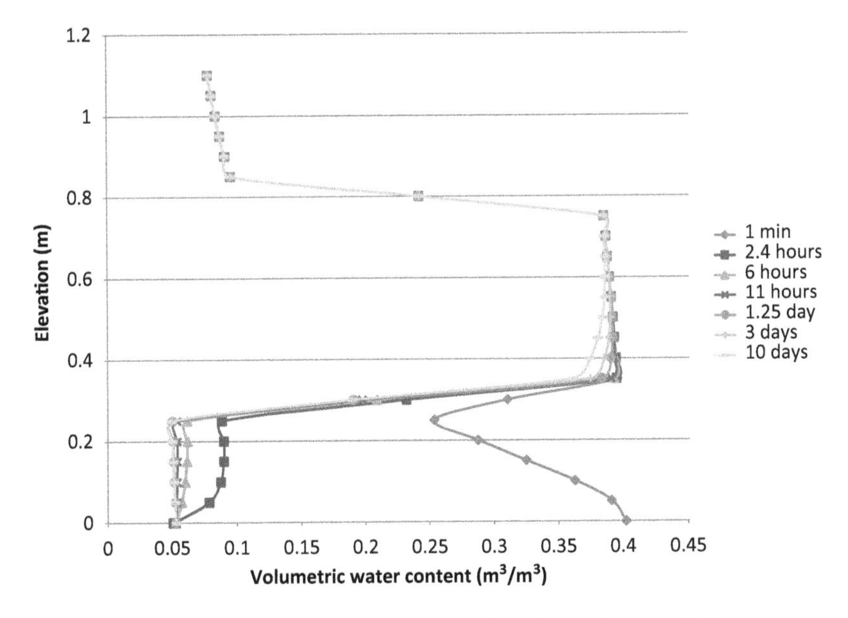

FIGURE 7.3 Numerical modeling results: coarse sand and fine silt used in the CCBE, water table at −2 m.

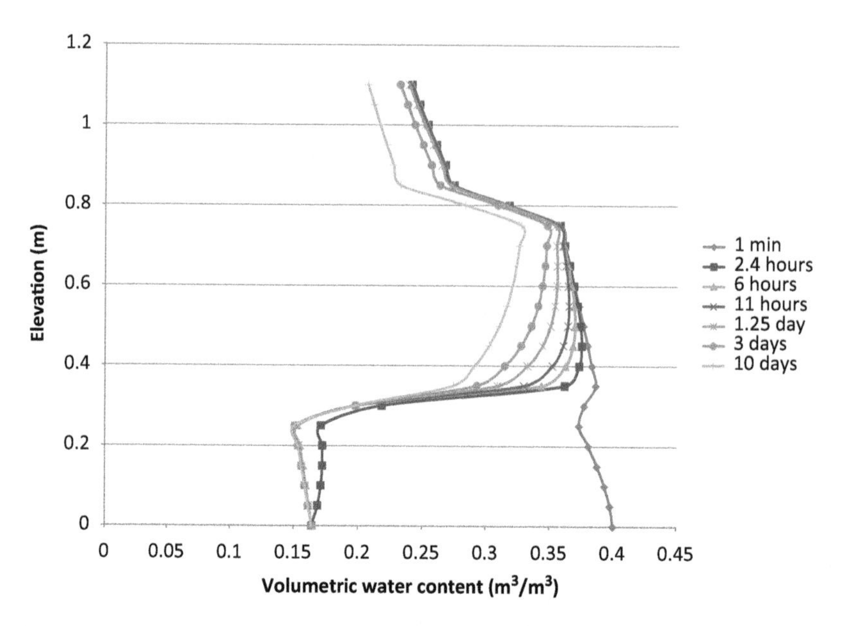

FIGURE 7.4 Numerical modeling results: fine sand and silt used in the CCBE, water table at −2 m.

a degree of saturation of 75%, and 0.33 (82.5%) near the top silt layer. A CCBE built with these materials would not be efficient in the long term; it would be wiser to select materials that have a larger difference in hydraulic properties.

If the cover material has a low sulfide content, these sulfides may oxidize in contact with oxygen and may reduce the oxygen flux that reaches the bottom of the cover (Aubertin et al. 2003; Mbonimpa et al. 2003; Demers et al. 2009). This effect can be beneficial to the cover performance as long as the cover itself does not generate AMD, and until all the sulfides are oxidized. Then the cover behaves like an inert material cover.

7.3.2 Site Conditions

Materials are not the only factor that can affect the performance of a CCBE; its environment can also significantly influence its performance. The site on which the CCBE is installed rarely has uniform hydrology and topography and can be subject to a variety of climatic conditions. The water budget of the tailings impoundment to be covered needs to be well understood, particularly precipitation and evaporation rates (Weeks and Wilson 2005). It is important to recall that a CCBE is usually chosen to reclaim unsaturated tailings, that is, on a site where the water table is low, typically below 2 m from the surface. The numerical model shown in Figure 7.2 was used to simulate different water table levels. The CCBE built with coarse sand and fine silt was shown to be efficient without water input for 10 days when the water table was placed 2 m underneath the cover. Figure 7.5 presents the same cover configuration but with a water table 5 m below the cover bottom. Again, both sand layers rapidly reach a low volumetric water content while the silt layer remains nearly saturated. After 10 days of drainage, the point just above the bottom sand–silt interface maintains a volumetric water content of 0.35, which corresponds to a degree of saturation of 87.5%. These results show that the performance of a well-designed CCBE is unaffected by the depth of the water table as long as the bottom capillary break remains at its residual water content.

However, if the contrast between the CBL and MRL is not large enough, the performance of the CCBE may be compromised, particularly when the water table is low. Figure 7.6 shows

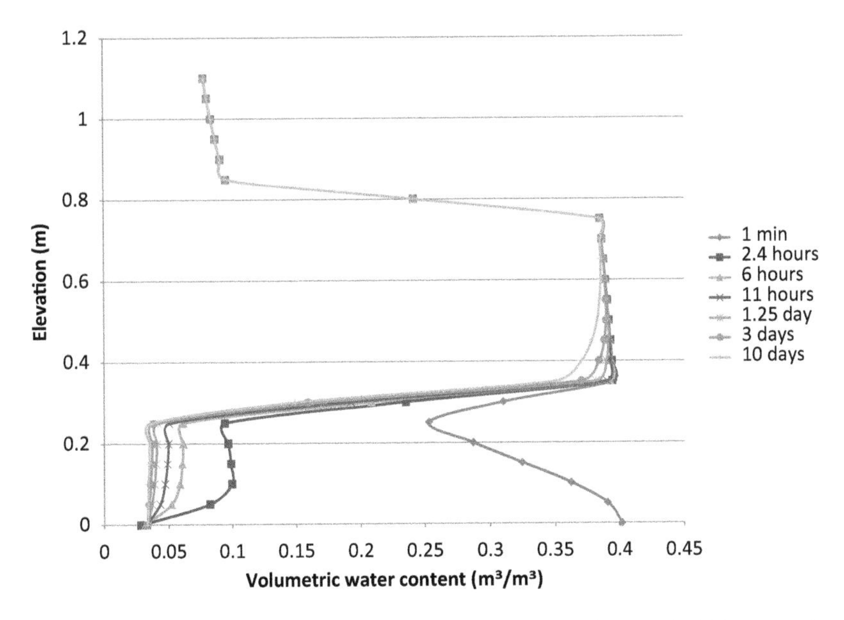

FIGURE 7.5 Numerical modeling results: coarse sand and fine silt used in the CCBE, water table at –5 m.

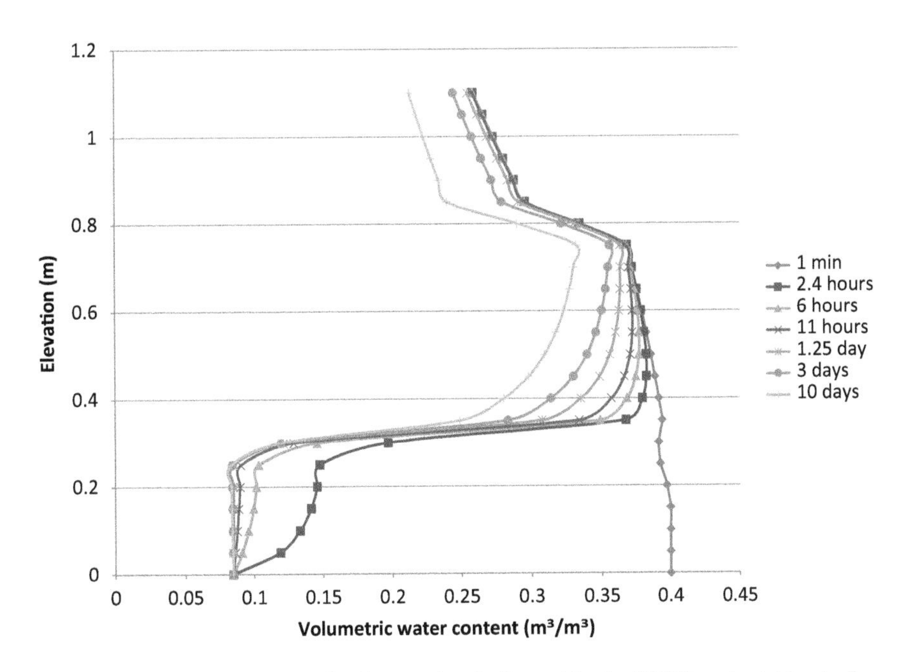

FIGURE 7.6 Numerical modeling results: fine sand and silt used in the CCBE, water table at –5 m.

a simulation performed with fine sand and silt as the CCBE constituents, with the water table placed 5 m below the cover. After 10 days, volumetric water content values around 0.28 are expected in the lower section of the MRL, while values up to 0.32 are predicted in the upper section. These water contents correspond to degrees of saturation of 70%–80%, which are not sufficient to provide long-term performance of the oxygen barrier.

Climatic conditions can affect the level of water table and its seasonal variations. In a temperate boreal climate, the water table tends to be at its highest after spring thaw, when recharge is high, while it reaches its lowest level during summer, when precipitation is low and evaporation is high. The CCBE needs to be designed to be efficient when the water budget is the least favorable, that is, during "dry" conditions. A well-designed CCBE will be able to maintain a nearly saturated MRL even during a dry period as long as the suction at the upper CBL–MRL interface remains below the AEV of the fine moisture-retaining material. If the suction becomes higher than the AEV, loss of water from the MRL could occur by drainage and evaporation. The thickness of the MRL ensures that, even if some desaturation occurs, a large portion of the layer remains at a high degree of saturation (>85%).

7.3.3 Slope Effect

The previous discussion focused on one-dimensional model of a CCBE. In the field, a CCBE covers a tailings impoundment that can have flat and inclined areas, and inclined areas can have anything from a very gentle slope to much steeper dike faces. The overall performance of the CCBE therefore includes a combination of its performance in all areas, including on inclined sections. Several researchers investigated water and gas movement in inclined covers (Ross 1990; Aubertin and Bussière 2001; Bussière et al. 2003). The general consensus is that volumetric water content lowers as one moves up the slope, and if the water content lowers below the target value, the cover may not be able to limit gas migration because of a degree of saturation lower than 85%–90% in a section of the cover. Three main factors influence CCBE performance on an inclined surface: slope angle, slope length, and material properties. Although beyond the scope of this chapter, physical stability aspects, especially for layers of fine materials, should also be considered in sloped covers.

7.3.3.1 Fieldwork

Desaturation caused by slope inclination was observed on the LTA site (Quebec, Canada) where slopes were instrumented with volumetric water content and suction sensors. Table 7.3 presents results from a fall sampling campaign. The target volumetric water content is 0.37 (degree of saturation of 85%), which was not reached at the upslope station, while it was exceeded in mid- and down slope. These data are consistent with other results obtained during monitoring of the site (Bussière et al. 2006).

7.3.3.2 Numerical Modeling

Numerical modeling was performed to enhance the comprehension of the slope desaturation phenomenon and to assess its impact on the performance of the CCBE to prevent oxygen migration. The numerical model is described in the study by Bussière et al. (2003). The model simulated the LTA cover, with a bottom CBL of 0.5 m, a MRL of 0.8 m, and an upper CBL of 0.3 m. Three slope angles were tested: 3H:1V (18.4°), 4H:1V (14°), and 5H:1V (11.3°). The main results are

TABLE 7.3
Results from a Sampling Campaign on a Slope Section of the LTA Site CCBE

Station Location	Volumetric Water Content in Moisture-retaining Layer	Degree of Saturation (%)	Volumetric Water Content in Bottom CBL
Upslope	34	78	25
Mid-slope	40	91	40
Downslope	40	91	32

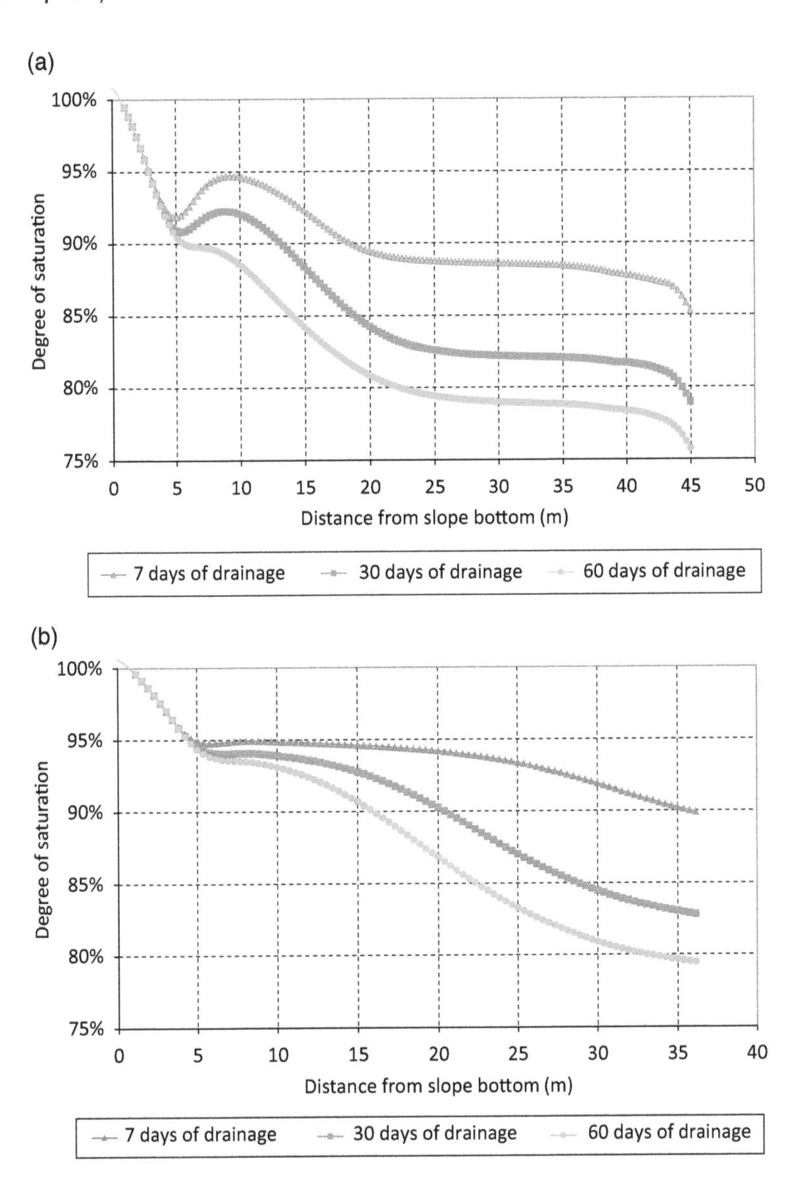

FIGURE 7.7 Evolution of the degree of saturation in the center (mid-depth) of the moisture-retaining layer as a function of the position along the slope (0 = bottom of the slope) for a slope inclination of 18° (a) and 11.3° (b) (adapted from the study by Bussière et al. 2003).

presented here. The slope length (50 m and 100 m were tested) had a minimal impact on the water content upslope. The slope angle effect is presented in Figure 7.7, where the degree of saturation in the middle of the MRL is expressed along the slope, from bottom to top, for three periods without water recharge. The slope effect is clearly seen as the 85% degree of saturation threshold is reached at 14 m for 60 days of drainage, at 19 m for 30 days of drainage, and around 45 m for 7 days of drainage with an 18° slope. With a slope inclination of 11.3°, slope lengths above 20 m can maintain a degree of saturation above 85% for the different drainage periods. For example, these results indicate that the degree of saturation 25 m upslope when inclined at 18° is 83% for 30 days of drainage, while for an inclination of 11.3° the same position on the slope would have a degree of saturation of 87%.

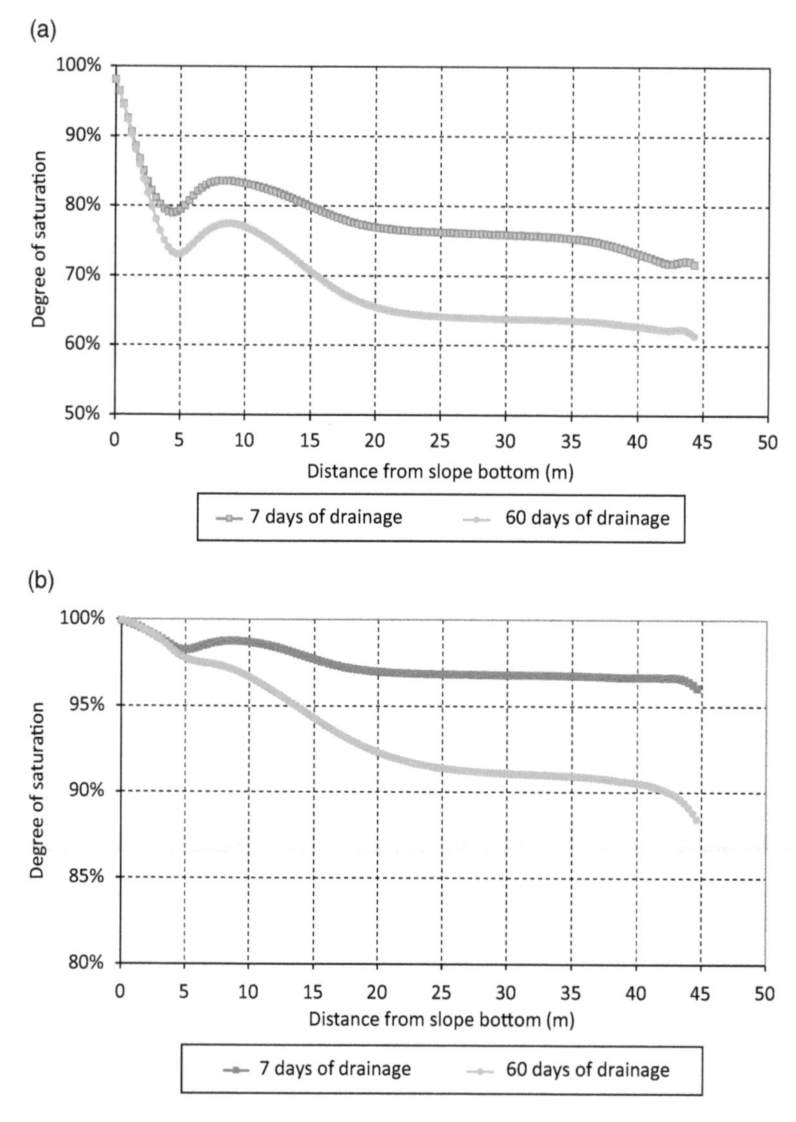

FIGURE 7.8 Evolution of the degree of saturation in the center (mid-depth) of the MRL as a function of the position along the slope (0 = bottom of the slope) when coarse Silt (a) and fine Silt (b) are used as fine-grained materials (adapted from the study by Bussière et al. 2003).

The effect of material properties was also evaluated numerically by replacing the silt used in the MRL with either a fine silt or a coarse silt. The fine silt has an AEV of 6.25 m and a k_{sat} of 9.9×10^{-6} cm/s, while the coarse silt has an AEV of 1.20 m and a k_{sat} of 1.0×10^{-4} cm/s. Modeling results presented in Figure 7.8 clearly show that a CCBE made with the coarse silt would desaturate below 85% less than 10 m upslope, whereas the CCBE made with the fine silt is much less sensitive to slope desaturation for a 18° degree slope. Indeed, it would keep a degree of saturation above 85% for a slope length over 40 m. The selection of a suitable material for the moisture-retaining layer should therefore include an evaluation of potential slope effects.

The primary objective of a CCBE is to limit oxygen diffusion by keeping the moisture-retaining layer at a high degree of saturation. Oxygen flux, calculated using Fick's laws, requires an effective diffusion coefficient, D_e, which is related to the degree of saturation (see Chapter 3). For a slope, the value of D_e changes along the length of the slope. The oxygen flux should therefore be calculated using numerical modeling tools, such as SEEP/W (Geo-Slope Int.). Bussière et al. (2003) developed an approach to make a preliminary estimate of the potential for desaturation of an inclined CCBE. The method proposed a new parameter called D_{e_Slope} that corresponds to the average D_e along the slope calculated from the degree of saturation in the center of the MRL. The parameter D_{e_Slope} is a function of the angle of the slope, the hydrogeological contrast between the MRL and bottom CBL, and the drainage time. The interested reader is referred to the study by Bussière et al. (2003) for more details.

7.3.4 Other Factors

The performance of a CCBE can also be influenced by other factors, such as previous oxidation of the tailings and vegetation. If sulfidic tailings were exposed in an unsaturated state before reclamation, for example, in abandoned mine sites, oxidation may have already started and pore water may be already contaminated. As discussed in Chapter 1, once the oxidation process begins, it may be difficult to stop because of the presence of Fe^{3+} and bacteria as oxidizing agents as well as oxygen. The installation of a CCBE reduces the source of oxygen from the atmosphere and reduces the rate of direct sulfide oxidation. However, the low amount of oxygen still present in the tailings pore space may be sufficient to oxidize Fe^{2+} into Fe^{3+} and continue the indirect oxidation of sulfides, even in near absence of oxygen (Pabst et al. 2017). Such considerations must be taken into account when evaluating the possibility of covering oxidized tailings with a CCBE. Water quality criteria will probably not be met in the short term, and passive treatment (Chapter 11) may be required to reach the water quality targets.

The final important factor that can influence CCBE performance is vegetation. In many jurisdictions, revegetation of reclaimed areas is mandatory. Plants, from small shrubs to trees, modify the site water budget, and their roots may create preferential pathways for air and water. A CCBE must be made robust enough to reduce the negative effects of vegetation on its long-term performance; for example, the proper selection of materials and a conservative design of protective layers may mitigate the long-term negative effects of vegetation. Chapter 14 examines the integration of vegetation in reclamation scenarios in more detail.

7.4 DESIGN METHODOLOGY

Each CCBE is site specific. There is no "one-size-fits-all" configuration or solution. The optimal CCBE for a given site is based on the tailings properties, site topography, available materials, and so on. Moreover, conditions will differ from one site to another. While CCBE designers may be inspired by other site designs and successes, there are specific, necessary steps in designing the right CCBE for a given site. The proposed design methodology, based on work by several authors (Aubertin et al. 2003; Yanful et al. 2006; INAP 2009), is illustrated as a flowsheet in Figure 7.9. The design process may take several years to complete because of the duration of laboratory and field simulations.

7.4.1 Site Characterization

The first step involves a thorough investigation of the tailings impoundment to be reclaimed. Information about the following aspects must be gathered:

- surface area and topography
- water table level and variations
- watersheds and water budget (precipitation, evaporation, percolation, and infiltration)
- typical and extreme climatic conditions (relative humidity, radiation, wind speed, and direction)
- plant species and distribution.

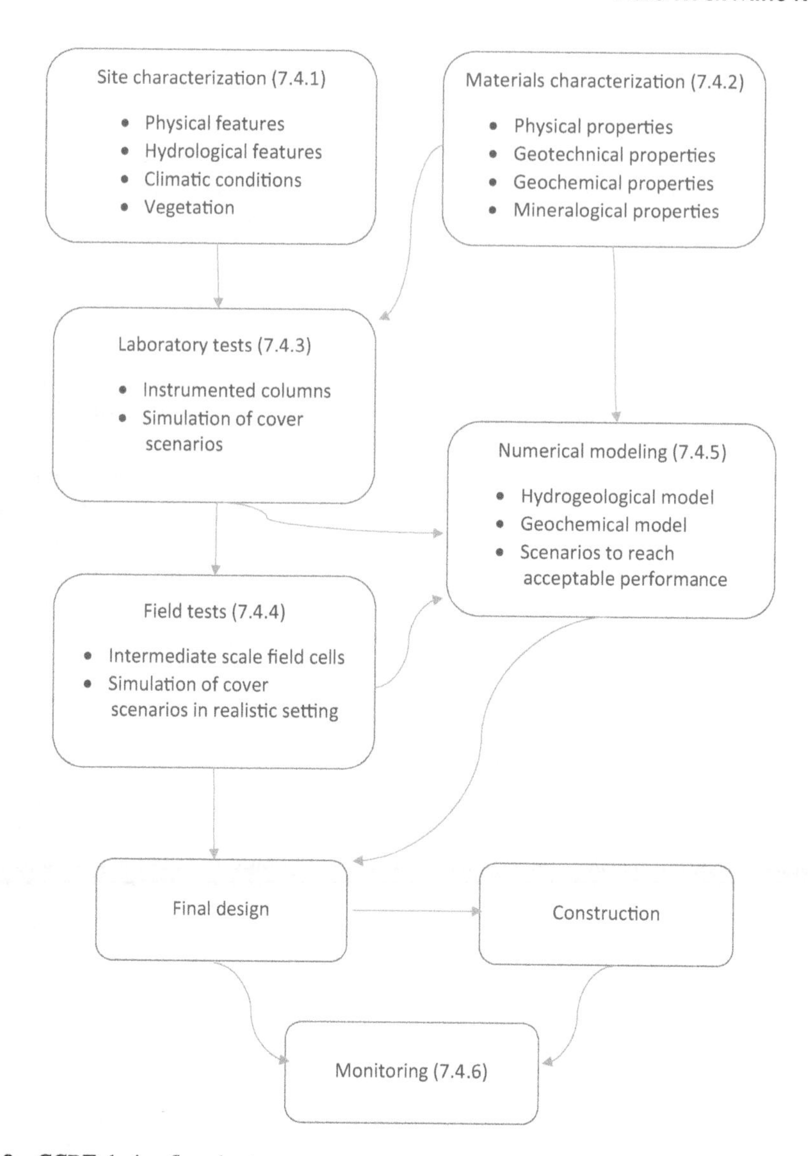

FIGURE 7.9 CCBE design flowsheet.

This information is useful to establish the baseline of the site before reclamation. Performance of the reclamation scenario may eventually be estimated according to the variation between the reclaimed state and baseline state.

7.4.2 Materials Characterization

The second step guides the selection of appropriate cover materials. Table 7.4 presents an overview of the main laboratory tests for material characterization, which should be performed in the CCBE design process. Specifically, for tailings, physical, chemical, geotechnical, and mineralogical properties must be determined with the appropriate laboratory and/or field tests. Indeed, the knowledge of tailings chemical and mineralogical composition is essential to interpret the laboratory and field simulation, the next two steps. This knowledge is also an input of numerical modeling (Section 7.4.5). Physical and geotechnical properties are also necessary to perform slope stability assessments (although beyond the scope of this book) and hydrogeological modeling (Section 7.4.5). Freeze–thaw cycle tests help to anticipate possible modification in materials properties.

TABLE 7.4
Main Laboratory Tests for Material Characterization

Parameter	Method	Standard or Reference
Acid generation potential	Acid–base accounting and kinetic tests	see Chapter 2
Chemical composition	X-ray fluorescence, atomic emission spectroscopy, or other techniques	
Mineralogical composition	X-ray diffraction and microscopy	Young (1995)
Particle size distribution	Sieve and laser particle size analyzer	ASTM D2487-06
Saturated hydraulic conductivity with and without freeze-thaw cycles	Rigid wall permeameter (coarse-grained samples) Flexible wall permeameter (fine-grained samples)	ASTM D5856-95, see Chapter 3 ASTM D5084-16, see Chapter 3
Water retention curve	Tempe cell, triaxial cell, column tests, and saline solutions	ASTM D6836-02, see Chapter 3
Oxygen diffusion coefficient and coefficient of reactivity	Diffusion-consumption tests	Mbonimpa, Aubertin et al. (2003) Elberling et al. (1994)
Compaction	Proctor test	ASTM D1557-07
Atterberg limits	Atterberg tests	ASTM D4318-17

7.4.3 Laboratory Physical Modeling

Once material characterization narrows down the possible types of material to use in the cover, small-scale representations of the CCBE are simulated in the laboratory. Instrumented column tests, used by several research groups (Aachib et al. 1994; Elberling et al. 1994; Yanful et al. 1999b; Bussière et al. 2004; Lewis and Sjöstrom 2010), are a necessary step to evaluate the performance of different configurations and materials. Columns provide a one-dimensional simulation of a real CCBE. Even if each column test has its specificity, the following general points are common:

- Columns can be a cylindrical (tube) or square-based prism. For cylindrical columns, a diameter of 15–20 cm is usually required for covers placed on tailings. For tests with coarse-grained materials (such as waste rock), the diameter of the column should be at least 6 times the size of the largest particle. For waste rock screened at 5 cm, a column with a 30 cm diameter is required. The height of the column depends on the thickness of the cover materials.
- For geochemical performance evaluation, sulfidic tailings are placed at the bottom of the columns. Tailings from the site to be reclaimed must be sampled and carried to the laboratory either submerged or in an air-tight container to avoid oxidation during handling and transport. The thickness of the tailings placed in the columns does not need to be equivalent to the tailings depth on the field; usually, oxidation occurs in the first tens of centimeters at the surface of the tailings, while the deeper portions tend to remain unoxidized for a long time. A typical tailings thickness used in column tests is 30–40 cm.
- To simulate realistic scenarios, the cover configuration is reproduced in a cross-section. For example, a CCBE may be composed of 30 cm of sand as the bottom capillary break, 75 cm of neutral tailings as the moisture-retaining layer, and 30 cm of sand as the upper capillary break. This configuration is installed on top of the sulfidic tailings in the column. A major advantage of column tests is the possibility to install several cover configurations to determine the most efficient for the given tailings and hydraulic conditions. Different types of materials may also be tested as a method to select the optimal match of coarse and fine materials.
- Placement of materials in the columns is a critical step. Again, the objective is to reproduce field conditions as much as possible, including the compaction and porosity of the materials. If the in situ tailings porosity is known, it can be reproduced in the column by placing a calculated mass of tailings in a given volume. For most tests, a typical porosity of 0.4–0.45 is

used for tailings (Bussière 2007), and for the cover materials, appropriate values are selected depending on the material type (neutral tailings, clay, sand, gravel, etc.). Compaction may be chosen based on optimum Proctor, or to simulate hydraulic deposition and self-weight consolidation. Installation of the materials in the column needs to be done conscientiously as it affects the results and repeatability of the test (Demers et al. 2011).
- Simulation of a water table can be achieved through the use of a porous plate that remains saturated at the base of the column, connected to a tube, which sets the water table level.
- Minimal instrumentation of the columns includes equipment to measure volumetric water content and suction at strategic locations in the different layers. A system to collect leachate is also necessary to evaluate geochemical evolution. Some column tests add gas sampling ports, pore water sampling lysimeters, oxygen consumption test equipment, or other specific devices depending on the objectives of the test. Depending on the type of sensors and dataloggers used, monthly, weekly, daily, or hourly values of volumetric water content and suction are recorded and used to evaluate the hydrogeological behavior of the cover system.
- Columns are periodically leached to force water movement in the pore space and collect the soluble elements. Leachate quality is an indicator of potential field water quality and, more importantly, gives information about the rate of release of metal species. Leachate analyses typically include determination of pH, Eh, electrical conductivity, chemical composition, acidity, and alkalinity. Column tests usually last around one year to give enough time for the system to stabilize (one to three months) and then to provide sufficient data to establish trends.
- At the end of the column test, dismantling by layers can be done to observe possible secondary mineral precipitation and to validate volumetric water content measurements by the probes.

Laboratory column tests are extremely useful in selecting cover configurations that have the greatest potential to be effective at a specific site, which leads to the next step: field tests. Data from the column tests are also essential for numerical modeling.

7.4.4 Field Tests

Laboratory tests should have narrowed down the CCBE configurations to a few options. After these configurations have proven efficient in controlled laboratory conditions, they must be tested in real conditions. Field tests are necessary to evaluate the behavior of the CCBE when subjected to precipitation, evaporation, and freeze–thaw cycles, and as a two-dimensional system. Field test plots can vary, from large barrels filled with the materials to large test plots of tens of meters (Yanful et al. 1999a; Bussière et al. 2007; Kalonji Kabambi et al. 2020).

Field test cell construction is executed in several steps, as illustrated in Figure 7.10:

- First, excavation is usually required to create a form for the cell. The cell should be in an area that has relatively easy access for monitoring, but it should not interfere with operations.
- Then, a geomembrane is placed in the cell form to isolate the cell content from the environment in order to recuperate all the percolating water from the cell.
- The drainage system, located at the bottom of the cell, is used to carry the water out of the cell to an outlet. This outlet may be equipped with a flow meter or automatic sampler.
- The materials are then placed in the cell, starting with the sulfidic tailings. The thickness of the tailings layer should be chosen according to the cell geometry, to provide sufficient material for hydrogeochemical interpretation. The CCBE should be placed as proposed for the site, in terms of layer thickness and compaction (porosity).
- Instrumentation is installed concurrently with material placement. Depending on the size of the cell, one or more monitoring stations can be installed. Instruments should measure volumetric water content and suction using robust equipment that will last several years on the field. Other instruments can be used for specific purposes, such as piezometers to measure the water level, gas sampling ports, oxygen consumption test tubes, etc.

FIGURE 7.10 Construction of the field cell. A: Excavation of the cell shape and drain. B: Placement of the impervious membrane. C: Installation of materials and instrumentation. D: Construction completed (Rey et al. 2016).

Field test cell monitoring needs to be done regularly and diligently over several seasons to obtain data for different seasonal conditions. With dataloggers, it is possible to obtain hourly or daily measurements of volumetric water content and suction, which will be used to determine the CCBE response to variations in climatic conditions. Suction and volumetric water content are also required to evaluate the performance of the CCBE to limit oxygen diffusion. Water samples are analyzed for the same parameters as laboratory column leachates. Water quality from the field cells are expected to be close to the future field water quality (Plante et al. 2014). Meteorological data from the site are required to relate to the cover performance. The installation of a portable weather station that can measure wind speed and direction, temperature, relative humidity, precipitations, and solar radiation is often warranted (see Chapter 10). The data obtained from the field test cells and column tests are useful for numerical modeling, the next step described below.

7.4.5 Numerical Modeling

Numerical modeling integrates data from the previous steps, including material and site characterization, to make a numerical representation of the site to be reclaimed. Because laboratory columns and field test cells cannot simulate all possible configurations and boundary conditions, numerical modeling is essential to obtain further information and to validate laboratory and field results. Several codes exist to solve the water and gas flow equations presented in Chapter 3, and other codes focus on geochemistry.

At this stage, a target oxygen flux is defined from the laboratory and field data, mostly using geochemical results. Then, a conceptual model of the laboratory tests is defined and validated using the laboratory data. The model can then be extended to a larger size, corresponding to the field test cell, and validated using the field cell results. Once the model validation is satisfactory, it can be used to test different cover configurations and different climatic conditions and to make predictions about the future behavior of the CCBE. It is important to keep in mind that the chosen cover system must be efficient for normal and extreme climatic conditions, and should take into account climate change (see Chapter 14).

7.4.6 Field Instrumentation and Monitoring

Once the final design is selected, it is essential to plan for field instrumentation and monitoring to evaluate the performance of the full-scale CCBE. Chapter 10 is dedicated to field monitoring. For a CCBE, the main parameters to monitor are volumetric water content and suction in the moisture-retaining layer and upper and lower CBLs, on flat surfaces and in slopes, because the performance is directly related to the unsaturated behavior of the cover.

7.5 CASE STUDIES

Table 7.5 presents the three cases of a full-scale implementation of CCBEs. Other sites used variations of the multilayer concept, but not with the same objective, and they were not trying specifically

TABLE 7.5
Summary of CCBE Case Studies

Site	Site Features	CCBE Configuration and Materials	References	Comments
Les Terrains Aurifères, Quebec, Canada	• 66 hectares • Tailings pond containing AMD-generating tailings • Non-AMD-generating tailings nearby • Humid climate	• 50 cm sand bottom capillary break • 80 cm non-acid-generating tailings as the MRL • 30 cm sand and gravel upper capillary break and protective layer	Bussière et al. (2006)	• Monitoring over 10 years confirmed adequate performance of the CCBE • Degree of saturation in MRL above 85%
Lorraine, Quebec, Canada	• 15 hectares • Abandoned site with AMD-generating tailings • Humid climate	• 30 cm sand bottom capillary break • 50 cm silt as the MRL • 30 cm sand and gravel upper capillary break and protective layer	Dagenais et al. (2005); Bussière et al. (2009)	• Periodic monitoring confirms adequate performance of the CCBE • Oxygen flux between 0.2 and 39 g $O_2/m^2/yr$ • Requires passive treatment to reach effluent quality norms because of previous tailings oxidation
Bouchard-Hébert, Quebec, Canada	• 68 hectares • Tailings pond with highly reactive tailings (59%–74% pyrite) • Humid climate	• 50 cm sand bottom capillary break • 60 cm clay or silt as the MRL • 30 cm sand and gravel upper CBL and protective layer	Bernier et al. (2005)	

to create capillary barrier effects. The three sites in Table 7.5 are located in Quebec, Canada, where the water budget is positive and temperatures drop below freezing for four–five months every year. For Lorraine and LTA sites, instrumentation was installed and regularly monitored to assess the performance of the cover. The covers are efficient at reducing the oxygen flux toward acid-generating tailings. In the case of Lorraine, effluent quality requires time to improve because the cover was installed over already oxidized tailings (see Section 7.3.4). The LTA site has an intermittent effluent that meets provincial water quality regulations.

7.6 APPLICABILITY OF CCBEs

7.6.1 ADVANTAGES

A CCBE has several advantages over single-layer covers – the first one being its efficiency in preventing AMD when the water table is below the surface of acid-generating tailings. Its efficiency rivals that of water covers but without the liability related to water-retaining dams. Because of the capillary barrier effect, the performance of the cover is generally stable over time and reacts very slowly to changing climatic conditions.

It is possible to build a CCBE with materials that have low susceptibility to freeze–thaw effects, which provide greater longevity and robustness. Furthermore, it was shown that mining materials, tailings, and waste rock can be used as cover materials following a laboratory and field demonstration (Aubertin et al. 1999; Bussière et al. 2004, 2007; Demers et al. 2008; Bussière and Demers 2020; Kalonji Kabambi et al. 2020). This use of mining materials is particularly interesting for large sites, which require significant volumes of cover material and would otherwise require the disturbance of large natural areas to obtain the appropriate material.

Finally, it is important to recall that the CCBE is a tested and proven method, at least for the short- and medium-term use. The earliest CCBEs were built in the 1990s. Thus, their performance can be confirmed for approximately two to three decades. The performance evaluated in the laboratory scale is replicated in the field, which validates the design methodology used.

7.6.2 LIMITS

Cost is a major limit in all reclamation work. A CCBE requires significant amounts of materials, and their transportation represents a major expense. If the appropriate materials are far from the site, costs rise rapidly. The use of mining materials can reduce the transportation costs if the materials are already on site. CCBEs require both coarse and fine materials, which may not be available in all locations. For example, in Northern Quebec, Canada, fine soil is minimal and is often not available in suitable quantities for any CCBE construction.

As discussed in Section 7.3.4, the installation of a CCBE over oxidized tailings is not enough to ensure proper effluent quality in the short term. Contaminated pore water contains Fe^{3+} and bacteria that oxidize sulfides when the pH is low even in the absence of oxygen. The CCBE performance to limit oxygen diffusion may meet the design objective, but indirect and bacterial oxidation affects effluent quality significantly. The designer must be aware of this limitation when reclaiming an abandoned mine site.

Finally, Section 7.3.3 showed that a CCBE placed on a slope may exhibit lower volumetric water contents upslope, which may compromise the overall performance. For a relatively gentle slope, such as dikes around a tailings impoundment, the slope effects can be mitigated by using a bench construction method or integrating suction breaks. These methods intend to cut the suction link between the top and the bottom of the slope, which reduces the generated suctions and desaturation. However, for large waste rock piles, slopes are steep and it would be difficult to maintain a high volumetric water content several tens of meters upslope to ensure performance of the CCBE. Research is ongoing on the use of CCBEs to reclaim large waste rock piles.

7.7 OPPORTUNITIES AND RESEARCH NEEDS

CCBEs may appear to be a known technology, but there is still room to improve our knowledge in this area. CCBEs will remain site specific, and new materials and numerical tools always need to be tested before implementation. The following list presents the main areas of research related to CCBEs:

- Use of mining materials: To reduce the amount of natural material used in the different layers, mine tailings and waste rock may be used. There are issues to be solved related to the use of mining materials, such as their long-term geochemical behavior, the hydrogeological behavior of coarse waste rock as a CBL, the compatibility between layers when using waste rock, and preferential flow in waste rock layers.
- Use of industrial waste: Other industrial wastes have the potential to be applied as part of a CCBE. The main issues are potential contamination from organic sources or other emerging contaminants, as well as long-term hydrogeological behavior.
- Vegetation: Chapter 14 examines the integration of vegetation in reclamation methods. For CCBE, where cover integrity is crucial for efficiency, vegetation (mainly roots) can have a detrimental impact. Until recently, operators would remove trees periodically from a reclaimed site with a CCBE to prevent deterioration of cover performance. Nowadays, it is recognized that vegetation is part of the environment and that the CCBE cannot be isolated from vegetation. Research focuses on the measurement of root impact on CCBE performance to specify the optimal cohabitation of CCBEs and vegetation.
- Evolution of material properties: Performance of a CCBE to prevent AMD is directly related to the hydrogeological behavior of the materials. With time, materials' properties may evolve, particularly when subjected to freeze–thaw and wet–dry cycles, so research is required to evaluate the extent of properties' evolution and their real impact on cover performance over the long term, including climate change effects. Chapter 14 is dedicated to long-term performance of reclamation.
- Waste rock pile reclamation: Finally, adaptation of the CCBE design for waste rock pile needs to be investigated. The steep slopes and very low water table are challenges for reclamation. Erosion effects in slopes, for both waste rock piles and tailings impoundment dikes, require further investigation to offer practical solutions and provide effective CCBE long-term performance.

REFERENCES

Aachib, M., M. Aubertin, and R. P. Chapuis. 1994. Column tests investigation of milling wastes properties used to build cover systems. *International Land Reclamation and Mine Drainage Conference*, Pittsburgh, USA.

Akindunni, F. F., R. W. Gillham, and R. V. Nicholson. 1991. Numerical simulations to investigate moisture-retention characteristics in the design of oxygen-limiting covers for reactive mine tailings. *Canadian Geotechnical Journal* 28:446–451.

Aubertin, M., and B. Bussière. 2001. Water flow through cover soils using modeling and experimental methods: Discussion: *Geotechnical and Geoenvironmental Engineering* 127, no. 9: 810–811.

Aubertin, M., R. P. Chapuis, M. Aachib, B. Bussière, J.- F. Ricard, and L. Tremblay. 1995. *Évaluation en laboratoire de barrières sèches construites à partir de résidus miniers. MEND report 2.22.2a.*

Aubertin, M., B. Bussière, M. Aachib, R. P. Chapuis, and J. R. Crespo. 1996. Une modélisation numérique des écoulements non saturés dans des couvertures multicouches en sol. *Hydrogéologie* 1:3–13.

Aubertin, M., J.- F. Ricard, and R. P. Chapuis. 1998. A predictive model for the water retention curve: Application to tailings from hard-rock mines. *Canadian Geotechnical Journal* 35, no. 1: 55–69.

Aubertin, M., M. Aachib, M. Monzon, A.- M. Joanes, B. Brussière, and R. Chapuis. 1999. *Étude de laboratoire sur l'efficacité de recouvrements construits à partir de résidus miniers. MEND 2.22.2b.*

Aubertin, M., B. Bussière, and L. Bernier. 2003. *Environnement et gestion des rejets miniers (CD-ROM).* Montréal, QC: Les Presses Internationales Polytechnique, École Polytechnique de Montréal.

Barbour, S. L., and E. K. Yanful. 1994. A column study of static nonequilibrium fluid pressures in sand during prolonged drainage. *Canadian Geotechnical Journal* 31, no. 2: 299–303.

Bernier, L., R. C. Bédard, J. Lemieux, and F. Latour. 2005. Plan de fermeture et restauration du parc à résidus miniers de la mine Bouchard-Hébert. *Symposium 2005 sur l'environnement et les mines*, Rouyn-Noranda, CIM.

Bussière, B. 2007. Colloquium 2004: Hydrogeotechnical properties of hard rock tailings from metal mines and emerging geoenvironmental disposal approaches. *Canadian Geotechnical Journal* 44, no. 9: 1019–1052.

Bussière, B., M. Aubertin, and R. P. Chapuis. 2003. The behavior of inclined covers used as oxygen barriers. *Canadian Geotechnical Journal* 40:512–535.

Bussière, B., M. Benzaazoua, M. Aubertin, and M. Mbonimpa. 2004. A laboratory study of covers made of low sulphide tailings to prevent acid mine drainage. *Environmental Geology* 45:609–622.

Bussière, B., A. Maqsoud, M. Aubertin, J. Martschuk, J. McMullen, and M. Julien. 2006. Performance of the oxygen limiting cover at the LTA site, Malartic, Quebec, *CIM Bulletin* 99, 1096.

Bussière, B., M. Aubertin, M. Mbonimpa, J. W. Molson, and R. P. Chapuis. 2007. Field experimental cells to evaluate the hydrogeological behaviour of oxygen barriers made of silty materials. *Canadian Geotechnical Journal* 44, no. 3: 245–265.

Bussière, B., R. Potvin, A.- M. Dagenais, M. Aubertin, A. Maqsoud, and J. Cyr. 2009. Restauration du site minier Lorraine, Latulipe, Québec: Résultats de 10 ans de suivi. *Déchets, sciences et technologies* 54:49–64.

Dagenais, A. M., M. Aubertin, B. Bussière, and J. Cyr. 2005. Performance of the lorraine mine site cover to limit oxygen migration. *SME Transactions* 318:190–200.

Demers, I., B. Bussière, M. Benzaazoua, M. Mbonimpa, and A. Blier. 2008. Column test investigation on the performance of monolayer covers made of desulphurized tailings to prevent acid mine drainage. *Minerals Engineering* 21, no. 4: 317-329.

Demers, I., B. Bussière, M. Mbonimpa, and M. Benzaazoua. 2009. Oxygen diffusion and consumption in low sulphide tailings covers. *Canadian Geotechnical Journal* 46: 454-469.

Demers, I., B. Bussière, M. Aachib, and M. Aubertin. 2011. Repeatability evaluation of instrumented column tests in cover efficiency evaluation for the prevention of acid mine drainage. *Water, Air and Soil Pollution* 219:113–128.

Elberling, B., R. V. Nicholson, E. J. Reardon, and P. Tibble. 1994. Evaluation of sulphide oxidation rates: A laboratory study comparing oxygen fluxes and rates of oxidation product release. *Canadian Geotechnical Journal* 31, no. 3: 375–383.

Geo-Slope. 2016. SEEP/W.

Hillel, D. 1998. *Environmental Soil Physics*. San Diego, CA: Academic Press.

INAP. 2009. *Global acid rock drainage guide (GARD guide).* http://www.gardguide.com.

Kalonji Kabambi, A., Bussière, B., Demers, I. (2020). Hydrogeochemical behavior of reclaimed highly reactive tailings, Part 2: Laboratory and field results of covers made with mine waste materials. *Minerals*, 2020, 10, 589, doi: 10.3390/min10070589.

Lewis, J. and J. Sjöstrom. 2010. Optimizing the experimental design of soil columns in saturated and unsaturated transport experiments. *Journal of Contaminant Hydrology* 115:1–13.

Mbonimpa, M., et al. 2003. Diffusion and consumption of oxygen in unsaturated cover materials. *Canadian Geotechnical Journal* 40:916–932.

Nicholson, R. V., R. W. Gillham, J. A. Cherry, and E. J. Reardon. 1989. Reduction of acid generation through the use of moisture-retaining cover layers as oxygen barriers. *Canadian Geotechnical Journal* 26:1–8.

Pabst, T., J. Molson, M. Aubertin, and B. Bussière. 2017. Reactive transport modelling of the hydro-geochemical behaviour of partially oxidized acid-generating mine tailings with a monolayer cover. *Applied Geochemistry* 78:219–233.

Plante, B., B. Bussière, and M. Benzaazoua. 2014. Lab to field scale effects on contaminated neutral drainage prediction from the Tio mine waste rocks. *Journal of Geochemical Exploration* 137:37–47.

Rey, N. J., I. Demers, B. Bussière, and M. Mbonimpa. 2016. Field experiments to test the elevated water table concept combined with a desulfurized tailings cover layer, *Geo-Chicago 2016: Sustainability, Energy, and the Geoenvironment*, Chicago, IL. https://ascelibrary.org/doi/10.1061/9780784480137.029

Ross, B. 1990. The diversion capacity of capillary barriers. *Water Resources Research* 28, no. 10: 2625–2629.
Weeks, B., and G. W. Wilson. 2005. Variations in moisture content for a soil cover over a 10 year period. *Canadian Geotechnical Journal* 42, no. 6: 1615–1630.
Woyshner, M. R.. and E. K. Yanful. 1995. Modelling and field measurements of water percolation through an experimental soil cover on mine tailings. *Canadian Geotechnical Journal* 32, no. 4: 601–609.
Yanful, E. K. 1993. Oxygen diffusion through soil covers on sulphidic mine tailings. *Journal of Geotechnical Engineering-Asce* 119, no. 8: 1207–1228.
Yanful, E. K., S. Morteza Mousavi, and L. P. DeSouza. 2006. A numerical study of soil cover performance. *Journal of Environmental Management* 81, no. 1: 72–92.
Yanful, E. K., P. Simms, R. K. Rowe, and G. Stratford. 1999a. Monitoring an experimental soil waste cover near London, Ontario, Canada. *Geotechnical and Geological Engineering* 17, no. 2: 65–84.
Yanful, E. K., P. Simms, and S. C. Payant. 1999b. Soil covers for controlling acid generation in mine tailings: A laboratory evaluation of the physics and geochemistry. *Water, Air, and Soil Pollution* 114, no. 3-4): 347–375.
Young, R. A. (1995). *The Rietveld Method*. New York: Oxford University Press.

8 Elevated Water Table with Monolayer Covers

Thomas Pabst

8.1 INTRODUCTION AND BACKGROUND

Oxygen barrier covers, such as water covers (Chapter 6) and covers with capillary barrier effects (Chapter 7), can efficiently limit oxygen ingress to the underlying reactive mine wastes and prevent acid mine drainage (AMD) generation. Previous chapters discussed some limitations of these techniques, especially technical challenges (including long-term physical stability) and costs. Under certain conditions, the elevated water table (EWT) technique, coupled with the construction of a monolayer cover, can be an efficient, robust, and affordable alternative to the other reclamation options. This relatively novel approach relies on the low diffusion of oxygen in water and saturated or nearly saturated materials. This technique was first proposed by SENES (1996) and since then has been validated in the laboratory by numerical models and in the field (Demers et al. 2008; Ouangrawa et al. 2009, 2010; Ethier et al. 2018).

This approach is principally adapted to relatively humid climates. It involves two components: first, elevating and controlling the water table position to maintain the reactive tailings saturated (Section 8.2) and, second, using a cover system to improve water balance on site (i.e., decrease evaporation and increase infiltration; Section 8.3). Selected case studies are presented in Section 8.4 to illustrate the challenges related to the implementation of the technique in situ. Main advantages and limitations of the technique are summarized in Section 8.5. Current research and future perspectives (Section 8.6) conclude this chapter.

The EWT technique with a monolayer cover applies primarily to tailings impoundments – that is, materials that have a substantial water retention capacity and where the water table can realistically be controlled. The technique is not directly applicable to waste rock piles in most cases because of their high permeability and low air entry value (AEV), the deep water table position, and the typical absence of confining dams. Therefore, the EWC approach will only be discussed in relation to tailings impoundments in this chapter.

8.2 ELEVATED WATER TABLE

8.2.1 Concept

The objective of the EWT technique is to maintain the reactive tailings saturated or close to saturation. A high degree of saturation helps to decrease oxygen diffusion (Chapter 3), thus reducing sulfide oxidation and AMD generation. The EWT approach is therefore relatively similar to water covers (Chapter 6), except that there is generally no free water above the surface of the tailings. The main challenge of this technique is to elevate the water table to keep the reactive tailings saturated all the time, regardless of seasonal variations and climate change (Chapter 14). The performance target is generally not set in terms of degree of saturation but of water table position (typically the depth of the water table with reference to the surface of the reactive tailings; Demers et al. 2008; Ouangrawa et al. 2009, 2010).

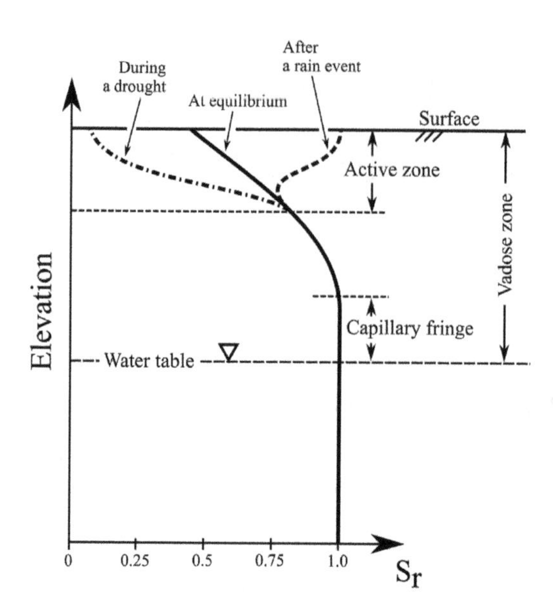

FIGURE 8.1 Variation of the degree of saturation with elevation in a tailings impoundment. Tailings remain saturated above the water table (capillary fringe) for suctions lower than their AEV. Above AEV, tailings are partially saturated. The degree of saturation in the active zone depends on the climatic conditions (infiltration, evaporation).

The water table does not necessarily need to reach the surface of the reactive tailings to keep them saturated. The tailings unsaturated hydrogeological properties, and especially their water retention capacity, imply that they will remain saturated above the water table (Figure 8.1; also see Chapter 3). The thickness of the capillary fringe can be estimated by the AEV. Typical AEV for tailings are comprised between 1 and 5 m (Bussière 2007). In other words, the water table could be located at a certain distance below the surface of the reactive tailings without significantly affecting the performance of the reclamation. A lower water table will significantly improve the short- and long-term geotechnical stability of the impoundment. It will also reduce the needs for high impervious dams. This is a strong argument in favor of the technique compared to a water cover.

An EWT can be considered independently of the geometry or the design of the impoundments, whether the impoundment is completely or partially surrounded by dams (Figure 8.2). However, it requires careful planning, preferably before the start of the operation and tailings disposal. An impermeable foundation is a significant advantage and can significantly contribute to elevating the water table (among other approaches; see Section 8.2). The construction of a protection layer (as pictured in Figure 8.2) will be discussed in Section 8.3.

8.2.2 Elevating the Water Table

Tailings are often transported and disposed of as slurry, and their initial water content is typically high. However, the water table in impoundments can vary with time, depending on climate and season, tailings properties, and impoundment design. Figure 8.3 shows inflows and outflows in a typical impoundment. Main inflows include infiltration from precipitation, surface runoff (in the particular case where the local topography is higher than the impoundment), and groundwater inflow (in the absence of an impermeable foundation). Outflows include evaporation (potentially coupled with transpiration if the site is covered by vegetation), seepage through dams (which are typically made of mine wastes and hence permeable), runoff through spillways, and percolation to the groundwater table (in the absence of an impermeable foundation).

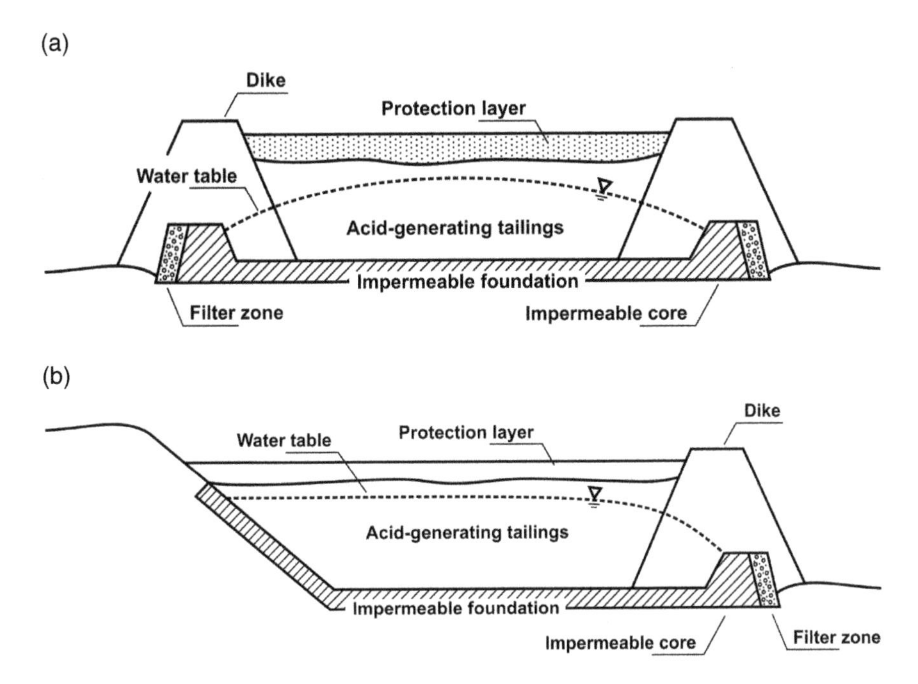

FIGURE 8.2 Conceptual models of an elevated water table within reactive tailings for reducing acid mine drainage (adapted from Aubertin et al. 1999). Two different impoundment geometries (a and b) are presented. A protection layer has been added at the surface of the reactive tailings to further improve the performance of the EWT technique (also see Section 8.3). Figures are not to scale.

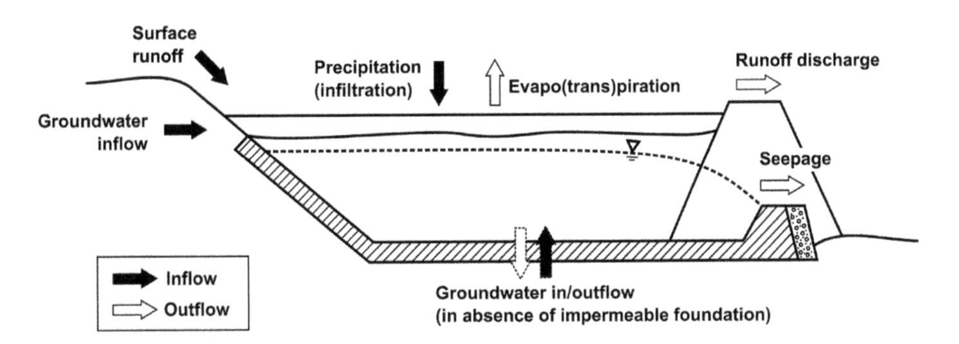

FIGURE 8.3 Water balance on a tailings impoundment reclaimed by using the EWT technique, with a monolayer cover.

The water table within a tailings impoundment usually decreases after closure and settles at an equilibrium located below the surface. The water table within the tailings is usually located above the regional water table, and the inflow is often limited to infiltration. Most of the water is lost by drainage through the dams and/or through the base of the impoundment (especially for old mine sites, where no measures have been taken to prevent percolation to the groundwater table). Evaporation can also induce significant water losses to the atmosphere. Finally, a substantial fraction of the water from large rain events can be lost by runoff through spillways. Therefore, some specific measures need to be undertaken to elevate the water table and keep it constant at the desired position.

Several approaches can be used to elevate the water table and modify the water balance within a tailings impoundment.

- Infiltration can be increased by limiting runoff and/or directing surface water toward the impoundment (where applicable). The installation of a coarse-grained layer at the surface, or leveling the surface of the tailings, can also help control runoff. Vegetation could contribute to reducing runoff but may in return increase water losses by transpiration, so its benefit is not clearly proven.
- Lateral seepage can be controlled by building impermeable or partially impermeable dams, where the core is at least as high as the targeted water table position. This method is very efficient in creating an artificial pond. It can, however, also be relatively costly, and it requires large quantities of materials with low permeability, especially in the case where the dams were initially built with tailings.
- The installation of an impermeable liner at the bottom of the impoundment to decrease seepage to the groundwater table is probably one of the most efficient methods to elevate the water table. However, it requires a careful construction and a careful monitoring to ensure the long-term integrity of the liner. The liner can be made of compacted clay (CCL), a geomembrane, a geosynthetic clay liner (GCL), or a combination of those. This approach can only be used if planned prior to disposal.
- Reducing tailings grain size through mineral processing could improve their retention capacity and increase the thickness of the capillary fringe. Sieving and controlled deposition could also be used to dispose of finer-grained tailings above the water table or in areas where the water table is lower. This requires careful planning and precise execution. Heterogeneities and operational challenges could also limit the applicability of such an approach. In addition, finer-grained materials usually have a larger specific surface, which could significantly increase the reactivity of sulfidic tailings, thus increasing the risk for contaminated drainage if the water table position is not strictly controlled.
- Densification can also reduce tailings saturated hydraulic conductivity and enhance their water retention capacity (i.e., their AEV). It is possible to increase tailings density by reducing their water content upon disposal (e.g., using filtration; tailings are then often referred to as thickened, paste, or filtered tailings depending on their water content), by speeding up drainage (e.g., using waste rock inclusions; Bolduc and Aubertin 2014), or by mechanical loading on site. Mixing is usually achieved prior to disposal. Even though densification can be costly, it increases the AEV and thus reduces the requirements for water table elevation. Densification can be complementary to other methods and could be specifically used for tailings that would be disposed above the water table. Tailings water retention capacity can also be increased by mixing tailings with binders and/or other mine wastes (e.g., paste rock).
- Vertical impermeable barriers can decrease lateral flow. Tailings are usually anisotropic in impoundments because of hydraulic disposal, which means that the vertical hydraulic conductivity can be one to two orders of magnitude smaller than the horizontal hydraulic conductivity (Blight 2010). By limiting lateral flow, the residence time of water in the tailings can be increased. The vertical barriers can be made of fine-grained materials or cement, which are usually mixed with tailings in situ. Numerical studies have indicated that the water table could be elevated by several meters using this method (SENES 1996), but practical and operational aspects have not been studied, and vertical impermeable barriers have so far not been used for the reclamation of a tailings impoundment.
- A cover system can also help to control evaporation and enhance infiltration; this solution is detailed in Section 8.3.

The choice of the most suitable method(s) is highly dependent on field conditions (climate, impoundment geometry, tailings properties, presence of an impermeable foundation) and many other technical, economic, and environmental constraints. Most of these solutions require some careful planning before disposal starts, which involves having a clear plan for the reclamation solution from the beginning of the production. It is much more challenging (and costly) to apply these solutions post facto. Integrated management approaches (like the concept of "Designing for closure"; Aubertin et al. 2002, 2016) are therefore strongly recommended to efficiently and durably elevate the water table.

8.2.3 Design Approach

Column tests, field investigations, and numerical modeling have shown that the EWT technique is very dependent on the position of the water table (Demers et al. 2008; Pabst et al. 2014). Conceptually, the water table should be no deeper than the AEV of the reactive tailings in order to maintain saturation. However, it is generally recommended that the water table should be kept at a depth less than half the AEV (Ouangrawa et al. 2009, 2010). This way, the design is slightly more conservative and can compensate (in part) for evaporation losses and overestimation of the AEV (graphical estimation, hysteresis; Maqsoud et al. 2012). Also, the prediction of in situ porosity (which has a direct influence on AEV) can be rather imprecise and can evolve over time.

The target of half AEV has been proposed based on analyses carried out for fresh unoxidized tailings. Recent studies have shown that the performance of an EWT for the reclamation of a pre-oxidized tailings impoundment can be different. In this case, it is recommended to elevate the water table even more, possibly to the surface of the tailings (Pabst et al. 2017a, 2017b; Ethier et al. 2018). One of the reasons for this difference may be the already acidic pore water that allows for indirect oxidation reaction. In this case, a small oxygen flux may be sufficient to maintain indirect oxidation reaction with Fe^{3+} (see Chapter 1). Maintaining the water table at the surface of the reactive tailings may be more efficient to prevent further oxidation and to increase the pH, so that Fe^{3+} would precipitate and the indirect oxidation would stop by itself.

The desaturation of the tailings was the main cause of most of the observed malfunctions and defaults for applications on fresh tailings. In several studies, the water table was located deeper than expected because of poor control of the water balance and material properties (Dobchuk et al. 2013). Therefore, other solutions should be considered when the minimum elevation of the water table cannot be guaranteed and long-term variations in the water balance are difficult to predict (e.g., due to climate change; see Chapter 14).

Piezometer and observation wells should be installed in situ to monitor the position of the water table. As the most critical parameter of the method's efficiency, the position of the water table must be monitored regularly and over a long time (Ethier et al. 2018). Automatic data loggers (in addition to water sampling and oxygen diffusion tests) may be useful to verify if the target is reached (also see Chapter 10).

8.2.4 Other Issues

Tailings are generally highly heterogeneous and anisotropic in impoundments. Some disposal methods (e.g., slurry) can create a natural segregation with coarser particles settling closer to the deposition point (typically close to the dams) and finer particles further away, toward the center of the impoundment (Bussière 2007; Blight 2010). A layering is therefore often observed in situ. In addition, mineralogical processes can evolve during the lifetime of the mine (typically several years), hence leading to different tailings properties (grain size, density, mineralogy). For all these reasons, tailings can have various properties throughout the site, and their AEV may vary significantly, laterally, and vertically. In this case, it is recommended to consider the lowest value of AEV (i.e., corresponding to the coarser tailings). Alternatively, it could be interesting, where feasible, to aim at disposing of more homogeneous and finer-grained tailings toward the end of the construction (e.g., filtered or paste tailings), when the site is near closure, to facilitate the application of the EWT technique.

The natural lateral segregation also usually creates coarser zones closer to the dams, which usually improves the general stability of the dams by decreasing interstitial pore water pressure and promoting a faster consolidation of the tailings. However, it can also induce a lowering of the water table closer to the dams, thus locally reducing the efficiency of the EWT technique. This effect has to be thoroughly studied to ensure a good performance of the reclamation everywhere on site. Less reactive materials (if available) could be placed in areas prone to desaturation. Otherwise, it may be necessary to increase the water table in these areas, or to lower the level of disposal.

Reclamation is intended to last for the lifetime of the impoundment, which can be very long and even "indefinite" (Vick 2001). Long-term variations, such as those caused by climate change, should therefore be taken into account in the design of the EWT. It is expected that climate change will have some (possibly significant) effect on the water balance and therefore on the water table position in the long term. Increased precipitation could contribute to elevating the water table, and should not be detrimental to geochemical stability aspects, but it could be more challenging to geotechnical stability. On the contrary, an increased evaporation could significantly alter the performance of the EWT technique. Climate change must be included in the design, for example, by using numerical models to simulate extreme events, like higher probable maximum precipitation or longer drought periods (Ethier et al. 2018; see also Chapter 14).

Keeping the reactive tailings saturated during the entire disposal may significantly improve the performance of the EWT technique by limiting pre-oxidation and reducing the risk of indirect oxidation. Underwater disposal, however, is not always possible for operational reasons, and in cases where tailings have already started to oxidize (this includes the particular case of abandoned mine sites), it is critical to consider the contamination already present in the pore water. Increasing the water table could release secondary precipitates and significantly alter water quality (Simms et al. 2000). Treatment or additional interventions may be required. Using slightly alkaline water may help to control the dissolution and mobility of metals and sulfates.

8.3 MONOLAYER COVERS

8.3.1 Concept

A cover can further add to the efficiency of the EWT technique by controlling the exchanges with the atmosphere. A cover can help to reduce evaporation from the tailings surface, improve infiltration, limit biological intrusions, and protect the surface of the impoundment against erosion. It can also act, in some cases, as an oxygen barrier. A cover will also help restore the site aesthetics and, under certain circumstances, promote revegetation (also see Section 8.6). Due to the high degree of saturation of the reactive tailings, a monolayer cover (also called a single-layer cover, i.e., made of one material only, contrary to multilayered covers) is generally sufficient (Dagenais et al. 2006). A monolayer cover is, however, only efficient when coupled with an EWT. As discussed below, a monolayer cover can have two different objectives, depending on the site and climatic conditions, and its design (composition, thickness) varies accordingly. Also, monolayer covers are usually installed on the flat portion on the surface of the tailings impoundment. A slope would make it more complex to elevate the water table.

8.3.1.1 Infiltration and Evaporation Control Layer

Evaporation is one of the main factors affecting the efficiency of the EWT technique. It is therefore usually critical to control evaporation by adding a monolayer cover so the water table level remains nearly constant, regardless of seasonal variations and climate change. The concept of such a monolayer cover is relatively similar to the top coarse-grained layer (drainage layer) in a cover with capillary barrier effects (CCBE) (also see Chapter 7). The objective is that the cover material loses its water as fast as possible (either by drainage and/or evaporation), so its (unsaturated) hydraulic conductivity becomes very small. Evaporation from the reactive tailings implies that the water is transported to the surface (where it changes phase), through the monolayer cover. If the cover material is unsaturated and its hydraulic conductivity is very low, evaporation will be slowed down and the monolayer cover will act as an evaporation barrier (Bussière et al. 2004). The development of such a capillary break requires a strong contrast (in terms of grain size, water retention capacity, and hydraulic conductivity) between the tailings and the cover material. Tailings are usually silty materials, so that monolayer covers used as evaporation control layers are often made of a coarser material such as sand or gravel. The choice of the cover material, however, is case specific and depends on tailings properties, climate, and site conditions (see Section 8.3.2 for more details).

A coarse material placed at the surface of the tailings can also contribute to decrease runoff and improve infiltration, thus increasing water balance on site and compensating for potential water losses. The objective of a monolayer cover made of a coarse-grained material is not to control oxygen flux. Its only purpose is to improve the water balance and help to maintain the EWT.

8.3.1.2 Water Retention Layer (Oxygen Barrier)

A monolayer cover can also be made of a fine-grained material, with a comparable or higher water retention capacity than the underlying tailings. The objective in this case is to maintain both the tailings and a part of the cover material saturated or close to saturation. The cover's high degree of saturation will contribute to further decreasing the oxygen flux to the reactive tailings, and therefore improve the performance of the EWT technique.

The grain size and hydrogeological properties of fine-grained monolayer covers are typically relatively similar to those of tailings and they generally do not develop a strong capillary barrier effect with the underlying reactive tailings. Consequently, an evaporation barrier made of a fine-grained cover will not be as efficient as that of coarser-grained materials to control water balance on site. The technique therefore usually requires a more rigorous control of the water table position, considering water losses to evaporation may be relatively high. It is particularly important to keep the cover close to saturation because fine-grained materials that desaturate can be prone to desiccation (depending on their plastic properties). Fractures can significantly affect the efficiency of a monolayer cover by reducing the cover's water retention capacity and creating preferential flow paths for oxygen. Deep fractures could also increase evaporation from the tailings and alter the position of the water table. Finally, such materials tend to generate more dust when they dry (if not revegetated). Adding another protection layer on the top (bi-layer cover system) to reduce dust generation may sometimes be required.

A monolayer cover will eventually desaturate (by drainage and evaporation) when the water table is too deep and will not prevent oxygen from oxidizing the underlying reactive tailings, hence leading to AMD generation. In other words, a fine-grained monolayer cover acting as oxygen barrier can only be efficient when coupled with an EWT.

There is no defined target for the choice of the water table position when a fine-grained monolayer cover is used. Generally, the water table should be elevated at a depth (below the surface of the tailings) that does not exceed half the smallest AEV between the reactive tailings and the cover material.

8.3.2 Monolayer Design Criteria

8.3.2.1 Monolayer Cover Materials

Various materials can be used to build monolayer covers. This section describes some of those materials. A critical aspect in the choice of a cover material is its long-term integrity and durability, as reclamation needs to remain efficient for an undetermined lifetime. Therefore, specific measures need to be undertaken to avoid erosion, cracking, and geochemical changes.

Natural soils, such as sand or gravel, are generally good materials for an evaporation barrier. Their homogeneity and stability, in addition to their hydrogeological properties, are significant advantages, but their availability (proximity to mine sites) can be limited.

Tills (also called moraine) are frequently used for the construction of cover systems, essentially due to their abundance close to many mine sites. However, till encompasses very different materials, and their performance as a cover is very dependent on their grain-size distribution. Some tills can be particularly suitable for cover systems. A reasonable amount of clay and silt particles can provide auto-sealing characteristics, reducing the effect of potential desiccation (Yanful et al. 1999). Coarser particles, on the other hand, tend to give tills a good resistance to erosion, evaporation, frost, and wetting–drying cycles. It is, however, critical to thoroughly characterize such materials, as other examples indicate that some type of moraines can be inefficient to play either the role of oxygen barrier or evaporation barrier (Pabst et al. 2018). Also, till materials are often relatively heterogeneous, because of their deposition history, so their grain size and properties can vary a lot spatially.

Natural materials often have good properties, and their long-term stability is usually not an issue. However, borrowing these materials in the vicinity of the mine can have a significant impact on the environment, especially for very large mine sites. It is, therefore, recommended to valorize waste materials (either from the mining or other industries) as much as possible (when applicable). In this case, however, the long-term geochemical stability must be thoroughly assessed.

Nonacid-generating tailings can be used in the construction of monolayer covers (oxygen barrier). Desulfurized tailings present similar hydrogeological properties and can be suitable too (Bussière et al. 2004, 2007; Demers et al. 2009; Dobchuk et al. 2013). In both cases, the small sulfide content (whether it is natural or residual) can contribute to a further decrease in the oxygen flux. However, despite their qualities, the use of desulfurized tailings can sometimes also lead to concentrations of metals and sulfates in the pore water that can exceed environmental criteria. This is particularly the case for metals that are highly mobile at near neutral pH (contaminated neutral drainage or CND).

Alkaline materials can contribute to reduce the acidity of the pore water in situ due to their natural buffering capacity. This effect, however, is limited, both in terms of neutralization capacity and durability, and the reclamation solution should not rely solely on this property. Alkaline materials can, nonetheless, sometimes be beneficial to neutralize the already acidic pore water.

8.3.2.2 Monolayer Cover Thickness

The thickness of a monolayer cover is generally comprised between 0.5 and 2 m. Cover performance increases with thickness, but so do costs. The objective of the design, therefore, is to find a balance between costs and performance.

Laboratory and field observation, together with numerical sensitivity analyses, have shown that the desaturation of a monolayer cover is often caused by evaporation. The portion of the cover affected by evaporation is strongly dependent on the cover material. For fine-grained materials, the affected depth can exceed 1 m. Evaporation usually affects only a few centimeters for coarser cover systems.

A site-specific assessment is generally required to design the cover thickness. The thickness will depend on the material chosen and the in situ climatic conditions. It is particularly important to take into account the climate change by using specific tests or numerical simulations to simulate the effect of long drought periods, for example.

8.3.3 Design Methodology

The coarse- or fine-grained criteria mentioned in the previous section is not an absolute characteristic, but must be understood in terms of hydrogeological contrast with the underlying reactive tailings. Experimental and numerical hydrogeological assessments, such as the following, are required to choose the most suitable cover material (Figure 8.4):

- Field investigations to assess specific site conditions (e.g., climatic conditions, geology, hydrogeology, geometry).
- Thorough characterization of both tailings and cover materials, including grain-size distribution, relative density of the grains, porosity, compaction properties, hydraulic conductivity,

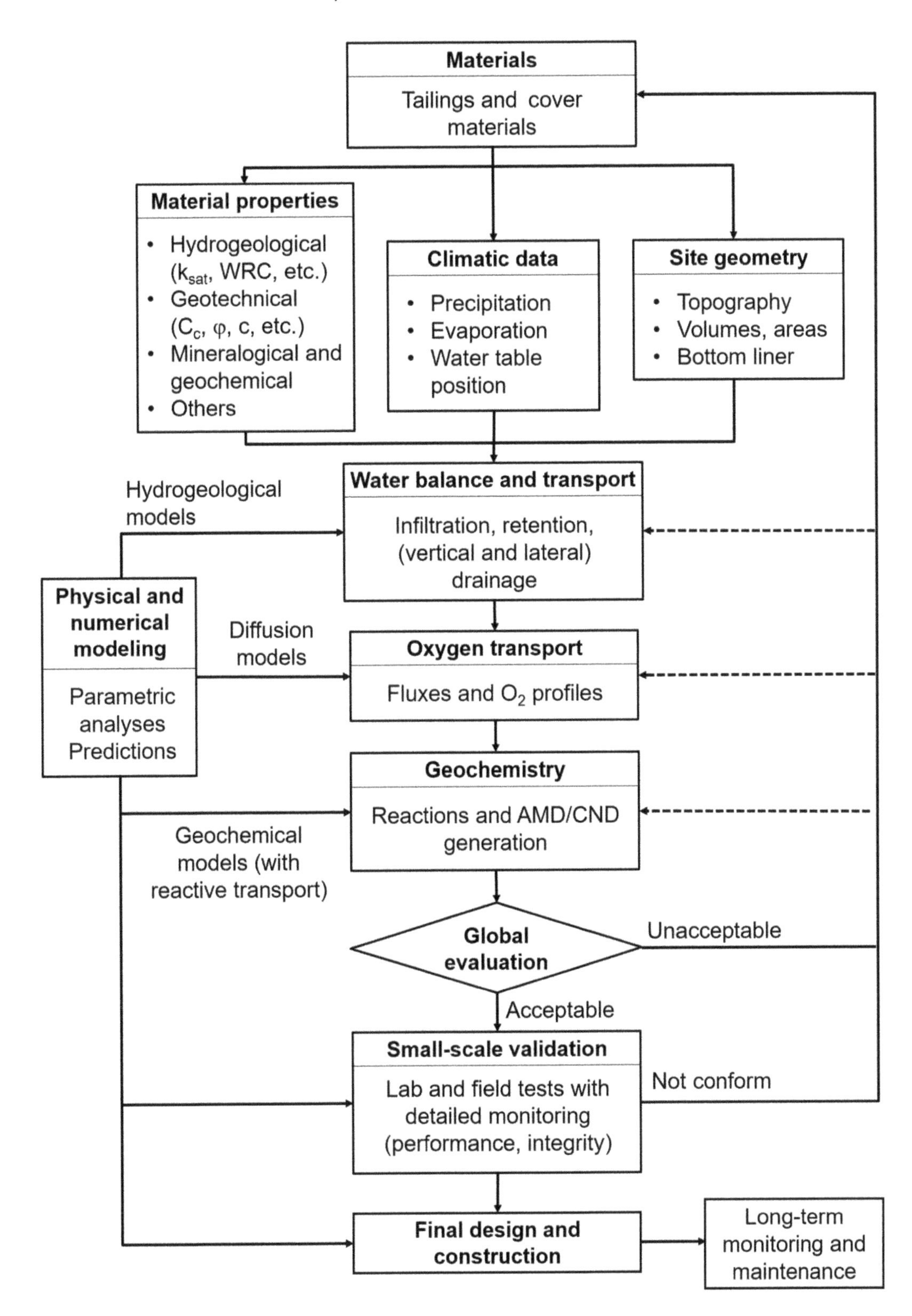

FIGURE 8.4 Design methodology for a monolayer cover coupled with an elevated water table. (Adapted from Aubertin et al. 2016).

and water retention curve. Additional testing (e.g., mineralogy or plasticity) may also be relevant, depending on the material properties. Depending on the climate in situ, susceptibility to freeze–thaw and wetting–drying cycles should also be assessed.

- Physical modeling at the laboratory scale (e.g., using column tests). Column tests must be carefully designed to be representative of field conditions, while also allowing for reasonable test duration and interpretation (Demers et al. 2011). The duration of such a test can exceed one year.
- Numerical simulations validated and calibrated on laboratory results. Models are used to extrapolate medium-scale tests to larger time and space scales. Alternative configurations, materials, and conditions (e.g., precipitation and evaporation rate, water table position) can be tested.
- Field tests. Field tests are used to validate the performance of the reclamation in real conditions at an intermediate scale (Bussière et al. 2007; Bossé et al. 2015). The hydrogeological behavior of the cover and the reactive tailings must be precisely monitored. Field cells are usually tested for 1–3 years, sometimes even longer.

Material costs and availability are also a decisive factor in this choice (Sjoberg et al. 2001; Aubertin et al. 2002).

The construction of the monolayer cover needs to be carefully supervised, like any other geotechnical infrastructure. Minimum thickness and materials' properties and homogeneity must be controlled during construction over the entire site to avoid localized oxidation zones. Parametric studies show that small variations in material grain size can indeed have significant impacts on the water retention curve and the permeability function. Numerical simulations have shown, for example, that a variation of less than an order of magnitude of the saturated hydraulic conductivity could significantly decrease the performance (Pabst et al. 2017a, 2017b).

When mine tailings are used as cover materials, hydraulic deposition may induce lateral segregation, which may cause local variation in grain size and porosity. Such heterogeneities can significantly alter the cover performance and need to be controlled by adjusting the sequence of deposition (multiple disposal points) or increasing the solid ratio of the pulp to limit segregation, for example.

Cover efficiency should also be assessed after construction. Piezometers should be installed at different locations throughout the impoundment, and in particular close to the dams. The water table should be monitored regularly. Water content probes should also be used to assess the degree of saturation of the tailings above the water table. Monitoring the degree of saturation is even more critical in the case of a monolayer cover whose role is to limit oxygen diffusion (Section 8.3.1). Water quality improvement is usually a good indicator of cover performance, and a good monitoring program should include regular water sampling, even though it may take some time to see improvements, especially in the case of an old pre-oxidized site. The most direct way to assess the performance of the reclamation may be to conduct oxygen consumption texts directly at the surface of the cover (Dagenais et al. 2012; see also Chapters 3 and 10). This approach directly and simultaneously takes into account both the effect of the EWT and the cover. It can, however, be complex to estimate very low oxygen fluxes using this technique because of experimental uncertainties, but a maximum target can be verified. More details on monitoring can be found in Chapter 10.

8.3.4 Other Challenges

A monolayer cover is directly exposed to atmospheric conditions, and measures need to be taken to protect it against erosion, bioturbation (roots, animals), or anthropic activities in the long term. Frost and heave as well as wetting and drying cycles could also alter the cover performance, especially with fine-grained materials. Desiccation cracks could create preferential paths for oxygen, thus decreasing the cover's efficiency as an oxygen barrier. Coarse-grained materials are usually more resilient.

In a few particular cases, when the water table is located deep under the tailings surface (i.e., deeper than half the AEV), and if the monolayer cover is made of a finer-grain material than tailings, a capillary break can develop at the interface with the tailings. The cover could then temporarily retain moisture and limit oxygen flux. However, in this scenario, evaporation can still significantly contribute to water losses (Pabst et al. 2018); it is therefore usually not possible to keep a monolayer saturated for a long time only through a capillary barrier effect. Thus, using a monolayer cover is not recommended unless the water table position is controlled and maintained high enough.

8.4 CASE STUDIES

The EWT technique, coupled with a monolayer cover, has been implemented in several sites in Canada. These sites were generally instrumented and closely monitored, which gave good information on the field performance of the technique in real conditions and at a large scale. This section describes a few examples of specific Canadian sites.

8.4.1 Manitou (Quebec, Canada)

The Manitou mine site (Figure 8.5) is located in the Abitibi region, in Quebec, 12 km east of Val d'Or. Between 1941 and 1979, the mine produced copper, zinc, silver, and gold. From 1970, the site also started to process ore from other mines. A total of 11 million tonnes of highly reactive tailings were disposed of in two impoundments. Dams were built in the 1950s to protect the installations. Several breaches occurred in the 1970s, and large quantities of tailings ended up in a nearby creek, covering several hectares along more than 6 km. When the reclamation work started around 2007, tailings were covering an area of approximately 190 ha, including the 42 ha of the original tailings impoundments. Tailings thickness varied from a few meters in the spill area to more than 12 m in the main impoundments.

The first studies on the site date back to the early 1990s. The reclamation plan for the lower part of the site consisted of an EWT covered by a 1-m thick monolayer cover made of nonacid-generating tailings produced at a nearby mine. The first step was to gather tailings and reduce the footprint of the spillway. Existing dams were reinforced and new ones were established to raise the water table. Local watercourses were altered to divert fresh water from the contaminated area. Several studies had shown that the water table had to be elevated to the surface of the reactive and pre-oxidized tailings to limit the risk of continued AMD generation due to indirect oxidation reactions.

The site is instrumented to monitor the water table position, assess oxygen fluxes to the tailings, and follow the evolution of water quality. Field monitoring indicates that the minimum water table

(a)

(b)

FIGURE 8.5 Manitou mine site before (a) and after (b) reclamation. Credit: RIME.

level target has, in general, been reached. Oxygen fluxes to the reactive tailings are also significantly lower than the target. Finally, water quality has significantly improved over the past few years, which suggests that the reclamation is efficient to control AMD generation (Ethier et al. 2018).

It was, however, considered unrealistic to try to elevate the water table to the surface of the old tailings impoundments, which were several meters higher than the rest of the site. Therefore, the reclamation option for this part of the site is to use a CCBE (see also Chapter 7). Reclamation work is ongoing on this portion of the site.

8.4.2 Detour Lake (Ontario, Canada)

Detour Lake Gold Mine is located 290 km northeast of Timmins, Ontario. The mine's operation began in 1983. The main tailings impoundment covers an area of approximately 300 ha and contains about 15 million tonnes of acid-generating tailings.

A large part of the impoundment is flooded and the water cover prevents the oxidation of the tailings (also see Chapter 6). The objective of the reclamation plan was to add a 1.0–1.5 m thick monolayer cover on top of the tailings where the water cover was too thin or nonexistent and unable to efficiently prevent AMD generation. Desulfurized tailings ($S < 1\%$ wt) were used as cover materials, and disposed of by end-pipe discharge. The desulfurized tailings are finer than the underlying reactive tailings, and it was expected that a capillary break effect could develop, maintaining the cover close to saturation to control oxygen diffusion to the acid-generating tailings.

Piezometers were installed, and several water content profiles were measured in this part of the impoundment to monitor water table and moisture fluctuations in the reactive tailings and the desulfurized tailings monolayer cover (Dobchuk et al. 2013). A thorough material characterization and numerical modeling were carried out to further assess the performance of the reclamation.

In practice, analyses revealed that the desulfurized tailings used as cover materials were coarser than initially expected and inadequate to create a capillary barrier effect with the underlying reactive tailings. The cover was consequently more prone to desaturation, especially where the water table was deeper. The cover was therefore not efficient as an oxygen barrier. Results also showed that oxygen diffusion was efficiently controlled where the water table was located less than 1 m below the surface of the reactive tailings (Dobchuk et al. 2013). These results confirm the importance of the water table position on the reclamation performance and how critical it can be to characterize both the reactive tailings and cover materials hydrogeological properties.

8.4.3 Aldermac (Quebec, Canada)

The Aldermac mine site is located in the Abitibi region of Quebec, 15 km west of Rouyn-Noranda. It was abandoned and is now the property of the Province of Quebec. The site was operated for zinc and copper from 1931 to 1943. Over 1.3 million tonnes of tailings were produced and disposed of over an area of 60 ha. No confining or reclamation measures were undertaken at that time, as a result, the highly reactive tailings (sulfur content can be up to about 50% by weight) were exposed to atmosphere and climatic conditions (Figure 8.6). As a result, the pore water on site was already very acidic and contaminated when the first mitigation begun in the 1990s. The first step was to build dams and berms to confine tailings within a 50-ha area. The remaining tailings were spread over 10 km along downstream streams, covering approximately 26 ha.

The site was divided in four areas. Various reclamation methods were used for the different zones. The North site, which covered approximately 26.5 ha, was reclaimed using an EWT coupled with a coarse-grained monolayer cover. New dams and berms were built and old dams were raised to increase the water table position in the tailings. Some tailings from other parts of the site were also disposed of in this area. Reactive tailings were then covered by a 1-m thick sand and gravel layer (Figure 8.6). The objective is to maintain the water table inside the coarse-grained layer to keep the reactive tailings saturated and add an additional oxygen barrier on top.

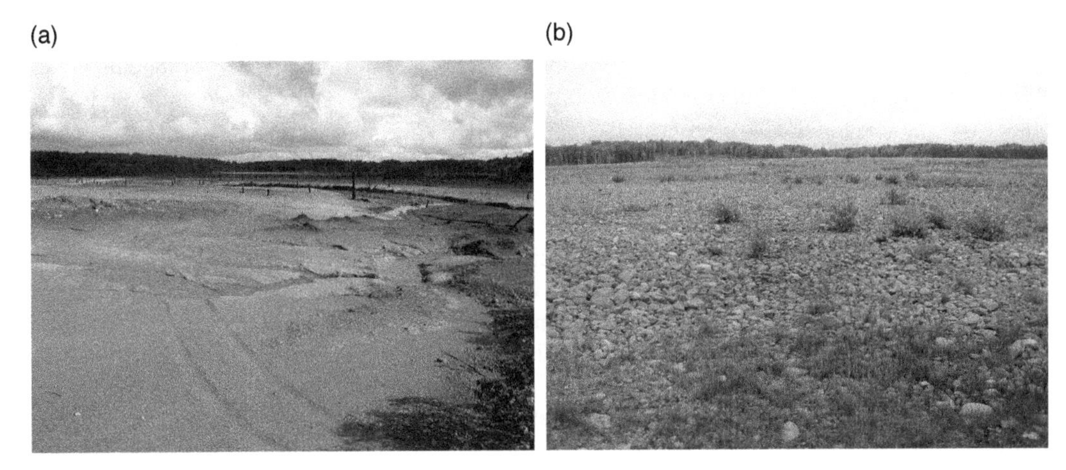

FIGURE 8.6 Aldermac mine site before (a) and after (b) reclamation. Credit: RIME.

The site is closely monitored to assess the efficiency of the reclamation (Maqsoud et al. 2015). Observation wells and piezometers were installed on the main impoundment and around the dams. They have two objectives: first, to monitor the water table position, and second, to assess the evolution of the groundwater quality and the possible connections with the impoundment. Various automatic sensors are installed to monitor water table position, and water is regularly sampled and analyzed for pH, electrical conductivity, and metal concentrations.

Preliminary results indicate that the position of the EWT is very dependent on climatic conditions. So far, however, the water table has remained above the target position – that is, in the sand-gravel monolayer cover. Results also indicate that the absence of bottom liner has caused a migration of contaminants to the aquifer. Water quality tends to slightly improve over time, but remains highly contaminated due to pre-oxidation and initial pore water quality (Maqsoud et al. 2015). Monitoring continues.

8.5 MAIN ADVANTAGES AND LIMITATIONS OF THE EWT TECHNIQUE

The main advantages and limitations of the EWT coupled with a monolayer cover are summarized in Table 8.1. The technique is usually less expensive than a CCBE because it requires less material for the construction of the monolayer cover. This implies, however, that cover material is

TABLE 8.1
Main Advantages and Limitations of the Elevated Water Table Technique Coupled with a Monolayer Cover

Advantages	Limitations
- Less expensive than other reclamation techniques if material is available and retention dams are impermeable - Efficient in the long term when well designed - Requires less material for the cover (reduced environmental footprint) - Better geotechnical stability than water covers because of lower pore water pressures on dikes - Possible use during operations (progressive reclamation)	- Very sensitive to climatic condition (evaporation, water balance) - Specific design criteria for pre-oxidized mining sites - Susceptible to failure by seismic events (liquefaction) - Limitations on future use of site - Applies only to tailings impoundments (not waste rock piles) - Not applicable everywhere (requires certain conditions in terms of geometry and climate)

available close to the site. Also, if impermeable dams need to be built after the construction of the impoundment, costs could significantly increase. Careful planning from the start of the operations ("Designing for Closure"; Aubertin et al. 2002, 2016) is therefore required for the method to be attractive. The technique also significantly improves the geotechnical stability of the dams compared to a water cover, because the pore water pressure is much lower with an EWT that remains below the surface of the tailings.

However, the EWT, coupled with a monolayer cover is very sensitive to water balance and climatic conditions, and thus not applicable everywhere. A decrease of the water table position below the AEV (or half the AEV) of the reactive tailings could significantly affect the performance of the method. The effects of climate change should be carefully assessed and reclamation design adapted accordingly. The technique is also not well suited for waste rock piles, and its application to pre-oxidized mine sites requires a more careful design and more severe performance targets. Finally, most of tailings remain saturated and susceptible to liquefaction following seismic events.

The choice and design of reclamation are site specific and require a thorough characterization of material properties (both mine wastes and cover material(s)), laboratory and field assessment, and a detailed numerical evaluation of influence parameters (water table position, cover thickness, climatic conditions). Also, like other reclamation techniques, the EWT coupled with a monolayer cover requires careful quality control during construction (heterogeneity, thicknesses) and long-term monitoring to ensure the performance of the reclamation. The site should also be protected and construction limited to avoid physical alteration to the cover properties.

8.6 PERSPECTIVES AND RESEARCH NEEDS

The EWT technique, coupled with a monolayer cover, is a relatively novel approach that has been used on several sites with good results. Yet several aspects could be improved, and the following aspects should be further studied in the near future (some are already under testing).

- Sand and gravel are not necessarily available in sufficient quantities close to the mine sites. Therefore, other types of materials should be considered for the monolayer cover. By-products (non-acid-generating of desulfurized tailings, industrial wastes) could be suitable, but their long-term durability still needs to be investigated and their availability depends on the mine location. In situ valorization of mine waste seems more promising. Waste rock in particular could constitute an interesting alternative to natural material. Their relative coarse grain size should allow them to be an efficient evaporation barrier, while also being available in large quantity on mine sites. This implies that they are not reactive and they do not generate AMD or CND. They haven't been used yet and some assessment is required. Also, all waste rocks may not be suitable. They should not be too fine, but also not too coarse; otherwise, wind could evaporate water directly from the underlying reactive tailings.
- Also, as mentioned in the introduction, the EWT technique is not well suited to reclaim waste rock piles. Some work is currently in progress to evaluate the possibility of using the technique coupled with the co-disposal of waste rock in tailings impoundments. Surrounding tailings could maintain a high degree of saturation in the waste rock. Further research is required to assess the effect of inclusions on water flow and oxygen transport. The use of reactive waste rock in the inclusions is also under consideration.
- Finally, the ultimate purpose of the reclamation is to bring the site to a state that is integrated in the natural environment. Thus, revegetation is a part of reclamation. The impact of vegetation on the performance of the EWT with monolayer technique needs to be investigated. Some of these aspects are discussed in Chapter 14.

REFERENCES

Aubertin, M., B. Bussière, M. Monzon, et al. 1999. *Étude sur les barrières sèches construites à partir des résidus miniers. Phase II: Essais en place*. Rapport de Recherche, Projet CDT P1899. NEDEM/MEND 2.22.2c.

Aubertin, M., B. Bussière, and L. Bernier. 2002. *Environnement et gestion des rejets miniers*. Montréal: Presses Polytechnique Int

Aubertin, M., B. Bussière, T. Pabst, M. James, and M. Mbonimpa. August 14–18, 2016. Review of reclamation techniques for acid generating mine wastes upon closure of disposal sites. *Geo-Chicago: Sustainability, Energy and the Geoenvironment*, Chicago. https://ascelibrary.org/doi/abs/10.1061/9780784480137.034.

Blight, G. 2010. *Geotechnical Engineering for Mine Waste Storage Facilities*. London: CRC Press.

Bolduc, F. L., and M. Aubertin. 2014. Numerical investigation of the influence of waste rock inclusions on tailings consolidation. *Canadian Geotechnical Journal* 51, no. 9: 1021–1032.

Bossé B., B. Bussière, R. Hakkou, A. Maqsoud, and M. Benzaazoua. 2015. Field experimental cells to assess hydrogeological behaviour of store-and-release covers made with phosphate mine waste. *Canadian Geotechnical Journal* 52, no. 9: 1255–1269.

Bussière, B. 2007. Colloquium (2004). Hydro-geotechnical properties of hard rock tailings from metal mines and emerging geo-environmental disposal approaches. *Canadian Geotechnical Journal* 44, no. 9: 1019–1052.

Bussière, B., M. Aubertin, M. Mbonimpa, J. Molson, and R. Chapuis. 2007. Field experimental cells to evaluate the hydrogeological behaviour of oxygen barriers made of silty materials. *Canadian Geotechnical Journal* 44, no. 3: 245–265.

Bussière, B., M. Benzaazoua, M. Aubertin, and M. Mbonimpa. 2004. A laboratory study of covers made of low-sulphide tailings to prevent acid mine drainage. *Environmental Geology* 45, no. 5: 609–622.

Dagenais, A.- M., M. Aubertin, and B. Bussière. 2006. Parametric study on the water content profiles and oxidation rates in nearly saturated tailings above the water table. *Proceedings of the 7th International Conference on Acid Rock Drainage (ICARD)*, March 26–30, 2006, St. Louis, Missouri, 405–420.

Dagenais, A. M., M. Mbonimpa, B. Bussière, and M. Aubertin. 2012. A modified oxygen consumption test to evaluate gas flux through oxygen barrier cover systems. *ASTM Geotechnical Testing Journal* 35, no. 1: 150–158.

Demers, I., B. Bussière, M. Benzaazoua, M. Mbonimpa, and A. Blier. 2008. Column test investigation on the performance of monolayer covers made of desulphurized tailings to prevent acid mine drainage. *Mineral Engineering* 21:317–329.

Demers, I., B. Bussière, M. Aachib, and M. Aubertin. 2011. Repeatability evaluation of instrumented column tests in cover efficiency evaluation for the prevention of acid mine drainage. *Water, Air, and Soil Pollution* 219, no. 1: 113–128.

Demers, I., B. Bussière, M. Mbonimpa, and M. Benzaazoua. 2009. Oxygen diffusion and consumption in low sulphide tailings covers. *Canadian Geotechnical Journal* 46:454–469.

Dobchuk, B., C. Nichol, W. Wilson, and M. Aubertin. 2013. Evaluation of a single-layer desulphurized tailings cover. *Canadian Geotechnical Journal* 50:777–792.

Ethier, M.- P., B. Bussière, S. Broda, and M. Aubertin. 2018. Three-dimensional hydrogeological modelling to assess the elevated water table technique for controlling acid generation from an abandoned tailings site. *Hydrogeology Journal* 26, no. 4: 1201–1219.

Maqsoud, A., B. Bussière, M. Aubertin, and M. Mbonimpa. 2012. Predicting hysteresis of the water retention curve from basic properties of granular soils. *Geotechnical and Geological Engineering* 30:1147–1159.

Maqsoud, A., M. Mbonimpa, B. Bussière, and M. Benzaazoua. 2015. The hydrochemical behaviour of the Aldermac abandoned mine site after its rehabilitation. *Canadian Geotechnical Conference, GéoQuébec 2015*, Québec, Canada.

Ouangrawa, M., M. Aubertin, J. Molson, B. Bussière, and G. J. Zagury. 2010. Preventing acid mine drainage with an elevated water table: Long-term column experiments and parameter analysis. *Water, Air, and Soil Pollution* 213:437–458.

Ouangrawa, M., J. Molson, M. Aubertin, B. Bussière, and G. J. Zagury. 2009. Reactive transport modelling of mine tailings columns with capillarity-induced high water saturation for preventing sulfide oxidation. *Applied Geochemistry* 24:1312–1323.

Pabst, T., M. Aubertin, B. Bussière, and J. Molson. 2014. Column tests to characterize the hydrogeochemical response of pre-oxidized acid-generating tailings with a monolayer cover. *Water, Air, and Soil Pollution* 225, no. 2:1841.

Pabst, T., M. Aubertin, B. Bussière, and J. Molson. 2017a. Experimental and numerical evaluation of single-layer covers placed on acid-generating tailings. *Geotechnical and Geological Engineering* 35:1421–1438.

Pabst, T., B. Bussière, M. Aubertin, and J. Molson. 2018. Comparative performance of cover systems to prevent acid mine drainage from pre-oxidized tailings: A numerical hydro-geochemical assessment. *Journal of Contaminant Hydrology* 214:39–53.

Pabst, T., J. Molson, M. Aubertin, and B. Bussière. 2017b. Reactive transport modelling of the hydro-geochemical behaviour of partially oxidized acid-generating mine tailings with a monolayer cover. *Applied Geochemistry* 78:219–233.

SENES. 1996. *Review of use of an elevated water table as a method to control and reduce acidic drainage from tailings. Mine Environment Neutral Drainage*, Report 2.17.1. Ottawa, ON: Ministry of Natural Resources.

Simms, P. H., E. K. Yanful, L. St-Arnaud, and B. Aubé. 2000. A laboratory evaluation of metal release and transport in flooded pre-oxidized mine tailings. *Applied Geochemistry* 15:1245–1263.

Sjoberg, B., G. W. Wilson, and M. Aubertin. 2001. Field and laboratory characterization of desulphurized tailings cover system. *Proceedings of the 54th Canadian Geotechnical Conference*, September 16–19, 2001, Calgary, AB.

Vick, S. G. 2001. Stability aspects of long-term closure for sulphide tailings. *Proceedings of the Safe Tailings Dams Constructions*, Gaellivare, Sweden.

Yanful, E. K., P. H. Simms, and S. C. Payant. 1999. Soil covers for controlling acid generation in mine tailing, a laboratory evaluation of the physics and geochemistry. *Water, Air, and Soil Pollution* 114:347–375.

9 Insulation Covers

Vincent Boulanger-Martel, Bruno Bussière, and Jean Côté

9.1 INTRODUCTION

Reclamation of mine sites located in permafrost regions started to gain interest from the scientific community during the mid-1990s (MEND 1993, 1996a, 1996b, 1997; Erickson 1995). In the late 1990s and early 2000s, researchers intensified their efforts to understand the processes controlling sulfide oxidation in permafrost environments (e.g., Elberling 1998, 2001, 2005; Godwaldt 2001; Meldrum et al. 2001). At the same time, some works were dedicated to the assessment of covers' effectiveness to control acid mine drainage (AMD) for tailings storage facilities (TSFs; Kyhn and Elberling 2001; Meldrum et al. 2001; MEND 2004). From then on, an increasing number of studies related to the reclamation of TSFs and waste rock storage facilities (WRSFs) were conducted at several mine sites, mainly in Canada (e.g., Coulombe et al. 2013; Pham et al. 2013; Smith et al. 2013a; Amos et al. 2015; Boulanger-Martel et al. 2016; Lessard et al. 2018). Several types of covers were proposed to reclaim TSFs and WRSFs located in permafrost regions (e.g., MEND 2004, 2009). However, thus far, the insulation cover (also known as thermal cover) is the most documented and the only cover option specifically designed to reclaim TSFs and WRSFs in such cold regions. This technique seeks to favor the aggradation of permafrost (i.e., an increase in the thickness of permafrost) in the reactive mine wastes and cover materials. The low temperatures maintained in the mine wastes control the oxidation kinetics (i.e., reactivity) of sulfide minerals (see Chapter 1), whereas freezing conditions reduce the potential for transport of contaminated water. Indeed, the thawing of the surface layer's interstitial ice can allow the release and transport of contaminants into the receiving environment (Elberling 2001).

As described in Chapter 1, the aggradation of permafrost conditions (the low temperature) in mine tailings and waste rocks during operations and after mine closure can help control AMD generation and the transport of oxidation products. Therefore, in the context of mine site reclamation in permafrost regions, the thermal regime of the ground layer near the surface governs the geochemical and thermal behavior of mine wastes. In these climatic conditions, AMD is mainly generated in the surface layer during summer because the temperatures (above 0°C) are more prone to sulfide oxidation (Elberling 2001). For these reasons, the thermal behavior of the ground layer near the surface is the most important feature with respect to cover design. Because this layer is significantly affected by climatic parameters, the thermal behavior and performance of insulation covers can also vary each year as a result of short-term fluctuations in climatic and hydrogeological conditions.

9.2 FUNDAMENTALS OF PERMAFROST AND IMPLICATIONS FOR COVER DESIGN

Permafrost is one of the key geomorphological attributes that characterizes the ground thermal regime of high latitudes around the world. Permafrost is a material (e.g., mineral soil, rock, ice, and organic matter) that observes temperatures equal to or lower than 0°C for at least two consecutive years (Van Everdingen 1998). Hence, permafrost is strictly defined by temperature and is independent of the presence of water and its state – water in permafrost can be unfrozen, partially frozen, or frozen (French 2007; Dobinski 2011). The existence of permafrost conditions is mostly related to the climatic and terrain conditions, whereas the thickness of the permafrost is mainly controlled

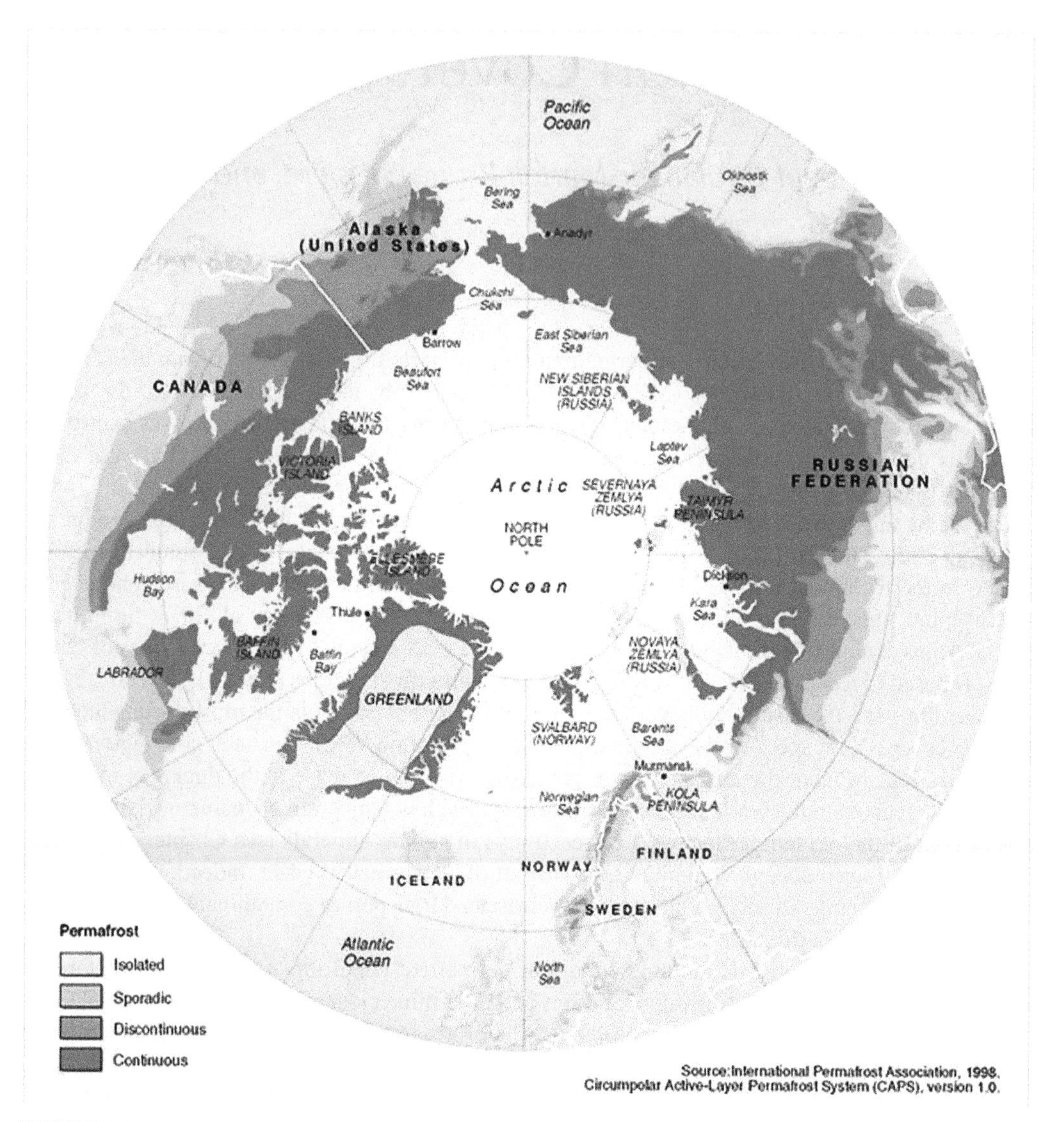

FIGURE 9.1 Distribution of isolated, sporadic, discontinuous, and continuous permafrost (Rekacewicz 2005). (source file: https://www.grida.no/resources/7000).

by the temperature at the surface (air and ground surface temperatures) and the local geothermal gradient (e.g., Andersland and Ladanyi 2004). In the northern hemisphere, the geographical occurrence of permafrost ranges from isolated patches, sporadic and discontinuous permafrost in the southernmost limit of the permafrost distribution to thick continuous permafrost conditions toward the north (French 2007; Figure 9.1). Discontinuous permafrost refers to the presence of thawed soil and permafrost coexisting in lateral extent, whereas continuous permafrost refers to a ground for which permafrost conditions are observed everywhere. The thickness of permafrost is highly variable but can range from less than a meter in the south and close to water bodies to more than 500 m toward the north and on the continent (French 2007; Smith et al. 2010, 2013b). A thermophysical model of permafrost is shown in Figure 9.2. The thickness of the permafrost represents the vertical distance between the permafrost table and permafrost base. The permafrost table is the upper boundary of permafrost and represents the depth at which ground temperatures become perennially below 0°C. The permafrost base is the lower boundary of permafrost and represents the deepest position at which ground temperatures are perennially below 0°C.

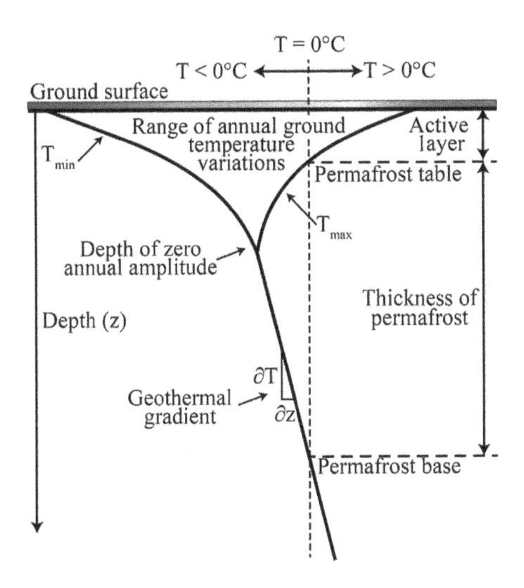

FIGURE 9.2 Simplified thermophysical model of permafrost (T_{max} and T_{min}: maximum and minimum annual ground temperatures).

The layer of the ground that is subject to seasonal thawing and freezing is defined as the active layer. The thickness of the active layer depends upon many factors, including the ground surface temperature, the subsurface moisture conditions and thermal properties, and the thickness of the snow cover (e.g., French 2007; Dobinski 2011). Because insulation covers require cold temperatures at the ground surface to work efficiently, their applicability is essentially limited to continuous permafrost regions (MEND 2004; Arenson and Sego 2007). The southernmost geographical extent of continuous permafrost corresponds to regions with mean annual air temperatures (MAATs) of about −8°C (French 2007).

9.3 DESIGN BASIS

9.3.1 Concept

Insulation covers essentially consist of a given thickness (z; Figure 9.3) of non-reactive materials placed over the reactive mine tailings or waste rocks to favor permafrost aggradation and low temperatures in the mine wastes (Elberling et al. 2000; Elberling 2005; Boulanger-Martel et al. 2020a). The design of insulation covers aims at optimizing the thickness of the cover in order to maintain the active layer within the cover materials and keep the mine wastes frozen year-round and below

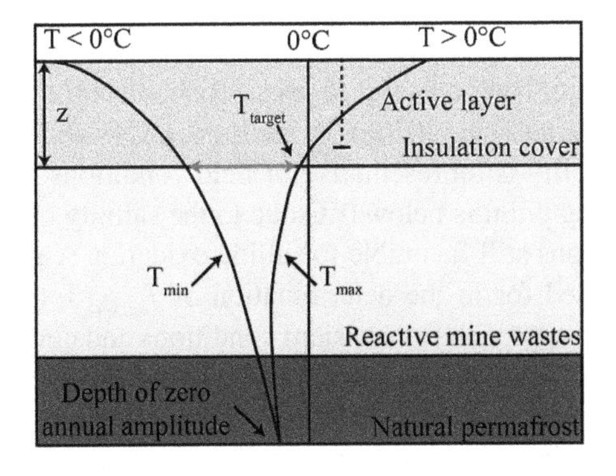

FIGURE 9.3 Working principle of an insulation cover for which the target temperature is slightly lower than 0°C.

a target temperature at which the oxidation of sulfide minerals is negligible (T_{target}; Figure 9.3). Accordingly, the performance of insulation covers mostly depends on the thermal regime established within the cover materials and more specifically on the temporal evolution of the temperature at the mine wastes-cover interface ($T_{interface}$) with respect to T_{target}.

Depending on the cover configuration, heat transfers in insulation covers can be governed by conduction and/or natural convection. In some cases, cover design can also include a layer with a high latent heat, which also affects heat transfers. These heat transfer processes were detailed in Chapter 3 but are discussed briefly in the following with respect to the design of insulation covers.

9.3.2 Reactivity of the Mine Wastes and Target Temperature

The temperature at which the oxidation of sulfide minerals becomes negligible is site specific (not necessarily 0°C) and must be determined prior to the design phase. T_{target} essentially depends on the relationship between the reaction rate (K_r) of the mine wastes and temperature (Kyhn and Elberling 2001; Meldrum et al. 2001; Coulombe 2012; Boulanger-Martel et al. 2020a). Few authors have assessed the impact of lowering temperatures on the reactivity of mine tailings and established T_{target}. For example, Meldrum et al. (2001) determined that the oxidation of tailings from Rankin Inlet mine (Nunavut, Canada) was minimized at −2°C and not measurable at −10°C; they used a T_{target} of −2°C for performance evaluation purposes. Elberling (2001) indicated that the reactivity of Nanisivik (Nunavut, Canada) mine tailings is only reduced by a factor of three to four when temperatures decrease from 15°C to −3°C and that sulfide oxidation was still noticeable at −4°C. Based on these results, Kyhn and Elberling (2001) used a T_{target} of −5°C to assess the performance of three insulation cover configurations. Coulombe (2012) showed that the reactivity of Raglan (Quebec, Canada) mine tailings is significantly reduced at temperatures varying from −2°C to −6°C and used both values as T_{target} to assess the performance of partial insulation covers. Boulanger-Martel et al. (2020a) assessed the effect of temperature on the K_r of Meadowbank mine (Nunavut, Canada) tailings and measured negligible K_r close to 0°C, resulting in a selected T_{target} of 0°C.

T_{target} can be determined based on results from oxygen consumption tests (OCTs) performed either in the laboratory or in the field, as detailed in Chapter 10. OCTs essentially involve measuring the decrease in oxygen concentration over time in a sealed air chamber located above the reactive materials. The magnitude of the decrease in oxygen concentration over time is usually expressed in terms of oxygen flux, which is related to the materials K_r and effective oxygen diffusion coefficient D_e (Elberling et al. 1994).

The value of T_{target} can be determined in the laboratory by performing OCTs at several different temperatures ranging from ambient air temperature to negative temperatures (Meldrum et al. 2001; Elberling 2005; Boulanger-Martel et al. 2020a). OCTs can be performed on mine wastes at different degrees of saturation (S_r) but should be performed at S_r between 40% and 60%, which are optimal for sulfide oxidation (Bouzahzah et al. 2010, 2014). The minimum testing temperature should be sufficiently low to capture the threshold temperature for which oxidation is no more quantifiable (near zero reactivity). A thermal bath can be useful to precisely control the temperature of the materials during testing (Boulanger-Martel et al. 2020a). In addition, OCTs should be performed on samples for which the pore-water quality is representative of field conditions. Factors such as the freezing point depression (i.e., freezing point is below 0°C due to the salinity of interstitial water) can result in hydrogeochemical conditions still favorable to sulfide oxidation reactions at temperatures below 0°C, which must be accounted for in the determination of T_{target}. Experimental results should be interpreted using a numerical approach for transient conditions and considering temperature-dependent conditions (Boulanger-Martel 2019). The $K_r - T$ relationship of the materials is then used to determine the temperature at which sulfide oxidation becomes negligible. Figure 9.4, from the work of Boulanger-Martel (2019), shows an example of a $K_r - T$ relationship obtained at the laboratory for reactive mine tailings. The results presented in Figure 9.4 illustrate the progressive decrease in K_r

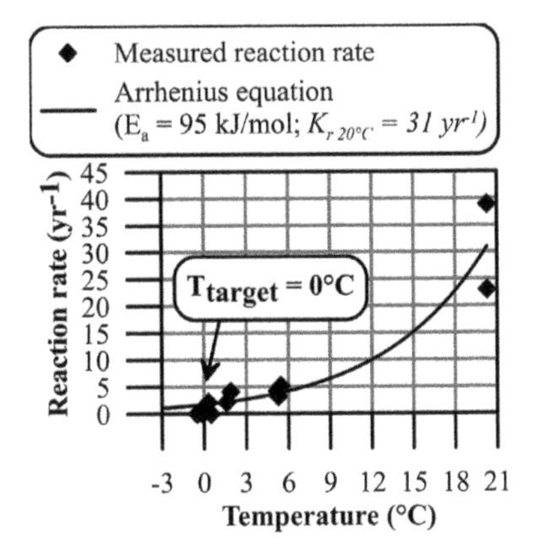

FIGURE 9.4 Effect of temperature on the reaction rate of some mine tailings and associated T_{target} (data from Boulanger-Martel 2019).

with decreasing temperature. In this example, K_r values of 39 to 23 yr^{-1} are observed at 20°C (average of 31 yr^{-1}), whereas negligible K_r values are observed at temperatures close to 0°C. In addition, no quantifiable reaction rates are observed below 0°C, suggesting a T_{target} of 0°C. Figure 9.4 also shows that the progressive reduction of K_r with decreasing temperatures can be described by matching the Arrhenius equation (see Equations 1.28 and 1.29; Chapter 1) to laboratory data. Fitting of the Arrhenius equation is done by fixing K_r at a reference temperature (e.g., 20°C; Figure 9.4) and by iteratively selecting the activation energy (E_a) that best describes the measured $K_r - T$ relationship. E_a essentially represents the amount of energy needed for a reaction to occur, which, in this case, is sulfide oxidation. The fitting process is performed in a way to best represent the mine waste's average measured K_r at any given temperature. So far, to the knowledge of the authors, this approach has been used strictly to assess the T_{target} of mine tailings, but the method could be adapted for waste rocks.

The field approach used to assess T_{target} is similar to that in the laboratory, except that OCTs are performed in situ on covered or uncovered tailings at different temperatures (Coulombe 2012; Boulanger-Martel et al. 2020a). Even though a numerical interpretation is recommended to increase precision (Mbonimpa et al. 2011), in situ OCTs are often interpreted using the standard method described by Elberling et al. (1994). In doing so, the relationship between the oxygen flux (i.e., reactivity) and temperature is obtained, and T_{target} can be determined.

9.3.3 Types of Cover Configurations

The literature reveals that several insulation cover configurations are used to control the temperature of mine wastes. The northern Canadian experience shows that insulation covers typically extend from 1 m up to 8 m in thickness and are made from materials of several origins and grain-size distributions such as:

- natural overburden materials (e.g., till, sand, and gravel esker materials);
- quarried rock fill (crushed rock); and
- non-potentially acid-generating (NPAG) waste rocks (e.g., Kyhn and Elberling 2001; MEND 2004; Smith et al. 2013a; Boulanger-Martel et al. 2020a).

Because the design of insulation covers must be site specific, several cover configurations can be developed to reach a given design criterion (i.e., preventing the active layer to reach the mine wastes and maintain $T_{interface}$ below T_{target}). Insulation covers can be made of one or several layers.

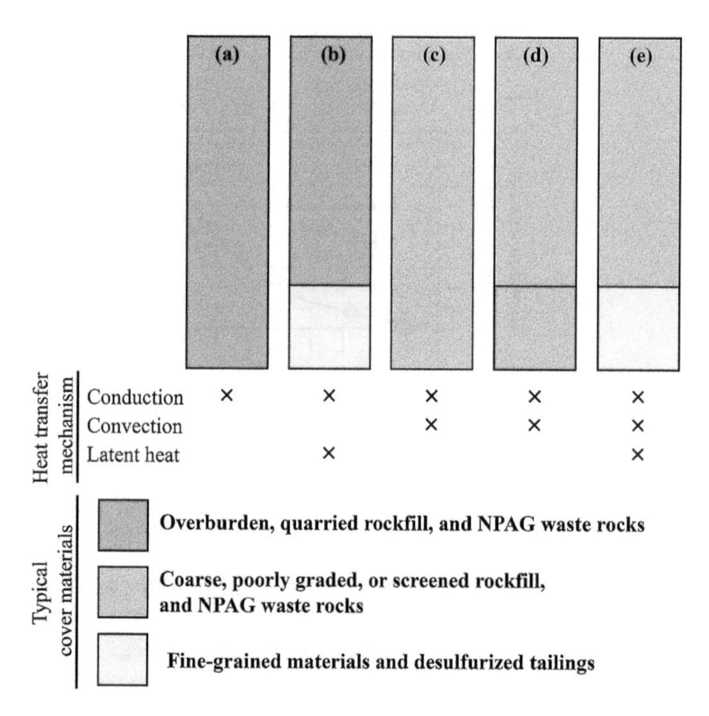

FIGURE 9.5 Schematic representation of several configurations of insulation covers and associated governing heat transport mechanisms.

Therefore, cover configuration will essentially depend on the materials that are available on site with considerations of their thermal and hydrogeological properties. A schematic representation of the main types of insulation cover configurations that are available is given in Figure 9.5 and described below.

The simplest form of insulation cover configuration consists of a single-layer cover made of granular materials. In such a cover, the temperature of the mine wastes is controlled by the conduction of heat through the cover layer (Figure 9.5a). Various relatively coarse-grained materials, such as overburden sand and gravel esker materials, quarried rockfill, or NPAG waste rocks, can be used for the construction of this type of cover. Due to their high porosity and low moisture-retention capacity, these materials usually provide good insulation properties – that is, low thermal conductivity.

In some cases, maintaining the $T_{interface}$ below T_{target} may require an excessively thick cover layer. Therefore, a layer of fine-grained, high moisture-retaining materials can be added to the cover design (Figure 9.5b). When placed underneath the drier insulating layer, a nearly saturated fine-grained layer acts as a latent heat reservoir that limits the progression of the thawing front (see Chapter 3). This latent heat layer aims to reduce the thickness of the active layer, thus preventing the thawing of the mine wastes, contributing to maintain $T_{interface}$ below T_{target}, and reducing the need for thick cover systems. In addition, such highly saturated layer can help reduce the migration and availability of oxygen to the oxidation reactions, which further contributes to control sulfide oxidation. A latent heat layer can be constructed with uncompacted or compacted fine natural materials such as tills or quarried fine-grained materials. Recent work on insulation covers has also indicated that fine compacted NPAG waste rocks or desulfurized tailings can potentially be used in the construction of a latent heat/moisture-retaining layer (Lessard et al. 2018; Boulanger-Martel et al. 2020b). The use of such materials can be beneficial to the efforts to reduce the footprint of the mine and to valorize the mine wastes (see Chapter 13). Although the latent heat layer can be placed at any position in the cover (e.g., MEND 2004; Claypool et al. 2009; Stevens et al. 2018; Boulanger-Martel et al. 2020b), it should ideally be placed at the base. In doing so, the effects of the latent heat layer on the thaw depths are optimized because the latent heat layer starts absorbing latent heat only when the thermal gradients are at the lowest within the cover. This contributes to further reduce the thaw depth.

The cooling potential of natural convection in rockfill embankment and earthwork can also be used in insulation covers (e.g., Arenson and Sego 2007; Pham 2013). In this case, the design aims to increase heat extraction rates by natural air convection during winter while limiting heat transfers to conduction alone during summer (Chapter 3). In doing so, generally colder ground temperatures are promoted than if conduction was the only year-round dominant heat transfer mechanism. Natural air convection in insulation covers is achieved by including a high air permeability (or high intrinsic permeability) layer made of coarse-grained materials (Figure 9.5c, d, and e). Insulation covers that use convective cooling can consist of a single layer of coarse materials (convection layer; Figure 9.5c) or multilayer systems that include a convection layer and a layer of finer materials. In the latter, the layer of finer materials further helps to control the temperatures because of its low thermal conductivity (Figure 9.5d) or latent heat (Figure 9.5e). Natural convection in cover materials is controlled by a combination of cover thickness and configuration, temperature gradient, and air/intrinsic permeability of the cover materials. However, for most field conditions, natural convection can occur in cover materials that have intrinsic permeabilities in the order of 10^{-7} m^2 or greater (Arenson and Sego 2007; MEND 2010; Pham 2013). Such intrinsic permeabilities are generally obtained for materials such as clean cobbles and other clean, coarse-grained materials (e.g., screened rockfill). Some waste rocks could also provide such high intrinsic permeabilities.

9.4 MAIN FACTORS OF INFLUENCE

The previous section highlighted that the design of insulation cover systems is based on optimizing the cover configuration in order to reach a T_{target} in the mine wastes. This optimization process is essentially based on the thermal properties of the cover materials that are available and in situ climatic and ground thermal conditions. In this section, more information about the influence of cover thickness and materials properties on the thermal behavior and performance of insulation covers is presented. Then, the main factors affecting the short- and long-term thermal behavior and performance of insulation covers are discussed. These factors include short-term fluctuations in climatic conditions, post-operation ground thermal conditions, and climate change.

9.4.1 Cover Thickness

Cover thickness (and configuration) is the most determinant factor of influence on the thermal behavior and performance of insulation covers. In general, an increase in cover thickness results in the reduction of the maximum and minimum temperatures observed at the mine waste-cover interface. Thus, a thicker cover generally promotes lower maximum $T_{interface}$. Figure 9.6a shows an example of the maximum $T_{interface}$ modeled at year 100 for insulation cover thicknesses ranging from 2 to 8 m. These results were obtained numerically by Boulanger-Martel (2019) from the modeling of the long-term thermal behavior of heat conduction-dominated, drained, coarse-grained insulation covers constructed over saturated tailings.

The thermal regime of the ground layer near the surface (i.e., insulation cover systems) is also often characterized in terms of the thickness of the active layer (or thaw depth). Accordingly, Figure 9.6b shows the impact of the cover thickness on the thaw depth for covers placed over saturated tailings. Based on this figure, it can be seen that, for insulation covers placed over saturated tailings, the relative impact of the cover thickness on the thaw depth progressively decreases with increasing cover thickness. Such behavior is observed because the tailings have an important dampening effect on the thaw depths when the cover is thin. This dampening effect progressively decreases with increasing cover thickness; thicker covers are less impacted by the presence of the saturated tailings, resulting in less important (or even negligible) variations in thaw depths with respect to cover thickness.

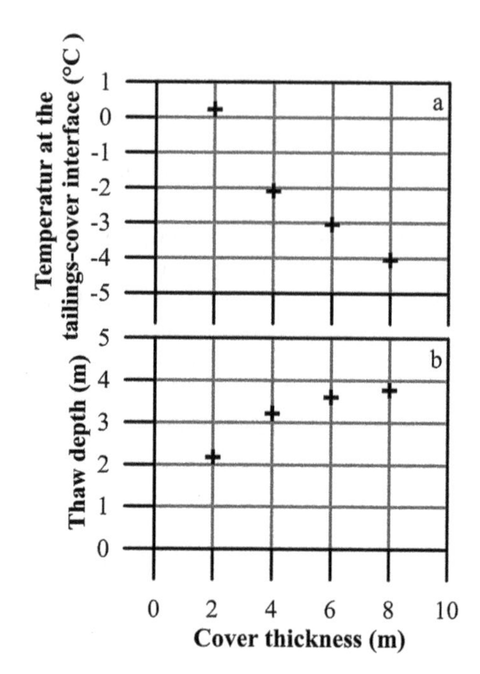

FIGURE 9.6 (a) Modeled maximum temperature at the tailings-cover interface and (b) thaw depths at year 100 as a function of cover thickness.

However, for cases where there is no significant difference in terms of thermal properties between the cover materials and the mine wastes (e.g., rockfill insulation cover on waste rocks), little to no variations in thaw depths with respect to cover thickness are to be expected.

For covers that include an air convection layer, an overall decrease of mine wastes temperatures is also observed with increasing cover thickness. This finding was highlighted by Pham (2013), who modeled the impact of varying the thickness of the air convection layer on the thermal behavior of insulation cover systems that use convective cooling. Pham (2013) considered insulation cover systems with thicknesses of the air convection layer varying from 2 to 6 m and overlaying a 1.5 m thick latent heat layer and 10 m of reactive waste rocks. The results of Pham (2013) demonstrated that for a given intrinsic permeability of the air convection layer, the freezing index of the ground at the base of the air convection layer increases with increasing cover thickness. The freezing index of the ground at the base of the air convection layer represents the sum of all the degree days observed at the base of the air convection layer when the average ground temperature is below 0°C during a calendar year. For example, Pham (2013) observed that for a cover's intrinsic permeability of 2×10^{-7} m^2, the freezing index of the ground at the base of the air convection layer increases from about 1250 to 3000°C days when the thickness of the cover increases from 2 to 6 m. This means that increasing the thickness of the air convection layer in an insulation cover that uses cold convection results in shallower thaw depths and colder $T_{interface}$.

The positive impact of a latent heat layer on the thaw depth and $T_{interface}$ has been recognized by several authors (e.g., MEND 2004; Pham 2013; Stevens et al. 2018). The impact of such latent heat layer becomes especially important when the thaw depth reaches the latent heat layer. For example, Boulanger-Martel (2019) numerically demonstrated that the addition of a 0.5 m thick latent heat layer at the base of a 2 m thick coarse-grained insulation cover (i.e., 0.5 m of saturated cover materials overlain by 1.5 m of drained cover materials) can help decrease the thaw depth and the maximum temperature at the interface by 0.3 m and 0.5°C, respectively, at year 100. However, if the same latent heat layer is added to a 4 m cover (i.e., 0.5 m of saturated cover materials overlain by 3.5 m of drained cover materials), the thaw depths and maximum temperature at the interface remain practically the same because the thaw depth doesn't reach the latent heat layer.

9.4.2 Cover Materials Properties

The main cover materials properties governing the thermal behavior and performance of insulation cover systems are thermal conductivity, intrinsic/air permeability (i.e., for covers using cold convection), and water content (latent heat).

Thermal conductivity is one of the main design parameters of insulation covers. In general, cover materials with lower thermal conductivities provide better insulation per unit of thickness than cover materials with higher thermal conductivities. Covers with lower thermal conductivity typically result in shallower thaw depths and lower $T_{interface}$ than covers with higher thermal conductivities. For example, Boulanger-Martel (2019) showed that increasing the thermal conductivity of the cover materials of a 4 m thick insulation cover by 25% increased the thaw depth by 0.2 m and the maximum $T_{interface}$ by 0.6°C. As described in Chapter 3, the thermal conductivity of geomaterials such as cover materials is a function of several factors, such as the thermal conductivity of the solid particles (i.e., related to the mineralogical composition), the degree of saturation and state of the interstitial water (i.e., unfrozen or frozen), and porosity (e.g., Côté and Konrad 2005a, 2005b). This means that, for a specific design (similar cover thicknesses made of the same geological unit), variations in cover materials' thermal conductivity is mainly attributed to in situ heterogeneities and variability in terms of mineral composition, density/porosity, moisture content, and grain-size distribution (which can affect materials' moisture-retention capacity and porosity) of the cover materials. However, for most field applications, the overall impact of in situ variations in thermal conductivity on the thermal behavior and performance of insulation covers resulting from in situ heterogeneities is considered low (Boulanger-Martel 2019).

For a similar cover thickness, covers that use an air convection layer theoretically show $T_{interface}$ that are lower than those from covers that control temperatures by conduction only. In turn, $T_{interface}$ tends to decrease with increasing intrinsic permeability of the air convection layer. Pham (2013) numerically assessed the impact of varying the intrinsic permeability of the air convection layer from 9×10^{-8} to 3×10^{-6} m^2 on the thermal behavior of an insulation cover that used convective cooling. Pham (2013) modeled the thermal behavior of an insulation cover system made of a 3 m thick air convection layer overlaying a 1.5 m thick latent heat layer and 10 m of potentially acid-generating waste rocks over 4 years. Pham (2013) also assessed the thermal behavior of this cover system considering conduction as the only heat transfer mechanism in the air convection layer (i.e., low intrinsic permeability; 2×10^{-9} m^2). The results of Pham (2013) show that natural convection can reduce the maximum $T_{interface}$ observed in summer by at least 2°C. They also suggest that increasing the intrinsic permeability of the convection layer from 2×10^{-7} to 8×10^{-8} m^2 can further reduce the maximum $T_{interface}$ by about 2°C. However, increasing the intrinsic permeability of the convection layer from 8×10^{-8} to 1×10^{-6} m^2 resulted in similar thermal behaviors, suggesting that there is a threshold intrinsic permeability at which increasing intrinsic permeability provides only small benefits in terms of cooling of the mine wastes-cover interface (Pham 2013).

The impact of cover materials' water content on the thermal behavior and performance of insulation covers was introduced earlier through the concept of the highly saturated latent heat layer. Despite having a high thermal conductivity, a cover layer at high S_r contributes to limit the thaw depths and reduces $T_{interface}$ via latent heat. The impact of cover materials' water content was demonstrated by the results of field experimental cells constructed at the Nanisivik mine (MEND 2004). Five 2 m thick insulation covers were constructed on Nanisivik TSF in order to assess the performance of several cover configurations, including the effects of covers' water content. The results of this study showed that the thaw depth varied with the covers' water content. The average thaw depth measured during the monitoring period was about 1.0 m for the cover that was placed at an initial water content of 33.5%, whereas it was about 1.4 m for the covers placed at an initial water content varying from 6.3% to 8.3% (MEND 2004).

9.4.3 Short-Term Fluctuations in Climatic Conditions

Air temperature is one of the main parameters affecting the thermal behavior of insulation cover systems. As shown in Figure 9.7, the thaw depth in insulation cover systems is often expressed as a function of the total air thawing index (I_{at}), which represents the sum of all degree days when the air temperature is above 0°C during a calendar year (Andersland and Ladanyi 2004). Accordingly, for a given site and cover configuration, the year-to-year fluctuations in air temperatures result in variations of thaw depths and $T_{interface}$. Because of latent heat, the impact of fluctuations in air temperatures is less important in saturated materials compared to drained materials (see the estimated thaw depths in unsaturated waste rocks and saturated tailings in Figure 9.7).

However, like other geotechnical infrastructures, it is in fact the temperature of the ground surface that governs the amount of heat that can be exchanged through insulation cover systems. The heat flux at the ground surface is best expressed in terms of the surface energy balance. Engineers have previously avoided the use of the energy balance to assess the ground surface temperature because of the complex nature of thermal exchanges between the ground and the atmosphere (conduction, radiation, convection/advection, and evaporation/condensation). Measurements of ground surface temperature have also been very challenging in several conditions. Therefore, because the air temperature has a strong effect on the ground surface and is much easier to measure than surface temperature, correlations have been established between the two mainly through thawing and freezing indices. On average, the ground surface temperature is higher than the air temperature, and this difference is attributed to the combined effects of factors such as the relief and the composition of the ground surface, the orientation and inclination of the surface, vegetation, albedo, net radiation, snow cover, subsurface drainage conditions, and the ground thermal properties and conditions (e.g., Andersland and Ladanyi 2004; Throop et al. 2012).

Among the factors affecting the ground surface temperature, snow cover thickness and density are known to have a strong influence as it insulates the ground surface during the cold winter temperatures and limits heat extraction (e.g., Goodrich 1982; Zhang 2005). Boulanger-Martel et al. (2020a) observed differences of 8°C to 10°C in average ground surface temperatures measured during the winter between an insulation cover covered with snow and another with negligible snow cover (see also Boulanger-Martel 2019). Kyhn and Elberling (2001) also showed that the presence of a snow thickness as thin as 5 cm can have a significant impact on the temperatures at the mine waste-cover interface.

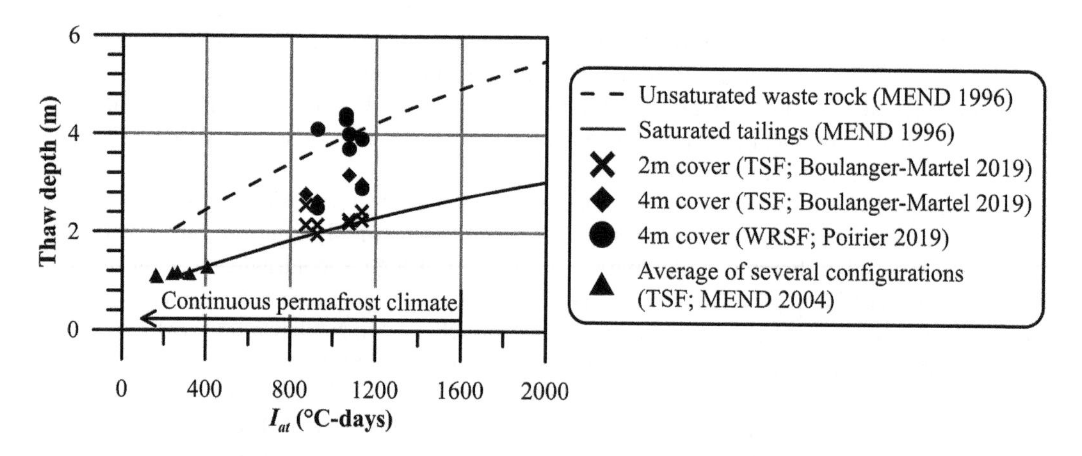

FIGURE 9.7 Estimated thaw depths in unsaturated waste rocks and saturated tailings and observed thaw depths in insulation covers for some tailings (TSF) and waste rock (WRSF) storage facilities as a function of the total air thawing index (I_{at}).

The advective heat flux associated with the infiltration of rain and snowmelt waters is another factor that could also accelerate heat transfers but is usually neglected due to the low net infiltration in continuous permafrost regions.

These factors are the main source of year-to-year fluctuations in cover performances. The magnitude of such fluctuations on the thermal behavior and performance of insulation covers should be assessed and incorporated during the cover design phases.

9.4.4 Post-Operations Ground Thermal Conditions

Waste rocks and tailings are stored at the surface at temperatures usually much greater than the natural ground temperatures (i.e., unfrozen mine wastes vs. frozen natural ground). Therefore, TSFs and WRSF disturb the near-surface ground thermal regime, and some period of time following the operations is required for the mine wastes to freeze-back and reach thermal equilibrium with the natural ground. This process occurs in both TSFs (e.g., Meldrum et al. 2001; Boulanger-Martel 2019) and WRSFs (e.g., Pham et al. 2013; Poirier 2019) and can take from a few years to several decades to achieve. The time required for freeze-back is strongly related to the water content of the mine wastes; due to latent heat, wet materials take more time to freeze and reach thermal equilibrium than dry materials. Therefore, the thermal behavior of TSFs is generally more significantly impacted by this process than WRSFs. The freeze-back time of TSFs also depends on the deposition methods; dense and dry tailings, such as filtered tailings, tend to freeze faster than wet and loose tailings (i.e., conventional tailings; see Bussière 2007 and Chapter 13 for tailings types definitions). At the same time, some operational practices can contribute to accelerating the freeze-back time of the mine wastes. For example, the deposition of tailings and waste rocks in thin layers can help favor a rapid freezing of the mine wastes. In addition, the use of deposition methods that reduce water, ice, and snow entrapment can also contribute to accelerate freeze-back of the mine wastes. In some cases, the presence of open or closed taliks underneath the storage facility can also affect the thermal behavior of the waste storage facility, resulting in longer freeze-back times. Taliks are zones of unfrozen ground within the permafrost occurring as a result of a local anomaly in thermal, hydrological, or hydrogeological conditions, and are often related to the presence of water bodies (French 2007).

An example of the impact of post-operation ground temperatures on the thermal behavior of a 4 m thick insulation cover constructed over 15 m of conventional tailings deposited as pulp and 85 m of rock is given in Figure 9.8a. The results were obtained from (1) the TEMP/W numerical software (GeoSlope International Ltd. 2017), (2) the materials' thermal properties provided in Table 9.1, and (3) simplified climatic and natural ground conditions representative of a low Arctic climate (see Boulanger-Martel 2019). The impact of freeze-back was modeled by running for 100 years a 1-D heat conduction model using air temperatures representative of the average 50-year climate and initial temperatures of −20°C for the cover materials and 1°C for the unfrozen tailings. A temperature profile at equilibrium with the geothermal gradient was used for the rock-foundation materials. Figures 9.8a indicates that, for this simplified case, subzero temperatures are reached at the tailings-cover interface within the first two years following cover construction, but it takes about nine years to freeze the center of the tailings layer and up to 50 years to reach thermal equilibrium. This example illustrates that freeze-back has a significant impact on the thermal behavior of storage facilities. However, because $T_{interface}$ is mostly affected in the first few years, this example also suggests that freeze-back mainly impacts the short-term performance of insulation covers to limit sulfide oxidation.

9.4.5 Climate Change

The largest source of uncertainties with respect to the performance of insulation covers over time is climate change (Boulanger-Martel 2019). The projected increase in air temperature (see Chapter 14)

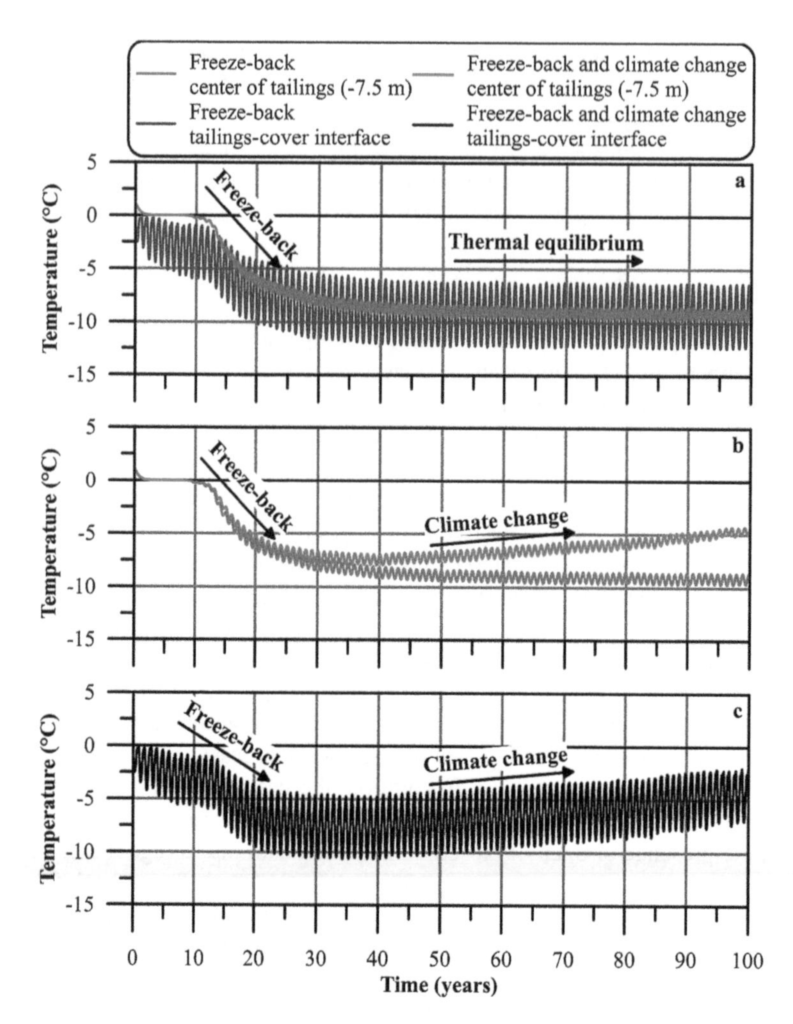

FIGURE 9.8 (a) Evolution of the temperatures at the center of the tailings and at the tailings-cover interface during freeze-back in current days temperatures; (b) evolution of the temperatures at the center of the tailings during freeze-back in current days temperatures and considering climate change; (c) evolution of the temperatures at the tailings-cover interface during freeze-back considering climate change.

TABLE 9.1

Thermal Model Parameters and Materials Properties

Parameters/Model Function	Drained Cover Materials	Saturated Tailings	Foundation
Materials thermal model	Full	Full	Simplified
Freezing curve	Temp/W gravel	Temp/W silt	–
$\lambda - T$ function	Temp/W gravel	Temp/W silt	–
Unfrozen thermal conductivity (W/m K)	1.00	2.28	2.70
Frozen thermal conductivity (W/m K)	1.14	3.89	2.70
Porosity (-)	0.38	0.40	0.005
Volumetric water content (m^3/m^3)	0.057	0.40	0.005
S_r (%)	15	100	100
Unfrozen volumetric heat capacity (MJ/m^3 K)	1.48	2.88	2.20
Frozen volumetric heat capacity (MJ/m^3 K)	1.37	2.17	2.20

could very well result in increased $T_{interface}$, ultimately exceeding T_{target} for a certain number of days, with that number of days increasing over time. This could lead to the oxidation of the covered sulfide minerals and the release of AMD.

To illustrate the potential impact of climate change on the long-term thermal behavior and performance of insulation cover, the numerical model described in Section 9.4.4 was used to assess the impact of an increase in temperature of 6°C over 100 years on a 4 m thick insulation cover. Such increase in temperature essentially corresponds to that of a medium-high radiative forcing climate change scenario (i.e., RCP6.0; Chapter 14) for Canada's low Arctic regions (Boulanger-Martel 2019). The rising temperature was implemented by linearly increasing the daily temperature of the yearly sinus air temperature function by 0.000164°C (or 0.06°C/year) over the 100 years of the model. This approach is a simplified way to assess the impact of climate change, which usually requires more complex climate models (see Chapter 14). However, using such a simplified approach is representative enough of average conditions and is adequate for demonstration purposes. The impact of climate change on the thermal behavior of the TSF is illustrated in Figure 9.8b by comparing the evolution of the temperatures modeled in the center of the tailings for cases that consider current day climate (blue line) and climate change (red line). Figure 9.8b shows that the tailings reach subzero temperatures after nine years for both scenarios, indicating similar thermal behaviors in the short term. For this time horizon, the increase in air temperature is not significant enough to affect the freezing time of the tailings. However, afterward, the effects of the increase in air temperature start to affect the thermal behavior of the TSF, a trend that increases with time. Temperatures within the TSF decrease during the first 30 to 40 years and then gradually increase due to climate change. This thermal behavior is also reflected at $T_{interface}$ (Figure 9.8c). A maximum temperature of −4.7°C is modeled for $T_{interface}$ at year 30, whereas a maximum temperature of −2.1°C is modeled at year 100 (thaw depth = 3.2 m). As a result, $T_{interface}$ shows a warming trend in the long term, which, depending on T_{target}, could result in insufficient cover performance after a certain amount of time (and for >100 years horizon). If an increase in air temperature of 8°C over 100 years is used instead, the maximum $T_{interface}$ would have reached −1°C, and the thaw depth increased by 0.43 m at year 100.

9.5 METHODOLOGY FOR THE DESIGN

The applicability of a cover option, including the insulation cover, is specific to each site, and its effectiveness must be demonstrated in the short- and long-term before being implemented at full scale. The design of an insulation cover essentially consists of optimizing the thickness of the cover in order to maintain the reactive mine wastes below a T_{target} in the long term. The minimum cover thickness required for a given site will mainly depend on the cover materials that are available and their properties, the properties of the mine wastes as well as the site-specific climatic and ground thermal conditions. Therefore, the design of such a cover system is an iterative process that includes several design stages, which form the generalized design methodology of insulation cover presented in Figure 9.9. This methodological approach is based on the work of several authors (Aubertin et al. 2002; Yanful et al. 2006; INAP 2009; MEND 2012; Boulanger-Martel 2019) and should be followed in conjunction with permitting and economic analyses. The different design steps are discussed in the following.

9.5.1 STEP 1: SITE CHARACTERIZATION

The first step is to obtain reliable and good quality data on the site conditions before, during, and after exploitation. Most of the conditions presented in step 1 in Figure 9.9 are determined during the baseline studies. However, some parameters such as the ground thermal regime, the hydrogeological behavior, and the hydrological conditions of the site should also be monitored throughout the operations. For example, monitoring the evolution of the internal temperatures in a TSF or WRSF

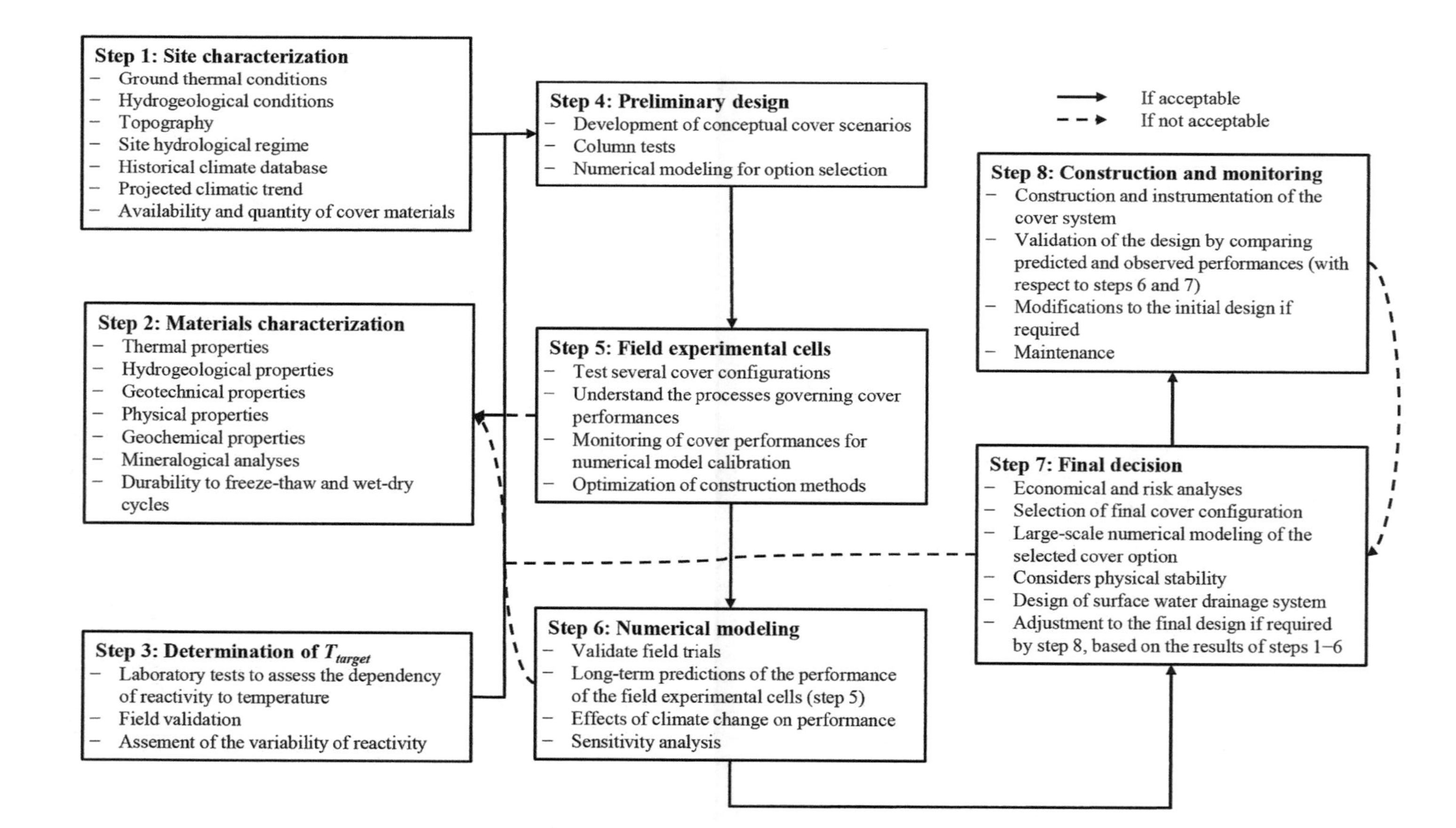

FIGURE 9.9 Methodology for the design of insulation covers.

can provide relevant information on the progress of freezing within these structures and allow the calibration of a numerical model prior to predicting the behavior of the structure. This step also requires the monitoring of the climate parameters on the site from the start of operations as well as the development of a historical climate database. The climate parameters that should be monitored include temperature, precipitation, snow depth, wind speed and orientation, relative humidity, and solar radiation. These data are useful at several levels (e.g., performance evaluation and inputs of numerical models such as a surface energy balance boundary condition), but above all will allow the scaling down of Global Climate Models, which will be used for developing a long-term climate database that considers several different climate change scenarios (see Chapter 14). At this stage, an inventory of available cover materials and their quantity must also be done.

9.5.2 Step 2: Materials Characterization

The next step in the design consists of characterizing the properties of potential cover materials and the mine wastes. The key thermal properties to be determined are the thermal conductivity of the solid particles and the thermal conductivity as a function of S_r at the frozen and unfrozen states (see Chapter 3). The intrinsic and air permeability should also be determined for covers using a convection layer (see Chapter 3). The hydrogeological properties of the materials are also key parameters for insulation cover designs, which include the saturated hydraulic conductivity and the water retention curve (see Chapter 3). In the preliminary design phase, these parameters can be estimated by the several different models presented in Chapter 3. However, laboratory-based characterizations are required for the more advanced design phases. The physical (relative density and particle size distribution) and geochemical (especially the potential for AMD generation) properties as well as mineralogical composition of the materials must also be determined following the methods provided in the previous chapters (Chapters 2 and 3). Although not necessary for thermal design, other geotechnical properties (e.g., consolidation and shear strength) are also necessary for the stability analyses of the structure. Ultimately, the durability of the materials with respect to freeze-thaw and wet-dry cycles must be assessed to ensure the performance of the structure in the long term (see Chapter 14 and Boulanger-Martel et al. 2020c).

9.5.3 Step 3: Determination of T_{TARGET}

As explained previously, the performance of an insulation cover is mainly based on its ability to maintain the mine wastes below a T_{target}. Therefore, this temperature must be determined in order to design according to the reactivity of mine wastes. It is recommended to determine T_{target} in the laboratory using OCTs performed at different temperatures as described in Section 9.3.2 of this chapter. T_{target} should also be validated using field experiments (i.e., during step 5). Finally, the variability in terms of reactivity of the mine wastes must be assessed to ensure that T_{target} is adequate for the entire surface of the storage structure.

9.5.4 Step 4: Preliminary Design

This step consists in developing preliminary conceptual insulation cover scenarios and their geometry based on the cover materials that are available and their associated properties. Insulation covers options include conduction-dominated insulation covers, convection/conduction-dominated insulation covers, and insulation covers incorporating a layer with a high latent heat. At this stage, several cover options and thicknesses can be considered, and preliminary numerical modeling is used to determine the most promising options. Column tests can be performed to test different cover configurations and materials, simulate different boundary conditions, calibrate numerical models (e.g., material properties), and obtain representative material properties (e.g., Boulanger-Martel et al. 2016; Lessard et al. 2018). The geochemical behavior of the covered and uncovered mine wastes can also be assessed in the laboratory at this stage.

9.5.5 Step 5: Field Experimental Cells

Once one or few cover options have been selected during the preliminary design phases, the performance of the selected covers should be assessed with respect to in situ climatic and ground thermal conditions. Field experimental cells are important to:

- evaluate the influence of several factors on cover performance (e.g., slope, location and scale effects);
- understand the climatic, hydrogeological, thermal, and geochemical processes governing the tested cover options;
- obtain field measurements for the calibration of numerical models; and
- test and optimize construction methods before full-scale implementation of the cover.

Field experimental cells are constructed in several steps, which can be different from one site to another, and from one cover configuration to another (e.g., MEND 2004; Smith et al. 2013a; Boulanger-Martel et al. 2020a, 2020b). The design can also aim to test one or several cover configurations at the same time. Instrumentation should be designed to monitor the evolution through time and space (depth and side effects) of temperature and unfrozen volumetric water contents (Figure 9.10; see also Chapter 10). Other relevant parameters related to the evaluation of the thermal behavior and performance of the cover systems should also be monitored, such as water quality, gas pressure, and oxygen consumption. Field experimental cells should be large enough to be able to avoid side effects and monitoring should be conducted for a minimum of three to five years.

9.5.6 Step 6: Numerical Modeling

The monitoring results of the field experimental cells should have narrowed down the number of cover options considered for the final design. Monitoring results from experimental cells can then be used to validate a numerical model and describe the behavior of the most promising scenarios tested in the field. This step also allows for the validation of the materials properties and boundary conditions that will be used to model the covers at larger scales. This numerical validation step will then

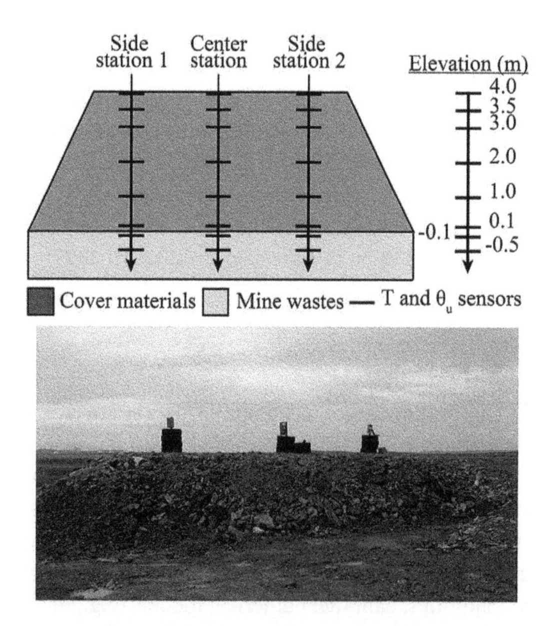

FIGURE 9.10 Example of an insulation cover field experimental cell instrumented to monitor the temperatures (*T*) and unfrozen volumetric water contents (θ_u) within the cover materials and close to the mine waste-cover interface.

make it possible to predict the long-term field behavior and performance of the tested configurations. In this case, the long-term climate database developed at step 1 is used as input to the numerical models. The performance of the cover systems should then be assessed for normal and extreme climatic conditions. Ultimately, a sensitivity analysis of the different cover configurations must be made to test their robustness. Sensitivity analyses should assess the influence of parameters such as variations in terms of materials properties, cover layers' thicknesses, and top (climatic function) and bottom (geothermal gradient and fixed temperature) boundary conditions. Several commercially available modeling software can be used to carry out this step.

9.5.7 Step 7: Final Decision

The optimization of cover thickness is based on the relation between the cover configuration, cover materials' properties, the in situ ground thermal properties, and moisture conditions. Freeze-back may result in a period of time for which $T_{interface}$ is greater than T_{target} after the construction of the cover. The expected climate change results in a gradual increase in the temperature at the mine wastes-cover interface, meaning that cover thickness must also be optimized to ensure the long-term performance of insulation cover systems. Thicker covers generally result in greater freeze-back time but provide a greater robustness to climate change (Boulanger-Martel 2019). In any case, cover thickness must maintain the $T_{interface}$ below T_{target} in the short- and long-term. Overall, the final configuration is selected based on all the results obtained in the previous stages as well as performance, economic, and risk analyses. In parallel to thermal design, aspects related to the physical stability of the selected option must be considered to ensure the long-term performance of the cover system.

9.5.8 Step 8: Construction and Monitoring

Once the final concept has been selected and constructed, the evolution of the system's performance must be monitored over time. Important parameters related to the performance of the cover systems include monitoring the climatic conditions at the site, the evolution the temperatures in the cover materials, at the wastes-cover interface and in the tailings or waste rocks (freeze-back) as well as the water contents in the cover materials and mine wastes. Monitoring the cover system's behavior allows for the validation of the design (comparison of observed performance vs predicted) and opportunities to make corrections if necessary.

9.6 CASE STUDIES

Table 9.2 presents studies of mine sites where insulation covers are planned or were tested at the field scale or implemented as their final reclamation cover. Table 9.2 indicates that insulation covers were considered for mine sites located in regions where the MAAT ranges from −8 to −15°C. These studies show that several insulation cover configurations were designed for reclaiming TSFs. These cover designs use various materials and at different S_r, ranging from fine silty sand to coarse rockfill materials and waste rocks. The thickness and configuration of covers are site specific and can vary from 1 to 8 m. Insulation covers for WRSF are typically between 3 and 5 m thick and generally use coarser materials than covers for TSFs. In most cases, NPAG waste rocks are used for cover construction.

Currently, insulation covers are implemented at large scale at three mine sites: Meadowbank, Nanisivik, and Rankin Inlet. Studies performed on the geochemical behavior of the Meadowbank, Nanisivik, and Rankin Inlet mines' tailings recommended T_{target} values of 0, −5, and −2°C, respectively. Meadowbank's and Rankin Inlet's insulation covers rely on the thermal conductivity and thickness of the cover to control sulfide oxidation. The design of Nanisivik's mine insulation cover also aimed to develop ice-rich permafrost at the base of the crushed shale fill. Insulation covers are also currently being considered as a reclamation method for other active mine projects in the Arctic (e.g., High Lake mine project (67°22'N) or Back River mine project (65°29'N); Stevens et al. 2018).

TABLE 9.2

Examples of Planned or Implemented Insulation Covers Configurations for Active and Inactive Mine Sites Across Canada

Mine Site, Location	Latitude	MAAT (°C)	Reclamation Work and Tested Covers (Configurations given from Bottom to Top)		Source
			TSF	WRSF	
Diavik – NWT, Canada	64°30'N	−8.9	–	Field experimental cell (2004–2010) Planned cover: 1.5 m till + 3.0 m waste rocks	Pham (2013); Smith et al. (2013a)
Ekati – NWT, Canada	64°43'N	−10.2	–	Planned covers: 5.0 m of NPAG waste rocks or 3.0 m of till and 1.0 m of NPAG waste rocks	Stevens et al. (2018)
Mary River – Nu, Canada	71°20'N	−14.4		Planned cover: 3.0 m of NPAG waste rocks	Stevens et al. (2018)
Meadowbank – NU, Canada	62°01'N	−11.2	Field experimental cells (2014–2019) Planned cover: 2–8 m NPAG waste rocks (progressive reclamation)	Final cover: 4.0 m NPAG waste rocks (progressive reclamation)	Boulanger-Martel (2019); Poirier (2019); Boulanger-Martel et al. (2020a)
Nanisivik – NU, Canada	73°02'N	−14.8	5 field experimental cells (1991–1992). Final cover of 1.25 m: 1.0 m crushed shale fill + 0.25 m sand and gravel	Final cover of 2.35 m: 2 m crushed shale fill + 0.35 m of sand, gravel and cobble fill	MEND (2004); Elberling (1998); Elberling (2001); Kyhn and Elberling (2001); Elberling (2005); Claypool et al. (2009); Cassie (2015)
Raglan – QC, Canada	61°41'N	−10.3	Field experimental cells constructed in 2001 and 2012 including Conduction covers: 1.2 m fine crushed rock + 1.2 m coarse crushed rock Convection cover: fine crushed rock + 2.0 m crushed rock 65–200 mm Final design not fixed	–	Coulombe (2012); Boulanger-Martel (2019)
Rankin Inlet – NU, Canada	62°48'N	−11.0	Final cover: 1.0 m insulation cover made of esker materials	–	Erickson (1995); Meldrum (1998); Meldrum et al. (2001)

MAAT: mean annual air temperature

9.7 ADVANTAGES AND LIMITS

There are several advantages related to insulation covers. First, insulation covers are relatively simple to design and construct. They usually consist of only a few layers that can be made from several materials of different origin. Second, the design of insulation covers is easy to adapt. If monitoring reveals that the cover thickness is insufficient to ensure the long-term performance of the cover system, it is relatively simple to increase its thickness by adding cover materials on top of the existing cover system. Third, depending on the availability of cover materials and cover thickness, insulation covers can be less expensive compared with multilayer systems, particularly those constructed with man-made materials (such as geomembranes or gecomposite clay liners; see Chapter 4). In addition, NPAG mine waste (waste rock or tailings) can be valorized as cover materials, thus reducing the overall footprint of the mine and potentially construction costs.

Despite all these advantages, the performance of insulation covers is highly sensitive to climate change. This brings significant uncertainties with respect to the long-term performance of these cover systems, which ultimately could jeopardize the full-size applicability of this reclamation method in the future. Finally, the uncertainties with respect to the long-term performance of insulation covers in the climate change context make social acceptability challenging.

9.8 PERSPECTIVE AND RESEARCH NEEDS

Insulation covers offer an attractive option to reclaim TSFs and WRSFs in continuous permafrost regions. However, there is still some additional research to perform to improve design insulation covers.

- The performance of insulation covers is strongly based on the value of T_{target}, which is currently determined based on reaction rate/oxygen flux measurements. In order to better relate T_{target} to the potential of the mine wastes for AMD generation, the relationships between the reaction rate, temperature, and water quality should be assessed.
- Climate change is one important factor with respect to the long-term performance of insulation cover. Best practice in integrating climate change into the design of insulation covers is by downscaling Global Climate Models based on representative concentration pathways (see Chapter 14). However, no generic approach currently exists to assess the impact of extreme events or periods (temperature) on the thermal behavior (and recovery from those events) on the performance of insulation covers.
- In a climate change context, the design of insulation covers could be adjusted in order to control more than one key driver of AMD generation (e.g., temperature). Insulation covers that also have a specific objective of controlling oxygen availability should be developed; these covers have been identified in the literature as insulation cover with capillary barrier effects (Boulanger-Martel et al. 2016). In such case, reactivity would be controlled when $T_{interface}$ is below T_{target} and oxygen fluxes would control sulfide oxidation for the time when the temperature at the waste-cover interface is above T_{target}.
- Based on the principle of the Seasonally Frozen Capillary Break Cover (Barbour et al. 2011), insulation covers could be adapted to reduce water infiltration along the slopes of TSFs and WRSFs. This concept relies on the presence of a highly saturated and frozen cover layer at the base of the cover, which could divert meltwater during thawing and act as a barrier to infiltration.
- Insulation covers are systems that work in complex systems that often involve coupled analyses of thermal, hydrogeological, and geochemical processes. In the presence of low temperatures and freezing soils, these processes become even more complex. Therefore, efforts to understand these processes and especially how they are related should be pushed forward.
- Understanding the long-term evolution of the thermal behavior of TSFs and WRSFs at large scale with respect to local ground thermal and hydrogeological conditions is also challenging. This is especially the case for WRSFs in which more complex processes such as wind-induced advection and convection could occur and affect the thermal behavior, especially during the warmer season.

REFERENCES

Amos, R. T., D. W. Blowes, B. L. Bailey, D. C. Sego, L. Smith, and A. I. M. Ritchie. 2015. Waste-rock hydrogeology and geochemistry. *Applied Geochemistry* 57:140–156.

Andersland, O. B., and B. Ladanyi. 2004. *Frozen ground engineering*. 2nd ed. Chichester: John Wiley & Sons.

Arenson, L., and D. Sego. 2007. Protection of mine waste tailing ponds using cold air convection. In *Assessment and Remediation of Contaminated Sites in the Arctic and Cold Climates (ARCSACC)*, ed. K. Biggar, G. Cotta, M. Nahir, A. Mullick, J. Buchko, A. Ho, and S. Guigard et al., 256–264. Edmonton, AB.

Aubertin, M., B. Bussière, and L. Bernier. 2002. *Environment and management of mine wastes*. Montreal: Presses Internationales Polytechnique.

Barbour, S. L., J. D. Zettl, Q. Song, M. O'Kane, and M. Nahir. 2011. *Evaluation of a seasonally frozen capillary break cover for mine waste in cold regions*. Paper presented at *Tailings and Mine Waste '11*, Vancouver, BC, Canada.

Boulanger-Martel, V. 2019. *Évaluation de la performance de recouvrements miniers pour contrôler le drainage minier acide en climat arctique*. PhD diss., Polytechnique Montréal, Montreal, Qc.

Boulanger-Martel, V., B. Bruno, and J. Côté. 2020a. Thermal behavior and performance of two field experimental insulation covers to control sulfide oxidation at Meadowbank mine, Nunavut. *Canadian Geotechnical Journal*, in press.

Boulanger-Martel, V., B. Bussière, and J. Côté. 2020b. Insulation covers with capillary barrier effects to control sulfide oxidation in the Arctic. *Canadian Geotechnical Journal*, in press.

Boulanger-Martel, V., B. Bussière, and J. Côté. 2020c. Resistance of a waste rock unit to freeze-thaw and wet-dry cycles: implications for use in a reclamation cover in the Canadian arctic. *Bulletin of Engineering Geology and the Environment*, in press.

Boulanger-Martel, V., B. Bussière, J. Côté, and M. Mbonimpa. 2016. Influence of freeze–thaw cycles on the performance of covers with capillary barrier effects made of crushed rock–bentonite mixtures to control oxygen migration. *Canadian Geotechnical Journal* 53, no. 5: 753–764.

Bouzahzah, H., M. Benzaazoua, and B. Bussière. 2010. A modified protocol of the ASTM normalized humidity cell test as laboratory weathering method of concentrator tailings. *Proceedings of International Mine Water and the Environment (IMWA), Mine Water and Innovative Thinking*, Sydney, NS, 15–18.

Bouzahzah, H., M. Benzaazoua, B. Bussière, and B. Plante. 2014. Prediction of acid mine drainage: importance of mineralogy and the test protocols for static and kinetic tests. *Mine Water and the Environment* 33, no. 1: 54–65.

Bussière, B. 2007. Colloquium 2004: hydrogeotechnical properties of hard rock tailings from metal mines and emerging geoenvironmental disposal approaches. *Canadian Geotechnical Journal* 44, no. 9: 1019–1052.

Cassie, J. 2015. *Mine closure design and performance monitoring, Nanisivik Mine, Nunavut, RPIC Federal contaminated sites regional workshop – Assessment and Remediation on Remote or Northern Sites*. Keynote presentation, Edmonton, AB.

Claypool, G., J. Cassie, and G. Carreau. 2009. *Reclamation of a tailing disposal area in the Canadian Arctic. Proceedings of 8th International Conference on Acid Rock Drainage*, Skelleftea, Suisse.

Côté, J., and J.-M. Konrad. 2005a. A generalized thermal conductivity model for soils and construction materials. *Canadian Geotechnical Journal* 42, no. 2: 443–458.

Côté, J., and J.-M. Konrad. 2005b. Thermal conductivity of base-course materials. *Canadian Geotechnical Journal* 42, no. 1: 61–78.

Coulombe, V. 2012. *Performance de recouvrements isolants partiels pour contrôler l'oxydation de résidus miniers sulfureux*. MScA thesis, Polytechnique Montréal, Montreal, QC.

Coulombe, V., B. Bussière, J. Côté, and M. Paradis. 2013. *Field assessment of sulfide oxidation rates in cold environment: case study of Raglan Mine. Northern Latitudes Mining Reclamation Workshop and 38th Annual Meeting of the Canadian Land Reclamation Association*, Whitehorse, YK, 32–42.

Dobinski, W. 2011. Permafrost. *Earth-Science Reviews* 108, no. 3–4: 158–169.

Elberling, B. 1998. *Processes controlling oxygen uptake rates in frozen mine tailings in the Arctic*. In *Ice in Surface Waters, Proceedings of the 14th International Conference on Ice*, July 28–30, A. A. Balkema, Rotterdam, New York, 183–188.

Elberling, B. 2001. Environmental controls of the seasonal variation in oxygen uptake in sulfidic tailings deposited in a permafrost-affected area. *Water Resources Research* 37, no. 1: 99–107.

Elberling, B. 2005. Temperature and oxygen control on pyrite oxidation in frozen mine tailings. *Cold Regions Science and Technology* 41, no. 2: 121–133.

Elberling, B., R. Nicholson, E. Reardon, and R. Tibble. 1994. Evaluation of sulphide oxidation rates: a laboratory study comparing oxygen fluxes and rates of oxidation product release. *Canadian Geotechnical Journal* 31, no. 3: 375–383.

Elberling, B., A. Schippers, and W. Sand. 2000. Bacterial and chemical oxidation of pyritic mine tailings at low temperatures. *Journal of Contaminant Hydrology* 41, no. 3: 225–238.

Erickson, P. 1995. *Reclamation of the North Rankin Nickel Mine tailings: Final report.* Report prepared for the Department of Indian Affairs and Northern Development.

French, H. M. 2007. *The Periglacial environment.* 3rd ed. West Sussex: John Wiley & Sons.

GEOSLOPE International Ltd. (2017). *Heat and mass transfer modeling with Geostudio 2018* (Second edition). Calgary, Alberta, Canada.

Godwaldt, R. 2001. *Acid mine drainage at sub-zero temperatures.* MSc thesis, University of Alberta, Edmonton, AB.

Goodrich, L. 1982. The influence of snow cover on the ground thermal regime. *Canadian Geotechnical Journal* 19, no. 4: 421–432.

INAP. 2009. *Global acid rock drainage guide (GARD guide).* www.gardguide.com.

Kyhn, C., and B. Elberling. 2001. Frozen cover actions limiting AMD from mine waste deposited on land in Arctic Canada. *Cold Regions Science and Technology* 32, no. 2: 133–142.

Lessard, F., B. Bussière, J. Côté, M. Benzaazoua, V. Boulanger-Martel, and L. Marcoux. 2018. Integrated environmental management of pyrrhotite tailings at Raglan Mine: Part 2 desulphurized tailings as cover material. *Journal of Cleaner Production* 186:883–893.

Mbonimpa, M., M. Aubertin, and B. Bussière. 2011. Oxygen consumption test to evaluate the diffusive flux into reactive tailings: interpretation and numerical assessment. *Canadian Geotechnical Journal* 48, no. 6: 878–890.

Meldrum, J. L. 1998. *Determination of the sulfide oxydation potential of mine tailings from Rankin Inlet, Nunavut, at sub-zero temperatures.* MSc thesis, Queen's University, Kingston, ON.

Meldrum, J., H. Jamieson, and L. Dyke. 2001. Oxidation of mine tailings from Rankin Inlet, Nunavut, at sub-zero temperatures. *Canadian Geotechnical Journal* 38, no. 5: 957–966.

MEND. 1993. *Preventing AMD by disposing of reactive tailings in permafrost, project report 6.1.* Mine Environment Neutral Drainage (MEND) Canada Center for Mineral and Energy Technology, Canada.

MEND. 1996a. *Acid mine drainage in permafrost regions: issues, control strategies and research requirements, project report 1.61.2.* Mine Environment Neutral Drainage (MEND) Canada Center for Mineral and Energy Technology, Canada.

MEND. 1996b. *Column leaching characteristics of Cullaton Lake B and Shear (S)-Zone Tailings, phase 2: cold temperature leaching final report, project report 1.61.3.* Mine Environment Neutral Drainage (MEND) Canada Center for Mineral and Energy Technology, Canada.

MEND. 1997. *Roles of ice, in the water cover option, and permafrost in controlling acid generation from sulfide tailings, project report 1.61.1.* Mine Environment Neutral Drainage (MEND) Canada Center for Mineral and Energy Technology, Canada.

MEND. 2004. *Covers for reactive tailings located in permafrost regions review, project report 1.61.4.* Mine Environment Neutral Drainage (MEND) Canada Center for Mineral and Energy Technology, Canada.

MEND. 2009. *Mine waste covers in cold regions, project report 1.61.5a.* Mine Environment Neutral Drainage (MEND) Canada Center for Mineral and Energy Technology, Canada.

MEND. 2010. *Cold regions cover research-phase 2, project report 1.61.5b.* Mine Environment Neutral Drainage (MEND), Canada Center for Mineral and Energy Technology, Canada.

MEND. 2012. *Cold regions cover system design technical guidance document, Report 1.61.5c.* Mine Environment Neutral Drainage (MEND), Canada Center for Mineral and Energy Technology, Canada.

Pham, H. N. 2013. *Heat transfer in waste-rock piles constructed in a continuous permafrost region.* PhD diss., University of Alberta.

Pham, N. H., D. C. Sego, L. U. Arenson, D. W. Blowes, R. T. Amos, and L. Smith. 2013. The Diavik Waste Rock Project: measurement of the thermal regime of a waste-rock test pile in a permafrost environment. *Applied Geochemistry* 36:234–245.

Poirier, A. 2019. *Étude du comportement thermique d'une halde à stérile en milieu nordique, département des génies géologique, civil et des mines.* MScA thesis, Polytechnique Montréal, QC.

Rekacewicz, P. 2005. *Permafrost distribution in the Arctic.* UNEP/GRID-Arendal Maps and Graphics Library. https://www.grida.no/resources/7000.

Smith, S. L., A. G. Lewkowicz, C. R. Burn, M. Allard, and J. Throop. 2010. *The thermal state of permafrost in Canada-Results from the International Polar Year. GEO2010, Proceedings of the 63rd Canadian Geotechnical Conference and the 6th Canadian Permafrost Conference*, Calgary, AB, 1214–1221.

Smith, L. J., M. C. Moncur, M. Neuner, M. Gupton, D. W. Blowes, L. Smith, and D. C. Sego. 2013a. The Diavik Waste Rock Project: design, construction, and instrumentation of field-scale experimental waste-rock piles. *Applied Geochemistry* 36:187–199.

Smith, S. L., D. W. Riseborough, M. Ednie, and J. Chartrand. 2013b. A map and summary database of permafrost temperatures in Nunavut, Canada. *Geological Survey of Canada, Open File* 7393. doi:10.4095/292615.

Stevens, C. W., T. Shapka-Fels, and M. Rykaart. 2018. Thermal cover design for mine waste facilities in cold regions. Paper presented at *Tailings and Mine Wastes*, Keystone, CO. https://www.asia-pacific.srk.com/sites/default/files/file/Chris_Stevens-Tailings_and_Mine_Waste_2018_Thermal_Cover_Design.pdf.

Throop, J., A. G. Lewkowicz, and S. L. Smith. 2012. Climate and ground temperature relations at sites across the continuous and discontinuous permafrost zones, northern Canada. *Canadian Journal of Earth Sciences* 49, no. 8: 865–876.

Van Everdingen, R. O. 1998. *Multi-language glossary of permafrost and related ground-ice terms in Chinese, English, French, German, Icelandic, Italian, Norwegian, Polish, Romanian, Russian, Spanish, and Swedish.* International Permafrost Association, Terminology Working Group.

Yanful, E. K., S. Morteza Mousavi, and L.-P. De Souza. 2006. A numerical study of soil cover performance. *Journal of Environmental Management* 81, no. 1: 72–92.

Zhang, T. 2005. Influence of the seasonal snow cover on the ground thermal regime: an overview. *Reviews of Geophysics* 43, no. 4: 1–23.

10 Monitoring the Performance of Mine Site Reclamation

Bruno Bussière, Thomas Pabst, Vincent Boulanger-Martel, Marie Guittonny, Benoît Plante, Carmen M. Neculita, Sylvette Awoh, Mamert Mbonimpa, Isabelle Demers, Abdelkabir Maqsoud, Adrien Dimech, and Pier-Luc Labonté-Raymond

10.1 INTRODUCTION

The final step to reclamation, after the cover system has been designed and built on site, is to ensure that reclamation meets the initial design objectives and maintains its performance over time. Even if most of the mine sites are regulated for water quality, other parameters must be included in monitoring program, given time lag between reactions leading to water contamination (sulfide oxidation) and the water quality at the final effluent. Instruments must then be installed to monitor performance parameters related to the functioning of the reclamation method applied (e.g., degrees of saturation for cover with capillary barrier effects [CCBE], elevated water table [EWT] with monolayer cover, water levels for water covers or temperatures for isolation covers, among others). Instrumentation and monitoring equipment are critical to ensuring the long-term performance of reclamation, but they also represent significant costs. They must therefore be carefully selected and installed in critical locations, and their data need to be regularly analyzed.

In this chapter, the most commonly used monitoring tools and approaches are presented, including meteorological, hydrological and hydrogeological, water quality, gas movement, vegetation, and soil temperature equipment and methods. The chapter ends with a discussion of different aspects of the monitoring strategy such as the selection criteria and instrument maintenance, the instruments' location and density, measurement frequency, and data treatment and reporting.

10.2 METEOROLOGICAL PARAMETERS

Previous chapters have highlighted that climate is one of the main parameters governing the behavior and performance of all reclamation covers. Meteorological parameters represent the main boundary condition governing the movement of water, air, and heat in cover systems. Therefore, the meteorological parameters relevant to each cover system must be evaluated in order to design adequately. In turn, these components must also be monitored to assess the actual field performance of a cover system and its evolution over time. In the context of mine site reclamation, the main meteorological parameters that are recorded are air temperature, relative humidity, precipitation (rain and snow), atmospheric pressure, solar radiation, and wind speed and direction. The thickness of snow can also be an important parameter to measure. Most of these parameters are usually recorded by governmental weather stations (see Figure 10.1), which, in some cases, are located close to the mine site. Weather data from those weather stations are generally publicly available. However, because site meteorological parameters can vary significantly for a given region, it is a best practice to install a monitoring station directly at the mine.

FIGURE 10.1 Meteorological station installed on a waste rock pile.

10.2.1 Weather Stations

Weather stations are widely used at mine sites to monitor climatic parameters (e.g. Swanson et al. 2003; Zhan et al. 2014; Bossé et al. 2015; Boulanger-Martel et al. 2020a). The configuration of weather stations varies from one site to another, meaning that the climatic variables that are measured at one given site can be different from another site. Basic weather stations usually measure air temperature, relative humidity, wind speed, and precipitations (i.e., rainfall and snowfall). However, more heavily equipped weather stations typically measure additional parameters such as the temperature of the ground surface, atmospheric pressure, solar radiation, and wind direction. The most relevant meteorological parameters to monitor at a given site should be selected by considering the

TABLE 10.1

Main Meteorological Parameters Measured by Weather Stations, Their Measurement Apparatus and Typical Recorded Values as Well as Their Frequency

Parameter	Apparatus	Typical Recorded Values	Typical Frequency of Recorded Values
Temperature	Resistance thermometer and thermistor	Average, maximum and minimum	Hourly, daily
Relative humidity	Electrical and dew-point hygrometer	Instantaneous measurement	Hourly, daily
Atmospheric pressure	Electronic barometer	Average, maximum and minimum	Hourly, daily
Rainfall	Rain gauge	Value in mm	Daily, annually
Snowfall	Rain gauge with snowfall adaptor	Value in mm of water	Daily, annually
Radiation	Pyranometer, pyrgeometer, net radiometer	Instantaneous measurement	Hourly
Wind speed	Anemometer	Average, maximum, and minimum	Hourly, daily
Wind direction	Wind vane, heated wire	Average	Hourly, daily

specific conditions of the site and be adapted to the selected reclamation option. A summary of the main apparatuses used to monitor the weather is provided in Table 10.1. Detailed description and working principles of these apparatuses can be found in the literature (EPA 2000; Aubertin et al. 2002). This section briefly discusses the instrumentation used to monitor the most relevant weather parameters to cover design and environmental monitoring.

Air temperature is an especially important parameter when designing and monitoring insulation covers but is also required in several estimation equations used to assess the water balance (see Chapters 3 and 9). For most weather stations, the temperature of the air is continuously measured. Accordingly, weather stations can provide an instantaneous air temperature. However, it is usually the maximum, minimum, and average hourly and daily temperatures that are recorded. Some weather stations are also equipped with an additional temperature sensor used to measure the temperature of the ground surface. The temperature of the air and ground surface is usually measured using a resistance thermometer (RTD) or a thermistor. Other types of older thermometers such as mercury thermometers and maxima and minima thermometers have generally been replaced by electronic temperature sensors.

Relative humidity is defined as the ratio of partial pressure of water vapor in the air to the saturated vapor pressure of water vapor at a given temperature, and is an important parameter related to evaporation (Fredlund et al. 2012). It is close to 0% when the air is very dry and close to 100% when the air is very humid. Relative humidity is measured by a hygrometer. Several types of hygrometer exist (mechanical, electrical, and dew-point hygrometers), but in weather stations, relative humidity is generally measured by capacitive or dew-point hygrometers. In many weather stations using capacitive hygrometers, relative humidity and temperature sensors are combined in a single probe. Air temperature and relative humidity sensors are usually housed in a casing that acts as a shield from weather and solar radiation.

Another important parameter to measure is precipitation. Liquid precipitation includes rain and freezing drizzle and is measured using rain gauges. Standard rain gauges are often used to measure rainfall. They essentially consist of a cylinder equipped with a funnel to collect rain. This type of rain gauge requires the intervention of a reader who must frequently record the amount of rain. The records of standard rain gauge represent the total amount of rain (expressed in mm) that occurred during the period since the last reading. One alternative to standard rain gauge frequently used in weather stations is the tipping bucket rain gauge. Tipping bucket rain gauges have the advantage of not requiring frequent visits to the weather station to record rainfall. Tipping bucket rain gauges essentially record (with a data logger) the number of tips that occur as a result of the filling of buckets of known volume due to rain. The number of tips is then converted to millimeter of water. This apparatus allows the continuous monitoring of rainfalls, and it provides information on the intensity and temporal distribution of precipitation. In cases where solid precipitations are involved (e.g., snow), rain gauges adapted for snowfall can be used to measure precipitation. Snowfall adaptors can consist of a heated sleeve or an antifreeze reservoir (to melt snow) added to the top of the funnel of the rain gauge (standard or tipping bucket).

Wind speed and direction play a role in the evaporation rate, in the creation of waves on water covers, and could also affect heat transfers in waste rock piles (Wilson et al. 1994; Mian and Yanful 2004; Pham 2013). There are two main approaches to measuring wind speed and direction. The first approach is to equip the monitoring station with a cup or propeller anemometer and a wind vane. The second is to use a hot wire anemometer to measure both wind speed and direction.

In weather stations, radiation is often measured in terms of total and net radiations, which are expressed as a solar radiation flux density (W/m^2). Total radiation can be entering (incident radiation) or being reflected from (reflected radiation) the Earth's surface. The net radiation is essentially the difference between the incident and reflected radiations. This definition of radiation is the most important in cover designs because it is used to calculate evaporation (see Chapter 3) and the soil energy balance. Total incident and reflected radiations are measured by pyranometers and pyrgeometers (usually thermopile pyranometers and pyrgeometers). Pyranometers are designed to measure the Earth's shortwave radiations (typically in the range of 0.3–3 μm), whereas pyrgeometers measure longer wavelengths (typically in the range of 4.5–100 μm). To measure the total incident radiation, pyranometers and pyrgeometers must face the sky. On the contrary, the total reflected radiation is measured by positioning these apparatuses to face the ground surface. Net radiometers are used to measure net radiations and consist of two pyranometers and two pyrgeometer (four-component configuration). In this configuration, a set of one pyranometer and one pyrgeometer faces upward (oriented toward the sun), whereas another set faces downward (oriented toward the ground surface).

10.2.2 Thickness of the Snow Cover

In some cases, it could be required to assess the thickness of the snow at specific locations. As snow reduces the intensity of freezing and reduces heat extraction from the soil, this is particularly important when dealing with insulation covers on tailings storage facilities (TSFs) and waste rock storage facilities (WRSFs) (e.g., Boulanger-Martel 2020a; see Chapter 9). Some direct methods are available to assess the thickness of the snow cover. However, the use of a snow ruler is probably the more practical way to measure the thickness of the snow cover at specific locations over a TSF or a WRSF. Using a sonic distance sensor can also provide measurements of the thickness of the snow cover. However, such setup is usually fixed in place and should be used to monitor the evolution of the thickness of the snow cover at a specific location. A sonic distance sensor measures the thickness of the snow cover based on time elapsed between the emission and return of an ultrasonic pulse (e.g., Fountain et al. 2010; Varhola et al. 2010).

10.2.3 Example of Meteorological Parameters Monitoring Results

Figure 10.2 shows an example of results obtained from the monitoring of meteorological parameters at the Meaowbank mine in Nunavut, Canada. Figure 10.2 presents the temporal evolution of the daily liquid and solid precipitations. Daily precipitation can also be used to calculate the yearly cumulative rainfall and snowfall (Figure 10.2a), 280 mm in this particular case. The distribution of precipitation between rain and snow each year is also important meteorological data. Minimum, maximum, and average daily air temperatures are also plotted as a function of time in Figure 10.2b, which allows for comparison of different years of monitoring. Sometimes it can be

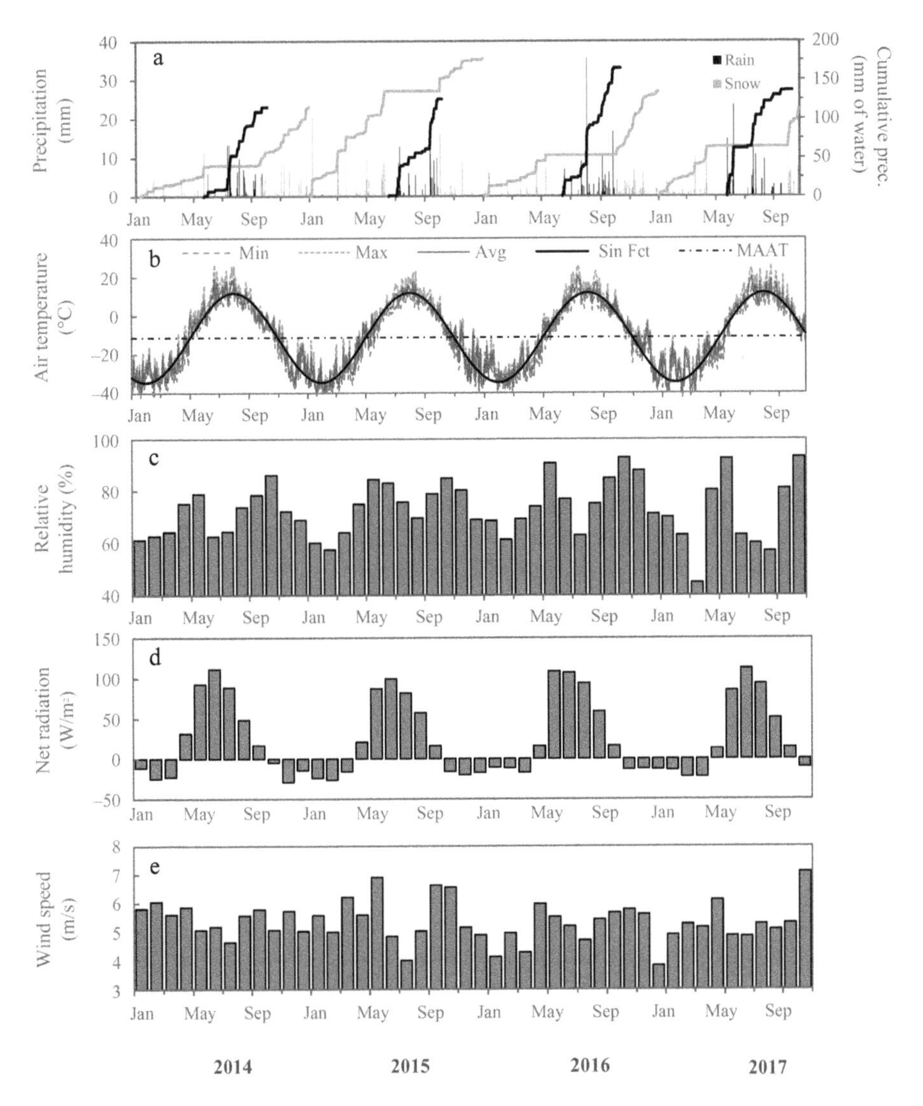

FIGURE 10.2 Weather data for January 2014 to October 2017 observed at the Meadowbank mine, Nunavut, Canada (from Poirier 2019). (a) Daily and cumulative rainfall and snowfall; (b) Minimum, maximum, and average daily air temperature, mean annual air temperature (MAAT), and sinusoidal air temperature function; (c) Monthly average relative humidity; (d) Monthly average net radiation; and (e) Monthly average wind speed.

useful to estimate the yearly evolution of the air temperature, which is done by fitting a sinusoidal function to the observed data (e.g., Andersland and Ladanyi 2004). Such function provides a simplified representation of the air temperature function that can be used in most geotechnical engineering software. Relative humidity measurements (Figure 10.2c) and net radiation (Figure 10.2d) are also key parameters to calculate evaporation. It is then important to obtain results for different periods of the year. Figure 10.2d shows that maximum net radiation occurs during summer when incident radiation is the most important. During winter (i.e., October to March), net radiation becomes negative mostly because the daylight time is shorter, solar radiation is less important than in summer, there is snow reflecting incident radiation, and the ground is freezing. This information related to net radiation is important in calculations of annual evaporation evolution. Finally, information on wind speed and wind direction is important for the monitoring of reclaimed mine sites. Information such as average wind speed and predominant wind direction is often required by reclamation professionals; in this example, wind speed was of 5.4 m/s on average (Figure 10.2e) and wind directions indicated a predominant wind from the northwest.

10.3 HYDROGEOLOGICAL AND HYDROLOGICAL PARAMETERS

10.3.1 Hydraulic Head

The hydraulic or total head (h) is one of the key parameters involved in all fluid transport equations (Freeze and Cherry 1979; Hillel 1998; see Chapter 3). In saturated porous media, h represents the energy per unit weight due to pore fluid (water), and is derived from Bernoulli's equation as the sum of the pressure head h_p and the elevation head z (see Equation 3.10 in Chapter 3) (Figure 10.3). Water movement in saturated conditions is a function of the saturated hydraulic conductivity k_{sat} and the hydraulic head gradient $\frac{\partial h}{\partial z}$.

Hydrogeological monitoring is important for some reclamation methods presented in the previous chapters, particularly the water cover, the EWT with monolayer cover, the CCBE, and the low saturated hydraulic conductivity cover (LSHCC) techniques. More specifically, knowledge of h at a specific location allows to determine the following information:

- Determining the hydraulic gradient between water cover and the surrounding groundwater to ensure that the water cover acts (or not) as a hydraulic trap.
- Determining the fluctuations of the water table level in the EWT technique to ensure that the depth of the water table always meets the design criteria.
- Determining the fluctuations of the water table level for sites reclaimed with CCBE because the water table position is one of the key boundary conditions in the numerical modeling used to assess the performance of CCBEs.
- Determining the drawdown position of the water table level inside the mine waste covered with LSHCC cover to ensure the efficiency of the cover to control water infiltration.

Measurements of h can be obtained using piezometers (Cassan 2005). There are two main types of piezometers: open-tube piezometers or open piezometers, and sealed piezometers. The sealed piezometers include hydraulic, pneumatic, electric, and vibrating wire piezometers. The sealed piezometers contain a sensitive part, which is closed and separated from pressurized water by a flexible diaphragm. The open-tube (perforated-pipe or standpipe) piezometer is directly in contact with the pressurized water.

10.3.1.1 Open-Tube Piezometers

Open-tube piezometers are the simplest form of piezometer. There are two types of open-tube piezometers: the perforated-pipe and the standpipe piezometer (McCarthy 2007; Price 2009).

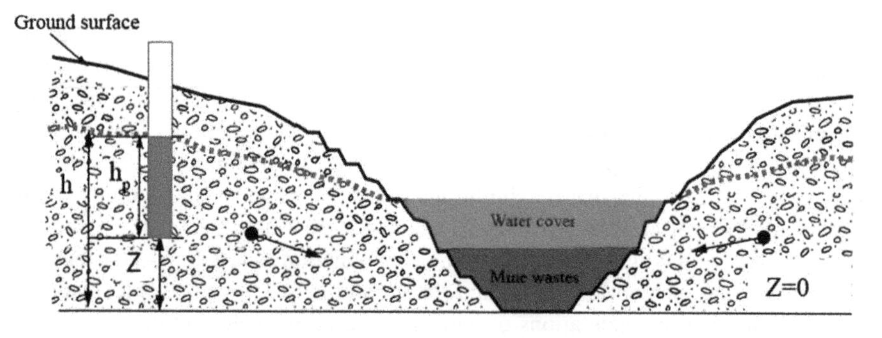

FIGURE 10.3 Illustration of pressure head calculation in a porous media.

The perforated-pipe piezometer consists of a pipe of 10 to 40 mm plugged at its bottom and perforated over a length of about 0.2 m (lower section). Coarse sand is used to fill the perforated pipe and acts as a filter element. It is usually used in soils with high permeability. The standpipe piezometer is the most common open piezometer used for measuring pressure head (McCarthy 2007; Yolcubal et al. 2004). It consists of a plastic pipe with a porous section at the bottom (filter tip) (McCarthy 2007; Price 2009). The annular space between the perforated tube and casing is filled with sand and the top surface of the sand is sealed with bentonite. The diameter of the tube may vary from 4 to 8 mm and should be selected based on soil permeability. The standpipe piezometer can be installed in sand, silt, and sometimes clays. In the latter case, one must wait until the water level stabilized before measurement.

Open piezometers are based on the principle that, when the water that is under pressure in the filter tip, is allowed to rise in a pipe, it will stabilize when the water level corresponds to the potential at the filter level (see Figure 10.4). The level to which water rises in the piezometer, considering an arbitrary reference datum, is h. The common way for estimating the h value in an open-tube piezometer in the field requires depth-to-water measurement in a piezometer (measured with electric sounding instrument or pressure sensor) and the knowledge of the elevation of the top of the piezometer casing (Yolcubal et al. 2004; Gorelick & GeS 2015). Hydraulic head in the piezometer is then calculated by subtracting depth to water (D) from the elevation of the top of the piezometer casing (E). Open-tube piezometers are more common than sealed piezometers described below and are generally used for the monitoring of mine reclamation scenarios.

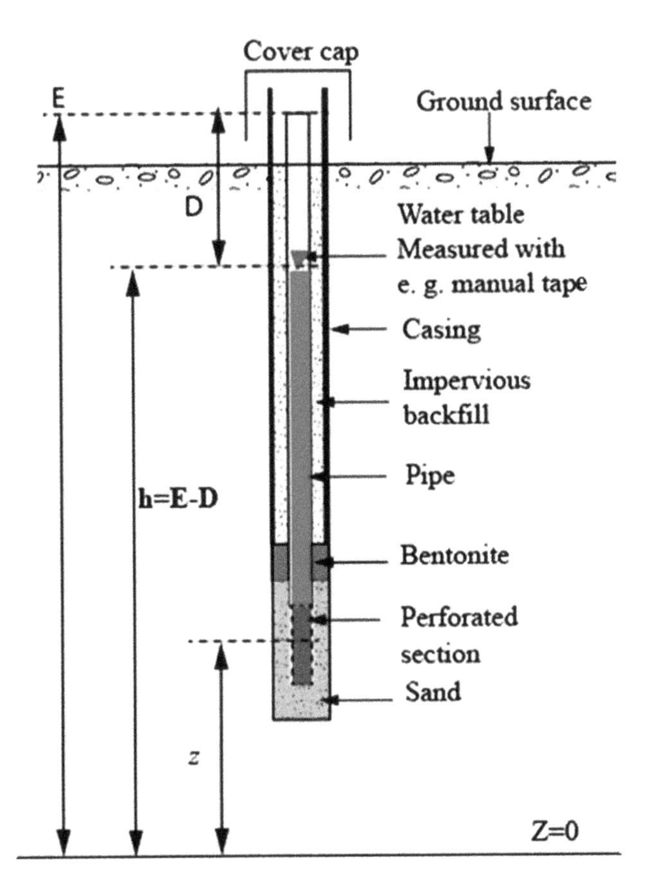

FIGURE 10.4 Schematic representation of an open-tube piezometer.

10.3.1.2 Hydraulic Piezometer

The hydraulic piezometer consists of a porous filter (sensing unit) connected to two flexible tubes at the end of which pressure gauge measurement devices are installed (Aubertin et al. 2002; Cassan 2005). The system is filled with an incompressible liquid such as deaerated water before installation. Any change of pressure at the porous filter is transmitted throughout the water in the tubes and to the pressure sensor. The water pressure and subsequently the pressure head of the sensing unit can be determined considering the elevation difference between the gauge and the sensing unit. The hydraulic head is obtained by adding the measurement read and the elevation of the pressure gauges (Aubertin et al. 2002). The two flexible tubes are used to circulate water through the system, removing air and ensuring that the reservoir remains full of water.

10.3.1.3 Pneumatic Piezometer

The pneumatic piezometer consists of two gas-filled tubes connected to a sealed cavity. The sealed cavity placed into the soil comprises a porous stone aimed at protecting the water-filled chamber from the entry of soils particles. The water in this chamber applies pressure on one side of a non-metallic flexible diaphragm. The sealed cavity is also connected to a measurement system placed at the surface. Dry gas such as nitrogen or CO_2 is injected under pressure to one of the tubes until equilibrium with the soil pore pressure is reached. There is then a separation of the diaphragm, which is expressed by a return of the gas at the end of the other tube. The pressure of the gas measured in the second tube corresponds to the pore pressure and is used to determine h (Aubertin et al. 2002; Cassan 2005; McCarthy 2007; Bureau of Reclamation 2014).

10.3.1.4 Electric Piezometer

The electric piezometer consists of a cylindrical probe inserted into the soil, equipped with a porous stone and an elastic diaphragm to which a strain gauge is attached. The pressurized water enters into the piezometer through the porous stone filter and pushes on the diaphragm, which causes the deflection of this diaphragm and induces a resistance variation of the strain gauge. A calibration curve between the pressure applied on the diaphragm and its deformation is needed to convert the obtained values into pore pressure (Aubertin et al. 2002; Cassan 2005; McCarthy 2007; Bureau of Reclamation 2014). Pressure measurements can then be converted into h values.

10.3.1.5 Vibrating Wire Piezometer

The vibrating wire piezometer consists a tight wire that is attached to an elastic diaphragm separated from a porous filter with water filled reservoir (Aubertin et al. 2002; McCarthy 2007; Bureau of Reclamation 2014). When the wire is excited by a magnetic field from a measuring station installed at the surface, the wire vibrates at a given frequency that is a function of the tension in the wire; this tension is affected by the pressure apply on the diaphragm and, consequently, by the water pore pressure. A calibration curve between vibration and wire tension is necessary to convert the measurements into pore pressure (Aubertin et al. 2002; McCarthy 2007). The pore pressure obtained is finally used to estimate h.

10.3.1.6 Advantages and Limits

Table 10.2 presents the main advantages and limits of each piezometer and some applications in the context of mine site reclamation. The open-tube piezometers are commonly used to monitor reclamation structure such as water cover, CCBE, and EWT because of their reliability, low costs, ease of installation, simplicity, and low maintenance (see Table 10.2). Sealed piezometers are mostly used for the monitoring of mine waste containment dikes (Bureau of Reclamation 2014). According to Bureau of Reclamation (2008), vibrating wire piezometer are now the piezometer of choice for tailings dams monitoring.

TABLE 10.2
Advantages and Limits of Each Piezometer and Some Applications in the Context of Reclamation (Modified from Aubertin et al. 2002; Bureau of Reclamation 2014)

	Piezometers			
Properties	**Open Tube**	**Hydraulic**	**Pneumatic**	**Electric and Vibrating Wire**
Longevity/Reliability	Long	Long	Short	Very short
Precision of data	Moderate	Low	Low	High
Time lag in impervious materials	Long	Short	Very short	Very short
Interference with new construction	Substantial	Moderate	Moderate	Moderate
Installation	Very simple	Moderate	Complex	Moderate
Installation cost	Not expensive	–	Expensive	More expensive
Maintenance requirements	Low	High (Significant complexity of maintenance)	Low	Low
Installation problems?	Quite low	High over time	High	Very low
Remote reading (telemetry)	No	Yes	Yes	Yes
Reading time	Long	Moderate	Long enough	Very short
Reading negative pressures	No	Yes	No	Yes
Other considerations	Freezing problem if high water level	Tubes can get clogged without regular maintenance	Must prevent moisture from entering tubes	Lightning damage can occur. Nearby electrical transmission, lines or equipment, can impact data.
Example of application in the reclamation context	Installation of open-tube piezometer to monitor groundwater around Don Rouyn pit lake. Used to evaluate the hydraulic connection between the Don Rouyn water cover and groundwater (QC, Canada, Awoh et al. 2013)	Mostly used in dam embankment (Bureau of Reclamation 2014)		
	Installation of open-tube piezometer on tailings impoundment where EWT are applied at Manitou (QC, Canada, Ethier et al. 2018); Detour Lake (ON, Canada, Dobchuk et al. 2013) and at Aldermac (QC, Canada, Maqsoud et al. 2013, 2015) Installation of open-tube piezometer to evaluate the drawdown of the water table in covered tailings with high density polyethylene (HDPE) at Normétal Mine (QC, Canada; Maurice 2012) and at Poirier site (Joutel, QC, Canada; Maurice 2002)	Installation of hydraulic piezometers to monitor pore pressure in several dams (Bureau of Reclamation 2014)	Installation of pneumatic piezometers to monitor phreatic surface within the dams (Poly Met Mining, Inc.'s (Poly Met's, Minnesota, USA; Radue 2017)	Installation of vibrating wire piezometers to monitor in pit dikes performance at Muskeg River mine (Fort McMurray, AB; SoeMoe et al. 2013) Installation of vibrating wire piezometers to monitor the dam of the Flotation Tailings basin (Poly Met Mining, Inc.'s (Poly Met's, Minnesota, USA; Radue 2017)

10.3.2 Surface Water Flux Monitoring

Surface water flux monitoring is necessary to evaluate water balance in mine waste storage facilities and is key information to validate reclamation performance. Such monitoring is important to all types of reclamation systems but are particularly critical to water covers, where the risk of overtopping is greater (see Chapter 6). Surface water flux monitoring systems must be designed based on local conditions (see below) but also need to sustain extreme climatic events and must therefore be resilient to climate change.

10.3.2.1 General Concept

Surface water management approaches on mine sites vary greatly depending on local climate conditions, seasonal weather variations, and water availability. In regions with limited water supply, the objective will generally be to collect, store, and reuse water. For example, collected surface water can be returned underground for extraction, recycled for ore processing, or transported to surface mine waste disposal sites to improve the performance of cover systems (e.g., water covers; see Chapter 6). In more humid regions with a positive water balance, excess water can cause a threat to dam stability, and the objective will usually be to collect, test, treat if necessary (to meet regulations), and discharge water into the environment (Garner et al. 2012). Surface water management infrastructures on a mine site therefore typically aim to segregate clean water (i.e., which has not been in contact with mine wastes or contamination sources) and contaminated water, collect seepage from waste rock piles and TSFs, and collect runoff from precipitations (Environment Canada 2009). Spillways are also built in TSFs (or tailings impoundments) to limit the risk of failure associated with overtopping (Szymanski and Davies 2004; Cacciuttolo and Tabra 2015).

Measurement approaches and equipment depend on the objective of monitoring. Direct and indirect measurements methods with various levels of precision and installation costs can be used. For example, the performance of water covers can be monitored using simple pressure transducers placed in monitoring wells (Freeman et al. 2004) to measure water levels above the surface, and by monitoring discharges in water channels, which often require more sophisticated installations to record water fluxes and elevations continuously. The required precision of the measurements also depends on the purpose of the monitoring. For example, regulators may require a high degree of precision to measure the discharge of treated water in the environment, whereas water heads may be measured with simple pressure transducer to monitor the performance of surface water infrastructure (basins, ditches, spillways) used to prevent uncontrolled discharges. Monitoring seepage from tailings dikes and rainfall runoff discharge, however, requires the direct measurement of water fluxes with a greater precision. Water balance calculations also usually require the installation of weather stations on the site to account for local rainfall variability (Zhou et al. 2019; also see Section 10.1). Some of the most commonly used surface water flux monitoring approaches are discussed below.

10.3.2.2 Direct Measurement Methods

The flow of an open-flowing stream can be directly estimated by multiplying the cross-sectional area of the flow by the average speed of the water in that stream. However, the speed of water in a cross section is usually not homogeneous, in particular, because of the friction between water and the surface of the ditch. Consequently, water speed is often faster at the center of the stream and close to the surface, and slower on the sides and deeper in the stream (Meals and Dressing 2008). Therefore, in practice, the flow is often measured punctually using direct measurements such as the velocity area method. This method consists of dividing the channel in segments where water velocity (considered homogeneous in the segment) is measured. The total discharge is then calculated by adding the contribution of the area of every segment multiplied by the water velocity. A number of individual measurements can be used to generate a rating curve (i.e., the relationship between water height and discharge in a channel), and the water flux is calculated in real time using a method such as a level bubbler, or an ultrasonic or pressure sensor (Ferland et al. 2011).

For small streams, the volumetric method can also be used to generate a rating curve. In this method, discharge rates are determined by measuring the time required to fill a containing vessel of known volume (Gore and Banning 2017).

10.3.2.3 Indirect Methods

Indirect methods are simpler and cheaper than direct measurements and can also be used to estimate the water flux in an open channel when a higher degree of precision is not needed. Manning's equation is commonly used to calculate streamflow based on water levels as well as channel geometry and roughness (NRCS 2004):

$$Q = \frac{1}{n} A R^{2/3} S^{1/2} \tag{10.1}$$

where
Q: Discharge [$L^3 \cdot T^{-1}$].
n: Manning's roughness coefficient [$L^{-1/3} \cdot T$].
A: Cross-section area of the channel [L^2].
R: Hydraulic radius [L].
S: Slope of the channel bed [–].

10.3.2.4 Hydraulic Measurement Structures

When a higher degree of precision over a long period is required, permanent infrastructures can be installed directly in a channel to measure discharge. The two most common installations include weirs and flumes.

A measuring flume requires the installation of an imposing structure in the watercourse. This structure comprises a first converging section, which gradually narrows the width of the flow section, followed by a control section (or throat section), where the flow is measured, and then a diverging section, where the stream recovers its initial width (Michalski 2000). The measurements obtained by this method are very precise (typically error < 5%; ASTM 2013) but it is an expensive structure that also needs to be designed according to a predetermined flow range (CEAEQ 2008).

Weirs are structures placed perpendicularly to the flow that act as a dam, creating a reservoir upstream, and allowing the water to flow through an opening (Figure 10.5). The geometry of the

FIGURE 10.5 Example of a V-notch weir measuring seepage from a tailings dam.

opening is variable and includes, for example, V, triangular, or trapezoidal shapes (Michalski 2000; USDI 2001). The height of water in this opening is measured and converted to a flow rate using precalibrated rating curves. Weirs are accurate but also need to be designed for a given flow range.

10.3.3 Volumetric Water Content

The in situ performance of some reclamation methods at the large scale involves the measurements of the unsaturated behavior of cover materials (e.g., store-and-release (SR) cover, EWT with monolayer cover, CCBE, LSHCC presented in Chapters 5, 8, 7, and 4, respectively). One key parameter related to unsaturated behavior of cover materials is the volumetric water content (θ). Various methods to evaluate in situ volumetric water content measurements (θ) have been developed. Some of these methods are simple, whereas others use advanced techniques. The main methods presented in the following can be divided into four categories: gravimetric method, methods based on high-energy ray sensors, dielectric methods, and hydrogeophysical methods. Some of their main characteristics are given in Table 10.3.

TABLE 10.3
Main Properties of Measurement Techniques for Volumetric Water Content

	Accuracy (cm^3/cm^3)	Installation method	Logging capability	Remarks	Case studies
Neutron probe	0.005	Access tube	No	Calibration curve is needed for each tested soil; involves a radioactive source and a specific permitting process; automatic logging not possible	Equity Silver (O'Kane et al. 1998)
TDR	0.03 and up to 0.01 with calibration curve	Permanent or inserted	Yes	Results can be affected by salinity and sulfide content; automatic logging feasible; a general equation can be used to deduct θ for standard soil, but precision can be improved by using a specific calibration curve	LTA (Ricard et al. 1997; Bussière et al. 2006); Lorraine (Dagenais et al. 2005), Health Steele (Yanful et al. 1993)
FD	0.03 and up to 0.01 with calibration curve	Permanent or inserted	Yes	Results can be affected by salinity and sulfide content; automatic logging feasible; a general equation can be used to deduct θ for standard soil, but precision can be improved by using a specific calibration curve; temperature and electrical conductivity can be measured with the same probe	LTA (Bussière et al. 2006); Lorraine (Bussière et al. 2009); Manitou (Ethier et al. 2018); Kidston (Williams et al. 2006)
Geoelectrical monitoring	Variable	Permanent or temporary	Yes	Large investigation volumes; complementary data to extend spatially local measurements	LTA (Maqsoud et al. 2011), Lac Tio WRP (Dimech et al. 2019)
Radar tomography	Variable	1-day investigation	Yes	Large investigation volumes, complementary data to extend spatially local measurements	LTA (Maqsoud et al. 2011)

10.3.3.1 High-Energy Emission (or Ray) Methods

These methods use a radioactive source for θ measurement; the most used method is the neutron probe. The neutron probe method is based on the measure of the mobile neutrons which are slowed down or thermalized by an elastic collision with hydrogen particles in the soil and water. The neutron probes use the property of preferential deceleration by the hydrogen atoms to determine θ of a soil. These hydrogen atoms are mostly associated to water molecules, and consequently the deceleration of the neutrons is related to the soil moisture. A calibration curve must be determined between the neutron probe count ratio and the volumetric water content. The main advantage of the neutron probe method is the accuracy of the measurements with a good calibration curve. However, the use of these techniques involves the handling of a radioactive source, the drilling of access tubes, and a specific calibration for each soil.

10.3.3.2 Dielectric Methods

Each soil component (solid, water, and gas) has its own electromagnetic property (commonly called relative dielectric constant ε_r); at 20 °C the relative dielectric constant of air is $\varepsilon_{air} = 1$, the solid relative dielectric constant ε_{solid} ranges between 2 and 5, and the relative dielectric constant of water ε_{water} is 80. Thus, the relative dielectric constant of the soil is mainly dominated by that of water. Consequently, it is possible to estimate θ by measuring ε_r. The two main techniques used to measure ε_r are briefly presented here. More information on these techniques can be found in the following literature: White and Zegelin (1995); Topp and Ferre (2002); Muñoz-Carpena (2004); Maqsoud et al. (2007, 2017); and Cooper (2016).

Time Domain Reflectometry (TDR) method is well known for θ measurements in soils (e.g., Topp et al. 1980; Robinson et al. 2005). The relative dielectric constant (ε_r) is determined by measuring the propagation time of an electromagnetic impulse along a transmission line placed in the soil. Topp et al. (1980) proposed a universal approach for soils and established a relationship between ε_r and θ. This relation is supposed to be valid for the majority of soils (independent of their composition and texture) and for θ values below 50% (Muñoz-Carpena 2004). However, measurements can be affected by factors such as salinity of the pore water, presence of sulfide minerals, and organic matter content. At higher volumetric water content, and for sulfidic, organic, or volcanic soils, specific calibration is required. The salinity and/or the clay composition of the soil (Skaling 1992; Zegelin et al. 1992; Jacobsen and Schjonning 1993) and/or the presence of iron and sulfide minerals (Robinson et al. 1994; Maqsoud et al. 2017) can affect measurements by attenuation of the reflected impulses.

It is then recommended to obtain a calibration curve for each material or to at least validate that the Topp et al. (1980) relationship give acceptable values for each material. The main advantages of using the TDR method to monitor covers are its logging capacity and the relatively good accuracy (Aachib 1997; Aubertin et al. 1997).

The Frequency Domain (FD) technique was proposed at the end of 1920s but remained unapplied until the development of a first probe for in situ measurements by Malicki (1983). The concept consists of measuring the time needed to charge a capacitor that influences the frequency of an oscillator. The capacitor is created using two rods, and the soil acts as the dielectric. By connecting this capacitor to an oscillator to form an electric circuit, changes in θ can be detected by the frequency variations of the circuit. In other words, the dielectric constant ε_r is determined by measuring the period of charge of the soil (capacitor). The main advantages and disadvantages of the FD method are similar to those of the TDR method and, as suggested for the TDR technique, it is recommended to obtain a calibration curve for each material (Topp et al. 1980; Dasberg and Hopmans 1992; Kup et al. 2011; Maqsoud et al. 2017).

10.3.4 Suction

Suction (ψ) is the other critical component that controls the unsaturated behavior of cover materials. Different types of equipment can be used for ψ measurements in geomaterials. For convenience, only the main techniques available on the market and that can be potentially used in the field to monitor reclamation methods are described in the following; more information on suction measurements can be found in Rahardjo and Leong (2006). Table 10.4 presents the main apparatus with their suction range and accuracy, the method of installation, the logging capability, and also some field examples where the instruments were used to measured suction in mine reclamation methods.

The tensiometer is a relatively simple device that consists of a plastic tube with a hollow ceramic tip filled with freshly boiled and cooled deionized water. The device is connected to a pressure gauge by a capillary tube. When a tensiometer is inserted into a soil mass, the ceramic tip transports water, via capillary action, from its interior to the exterior, creating a pressure in the tensiometer tube. This pressure is referred to as the suction or matric potential. The flow of water from the tensiometer tube into the soil mass through the ceramic tip continues until equilibrium is reached between the water in the tensiometer tube and that in the soil mass. At equilibrium, the water tension in the tensiometer is equal to the suction in the soil (Singh and Kuriyan 2003). The main advantage of the tensiometer is the simplicity of the device and the accuracy of the technique. However, the maintenance of the

TABLE 10.4

Comparison Between the Different Techniques Used for Suction Measurements

	Measurements Range (kPa)	Accuracy	Installation Method (kPa)	Logging Capability	Remarks	Case Studies
Tensiometer	0–100	1	Permanently	Yes	Precise measurements for low suction; cannot measure ψ values >80 kPa in the field; maintenance is necessary before winter in cold climate region	LTA (Bussière et al. 2006)
Gypsum block	30–1000	1	Permanently	Yes	Can dissolve in aggressive water; cannot measure low (<30 kPa) suction values	Health Steele (Yanful et al. 1993)
Granular matrix sensor	5–200	1	Permanently	Yes	High conductivity water can affect readings; less precise at suction values <5–10 kPa; no need for a calibration curve	LTA (Ricard et al. 1997; Bussière et al. 2006), Lorraine (Dagenais et al. 2005)
Heat dissipation sensor	10–1000		Permanently	Yes	Development of specific calibration curves is recommended; influenced by temperature and hysteresis; less precise at suction values <10 kPa	Equity (O'Kane et al. 1998); Health Steele (Yanful et al. 1993)

equipment is relatively complex: for instance, the water must be "pure" and changed regularly, the ceramic tip must be filled with nonfreezing liquid before the frost period, automatic monitoring is difficult, etc. The measurement range of standard tensiometer is limited to 0–100 kPa because of possible water cavitation. In short, the tensiometer is not well adapted for measurements close to the surface influenced by the evaporation phenomenon. However, a new type of tensiometer has been developed (high capacity tensiometer) that can measure suction values as high as 1500 kPa. More information on tensiometers can be found in Marinho et al. (2008).

Indirect methods for suction measurement used some properties of a controlled porous medium and relate them to soil suction. The two main sensors that use indirect methods are electrical conductivity and thermal conductivity sensors (Bulut and Leong 2008). Electrical resistance between electrodes can be used to evaluate the suction. When two electrodes are incorporated into a porous matrix (block), an electrical resistance between them is created, which is proportional to the water content of the matrix. This water content is directly related to the soil suction through the water retention curve of the media forming the matrix. The electrical resistance decreases as the medium and the block become dryer. This technique has been used for more than 60 years. The gypsum block was developed first (Stenitzer 1993), but because of dissolution problem with such matrix, other granular matrices have since been developed (e.g., Silica) and are preferred for field applications (Leib et al. 2003). Also, gypsum blocks must be calibrated on a regular basis for accurate measurements. As the newer granular matrices are more stable, the sensors do not need to be calibrated as often. Electrical resistance blocks have several advantages: they are simple to use, require little maintenance, provide easy logging capacity, and are not subjected to damage by freezing. The main disadvantage of this type of sensor is related to the fact that suction between 0 and 10 kPa cannot be measured with accuracy (range measurements; see Table 10.4); this suction range can be important is some reclamation applications.

The high thermal conductivity of water influences heat flow and/or dissipation from a porous media (a dry material is heated more quickly than a wet material). Heat transfer in a porous medium is proportional to its water content (see Chapter 3). A heat dissipation sensor is composed of a porous block containing a heat source and an accurate temperature sensor. The temperature of the block is measured before and after the ignition of the heat source for a few seconds. The water content of the block is obtained from the temperature variation (Muñoz-Carpena 2004). The porous block being placed in contact and in equilibrium with the soil, its water retention curve will give the matric suction of the soil (Zhang et al. 2001; Nichol et al. 2003). Consequently, the probe must be calibrated before its use by measuring its water retention curve. The main advantages of this sensor are the wide measurements range (10–1000 kPa), its logging capability, its low periodic maintenance requirement, and its relatively good precision. The main disadvantages are related to the environmental effect on suction measurements. Hysteresis effects associated with drying and wetting of the sensors (Feng and Fredlund 2003) and the effect of the ambient temperature on the sensors were found to influence the matric suction measurements (Tan et al. 2007).

Figure 10.6 shows, in a mosaic of photos, the installation of suction (electrical resistance block; Figure 10.6e) and volumetric water content sensors (TDR probe; Figure 10.6f) in an existing CCBE. After digging a trench in the cover, the different sensors are inserted in the cover layers with care and by making sure that the contact between the probe and the soil is good (Figure 10.6a, b, c). The trench is then filled and the cables (with the data acquisition system) are placed in a plastic reservoir to protect them against vandalism and animals (Figure 10.6d).

10.3.5 Lysimeters

A lysimeter is defined as a buried container exposed to the atmosphere that isolates a volume of soil and that allows for the measurement of water storage in the soil and the quantity and quality of percolation water (Muller 1996; Benson et al. 2001). Lysimeters are often used as part of the monitoring of engineering cover system to estimate the different components of the water balance,

FIGURE 10.6 Mosaic of photos representing the installation of volumetric water content and suction sensors in an existing CCBE; (a) trench digging; (b) placement of instruments in the soil layers; (c) instruments installed with cable protected in a plastic pipe; (d) cables and data acquisition system protected in a plastic reservoir; (e) electrical resistance block to measure suction; (f) TDR probes to measure volumetric water content.

particularly water percolation (e.g., Swanson et al. 2003), and for sampling of the percolated water for an eventual water quality assessment (e.g., Aboukhaled et al. 1986; Swanson et al. 2003; Howel 2005). In contrast to the indirect methods using the other components of the water balance to deduct percolation rate, lysimetry provides a direct measurement. Two main types of lysimeter exist: volumetric lysimeter and weighing lysimeters.

10.3.5.1 Volumetric Lysimeters

Volumetric lysimeters correspond to a reservoir that mimics the soil profile to be investigated, with a collection system at the bottom (see Figure 10.7). The main components of lysimeters are the buried containers with open tops that collect and measure soil water. The material used as backfill material should be similar (if not the same) to the one in which it is buried and should be placed in a similar state (or porosity). Water that infiltrates the cover reaches the buried container, accumulates, and then is transported into a measuring facility. Changes in soil water storage are obtained by integrating profiles of volumetric water content measured using probes (see Section 10.3.3) placed in the soil profile. The literature shows different variants of volumetric lysimeters that mainly aim to control the bottom boundary condition (e.g. Aboukhaled et al. 1986; Stenidzer et al. 2007): lysimeters with drainage without a water table (as shown in Figure 10.7); compensation lysimeters with constant level of water; compensation lysimeters with the water table placed above the ground surface.

Monitoring water flux through covers is often complicated by the fact that unsaturated conditions usually prevail. When designing such percolation lysimeters, a number of issues must be addressed, such as the geometry of the lysimeter, the anticipated infiltration rate, and the hydraulic properties of the backfill material (Bews et al. 1999). The lysimeter must be deep enough to develop a constant pressure head for the anticipated rates of infiltration. The length and width are usually five times larger than depth to ensure that preferential flow processes are captured in the test and that the lysimeter mimics full-scale conditions (Bews et al. 1999; Chiu and Shackelford 2000). The drainage

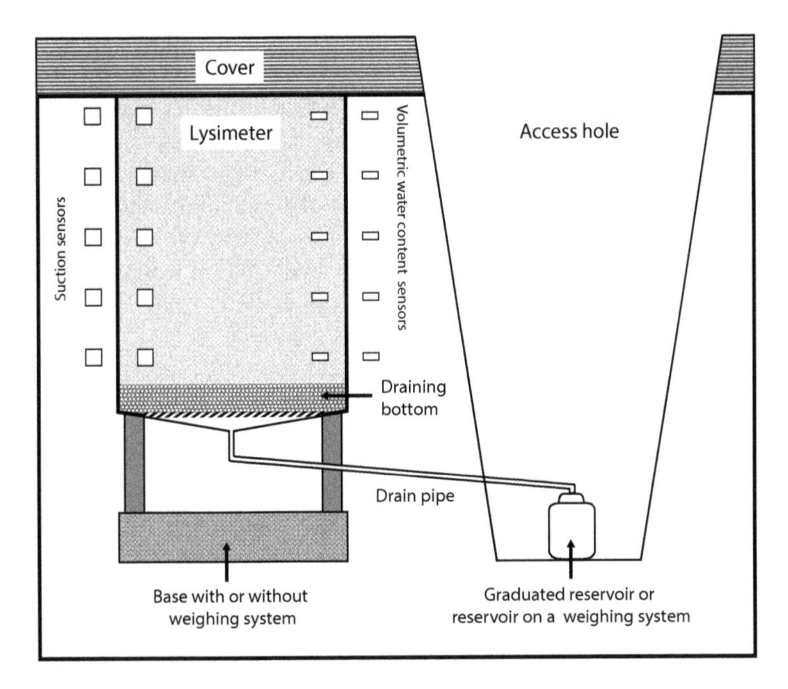

FIGURE 10.7 Schematic representation of a lysimeter to measure water percolation (volumetric lysimeter without weighing system, and weighing lysimeter with a base with a weighing system).

material used at the bottom must not create artificial capillary barrier effects that would modify the water storage (Khire et al. 1999). If the soil thickness in the lysimeter is not sufficient, the no-flow bottom boundary condition could modify water and vapor movement upward, and give erroneous measurements (Coons et al. 2000). Finally, monitoring volumetric water content and/or suction outside the lysimeter (at the same elevation) to validate that hydrogeological conditions in the lysimeters is close to those of the natural system is recommended (Figure 10.7). Unsaturated flow modeling is recommended to validate the design of a lysimeter.

10.3.5.2 Weighing Lysimeter

Changes in the mass of a lysimeter can be directly linked to the volume of water entering and leaving the tank; an increase in weight corresponds to the entry of water from precipitation or irrigation, whereas a decrease corresponds to the loss of water through evapotranspiration or drainage. Drainage water or percolation water through the lysimeter can be collected in containers connected to the lysimeter and measured periodically.

The main difference between the different types of weighing lysimeters is mainly related to weighing system used in their conceptions. Based on the weighing systems, four types of lysimeters are identified (Aboukhaled et al. 1986; Howell et al. 1991; Fisher 2012): mechanical, electronic, hydrostatic, and floating weighing systems. Only the two first will be described in the following sections due to their more frequent use in the engineered cover context.

For the mechanical system, the lysimeter is equipped with a retaining wall, allowing a free movement of the inner container, which contains the soil mass. The variation of the container mass can be evaluated by periodically lifting the inner container from its support, or placing the lysimeter on a weighing system to continuously record changes in the weight of the container. In the case of lifting the inner container from its support, the capacity of the weighing scales is limited and cranes have to be used. Therefore, the size of the lysimeters must be adapted to these weighing scales. For the second system, the mass changes in the lysimeter are evaluated using strain gauges or electronic load cells; the design improvement of load cells, their cost, and the development of different logger systems make this option the most attractive (Lorite et al. 2012; Schmidt et al. 2013; Meena et al. 2015; Bello and Van Rensburg 2017).

Difficulties that can be encountered with weighing lysimeters are essentially the same as those identified for volumetric lysimeters, but weighing lysimeters can also have problems related to the accuracy of the measurement and the detection of small mass variations that can occur inside the lysimeter. It is also worth mentioning that the size of weighing lysimeters is usually smaller (1–2 m^2) than volumetric lysimeters because of the limited capacity of scales (Benson et al. 2001). Several lysimeters in engineered covers such as CCBE and SR cover systems have been used successfully (Waugh 2004; Smith and Waugh 2004; Evett et al. 2009; Fisher 2012; Ruth et al. 2018; Padrón et al. 2020).

10.3.6 Hydrogeophysical Methods

Traditional hydrogeological sensors described previously are usually considered precise (with proper calibrations) and easy to install. They provide quasi-continuous measurements of key hydrogeological parameters (such as volumetric moisture content θ or suction ψ) in a volume of approximately 1 liter (roughly 15 cm around the sensor) (Robinson et al. 2008). This limited spatial extension can be considered a limitation when they are used to monitor large structures such as reclaimed WRSFs or TSFs. Many sensors are then necessary to have vertical profiles of saturation since these profiles must be located at several locations to monitor reclamation performance under different conditions (e.g., water table elevation, slope angle, or vegetation). For example, 120 and 160 sensors have been buried in the reclamation covers at Les Terrains Aurifères and Lorraine mine sites (Québec), respectively. This represents important costs for monitoring programs and large volumes of data to collect, process, and analyze (Bussière et al. 2006). Finally, the lack of lateral extension of these measurements makes it difficult to take into account material heterogeneity and to identify local failures of the reclamation designs as described by Rykaart et al. (2006).

In this context, hydrogeophysical methods have been suggested since they provide estimations of physical parameters with large spatial extensions (Rubin and Hubbard 2006). Radar tomography or electrical resistivity tomography (ERT) is frequently used to image the spatial distribution of dielectric constant ε or electrical conductivity σ, respectively. The investigated volume usually ranges from one to several hundred cubic meters (Binley et al. 2015). The distribution of ε or σ can be translated into volumetric water content θ in the medium according to empirical relationships similar to Topp's equation mention previously to interpret TDR results. Time-lapse ERT (also referred as geoelectrical monitoring) is one of the most promising approaches among hydrogeophysical methods for performance monitoring because permanent systems can be installed at reclaimed mine sites where autonomous monitoring can be carried out remotely (see Falzone et al. 2019 for a review of time-lapse ERT applications). Such monitoring systems can provide hydrogeophysical data with high spatio-temporal resolution, which can be used as complementary data to extend spatially local hydrogeological measurements for performance monitoring. Finally, the continuous monitoring can also be complemented with larger geophysical radar or ERT investigations on a yearly basis, for example.

From a theoretical point of view, time-lapse ERT reconstructs the spatio-temporal distribution of bulk electrical conductivity in the investigated volume from a resistance database. A resistance value can be obtained with four electrodes by measuring the potential difference created in the medium by a current injection using Ohm's law (Rubin and Hubbard 2006). Each configuration of four electrodes provides information about a specific location, and a single dataset can contain up to several thousands of configurations. A complete dataset can usually be measured in less than 1 hour, and repeated measurements allow for the monitoring of the spatio-temporal evolution of electrical conductivity (Wilkinson et al. 2006). Electrical conductivity in a porous medium can be affected by various key hydrogeological parameters such as water content, ionic content of interstitial water, porosity, and temperature. Archie's law is the most common empirical relationship used to convert σ into θ. Archie's law can be calibrated in the laboratory for each material and for specific water conductivities and temperatures (Merritt et al. 2016). The measured database can be processed to recover the most probable distribution of electrical conductivity in the investigated area through geophysical inversion, which can be converted into θ distribution (Chambers et al. 2014). The geoelectrical database can also be used as input for data assimilation techniques to optimize hydrogeological models in order to refine θ estimations in the covers (Bouzaglou et al. 2018). Time-lapse ERT can then be used to track the infiltration of a tracer or to monitor water content changes in a material, which is valuable information for performance monitoring of CCBE, EWT with monolayer cover, and SR covers.

From a practical point of view, several steps should be followed to design, install, and perform geoelectrical monitoring in reclamation covers:

1) The first step includes an identification of several zones of the covers to be monitored – for instance, where the performance is expected to be limited because of important slopes, low phreatic surface, dense vegetation, or heterogeneous material. Typical dimensions of monitored zones could be 100-m-long profiles or 10 m × 50 m rectangular grids.
2) The second step is the design of geoelectrical monitoring systems. Electrode position should be chosen according to this basic rule of thumb: the spatial resolution can be approximated to the electrode spacing, typically between 50 cm and 5 m (Rubin and Hubbard 2006). Most geoelectrical monitoring systems use between 50 and 500 stainless-steel electrodes. Buried electrodes below and above the layers of interest should be preferred to maximize coupling with the materials, measurement precision, and electrode lifetime (Kiflu et al. 2016).
3) The third step consists of installing the geoelectrical system in the cover during its construction to minimize the impacts of the instrumentation on cover geometry (as opposed to post-construction installation with trenches). Electrodes installation can be as fast as the cover construction if the electrode cables are well prepared. Buried electrodes and cover internal geometry should be properly surveyed for topographic corrections. Material sampling and in situ characterization of porosity are also recommended to constrain geoelectrical results and to calibrate hydrogeophysical relationships in the laboratory (Tso et al. 2019). The installation of some vertical profiles of traditional volumetric water content sensors, typically every 25 m along the electrodes grids to allow calibration and validation of geoelectrical results, is also recommended.
4) Finally, the last step includes autonomous measurements on a weekly, daily, or hourly basis depending on the needed time resolution. Measurements should be carried out with optimized protocols that minimize the number of configurations (i.e., the measurement time) while ensuring the needed spatial resolution. Data can usually be transferred to remote servers and autonomous processing could allow for the recovery of volumetric water content in the covers and extend spatially hydrogeological data to monitor reclamation performance (see Holmes et al. 2020 for details on autonomous remote geoelectrical monitoring).

Time-lapse ERT has become a recognized and valuable technique to monitor hydrogeological processes in various fields over the last 20 years. Permanent geoelectrical systems have been used to (1) monitor moisture content in railways embankments, (2) quantify permafrost retreat, (3) predict moisture-induced landslides, (4) track contaminant migration or coastal saline intrusion, and (5) monitor root propagation, CO_2 storage, or geothermal operations (see De Franco et al. 2009; Schmidt-Hattenberger et al. 2016; Mollaret et al. 2019; and Whiteley et al. 2019). However, hydrogeophysical applications for mining reclamation are mostly limited to one-day investigations to image the internal structure of waste rock piles or to delineate oxidation in tailing storage facilities (Anterrieu et al. 2010; Shokri et al. 2014; Power et al. 2018). Several time-lapse studies using geoelectrical monitoring to evaluate the performance of reclamation covers are nonetheless available in the literature. For example, Maqsoud et al. (2011) used time-lapse ERT to monitor changes of water content in a sloped CCBE built at the Les Terrains Aurifères (LTA) tailing storage facility. Dimech et al. (2019) also carried out geoelectrical monitoring during 10 days to track water infiltration in an experimental waste rock pile during a controlled infiltration test to assess the performance of a cover that includes a flow control layer (FCL) built on the top of the pile.

10.4 WATER QUALITY MONITORING

Water quality monitoring at reclaimed mine sites allows for the assessment of environmental contamination. Whenever water quality does not meet the enforced criteria, water treatment (active or passive) must be implemented (see Chapter 11) to ensure compliance with regulatory requirements. Several precautions must be taken for accurate and representative sampling of both underground and surface water collection and analysis. For an integrative and pertinent evaluation of water quality, assessment of monitoring results should combine physicochemical parameters and their potentially adverse biological effects (Altenburger et al. 2019).

10.4.1 Sampling Methods for Ground and Surface Waters

General precautions must be taken when preparing for a ground and surface water sampling campaign. Great care should be put in planning and preparation. Planning for the equipment needs is crucial: getting the right equipment (sampling notes, probes, filters, relevant types of bottles with and without preservatives for specific parameters, coolers, filtration apparatus, etc.) is obviously important, but the sampling points should be organized in increasing order of contamination (from less to more contaminated sampling points) to minimize the effects of eventual carryovers. Cleaning the equipment that is used in multiple samplings is also of utmost importance. Sampling notes need are recorded for the relevant information of each sample, including time and date, sample and sampler identification, equipment used, depth, volume, purging rate, and weather conditions. Finally, samples need to be stored in the dark at 4 °C until analysis. These general precautions can be found in many standards, some of which are listed in Tables 10.5 (groundwater) and 10.6 (surface water).

When sampling groundwater, each well must be inspected for casing integrity and to make sure no water is present at the surface. When purging of the well is necessary, the purging method needs

TABLE 10.5
Typical Standards and Guides for Groundwater Sampling

Standard	References
Standard guide for sampling groundwater monitoring wells	ASTM (2019a)
Planning and preparing for groundwater sampling	ASTM (2017b)
Standard guide for purging methods for wells used for groundwater quality investigations	ASTM (2018e)
Standard guide for the selection of purging and sampling devices for groundwater monitoring wells	ASTM (2014c)
Standard practice for low-flow purging and sampling for wells and devices used for groundwater quality investigations	ASTM (2018g)
Standard guide for documenting a ground-water sampling event	ASTM (2019c)
Standard guide for field preservation of groundwater samples	ASTM (2018f)
Guide for selection of purging and sampling devices for groundwater monitoring wells	ASTM (2014c)
Guide for field filtration of groundwater samples	ASTM (2017c)

TABLE 10.6
Typical Standards and Guides for Surface Water Sampling

Standard	References
Guide for quality planning and field implementation of a water quality measurement program	ASTM (2018c)
Practice for decontamination of field equipment used at waste sites	ASTM (2020)
Standard practice for sampling liquids using grab and discrete depth samplers	ASTM (2016c)
Practice for sampling with a dipper or pond sampler	ASTM (2019b)
Practice for sampling liquids using bailers	ASTM (2016b)
Guide for sampling wastewater with automatic samplers	ASTM (2019d)

TABLE 10.7
Stabilization Criteria for Groundwater for Three Consecutive Samples Taken 3–5 Minutes Apart (From Environmental Protection Agency (EPA) 2002 and Cloutier et al. 2015)

Parameter	Stabilization criteria
Temperature (°C)	±0.1
pH (pH units)	±0.1
Oxidation–reduction potential (ORP) (mV)	±10
Electrical conductivity (μS/cm)	±2–3%
Dissolved oxygen (mg/l)	±0.2–0.3
Turbidity (NTU)	±10% or visual estimate

to be selected based on several factors, such as the sampling objectives, the physicochemical parameters of interest, and the permeability of the media that is sampled, among others. Again, many guidelines are available to help select the most appropriate purging method (Table 10.5, groundwater). The purged water needs to be monitored for key parameters to stabilize before sampling; generally, three consecutive samples, separated by 3–5 minutes, need to be within an acceptable range before sampling, as given in Table 10.7 (stabilization parameters). When sampling surface waters such as streams, rivers, or lakes, the sampling equipment (e.g., telescopic pole, bailer, and bottle) needs to be rinsed downstream from the chosen sampling point. Sampling from a boat must be done at least 1 m away, upstream from the wind and wave directions.

10.4.2 Physicochemical Analyses

Water physicochemical parameters to be analyzed must be selected according to each site and legislative requirement, as well as according to the objectives (such as water treatment planning). Sometimes, even though some parameters are not specifically identified as target in the legislation, their indirect effects might identify them as a potential target. For example, some oxidizable species are not toxic, but their oxidation consumes dissolved oxygen (DO) and can cause water anoxia, which could be toxic, before anaerobic processes make the water toxic.

A comprehensive list of basic physicochemical parameters, methods of analysis of each type of equipment needed, precautions to be taken after sampling (e.g., filtration, addition of a preservative), conservation time, and standards to be used are detailed in Table 10.8 (Physicochemical analysis).

TABLE 10.8

Physicochemical Analysis on Mine Effluents

Analysis	Method	Equipment needed	Filtration	Preservative agent	Conservation time	References
pH	Electrode	Probe/m	No	No	2 h	APHA (2017t)
Oxidation–reduction potential (ORP)	Electrode	Probe/m	No	No	Immediately	ASTM (2018b) APHA (2017a)
Electrical conductivity (EC)	Electrode	Probe/m	No	No	28 days	ASTM (2014b) APHA (2017b)
Hardness	Titration	Titrator	No	No	6 months	ASTM (2014a) APHA (2017c)
Total dissolved (TDS) and suspended (TSS) suspended solids	Electrode/Gravimetry	Probe/m	TDS: 0.45 µm TSS: No	No	7 days	ASTM (2017a) APHA (2017d)
Turbidity	Nephelometry	Turbidity meter	No	No	48 h	ASTM (2018d) APHA (2017e)
Dissolved oxygen (DO)	Electrode	Probe/m	No	No	Immediately	ASTM (2014d) APHA (2017f)
Temperature	Electrode/ Thermometer	Probe/Meter	No	No	3 minutes	ASTM (2018a) APHA (2017g)
Acidity/Alkalinity	Titration	Titrator	No	No	14 days	ASTM (2014e) APHA (2017h)
Dissolved (filtered) and total (unfiltered) metals	Absorption/Emission spectroscopy	AAS (ICP-AES/-MS	Dissolved: 0.45 µm Total: no	Conc. HNO_3 (pH < 2)	28 days for Hg (6 months for other metals	ASTM (2016a) APHA (2017j, k)
Anions	Ionic chromatography	IC	0.45 µm	No	6 months	APHA (2017l)
Ammonia nitrogen NH_3-N	Electrode	Probe/Meter	No	Conc. H_2SO_4 (pH < 2)	28 days	APHA (2017m)
Sulfide	Colorimetry	Spectrophotometer	No	No		APHA (2017n)
Dissolved (DOC) and total (TOC) organic carbon	Combustion	Carbon Analyzer	DOC: 0.45 µm TOC: no	HCl or H_2SO_4 or H_3PO_4 (pH < 2)	28 days	APHA (2017o)
Fe^{2+}/Fe^{3+}	Colorimetry	Spectrophotometer/ ICP-AES/-MS	Fe^{2+}: 0.45 µm Fe^{3+}: no	Conc. HNO_3 (pH < 2)	Immediately/28 days	APHA (2017p)
Mn^{2+}	Absorption/Emission spectroscopy	Spectrophotometer/ ICP-AES/-MS	0.45 µm	Conc. HNO_3 (pH < 2)	28 days	APHA (2017q)
As^{3+}/As^{5+}	High-performance liquid chromatography	HPLC	0.45 µm	Conc. HNO_3 (pH < 2)	28 days	APHA (2017r)
Cyanides and derivatives: CN^-, SCN^-, CNO^-	Colorimetry	Spectrophotometer/IC	No	NaOH (pH > 12)	14 days	APHA (2017s)
Thiosalts (ex. $S_2O_3^{2-}$, $S_3O_6^{2-}$, $S_4O_6^{2-}$)	Ion chromatography	IC	0.45 µm	Freezing	14 days	

10.4.3 Ecotoxicological Analyses

Water toxicity testing is mandatory in many countries (e.g., Belgium, Brazil, Canada, Germany, Italy, Netherlands, Spain, Sweden, Switzerland, UK, USA) and is part of mine water monitoring requirements. The types of organisms and testing duration of toxicity tests vary from one test to the other. The most commonly used organisms to properly cover the key trophic levels (producers, primary consumers, and secondary consumers) are algae, invertebrate, and fish (Norberg-King et al. 2018). A comprehensive review of comparative regulatory and biological monitoring programs for wastewater and industrial effluents, especially for Canada, the EU, and the United States, is presented by Norberg-King et al. (2018). Moreover, Altenburger et al. (2019)'s review identifies potential solutions for future water quality monitoring, and improving the balance between exposure and toxicity assessments of real-world pollutant mixtures.

10.5 GAS MOVEMENT

10.5.1 Introduction

Water covers (Chapter 6), covers with capillary barrier effects (CCBE) (Chapter 7), and the EWT with a monolayer cover (Chapter 8) are the main oxygen barriers used in humid climate regions to control acid mine drainage (AMD) generation. Low saturated hydraulic conductivity covers (LHSCC), such as those that integrate geomembranes (Chapter 4), can also help to control the migration of oxygen even though their main objective is to control water infiltration. The performance of oxygen barriers can be evaluated by comparing the quantity of oxygen consumed by the mine wastes exposed to the air (Q_{before}) during a given period t (example t = 1 year) to that crossing the cover and consumed by the mine wastes during the same period (Q_{after}) (see Figure 10.8). The cover performance can also be assessed by limiting the amount of oxygen Q_{after} to a target value defined at the design stage.

The amount of oxygen consumed by the mine wastes Q_{before} and Q_{after} can be calculated using numerical codes considering transient conditions (to account for changing conditions due to soil–atmosphere exchanges) based on the oxygen fluxes consumed at the surface of the wastes before F_{before} and F_{after} reclamation. These oxygen fluxes consumed by the wastes (with and without cover) can be determined experimentally using oxygen consumption test (see Section 10.5.2). They can also be determined from oxygen concentration profiles measured in the cover and in the underlying tailings (see Section 10.5.3) by applying Fick's first law (see Chapter 3; Section 3.5.1).

10.5.2 Oxygen Consumption Test

Initially, the main objective of the oxygen consumption test (OCT) was to estimate a quasi-instantaneous steady-state oxygen flux through reactive mine wastes exposed to air at a given time (Elberling et al. 1994). The test usually consists of the insertion of a cylinder directly in the mine

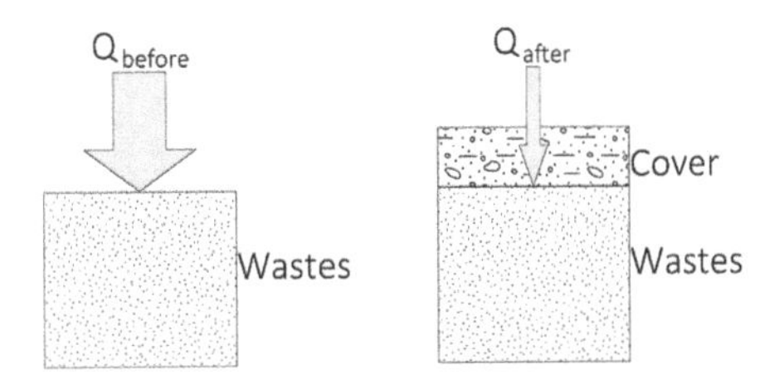

FIGURE 10.8 Principle for evaluating the oxygen barrier cover performance with the mass of oxygen consumed by tailings without cover (Q_{before}) and the mass of oxygen consumed by tailings with cover (Q_{after}).

waste, leaving an headspace of approximately 10 cm. A cap with an oxygen sensor is placed on top of the cylinder for approximately 3–5hours. The decrease in oxygen with time in the headspace allows for the calculation of the oxygen surface flux that is representative of the waste's reactivity. The concept and the method of interpretation are presented in Section 3.5.2 of Chapter 3.

The test was modified later to measure oxygen fluxes through cover systems (mainly CCBE and EWT with monolayer cover) at a given time (Mbonimpa et al. 2002; Dagenais et al. 2012). Figure 10.9 illustrates the measurement approach for the case of a CCBE. The drainage and top layer of the CCBE is excavated, and the cylinder is pushed into the moisture-retaining layer (MRL) and capillary break layer (CBL). Typically, the test is performed inside a 0.50 to 1-m-long, 10 to 20-cm-diameter aluminum cylinder, leaving a top headspace usually less than 10 cm to create the oxygen reservoir. The modified OCT last 5–9 days to obtain a significant variation (3% or more) of the oxygen concentration in the reservoir. The test interpretation is based on a comparison between measured oxygen concentrations in the top chamber and simulated results obtained from the numerical solution of the modified Fick's laws. This interpretation requires the porosity n, volumetric water content θ of the different materials, and the reaction rate coefficient of the different materials (Mbonimpa et al. 2002; Ethier et al. 2018). When oxygen fluxes with and without cover are available, the performance of the cover system can be calculated. Examples of modified OCT performed and interpreted to estimate oxygen flux through covers can be found in Dagenais et al. (2012) for CCBE and in Ethier et al. (2018) for an EWT with monolayer cover.

Ideally, OCT are performed at different locations and at different seasons on reclaimed sites to assess the spatial and temporal performance variability. Cylinders are usually removed after the test because water flow in the cylinder could not be representative of the long-term reality. Finally, it is worth mentioning that OCT can also be performed below insulation covers (see Chapter 9) to assess the tailings reactivity after cover placement. This test measures the influence of tailings temperature on sulfide reactivity. More information on these OCT can be found in Coulombe et al. (2013).

10.5.3 Gas Concentration Measurements

Table 10.9 presents the different methods for oxygen monitoring in water and soil covers. In the case of water covers, DO concentration can be measured using DO probes at different depths in the water column directly in situ or in water samples collected with a horizontal water sampling

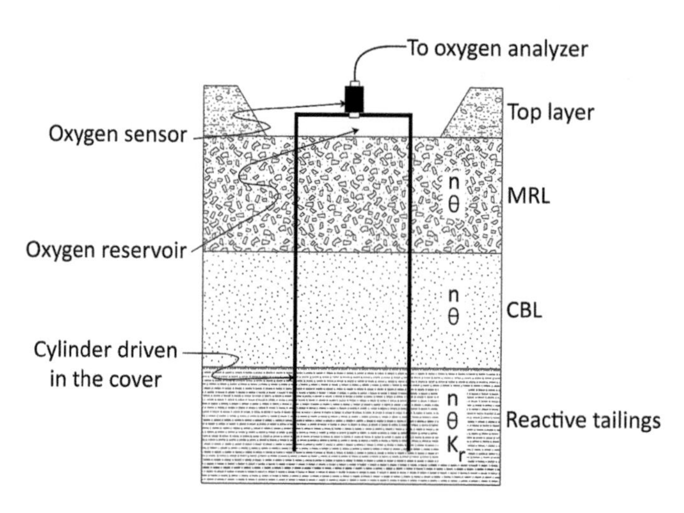

FIGURE 10.9 Illustration of an OCT performed through a CCBE (n: porosity, θ: volumetric water content, K_r: reaction rate coefficient).

TABLE 10.9
Methods for Oxygen Monitoring in Water and Soil Covers

Cover type	Principle	Measurement	References – mine site
Water cover–water column	DO probes or micro-electrodes	Direct on site or in water sampled	Vigneault et al. (2001) – Louvicourt site, Canada Awoh et al. (2013) – Don Rouyn site, Canada
Water cover–pore water	DO micro-electrodes	In situ	Vigneault et al. (2001) – Louvicourt site, Canada
Engineered covers	Active sampling	Sub-slab probe, well with annular seal, direct-push screened interval	ASTM (2018)
		Direct-push	ASTM (2018) Elberling et al. (1993) – Falconbridge Limited's East Mine, Canada Elberling (2005) – Nanisivik Mine site, Northern Canada Mbonimpa et al. (2008) – Les Terrains Aurifères (LTA) site, Canada
		Perforated bottles	Elberling (2005) – Nanisivik Mine site, Northern Canada
		Chainsaw gas filter	Adu-Wusu and Yanful (2006) – Whistle Mine, Canada
		Lysimeters and balls	Alakangas et al. (2008) – Kristinberg site, Sweden
		Slotted/perforated tubings screens	Vriens et al. (2019) – Antamina Mine, Peru Smith et al. (2013) – Diavik Mine, Canada
	Passive sampling	Sorbents	ASTM (2017)
	Direct measurement	Electrochemical (galvanic) oxygen sensor	Vriens et al. (2019) – Antamina Mine, Peru
		Zirconia oxygen sensor	Ishii and Kadoya (1991)
		Optical oxygen sensor	Klimant et al. (1995, 1997); Mbonimpa et al. (2008) – Les Terrains Aurifères (LTA), Canada
		Infrared oxygen sensor	Hanslin et al. (2005)

bottle (Awoh et al. 2013). During the sample recovery from the bottle and the measurement, any mixing of the water sample should be avoided to control possible dissolution of oxygen from air. Interstitial DO measurements in the wastes can be done using a DO micro-electrode attached to submersible micro-manipulator to allow measurements at very small depth intervals (e.g., each 0.5 mm) because oxygen disappears within only a few millimeters of the tailings–water interface (Vigneault et al. 2001).

In the case of soil covers, gas concentration within the cover materials and the underlying wastes can be performed in two ways: by measuring the gas concentration in sampled interstitial gas or by direct monitoring of the gas concentration with gas probes. Sampling of interstitial gas within tailings and soil covers can be performed following active and passive techniques as described in ASTM (2018h) and ASTM (2017d), respectively. Active sampling aims to provide the instantaneous environment of the soil gas at a particular time and at a specific depth. The gas is pumped from sampling probes installed at different depths and the gas concentration can be analyzed on site using an appropriate sensor or in the laboratory using gas-phase chromatography. Sub-slab probe, direct-push drive point, direct-push screened interval, and well with annular seal are the installation options according to ASTM (2018h) (see Table 10.9). For the direct-push installation, a probe tip through which the soil gas is sampled is driven in the cover or wastes using a probe tube that extends from the tip to the ground surface. Instead of installation options given by ASTM (2018h), perforated bottles, sampling ports consisting of a chainsaw gas filter covered with aluminum window screening, lysimeters, and balls buried into the ground can also be used (see Table 10.9). The sampler can also be placed into a hole after excavation where the annular space is sealed with an inert impermeable material. Furthermore, the probe installation can be done during the construction, being careful not to damage them. In the case of covers with geomembrane (GM), the gas concentration in the wastes beneath the GM should be monitored using other than the direct-push method dedicated samplers (to avoid destroying the GM).

For the passive soil gas sampling, an equilibrium is established (by diffusion and adsorption of interstitial gas) over a period of a few days to a few weeks between the soil gas and the passive samplers (appropriate sorbents). Once the equilibrium is reached, the samplers are retrieved from the sample holes, capped/resealed, and returned to the laboratory for analysis. The presence of oxygen and CO_2 in air can lead to contamination, making this approach cumbersome. The direct monitoring of the soil gas concentration can be performed in situ with gas electrochemical (galvanic), zirconia, optical, and infrared sensors (see Table 10.9). These probes allow a continuous measurement of the oxygen concentration, which is not possible when the concentration is measured in sampled interstitial gas. A calibration of the sensors is required.

In most cases, active methods are preferred to measure oxygen concentration in and below mine covers. Once oxygen concentrations are measured at different elevation in the oxygen barrier, oxygen flux can be calculated with the first Fick's law. The use of this law also requires knowledge of the effective diffusion coefficient D_e, which can be estimated if the porosity and the volume water content in the cover layers of materials and in the wastes are known (see Chapter 3; Section 3.5.3). The methods for measuring the volumetric water content are given in Section 10.3. The oxygen concentration profiles measured at different periods can also be used to validate the numerical models used to solve Fick's second law under transient conditions. For this, the measured and calculated oxygen profiles are compared (Demers et al. 2009; Ethier et al. 2018).

The frequency of oxygen concentration measurements is usually low, typically from one to three times per year. The oxygen concentration measurements are usually used as a complementary method to assess the performance of oxygen barriers to control AMD; the primary one consists of measuring volumetric water content in the oxygen barrier and calculating oxygen flux based on the first Fick's law.

10.6 VEGETATION PARAMETERS

Many countries have regulations requiring that reclaimed mine sites are revegetated (see Chapter 12). Revegetation helps to give the former mine site an appropriate land use for adjacent communities. After revegetation is implemented, it is important to measure several vegetation parameters in order to evaluate the revegetation success and eventually to plan maintenance. Moreover, if revegetation is performed over a reclamation cover, as it can be the case for LSHCC (Chapter 4), SR covers (Chapter 5), CCBE (Chapter 7), and EWT with monolayer covers (Chapter 8), the measurement of appropriate vegetation parameters will include the vegetation contribution to evapotranspiration in the numerical prediction of the reclamation performance (see Chapters 3 and 14).

The revegetation success is basically evaluated through the measurement of two types of parameters related to plant community composition and plant productivity. On a reclaimed mine site, plant community composition is indicative of the ecological value of the revegetated area. It is used to follow up the evolution of the plant community toward the plant community characteristic of the natural or targeted ecosystem. The evaluation of plant community composition also checks that no exotic or invasive species has established on the reclaimed mine site. Plant productivity is defined as the production of living organic matter (or plant biomass), thanks to photosynthesis. It is related to the greening and erosion control of the revegetated site. It is usually evaluated at the aboveground level. Finally, vegetation can increase or decrease the performance of reclamation covers by influencing the water balance of the site. It is thus useful to include some vegetation parameters as inputs in the numerical models used to predict cover performance for robust design and accurate long-term performance evaluation. Several numerical models can include vegetation inputs, mainly used to adjust the evapotranspiration component of the water balance. The most used vegetation parameters in these models, such as plant cover, leaf area index (LAI), maximal rooting depth, and root length density (RLD), as well as some measurement technics of these parameters, will be presented in the following.

10.6.1 Plant Community Composition

The plant community composition is simply a qualitative list of the different plant species occurring on the revegetated mine site. It implies achieving a vegetation inventory on a determined area. The minimum surface likely to include most of the species present on the reclaimed site can be determined by plotting the species number against the sampled area. When the curve is reaching a plateau, the corresponding area can be considered as the appropriate surface for conducting the inventory (Guinochet 1973). If the sampled area is too small, the inventory may be not representative since it may contain only a portion of the plant species occurring on the reclaimed site. If the sampled area is too large, the risk is to include plant species from the plant communities from adjacent ecosystems to the reclaimed mine site, increasing the heterogeneity. The required area for the inventory increases with the size of the plants present on the reclaimed mine site. For young plant communities evolving on reclaimed mine sites, typically herbaceous or shrub vegetation types, the required area may vary between 25 and 200 m^2. Forest vegetation rather requires 300–800 m^2 inventory areas. It is easier to identify plant species when they carry flowers. Thus, a plant inventory may imply several sampling times during the growing season to account for different flowering periods of species. Plant species diversity or, in other words, plant species variety can then be evaluated from the species inventories. Diversity indexes are used to quantify the plant species diversity. The most used index is the total species richness index. It is calculated as the number of different species by sampled surface. It is also possible to focus on the number of rare, patrimonial, or protected species by sampled surface. Other diversity indexes take into account the evenness of species in the community. For example, the Shannon–Weaver index (Shannon 1948) is calculated with the relative cover or density of each species. It is smaller when the number of species decreases and when a few species are dominant in terms of cover or density (see Section 10.6.2 on vegetation productivity). At the scale of a revegetated mine site, plant diversity could also be evaluated at the genetic or community levels complementary to the species level, but this will not be dealt with in this book.

10.6.2 Vegetation Productivity

The parameters used to measure plant productivity change according to the type of plants considered.

10.6.2.1 Small Plants

For numerous small plants (like herbaceous plants), plant cover or biomass are the most relevant parameters. Plant cover is simply the soil percentage area that is covered by the plant aboveground parts after being vertically projected on the soil. Plant cover can be estimated visually, by intervals (usually <5%, 5–25%, 25–50%, 50–75%, and 75–100%), for the whole plant community (total plant cover), or by species or groups of species. A useful method to quantitatively measure the plant cover is the point intercept method (Jonasson 1983) on transects or quadrats. Transects are a straight line of measurement of a given length, whereas quadrats are a squared surface of a given area (see Figure 10.10). A thin rod is moved along the transect line or along a grid in the quadrat at a fixed measurement interval (see Figure 10.10). For each position of the rod, if a plant touches the rod, the corresponding plant species is noted present in the survey. This species cover is then calculated as the ratio between the number of positions where the species is present to the total number of positions in the survey multiplied by 100 (see Table 10.10). The size of the measurement interval needs

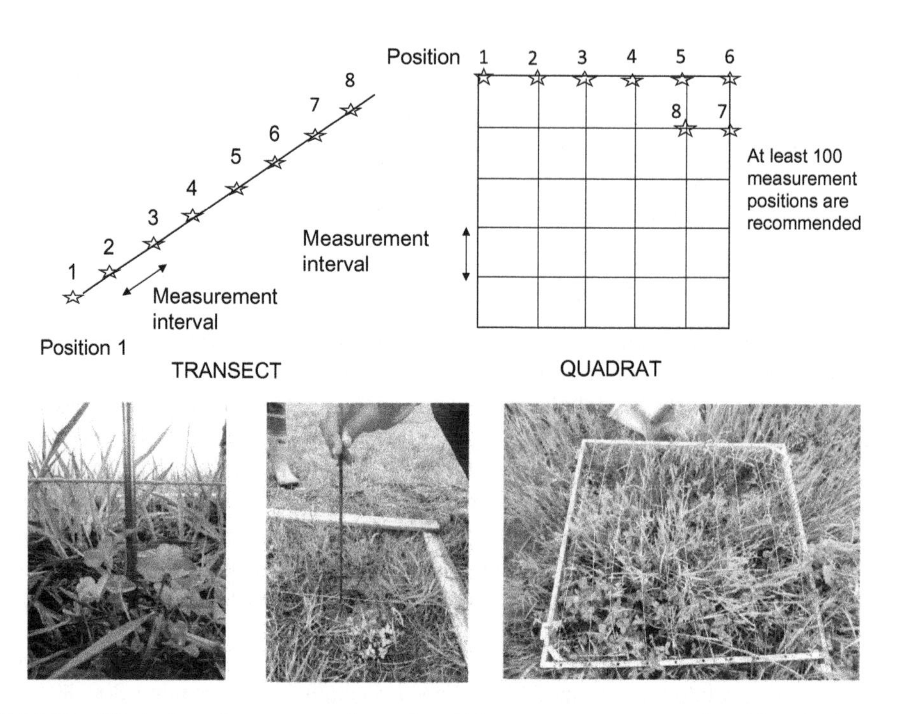

FIGURE 10.10 Illustration of a transect and a quadrat with measurement at several positions in the vegetation to quantify plant cover.

TABLE 10.10

Example of Point Intercept Survey on a Transect of 100 cm With a Measurement Interval of 10 cm. When a Species is Present at the Measurement Position, it is Noted With a 1, When it is Absent, it is Noted with a 0. The Percentage of Cover is Then Calculated for Each Species and for the Whole Vegetation

Species/Measurement position	10	20	30	40	50	60	70	80	90	100	% Cover
1- *Phleum pratense*	0	0	0	1	1	0	1	0	0	0	30
2- *Trifolium repens*	1	0	0	1	0	0	0	0	0	0	20
3- *Melilotus alba*	0	0	0	0	0	0	0	0	0	0	0
4- *Salix sp.*	0	0	1	0	0	0	0	0	0	0	10
5- *Carex sp.*	0	0	0	0	0	0	0	0	0	0	0
Total cover (whole vegetation cover)[a]	1	0	1	1	1	0	1	0	0	0	50

[a] If at least one species is present at a measurement position (i.e., in a column), the value of 1 is given. If no species is present, the value 0 is given. Then, the total cover = number of 1 in the total cover line total number of measurement positions (i.e., number of columns) × 100.

to be adjusted to the size and colonization intensity of the vegetation. For example, for herbaceous vegetation, a measurement interval of 10 cm is generally appropriate. Plant biomass (g.m^2) is the dry mass of aboveground parts collected on a given area of soil. For small size vegetation, direct measurement through cutting of aboveground parts of vegetation on a given soil surface is possible. Then the vegetation is dried until constant weight. The plant biomass can also be estimated nondestructively with the point intercept method. In this case, the number of contacts of each plant species is noted at each position of the rod. The linear relationship between the number of contacts and the biomass of each plant must be calibrated, however, in quadrats where point intercept surveys are coupled with direct biomass measurements.

10.6.2.2 Large Plants

For well individualized plants or large plants (like trees), density (i.e., number of individuals by surface) measurements evaluate the intensity of colonization, whereas growth measurements (like tree maximal height and basal diameter) for a given age accounts for productivity. Plant density is measured by counting the number of individuals in a plot. The size of the plot increases with the size of the considered plant. For example, for small trees or shrubs, an area of 6 m^2 is appropriate. In plantations, the survival rate is measured as the number of living individuals to the total number of planted individuals multiplied by 100. Tree growth is measured by the height of the tree (m) between the soil and the alive bud further from the soil, and also by the diameter of the main shoot at the collar (zone between the beginning of the shoots and the roots) or at breast height (1.3 m from the soil). Survival, height, and diameter measurements are usually performed at the end of each growing season after planting, on all planted trees if possible, or on a representative sample. Shrubs usually develop several shoots from the base rather than one dominant main shoot; thus, measurements of maximal height and diameter are not always easy. If height and diameter measurements are performed on naturally colonizing individuals, it is important to get their age to be able to compare their growth to appropriate references.

10.6.3 Leaf Area Index

The leaf area index (LAI) is the ratio of the total leaf surface (one sided) to the soil surface where the canopy is perpendicularly projected. It is a dimensionless value, generally lower than 19; higher values are usually reached for conifers. LAI is used to calculate the potential transpiration component of the water budget (see Fredlund et al. 2012 and Chapter 3). LAI changes over time because it depends on plant physiology, development stage, sanitary state, and growing season period. For small vegetation, LAI can be directly measured by collecting the leaves of plants growing on a given soil surface and measuring the total leaf area. The total leaf area is measured using a leaf area meter or an image analysis software on the scanned leaves. If the relationship between leaf biomass and leaf area is known, LAI can also be obtained from leaf biomass measurements on a given soil surface (see Section 10.6.2). For vegetation with well-developed trees, indirect measurement methods of LAI are preferred. Some of these methods are based on light transmittance through canopy. The incident flux of light above the canopy and the light flux reaching the soil under the canopy are measured with optical sensors. Then the LAI is obtained by using a modified Beer–Lambert extinction law (Bréda 2003). At the regional scale, remote sensing can be used to obtain LAI. The methods to measure or estimate LAI are reviewed in Bréda (2003).

10.6.4 Rooting Depth and RLD

Maximal rooting depth (m) is used in numerical models to include the depth up to which root pumping and water losses by transpiration will occur. The easiest method to measure maximal rooting depth from the soil surface is direct measurement on vertical trenches dug under the vegetation (Figure 10.11). The depth of the trench can reach several meters under mature trees. The RLD represents the colonization intensity of roots in terms of length. It is calculated as the total length of roots in a given soil volume ($m.m^{-3}$). The distribution of RLD along soil depth is used by numerical models to distribute potential transpiration in the soil (see Chapter 3, Section 3.6). An appropriate method to measure RLD is the root auger method that implies soil coring followed by root extraction from the soil matrix, root digitalizing, and then root image analysis in 2D using a software (see Figure 10.12). Dedicated software provides cumulative root length from an image

FIGURE 10.11 Photographs of a vertical trench used to measure maximal rooting depth of the vegetation.

FIGURE 10.12 Illustration of the procedure for the root length density measurement, from sampling to digitalization of roots.

of roots. Soil cores with a diameter of 7–10 cm and 10–20 cm long are typically used. Vegetation heterogeneity and evolution (age, species, density) should be taken into account to repeat and position trenches and cores. In particular, the spatial variability of RLD around a plant is high. It is thus recommended to sample several cores at an increasing distance of the plant. More information on the available methods to measure root parameters can be found in the specialized book from Smit et al. (2000).

10.6.5 Complementary Remarks

The success of revegetation could also be evaluated through parameters related to ecosystem functions and ecosystem services measurements. The quantity of water potentially transpired by vegetation, calculated thanks to LAI, is one of these parameters. Other parameters could be of interest, such as the quantity of sequestered carbon, or the quantity of wood production.

Vegetation is heterogeneous and only a portion of the vegetation will likely be measured on revegetated mine sites (Albright et al. 2010). Thus, a sampling design based on statistical methods is usually required to ensure that the collected data are representative of the whole plant community. For this purpose, randomized, stratified, or systematic sampling with repetitions is recommended. Then the analysis of geometric mean or median values, percentile values, and confidence intervals can be used to evaluate the representativeness of the data. The comparison of mean values of parameters measured with ranges of targeted values can be used to evaluate the success of the establishment and progress of the revegetated plant community. Target values of measured parameters generally come from reference zones, where ecosystems similar to the one desired on the revegetated mine site are present. These target values may be different in the short and the long term since the vegetation of revegetated zones is dynamic and will likely evolve through plant succession (Chapter 12).

10.7 GROUND TEMPERATURE

The monitoring of ground temperatures is usually required for all TSFs and WRSFs as well as most cover configurations constructed in permafrost environments. Such monitoring is conducted to assess the thermal behavior and performance of these structures throughout most of their service life. Among the several covers discussed in this book, the insulation cover (Chapter 9) is the only cover option that uses temperature as a control strategy for limiting sulfide oxidation. The performance of insulation covers essentially relies on maintaining the temperature at the mine wastes–cover interface ($T_{interface}$) below a target temperature (T_{target}; temperature at which sulfide oxidation becomes negligible). Therefore, the monitoring of ground temperatures is critical to evaluate the performance of this type of cover. However, other types of covers, which, in temperate climates, do not typically require the monitoring of ground temperatures may use temperature as an additional design criterion and performance indicator in permafrost environments (e.g., Boulanger-Martel et al. 2016, 2020b; MEND 2004, 2009, 2012).

In addition, monitoring the evolution of the temperatures within TSFs and WRSFs as well as their foundations is important to assess the freeze-back time of the mine wastes and the evolution of the ground thermal regime (i.e., thermal interactions between the cover, mine wastes, and natural ground). In this context, the main temperatures that should be monitored include ground surface, internal (cover and mine wastes), and natural ground temperatures. The temperature of the ground surface is mainly used to characterize the amount of heat that can be exchanged through a cover system, whereas internal and natural ground temperatures are mainly measured to assess the thermal behavior of the cover – mine waste – natural ground system. It should be noted that a detailed characterization of the natural ground temperatures is also done as a part of the baseline studies (using similar monitoring methods). However, in this case, temperatures are measured at great depth, often reaching over several hundred meters. Even though such thermal monitoring is often continued during the operations, it is outside of the scope of this book to discuss that monitoring in detail, and the following sections will focus on the instruments and methods used to measure the temperature in the cover materials, mine wastes as well as the top of the foundation materials.

10.7.1 Temperature Sensors

Many types of temperature sensors are available for the measurement of ground temperatures (e.g., Andersland and Ladanyi 2004). However, in the context of the reclamation of TSFs and WRSFs, ground temperatures are generally measured using thermistors (e.g., Meldrum et al. 2001; Smith et al. 2013; Poirier 2019; Boulanger-Martel et al. 2020a). For a given monitoring station (or location), ground temperatures are generally measured at several depths below the ground surface. Depending on the intended usage and depth range, temperature measurements can be achieved using whether multi-point thermistor strings (thermistor string; Meldrum et al. 2001; Smith et al. 2013; Poirier 2019) or multiple single-point thermistors (Coulombe 2012; Boulanger-Martel et al. 2020a, 2020b).

Thermistor strings are the most widely used and consist of a chain of individual temperature sensors (thermistor) that are encapsulated along a multiconductor cable jacketed with PVC or polyurethane. Thermistor strings are manufactured to the custom needs of the user: the user can specify the overall cable length, number of thermistors, their accuracy, and location along the string. Most thermistors have an accuracy between ±0.1 and ±0.5 °C, but an accuracy of ±0.2 °C or better is desirable for most engineering purposes (Andersland and Ladanyi 2004). The versatility of multi-point thermistor strings mean that they can be installed in shallow structures such as cover systems (i.e., cover only) and field experimental cells (e.g., Smith et al. 2013; Pham 2013) or thicker (deeper) structures such as in TSFs and WRSFs (e.g., Meldrum et al. 2001; Poirier 2019). For such applications, thermistor strings are typically installed in boreholes (Figure 10.13a), which can run from the ground surface to several meters (10–50 m) into the foundation materials. Thermistor installation is done by inserting the thermistor string directly in the borehole and by back filling the borehole. For most granular materials, best practice suggests the borehole to be fully grouted with a cement–bentonite grout. However, in some case the borehole can be filled by other available materials such as sand, gravel, tailings, or fresh water (depending in which materials the thermistor sting is being installed in). Some installation may not require a casing, whereas a PVC or steel guide-pipe could be required for another installation to help reach the instruments' intended depths. Depending on the amount of thermal disturbance caused by the drilling operations, it may take several days or weeks before thermal equilibrium is reached within the boreholes (Andersland and Ladanyi 2004).

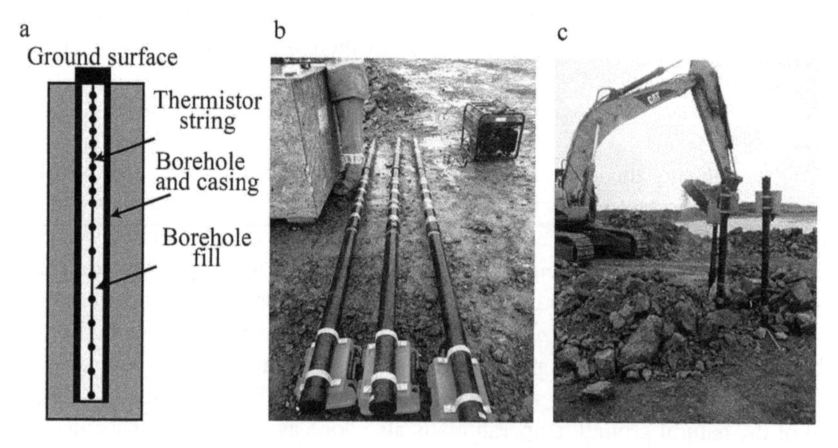

FIGURE 10.13 (a) Multi-point thermistor string, and (b, c) multiple single-point thermistors setup (Boulanger-Martel 2019) for measurement of ground temperatures in mine wastes and cover.

Thermistor strings can also be installed during construction (e.g., Smith et al. 2013; Pham 2013). Smith et al. (2013) installed thermistors strings along the construction of covered and uncovered field experimental cells for waste rock piles. In this case, they housed the thermistor strings in flexible PVC pipes for protection. PVC pipes were insulated between each sensor in order to prevent airflow within the setup. Thermistor strings are usually connected to a data logger but can also be read manually.

Multiple single-point thermistors can also be used to measure ground temperatures (Figure 10.13a). However, in this case, the intended usage is limited in terms of depth and is restricted to shallow applications (i.e., less than 4–5 m) such as insulation covers (Coulombe 2012; Boulanger-Martel et al. 2020a, 2020b). This monitoring setup essentially consists of installing several single-point thermistors at selected elevations and relaying them to a data logger at the surface. Prior to being installed in the field, sensors can be attached to a PVC pipe, which is also used to protect the cables and route the cable to the surface (Figure 10.13b). This type of instrumentation setup is usually installed in the top portion of the mine wastes (<1 m) and in cover materials during cover construction (Figure 10.13c). Multiple single-point thermistors can also be installed by excavating an existing cover (Coulombe 2012).

10.7.2 Measurement Depth and Density

Whether thermistor strings or multiple single-point thermistors are used to monitor ground temperatures, there are two main aspects to consider when planning the installation of thermistors: the overall measurement depth and the spacing between each temperature measurements. For most applications, the overall depth of the installation should allow for measuring the temperatures within the cover materials, the mine waste, and ideally the top portion of the foundation (Figure 10.14). Thermistor strings installed in waste storage facilities should reach depths ≥10 m within the foundation materials. Temperature measurements in the foundation can provide useful information on the thermal behavior of the storage structure and provide reliable boundary condition for heat transfer analyses (Pham 2013; Boulanger-Martel 2019). However, for some applications such as field experimental cells, the ground temperature measurement can be focused on the cover materials and the surface portion of the mine wastes (e.g. Boulanger-Martel et al. 2020a, 2020b).

The density of temperature measurements should be high in the cover system and in the first meter(s) of mine wastes (small sensor spacing; Figure 10.14). In insulation cover systems, a high density of temperature measurements is especially required near the anticipated position of the 0 °C isotherm and close to the mine waste–cover interface. As shown in Figure 10.14, lower density of temperature measurements (greater sensor spacing) can be used with increasing depths in the mine wastes and the foundation (see Table 10.11). Ground temperature monitoring stations are usually installed vertically (Table 10.11). However, in some cases, it may be necessary to monitor the ground temperatures along an inclined plane. For example, an inclined installation can be useful to measure the ground temperatures near the surface and assess side effects along the slope of a WRSF (Poirier 2019). It could also be used to monitor the thermal conditions along a specific cover layer (Pham 2013; Smith et al. 2013).

Ground temperatures near the surface continuously vary with climatic conditions, meaning that temperature measurements should be made frequently at shallow depths (Andersland and Ladanyi 2004). Hence, ground temperatures in shallow structures such as insulation covers are typically recorded every 4–12 hours. However, ground temperatures within thicker structures such as TSF and WRSF are logged daily but are sometimes recorded manually on a weekly or monthly basis. The amount and spatial distribution of temperature monitoring stations across a mine site will depend on several site-specific factors, but generally the monitoring strategy must assess the overall thermal behavior of the structure as well as that of critical zones. More details with respect to the monitoring strategy are discussed in Section 10.8.

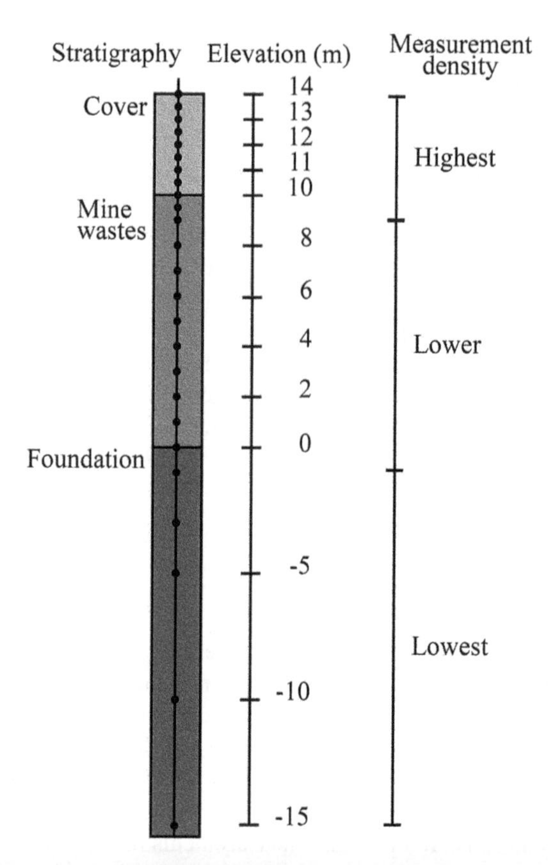

FIGURE 10.14 Density of temperature measurements (sensor spacing) required to assess the thermal behavior and performance of a cover – mine waste – natural ground system.

TABLE 10.11
Typical Type of Installation and Range of Temperature Sensor Spacing

	Typical type of installation	Typical range of sensor spacing (m)
Cover materials and near the ground surface	Vertical and inclined	≤0.5–1
Tailings Storage Facility	Vertical	1–5
Waste Rock Storage Facility	Vertical and inclined	1–5
Foundation	Vertical	1–5

10.8 MONITORING STRATEGY

10.8.1 Selection Criteria and Instruments Maintenance

As mentioned previously, monitoring the long-term performance of AMD control methods is key to ensuring that the cover functions as intended and that the performance is maintained for the nearly infinite lifetime of the closed mine site. Monitoring will also help determine corrective measures if necessary and calibrate and validate numerical simulations. This chapter has briefly presented different techniques to monitor the key parameters associated with the performance of the main AMD reclamation methods. The choice of the most appropriate equipment for each parameter should be based on selection factors such as the range and accuracy of measurements, the need for continuous measurements, environmental risk, soils properties, site configuration and exposure conditions, geographical location and susceptibility to climate change, durability, accessibility, and available budget. Monitoring equipment should also be regularly tested, calibrated (at least once before installation), and replaced, depending on its lifetime. AMD can often reduce equipment lifetime. It is recommended to measure properties manually with destructive methods (from trenches or boreholes samples) or using another method at least once a year to ensure that automatic systems are functioning well.

10.8.2 Instrument Location and Density

The critical aspects of AMD reclamation methods instrumentation are the location and number of monitoring stations. There are no rules to decide the number of instruments per area of the storage facilities or stored volumes of waste. Monitoring design requires experience, common sense, and a strong comprehension of the site and its particularities. The first step in determining these two parameters is to divide the reclaimed sites in zones where the system (e.g., cover system, tailings, boundary conditions) can be considered uniform. For example, the inclined zones should be considered a distinct zone because of their different hydraulic behavior (e.g., Bussière et al. 2003, 2006, and Chapters 5 and 7). For CCBE and EWT with monolayer cover, zones where the water table is deep should be monitored differently and more intensively than areas where it is close to the surface because of the risk of desaturation. The surface area but also the depth (profile monitoring) should be adequately covered. As different materials can be used to reclaim a mine site, these must be monitored in a distinct and appropriate manner. The choice of equipment also often depends on the material properties (e.g., particle size distribution, expected degree of saturation). Ideally, all materials used in the covers should be monitored, with each specific zone adequately instrumented (sometimes several sensors can be placed vertically in the same layer – for example, to monitor suction or water content profiles) and analyzed separately. The replication of sensors at some locations (with similar or different equipment) is also important (Martin et al. 2017), particularly for locations where measurements are crucial to conclude on the performance of a given reclamation methods (e.g., suction and volumetric water content in the MRL of a CCBE). There is no clear rule concerning the number of monitoring stations to install on a reclaimed site. Different parameters, such as the number of zones, budget, regulations, and information needed to validate the design assumptions and objectives, affect the required number and location of the stations on a given site. Examples of monitoring programs for existing reclaimed sites can be found in the following literature: Ricard et al. (1997), O'Kane et al. (1998), Adu-Wusu et al. (2001); Dagenais et al. (2005); Bussière et al. (2006); Maurice, (2012); Zhan et al. (2014); Ethier et al. (2018); Poirier (2019).

10.8.3 Measurements Frequency and Data Treatment

Measurement frequencies for all parameters used on these and other sites are also variable. Typically, the θ, ψ, and T measurement frequencies vary during the cover life, being mere frequent during the first 3–5 years. As new systems available automatically recover data, a daily frequency can be applied to most sites. The frequency of water quality sampling and analysis of surface and underground waters are usually driven by regulatory requirements and can vary from one country to the other, but typically occurs two to four times a year at the post-closure stage. The measurement frequencies for meteorological parameters are typically at least once a day, but measurement every 6 hours strongly improves the quality of the data, and OCT are performed at least once per year. Vegetation on reclaimed mine sites should also be monitored with a frequency of once per 3–5 years. It is also important to integrate into the monitoring strategy more intensive periods to better understand the short-term behavior (especially system response to extreme climate events) and to evaluate the variability of typical measurements.

10.8.4 Data Analysis and Reporting

With the new capacity of data acquisition systems, significant volumes of data are generated annually. It is important to rapidly treat the data to avoid an accumulation of untreated data in files and data storage systems. A clear monitoring plan must integrate a quality-control stage, followed by a preliminary and a detailed analysis. Preliminary reports must be produced monthly, while a complete report must be written every year. This report should integrate a comparison with the previous years to follow the evolution of the performance closely.

10.9 FINAL REMARKS

The aim of this chapter was to present well-known and widely used monitoring equipment and techniques that have been tested on existing mine sites. However, some recent and emerging techniques are promising to monitor reclamation methods. For example, innovative equipment proposes to measure the actual evapotranspiration (AET) components directly instead of predicting the AET using numerical codes. These techniques include latent and sensible heat flux measurements using micrometeorological methods such as eddy covariance (EC). Although some researchers have used the EC technique to study water balance components and surface–atmosphere interactions (Aubinet et al. 1999; Carey et al. 2005; Scott 2010), it has rarely been utilized for assessing the performance of mine site reclamation strategies (Carey 2008). Satellite-based techniques are also used more and more to remotely monitor mine sites, and could be used to detect contamination of water (Riaza et al. 2011), vegetation evolution (Raval et al. 2013), or infrastructures displacement (Hu et al. 2017).

The monitoring approaches presented in this chapter should also be coupled with geotechnical and geomechanical monitoring tools to assess the physical stability of mining infrastructures in the short and long term. However, physical stability of mining infrastructures (dikes, pit walls, etc.) is outside the scope of this book. The interested reader can refer to the following references regarding the monitoring of tailings dams (International Commission on Large Dams (ICOLD) 1996; Aubertin et al. 2002; Blight 2010).

Finally, monitoring of reclamation methods must continue for the entire lifetime (often centuries) of these structures. However, the lifetime of equipment is typically 5–15 years, so frequent replacement must be planned. Hence, funds should be prepared and guarantied for the future replacement of monitoring equipment. Local communities should also be involved in the monitoring process to maximize the benefits of mining operations for these communities at the post-closure stage. After appropriate training, they could participate in sampling and results interpretation. Ideally, long-term monitoring results should be shared with surrounding communities to keep them informed on the overall reclamation performance and to reassure them regarding the protection of their environment.

REFERENCES

Aachib, M. 1997. *Étude en laboratoire de la performance de barrières de recouvrement constituées de rejets miniers pour limiter le DMA. Ph.D Thesis*, Department of Civil and Geological and Mining Engineering, École Polytechnique de Montréal, Montréal.

Aboukhaled, A., J.F. Alfaro, and M. Smith. 1986. *Les lysimètres*. Bulletin FAO d'irrigation et de drainage, 39, M-56 ISBN 92-5-201186-2, Food and Agriculture Organization of the United Nations, FAO, Rome.

Adu-Wusu, C., Yanful, E.K. 2006. Performance of engineered test covers on acid-generating waste rock at Whistle mine, Ontario. *Canadian Geotechnical Journal*. 43: 1–18.

Adu-Wusu, C., Yanful, E.K., Mian, M.H. 2001. Field evidence of resuspension in a mine tailings pond. *Canadian Geotechnical Journal*, 38: 796–808.

Alakangas, L. Lundberg, A., Öhlander, B. 2008. Pilotscale Studies of Different Covers on Unoxidised Sulphide-rich tailings, Northern Sweden: Oxygen Diffusion. In Mine Water and the Environment: Proceedings of the 10th IMWA Congress 2008, Karlsbad, Czech Republic, 2–5 June, 2008.

Albright, W. H., C. H. Benson, and W. J. Waugh. 2010. *Water balance covers for waste containment – Principles and practice*. Reston, VA: American Society of Civil Engineers (ASCE) Press.

Altenburger, R., W. Brack, R.M. Burgess, et al. 2019. Future water quality monitoring: Improving the balance between exposure and toxicity assessments of real-world pollutant mixtures. *Environmental Sciences Europe* 21: 12. https://doi.org/10.118i6/s12302-019-0193-1.

Andersland, O. B., and B. Ladanyi. 2004. *Frozen Ground Engineering*. 2nd ed. Chichester: John Wiley & Sons.

Anterrieu, O., M. Chouteau, and M. Aubertin. 2010. Geophysical characterization of the large-scale internal structure of a waste rock pile from a hard rock mine. *Bulletin of Engineering Geology and the Environment*, 69(4), 533–548.

APHA. 2017a. *2580 Oxidation-reduction potential (ORP). Standard Methods for the Examination of Water and Wastewater*, 23rd ed., Washington DC, USA. American Public Health Association. DOI: 10.2105/SMWW.2882.034

APHA. 2017b. *2510 Conductivity. Standard Methods for the Examination of Water and Wastewater*, 23rd ed., Washington DC, USA. American Public Health Association. DOI: 10.2105/SMWW.2882.027

APHA. 2017c. *2340 Hardness. Standard Methods for the Examination of Water and Wastewater*, 23rd ed., Washington DC, USA. American Public Health Association. DOI: 10.2105/SMWW.2882.025

APHA. 2017d. *2540 Solids. Standard Methods for the Examination of Water and Wastewater*, 23rd ed., Washington DC, USA. American Public Health Association. DOI: 10.2105/SMWW.2882.03

APHA. 2017e. *2130 Turbidity. Standard Methods for the Examination of Water and Wastewater*, 23rd ed., Washington DC, USA. American Public Health Association. DOI: 10.2105/SMWW.2882.018

APHA. 2017f. *4500-O Oxygen. Standard Methods for the Examination of Water and Wastewater*, 23rd ed., Washington DC, USA. American Public Health Association. DOI: 10.2105/SMWW.2882.091

APHA. 2017g. *2550 Temperature. Standard Methods for the Examination of Water and Wastewater*, 23rd ed., Washington DC, USA. American Public Health Association. DOI: 10.2105/SMWW.2882.031

APHA. 2017h. *2310 Acidity. Standard Methods for the Examination of Water and Wastewater*, 23rd ed., Washington DC, USA. American Public Health Association. DOI: 10.2105/SMWW.2882.022

APHA. 2017i. *2320 Alkalinity. Standard Methods for the Examination of Water and Wastewater*, 23rd ed., Washington DC, USA. American Public Health Association. DOI: 10.2105/SMWW.2882.023

APHA. 2017j. *3120 Metal by plasma emission spectroscopy. Standard Methods for the Examination of Water and Wastewater*, 23rd ed. ed., Washington DC, USA. American Public Health Association. DOI: 10.2105/SMWW.2882.047

APHA. 2017k. *3120 Metal by inductively coupled plasma – mass spectrometry. Standard Methods for the Examination of Water and Wastewater*, 23rd ed., Washington DC, USA. American Public Health Association. DOI: 10.2105/SMWW.2882.048

APHA. 2017l. *4110 Determination of anions by ion chromatography. Standard Methods for the Examination of Water and Wastewater*, 23rd ed., Washington DC, USA. American Public Health Association. DOI: 10.2105/SMWW.2882.070

APHA. 2017m. *4500-NH3 Nitrogen (ammonia). Standard Methods for the Examination of Water and Wastewater*, 23rd ed., Washington DC, USA. American Public Health Association. DOI: 10.2105/SMWW.2882.087

APHA. 2017n. *4500-S2- Sulfide. Standard Methods for the Examination of Water and Wastewater*, 23rd ed., Washington DC, USA. American Public Health Association. DOI: 10.2105/SMWW.2882.096

APHA. 2017o. *5310 Total Organic Carbon (TOC). Standard Methods for the Examination of Water and Wastewater*, 23rd ed., Washington DC, USA. American Public Health Association. DOI: 10.2105/SMWW.2882.104

APHA. 2017p. *3500-Fe Iron. Standard Methods for the Examination of Water and Wastewater*, 23rd ed., Washington DC, USA. American Public Health Association. DOI: 10.2105/SMWW.2882.055

APHA. 2017q. *3500-Mn Manganese. Standard Methods for the Examination of Water and Wastewater*, 23rd ed., Washington DC, USA. American Public Health Association. DOI: 10.2105/SMWW.2882.059

APHA. 2017r. *3500-As Arsenic. Standard Methods for the Examination of Water and Wastewater*, 23rd ed., Washington DC, USA. American Public Health Association. DOI: 10.2105/SMWW.2882.051

APHA. 2017s. *4500-N Nitrogen. Standard Methods for the Examination of Water and Wastewater*, 23rd ed., Washington DC, USA. American Public Health Association. DOI: 10.2105/SMWW.2882.086

APHA (American Public Health Association). 2017. *4500-H+ pH value. Standard Methods for the Examination of Water and Wastewater*, 23rd ed., Washington DC, USA. American Public Health Association. DOI: 10.2105/SMWW.2882.082

ASTM. 2013. Standard Test Method for Open Channel Flow Measurement of Water with the Parshall Flume. *Designation: D-1941-91*, West Conshohocken, PA: ASTM International.

ASTM. 2014a. Standard Test Methods for Electrical Conductivity and Resistivity of Water. *Designation: D1125–14*, West Conshohocken, PA: ASTM International.

ASTM. 2014b. Standard Test Method for Oxidation-Reduction Potential of Water. *Designation: D1498–14*, ASTM International

ASTM. 2014c. Standard Guide for the Selection of Purging and Sampling Devices for Groundwater Monitoring Wells. *Designation: D6634/D6634M*, West Conshohocken, PA: ASTM International

ASTM. 2014d. Standard Test Method for On-Line Measurement of Turbidity Below 5 NTU in Water. *Designation: D6698–14*, West Conshohocken, PA: ASTM International

ASTM. 2014e. Standard Specification for ASTM Liquid-in-Glass Thermometers. *Designation: E1–14*, West Conshohocken, PA: ASTM International

ASTM. 2016a. Standard Test Methods for Acidity or Alkalinity of Water. *Designation: D1067–16*, West Conshohocken, PA: ASTM International

ASTM. 2016b. Standard Practice for Sampling Liquids Using Bailers. *Designation: D6699–16*, West Conshohocken, PA: ASTM International

ASTM. 2016c. Standard Practice for Sampling Liquids Using Grab and Discrete Depth Samplers. *Designation: D6759–16*, West Conshohocken, PA: ASTM International

ASTM. 2017a. Standard Test Method for Hardness in Water. *Designation: D1126–17*, West Conshohocken, PA: ASTM International

ASTM. 2017b. Standard Guide for Planning and Preparing for a Groundwater Sampling Event. *Designation: D5903-96e1* (Reapproved 2017). West Conshohocken, PA: ASTM International

ASTM. 2017c. Standard Guide for Field Filtration of Groundwater Samples. *Designation: D6564/D6564M-17*, West Conshohocken, PA: ASTM International

ASTM. 2017d. Standard Practice for Passive Soil Gas Sampling in the Vadose Zone for Source Identification, Spatial Variability Assessment, Monitoring, and Vapor Intrusion Evaluations. *Designation: D7758–17*, West Conshohocken, PA: ASTM International

ASTM. 2018a. Standard Test Methods for Dissolved Oxygen in Water. *Designation: D888–18*, West Conshohocken, PA: ASTM International

ASTM. 2018b. Standard Test Methods for pH of Water. *Designation: D1293–18*, West Conshohocken, PA: ASTM International

ASTM. 2018c. Standard Guide for Quality Planning and Field Implementation of a Water Quality Measurement Program. *Designation: D5612–94* (Reapproved 2018). West Conshohocken, PA: ASTM International

ASTM. 2018d. Standard Test Methods for Filterable Matter (Total Dissolved Solids) and Nonfilterable Matter (Total Suspended Solids) in Water. *Designation: D5907–18*, West Conshohocken, PA: ASTM International

ASTM. 2018e. Standard Guide for Purging Methods for Wells Used for Ground Water Quality Investigations. *Designation: D6452–18*, West Conshohocken, PA: ASTM International

ASTM. 2018f. Standard Guide for Field Preservation of Ground Water Samples. *Designation: D6517–18*, West Conshohocken, PA: ASTM International

ASTM. 2018g. Standard Practice for Low-Flow Purging and Sampling for Wells and Devices Used for Ground-Water Quality Investigations. *Designation: D6771–18*, West Conshohocken, PA: ASTM International

ASTM. 2018h. Standard Practice for Active Soil Gas Sampling in the Vadose Zone for Vapor Intrusion Evaluation. *Designation: D7663–12* (Reapproved 2018). West Conshohocken, PA: ASTM International

ASTM. 2019a. Standard Guide for Sampling Ground-Water Monitoring Wells. *Designation: D4448–01* (Reapproved 2019). West Conshohocken, PA: ASTM International

ASTM. 2019b. Standard Practice for Sampling With a Dipper or Pond Sampler. *Designation: D5358–93* (Reapproved 2019). West Conshohocken, PA: ASTM International

ASTM. 2019c. Standard Guide for Documenting a Groundwater Sampling Event. *Designation: D6089–19*, West Conshohocken, PA: ASTM International

ASTM. 2019d. Standard Guide for Sampling Wastewater With Automatic Samplers. *Designation: D6538–12* (Reapproved 2019). West Conshohocken, PA: ASTM International

ASTM. 2020. Standard Practice for Decontamination of Field Equipment Used at Waste Sites. *Designation: D5088–20* West Conshohocken, PA: ASTM International.

Aubertin, M., Bussière, B., Bernier, L. 2002. *Environnement et gestion des rejets miniers.* CD-ROM, Les Presses Internationales Polytechnique, École Polytechnique de Montréal, Québec, Canada.

Aubertin, M., B. Bussière, J.M. Barbera, R.P. Chapuis, M. Monzon, and M. Aachib, 1997. *Construction and instrumentation of in situ plots to evaluate covers built with clean tailings. Proceeding of the Fourth International Conference on Acid Rock Drainage (ICARD),* Vancouver, Vol. II: 715–729.

Aubinet, M., Grelle, A., Ibrom, A. et al. 1999. Estimates of the annual net carbon and water exchange of forests: The EUROFLUX Methodology, *Advances in Ecological Research,* 30 113–175.

Awoh, A.S., Mbonimpa, M., Bussière, B. 2013. Field study of the chemical and physical stability of highly sulphide-rich tailings stored under a shallow water cover. *Mine Water and the Environment,* 32(1): 42–55.

Bello, Z. A. and L. D. Van Rensburg. 2017. Development, calibration and testing of low-cost small lysimeter for monitoring evaporation and transpiration. *Irrigation and Drainage,* 66: 263–272.

Benson, C., T. Abichou, W. Albright, G. Gee, and A. Roesler. 2001. Field evaluation of alternative earthen final covers. *International Journal of Phytoremediation,* 3(1), 105–127.

Bews, B., B. Wickland, and S. Barbour. 1999. *Lysimeter design in theory and practice. Proceeding of Tailing and Mine Waste* '99, Fort Collins, Colorado, USA, 24–27 January 1999, Balkema, Rotterdam, 13–21.

Binley, A., S.H. Hubbard, J. A. Huisman, A. Revil, D.A. Robinson, K. Singha, and L. D. Slater. 2015. The emergence of hydrogeophysics for improved understanding of subsurface processes over multiple scales. *Water Resources Research,* 51(6), 3837–3866.

Blight, G.E. 2010. *Geotechnical Engineering for Mine Waste Storage Facilities.* CRC Press, Boca Raton, FL.

Bossé, B., Bussière, B., Hakkou, R., Maqsoud, A., Benzaazoua, M. 2015. Field experimental cells to assess the hydrogeological behaviour of store-and-release covers made with phosphate mine waste. *Canadian Geotechnical Journal,* 52(9): 1255–1269.

Boulanger-Martel, V. 2019. *Évaluation de la performance de recouvrements miniers pour contrôler le drainage minier acide en climat arctique.* PhD diss, Polytechnique, Montréal.

Boulanger-Martel, V., B. Bussière, J. Côté, and M. Mbonimpa. 2016. Influence of freeze–thaw cycles on the performance of covers with capillary barrier effects made of crushed rock–bentonite mixtures to control oxygen migration. *Canadian Geotechnical Journal* 53, no. 5: 753–764.

Boulanger-Martel, V., B. Bussière, and J. Côté. 2020a. Insulation covers with capillary barrier effects to control sulfide oxidation in the Arctic. *Canadian Geotechnical Journal,* doi.org/10.1139/cgj-2019-0684.

Boulanger-Martel, V., B. Bruno, and J. Côté. 2020b. Thermal behaviour and performance of two field experimental insulation covers to control sulfide oxidation at Meadowbank mine, Nunavut. *Canadian Geotechnical Journal,* doi.org/10.1139/cgj-2019-0616.

Bouzaglou, V., Crestani, E., Salandin, P., Gloaguen, E. and Camporese, M. 2018. Ensemble Kalman Filter Assimilation of ERT Data for Numerical Modeling of Seawater Intrusion in a Laboratory Experiment. *Watermark,* 10(4), 397.

Bréda, N., 2003. Ground-based measurements of leaf area index: a review of methods, instruments and current controversies. *Journal of Experimental Botany* 392, 2403–2417.

Bulut, R., and E.C. Leong. 2008. Indirect Measurement of Suction. *Geotechnical and Geological Engineering,* 26(6):633–644.

Bureau of Reclamation. 2008. History Essays from the Centennial symposium. *Reclamation managing water level in the west,* Dept. of Interior Bureau of Reclamation Volume 1, 962p.

Bureau of Reclamation. 2014. Embankment Dams Chapter 11: Instrumentation and Monitoring. *Reclamation managing water level in the west,* Dept. of Interior Bureau of Reclamation, DS-13(11)-9:1 Phase 4 Final, 113p.

Bussière, B., Dagenais, A.-M., Mbonimpa, M., Aubertin, M. 2002. *Modification of oxygen-consumption testing for the evaluation of oxygen barrier performance. Proc. 55th Can. Geotech. Conf. –3rd Joint IAH – CNC– Can. Geotech. Society Conf.: Ground and Water: Theory to Practice,* Niagara Falls, pp. 139–146. CD-ROM.

Bussière, B., Aubertin, M., Chapuis, R.P. 2003. The behaviour of inclined covers used as oxygen barriers. *Canadian Geotechnical Journal,* 40: 512–535.

Bussière, B., A. Maqsoud, M. Aubertin, J. Martschuk, L. McMullen, and M. Julien. 2006. Performance of the oxygen limiting cover at the LTA site, Malartic, Quebec. *CIM Bulletin,* 1(6), 1–11.

Bussière, B., Potvin, R., Dagenais, A-M., Aubertin, M., Maqsoud, A., Cyr, J. 2009 Restauration du site minier Lorraine, Latulipe, Québec: résultats de 10 ans de suivi. *Déchets, Sciences et Techniques* 54: 49–64.

Cacciuttolo, C., & Tabra, K. 2015. *Water Management in the Closure of Tailings Storage Facilities.* Santiago: 10a ICARD IMWA

Carey, S.K. 2008. Growing season energy and water exchange from an oil sands overburden reclamation soil cover, Fort McMurray, Alberta, Canada. *Hydrological Processes* 22, 2847–2857.

Carey, S.K., Barbour, S.L., Hendry, M.J. 2005. Evaporation from a waste-rock surface, Key Lake, Saskatchewan. *Canadian Geotechnical Journal* 42, 1189–1199.

Cassan, M. 2005. *Essais de perméabilité sur site dans la reconnaissance des sols*. Presses de l'École nationale des ponts et chaussées. ISBN 2-859e78-396-2, Paris, France.

CEAEQ. 2008. Cahier 7 Méthodes de mesure du débit en conduit ouvert. In: *Guide d'échantillonnage à des fins d'analyses environnementales* (248 p). (http://www.ceaeq.gouv.qc.ca/analyses/index_en.htm)

Chambers, J. E., D. A. Gunn, P. B. Wilkinson, et al. 2014. 4D electrical resistivity tomography monitoring of soil moisture dynamics in an operational railway embankment. *Near Surface Geophysics*, 12(1), 61–72.

Chiu, T. and C. Shackelford. 2000. Laboratory evaluation of sand underdrains, *Journal of Geotechnical and Geoenvironmental Engineering*, ASCE, 126(11), 990–1002.

Cloutier, V., E. Rosa, S. Nadeau, P.L. Dallaire, D. Blanchette, and M. Roy. 2015. *Projet d'acquisition de connaissances sur les eaux souterraines de l'Abitibi-Témiscamingue (partie 2)*. Groupe de recherche sur l'eau souterraine, Institut de recherche en mines et en environnement, Université du Québec en Abitibi-Témiscamingue. Rapport de recherche no. P002.R3, 347p.

Coons, L., M. Ankeny, and G. Bulik. 2000. *Alternative earthen final covers for industrial and hazardous waste trenches in southwest Idaho, Proceeding of the 3rd Arid Climate Symposium*, Solid Waste Association of North America, Silver Springs, MD, 14-1–14-16.

Cooper, J.D. 2016. *Soil Water Measurement: Practical Handbook*. John Wiley & Sons, Ltd, Chichester, UK.

Coulombe, V. 2012. *Performance de recouvrements isolants partiels pour contrôler l'oxydation de résidus miniers sulfureux*. MScA Thesis, Polytechnique Montréal, Montréal.

Coulombe, V., Bussière, B., Côté, J., Paradis, M. 2013. *Field assessment of sulphide oxidation rates in cold environment: case study of Raglan Mine. 38th Annual Meeting of the Canadian Land Reclamation Conference*, Whitehorse, Yukon September 9–12, 2013.

Dagenais, A. M., Aubertin, M., Bussière, B., and Cyr, J. 2005. *Performance of the Lorraine site cover to limit oxygen migration. Proceeding SME annual meeting*, Salt Lake City, UT, 1–15, Preprint 05–104.

Dagenais, A.M., Mbonimpa, M., Bussière, B. and Aubertin, M. 2012. The modified oxygen consumption test to evaluate gas flux through oxygen barrier covers. *Geotechnical Testing Journal*, ASCE, 35(1): 150–158.

Dasberg S., J.W. Hopmans. 1992. Time domain reflectometry calibration for uniformly and nonuniformly wetted sandy and clayey loam soils. *Soil Science Society of America Journal*, 56:1341–1345.

De Franco, R., Biella, G., Tosi, L., et al. 2009. Monitoring the saltwater intrusion by time lapse electrical resistivity tomography: The Chioggia test site (Venice Lagoon, Italy). *Journal of Applied Geophysics*, 69(3–4), 117–130.

Demers, I., Bussiere, B., Mbonimpa, M. and Benzaazoua, M. 2009. Oxygen diffusion and consumption in low-sulphide tailings covers. *Canadian Geotechnical Journal*, 46(4): 454–469

Dimech, A., M. Chouteau, M., Aubertin, B. Bussière, V. Martin, and B. Plante. 2019. Three-dimensional time-lapse geoelectrical monitoring of water infiltration in an experimental mine waste rock pile. *Vadose Zone Journal*, 18(1) https://doi.org/10.2136/vzj2018.05.0098

Dobchuk, B., Nichol, C., Wilson, W., and Aubertin, M. 2013. Evaluation of a single-layer desulphurized tailings cover. *Canadian Geotechnical Journal*, 50: 777–792.

Elberling, B. 2005. Temperature and oxygen control on pyrite oxidation in frozen mine tailings. *Cold Regions Science and Technology* 41 (2005) 121–133.

Elberling, B., Nicholson, R.V., David, D.J. 1993. Field evaluation of sulphide oxidation rates. *Nordic Hydrology*, 24, 1993, 323–338.

Elberling, B., Nicholson, R.V., Reardon, E.J., Tibble, P. 1994. Evaluation of sulphide oxidation rates: a laboratory study comparing oxygen fluxes and rates of oxidation product release: *Canadian Geotechnical Journal*, 31: 375–383.

Environment Canada. 2009. *Environmental Code of Practice for Metal Mines*. https://www.ec.gc.ca/lcpe-cepa/default.asp?lang=En&n=CBE3CD59-1&offset=2

Environmental Protection Agency (EPA). 2000. *Meteorological monitoring guidance for regulatory modeling applications*. Report EPA-454/R-99-005, *Office of Air and Radiation, Office of Air Quality Planning and Standards Research Triangle Park*, NC 27711.

Environmental Protection Agency (EPA). 2002. Ground-water sampling guidelines for superfund and RCRA project managers. Ground water forum issue paper. Written by D. Yeskis and B. Zavala. *Office of Solid Waste and Emergency Response*, Report no: EPA 542-S-02-001, Washington, DC, 53p.

Ethier, M.-P., Bussière, B., Aubertin, M., Maqsoud, A., Demers, I., and Broda, S. 2018. In situ evaluation of performance of reclamation measures implemented on abandoned reactive tailings disposal site. *Canadian Geotechnical Journal*, 55: 1742–1755.

Evett, S.R., N.T. Mazahrih, M.A. Jitan, M.H. Sawalha, P.D. Colaizzi, and J.E. Ayars. 2009. A weighing lysimeter for crop water use determination in the Jordan valley, Jordan. *Transactions of the ASABE*, 52: 155–169.

Falzone, S., J. Robinson, and L. Slater. 2019. Characterization and monitoring of porous media with electrical imaging: a review. *Transport in Porous Media*, 130(1), 251–276.

Feng. M., and D.G. Fredlund. 2003. Calibration of thermal conductivity sensors with consideration of hysteresis. *Canadian Geotechnical Journal*, 44(2): 113–125.

Ferland, J., Cliche, B., Vigneault, R., Rochefort, J., Lalonde, A., Benanteur, A., & Demard, H. 2011. *Guide de Soutien Technique pour la Clientèle: Règlement sur la Déclaration des Prélèvements D'eau*: MDDEP, Québec, Canada.

Fisher, D.K. 2012. Simple weighing lysimeters for measuring evapotranspiration and developing crop coefficients. *International Journal of Agriculture and Biological Engineering* 5: 35–43.

Fountain A. G., Nylen T.H., Monaghan A., Basagic H. and Bromwich D. 2010. Snow in the McMurdo dry valleys, Antarctica. *International Journal of Climatology* 30, no. 5: 633–642.

Fredlund, D.G., H. Rahardjo, M.D. Fredlund. 2012. *Unsaturated soil mechanics in engineering practice*. John Wiley & Sons, Hoboken, New Jersey.

Freeman, L. A., Carpenter, M. C., Rosenberry, D. O., Rousseau, J. P., Unger, R., & McLean, J. S. 2004. Use of submersible pressure transducers in water-resources investigations. *US Geological Survey, Techniques of Water-Resources Investigations*, 8, A3.

Freeze, R.A. and Cherry, J.A. 1979. *Groundwater*. Prentice-Hall Inc., Englewood Cliffs, Vol. 7632, 604.

Garner, R., Naidu, T., Saavedra, C., Matamoros, P., & Lacroix, E. 2012. *Water Management in Mining: A Selection of Case Studies*. International Council on Mining & Metals, London.

Gore, J. A., & Banning, J. 2017. Discharge measurements and streamflow analysis. In: *Methods in Stream Ecology*, Eds, F. Richard Hauer and Gary Lamberti, Volume 1 (p. 49–70): Academic Press, Elsevier, Burlington, MA.

Gorelick, S., GeS, M. 2015. Hydrology, floods, and droughts: Groundwater and Surface Water. *Encyclopedia of Atmospheric Sciences (Second Edition)*, pp. 209–216.

Guinochet, M. 1973. *Phytosociologie* (Vol. 1). Masson, Paris.

Hillel, D. 1998. *Environmental Soil Physics Fundamentals, Applications, and Environmental Considerations*. Academic Press, Waltham.

Holmes, J., J. Chambers, P. Meldrum, P. Wilkinson, J. Boyd, P. Williamson, D. Huntley, K. Sattler, D. Elwood, V. Sivakumar, H. Reeves and S. Donohue. 2020. Four-dimensional electrical resistivity tomography for continuous, near-real-time monitoring of a landslide affecting transport infrastructure ein British Columbia, Canada. *Near Surface Geophysics*. doi:10.1002/nsg.12102.

Howell, T. A., A. D. Schneider, M.E. Jensen 1991. History of lysimeter design and use for evaporation measurements. P 1–9. In R.G. Allen et al. ed. *Lysimeter for Evapotranspiration and Environmental Measurements. Proceedings of the International Symposium on Lysimetry*, ASCE, New York, NY, USA.

Hu, X., T. Oommen, Z. Lu, T. Wang, J.-W. Kim. 2017. Consolidation settlement of Salt Lake County tailings impoundment revealed by time-series InSAR observations from multiple radar satellites, *Remote Sensing of Environment*, 202, 199–209.

International Commission on Large Dams (ICOLD) 1996. *Monitoring of Tailings Dams – Review and Recommendations*. Bulletin 104, Published by International Commission on Large Dams, Paris, France.

Ishii, T., Kadoya, K. 1991. Continuous measurement of oxygen concentration in citrus soil by means of a waterproof zirconia oxygen sensor. *Plant and Soil*, 131: 53–58.

Jacobsen, O.H., and P. Schjonning. 1993. Field measurements of soil water content during a drying period using time domain reflectometry and gravimetry. *Journal of Hydrology*, 151, 159–172.

Jonasson, S., 1983. The point intercept method for non-destructive estimation of biomass. *Phytocoenologia* 11, 385–388.

Khire, M., C. Benson, and P. Bosscher. 1999. Field data from a capillary barrier in a semiarid climate and model predictions with UNSAT-H. *Journal of Geotechnical and Geoenvironmental Engineering, ASCE*, 125(6), 518–528.

Kiflu, H., S. Kruse, M. H. Loke, P. B. Wilkinson and D. Harro. 2016. Improving resistivity survey resolution at sites with limited spatial extent using buried electrode arrays. *Journal of Applied Geophysics*, 135, 338–355.

Klimant, I., Meyer, V., Kühl, M. 1995. Fiber-optic oxygen microsensors, a new tool in aquatic biology, *Limnology and Oceanography*, 40, 1159–1165.

Klimant, I., Kühl, M., Glud, R.N., Holst, G. 1997. Optical measurement of oxygen and temperature in microscale: strategies and biological applications. *Sensors and Actuators B*, 38–39, 29–37.

Kup, F., R. Saglam, I. Tobi, H. Sahin, and C. Saglam. 2011. Calibration of time domain reflectometry (TDR) on the basis of torf sand and its optimisation for irrigation automations. *African Journal of Agricultural Research*, 6(10), 2386–2393.

Leib, B.G., J.D. Jabro, and G.R. Matthews. 2003. Field evaluation and performance comparison of soil moisture sensors. *Soil Science*, 168:396–408.

Lorite I.J., M. Garcià-Vila, M.A. Carmona, M.A. C. Santos and M.A. Soriano. 2012. Assessment of the irrigation Advisory Service recommendation and farmers irrigation management: a case study in Southern Spain. *Water Ressources Management*, 26, 2397–2419.

Malicki, M. 1983. A capacity meter for the investigation of soil moisture dynamics. *Zesty problemowe Postepow Nauk Rolniczych*, 201–214.

Maqsoud, A., B. Bussière, M. Mbonimpa, M. Aubertin, W.G. Wilson, W.G. 2007. *Instrumentation and monitoring of covers used to control Acid Mine drainage. Proceeding of the Mining Industry Conference, CIM*, Montréal, CD-rom.

Maqsoud, A., B. Bussiere, M. Aubertin, M. Chouteau and M. Mbonimpa. 2011. Field investigation of a suction break designed to control slope-induced desaturation in an oxygen barrier. *Canadian Geotechnical Journal*, 48(1), 53–71.

Maqsoud, A., Mbonimpa, M., Bussière, B., Dionne, J. 2013. *Réhabilitation du site minier abandonné Aldermac, résultats préliminaires du suivi de la nappe surélevée. 66th Canadian Geotechnical Conference and the 11th Joint CGS/IAH-CNC Groundwater Conference*, September 29 to Thursday October 3, 2013, Montréal.

Maqsoud, A., Mbonimpa, M., Bussière, B., Benzaazoua, M. and Dionne, J. 2015. *The hydrochemical behavior of the Aldermac abandoned mine site after its rehabilitation. 68th Canadian Geotechnical Conference, Québec, GeoQuébec 2015*, September 20 to 23, 2015, Québec.

Maqsoud, A., P. Gervais, B. Bussière, V. Le Borgne. 2017. Performance evaluation of equipment used for volumetric water content measurements. *WSEAS Transactions on Environment and Development*, 13, 27–32.

Marinho, F.A.M., W.A. Take, and A. Tarantino. 2008. Measurement of matric suction using tensiometric and axis translation techniques. *Geotechnical and Geological Engineering*, 26(6), 615–631.

Martin, V., Plante, B., Bussière, B. et al. 2017. *Controlling Water Infiltration in Waste Rock Piles: Design, Construction, and Monitoring of a Large-Scale In-situ Pilot Test Pile.* GeoOttawa 2017–70th Canadian Geotechnical Conference, Ottawa, ON, Canada.

Maurice, R. 2002. *Restauration du site minier Poirier (Joutel) — expériences acquises et suivi des travaux. Proc. of Defis & Perspectives: Symposium 2002 sur l'environnement et les mines*, Rouyn-Noranda, *on CD-ROM*, paper s32 a1021 pp.545.

Maurice, R. 2012. *Normétal mine tailings storage facility HDPE cover: design considerations and performance. Proc. of 9th International Conference on Acid Rock Drainage*, Ottawa, Canada, 20-26 May, Volume 1 and 2, paper 430.

Mbonimpa, M., Aubertin, M., Dagenais, A.-M., Bussière, B., Julien, M., Kissiova, M. 2002. *Interpretation of field tests to determine the oxygen diffusion and reaction rate coefficients of tailings and soil covers. Proc. 55th Can. Geotech. Conf.–3rd Joint IAH – CNC– Can. Geotech. Society Conf.: Ground and Water: Theory to Practice*, Niagara Falls, pp. 147–154. CD-ROM.

McCarthy, D. F. 2007. *Essentials of Soil Mechanics and Foundations: Basic Geotechnics.* Upper Saddle River, New Jersey: Pearson Prentice Hall, 850p.

Meals, D. W., & Dressing, S. 2008. Surface water flow measurement for water quality monitoring projects. *Tech Notes*, 3, Developed for U.S. Environmental Protection Agency by Tetra Tech, Inc., Fairfax, VA, 16 p.

Meena, H.M., R.K. Singh, and P. Santra. 2015. Design and development of a load-cell based cost effective mini-lysimeter. *Journal of Agricultural Physics*, 15, (1), 1–6.

Meldrum, J., H. Jamieson, and L. Dyke. 2001. Oxidation of mine tailings from Rankin Inlet, Nunavut, at subzero temperatures. *Canadian Geotechnical Journal* 38, no. 5: 957–966.

MEND. 2004. *Covers for Reactive Tailings Located in Permafrost Regions Review, Project Report 1.61.4.* Mine Environment Neutral Drainage (MEND) Canada Center for Mineral and Energy Technology, Canada.

MEND. 2009. *Mine Waste Covers in Cold Regions, Project Report 1.61.5a.* Mine Environment Neutral Drainage (MEND) Canada Center for Mineral and Energy Technology, Canada.

MEND. 2012. *Cold Regions Cover System Design Technical Guidance Document, Report 1.61.5c*, Mine Environment Neutral Drainage (MEND). Canada Center for Mineral and Energy Technology, Canada.

Merritt, A. J., J. E. Chambers, P. B. Wilkinson, L.J. West, W. Murphy, D. Gunn, and S. Uhlemann. 2016. Measurement and modelling of moisture – electrical resistivity relationship of fine-grained unsaturated soils and electrical anisotropy. *Journal of Applied Geophysics*, 124, 155–165.

Mian, M.H., Yanful, E.K. 2004. Analysis of wind-driven resuspension of metal mine sludge in a tailings pond. *Journal of Environmental Engineering and Science*, 3:119–135.

Michalski, A. 2000. Flow measurements in open irrigation channels. *IEEE Instrumentation and Measurement Magazine*, 3(1), 12–16.

Mollaret, C., C. Hilbich, C. Pellet, A. Flores-Orozco, R. Delaloye, and C. Hauck. 2019. Mountain permafrost degradation documented through a network of permanent electrical resistivity tomography sites. *The Cryosphere*, 13(10), 2557–2578.

Muller J.C. 1996. Un point sur trente ans de lysimétrie en France (1960–1990). In: *Une technique, un outil pour l'étude de l'environnement*, Ed. INRA, Co-Ed. Comifer, Paris.

Muñoz-Carpena, R. 2004. *Field Devices For Monitoring Soil Water Content Bulletin N° 343*, Department of Agricultural and Biological Engineering, Florida Cooperative Extension Service, Institute of Food and Agricultural Sciences, University of Florida, Gainesville, FL.

Nichol C., L. Smith, and R. Beckie. 2003. Long term measurement of matric suction using thermal conductivity sensors. *Canadian Geotechnical Journal* 40(3): 587–597.

Norberg-King, T.J., M.R. Embry, S.E. Belanger. et al. 2018. An international perspective on the tools and concepts for effluent toxicity assessments in the context of animal alternatives: reduction in vertebrate use. *Environmental Toxicology and Chemistry* 37 (11): 2745–2757. DOI: 10.1002/etc.4259.

NRCS. 2004. *National Engineering Handbook: Part 630—Hydrology*. USDA Soil Conservation Service: Washington, DC, USA.

O'Kane, M., G.W. Wilson, S.L. Barbour. 1998. Instrumentation and monitoring of an engineered soil cover system for mine waste rock. *Canadian Geotechnical Journal*, 35: 828–846.

Padrón, R.S., L. Gudmundsson, D. Michel, and S.I. Seneviratne. 2020. Terrestrial water loss at night: global relevance from observations and climate models. *Hydrololgy and Earth System Sciences*, 24, 793–807,

Pham, H. N. 2013. *Heat Transfer in Waste-Rock Piles Constructed in a Continuous Permafrost Region*. PhD diss., University of Alberta, Edmonton, Canada.

Power, C., P. Tsourlos, M. Ramasamy, A. Nivorlis, and M. Mkandawire. 2018. Combined DC resistivity and induced polarization (DC-IP) for mapping the internal composition of a mine waste rock pile in Nova Scotia, Canada. *Journal of Applied Geophysics*, 150, 40–51.

Price, D. G. 2009. Chapter 6: Field Tests and Measurements. In *Engineering Geology: Principles and Practice*. M. H. de Freitas (ed.), Springer–Verlag, Berlin, Heidelberg, 450p.

Radue, T. J. 2017. *Tailings Basin Geotechnical Instrumentation and Monitoring Plan Cell 2E*. Report Prepared for Poly Met Mining, Inc. Version 3, Minneapolis, Minnesota, 59p.

Rahardjo, H., and E.C. Leong. 2006. *Suction Measurements. Proceedings of the 4th International Conference on Unsaturated Soils (UNSAT'2006)*, Carefree, Arizona, USA. Vol. 1, pp. 81–104.

Raval, S., R. N. Merton, D. Laurence. 2013. Satellite based mine rehabilitation monitoring using WorldView-2 imagery, *Mining Technology*, 122:4, 200–207.

Riaza, A., Buzzi, J., García-Meléndez, E., Carrère, V., Müller, A., 2011. Monitoring the extent of contamination from acid mine drainage in the Iberian pyrite belt (SW Spain) using hyperspectral imagery. *Remote Sensing*, 3: 2166–2186.

Ricard, J.F., Aubertin, M., Firlotte, F.W., Knapp, R., and Mcmullen, J. 1997. *Design and Construction of a Dry Cover Made of Tailings for the Closure of Les Terrains Aurifères site, Malartic, Québec, Canada*. In *Proceedings of the 4th International Conference on Acid Rock Drainage (ICARD)*, Vancouver, B.C., 31 May–6 June, Vol. 4, pp. 1515–1530.

Robinson D.A., J.P. Bell and C.H. Batchelor 1994. The influence of iron minerals on the determination of soil water content using dielectric techniques. *Journal of Hydrology* 161:169–180.

Robinson, D. A., M.G. Schaap, D. Or, and S.B. Jones. 2005. On the effective measurements frequency of time domain reflectometry in dispersive and nonconductive dielectric materials. *Water Resources Research*, 41, W02007, doi:10.1029/2004WR003816.

Robinson, D. A., C.S. Campbell, J.W. Hopmans, et al. 2008. Soil moisture measurement for ecological and hydrological watershed-scale observatories: a review. *Vadose Zone Journal*, 7(1), 358–389.

Rubin, Y., & Hubbard, S. S. (Eds.). 2006. *Hydrogeophysics* (Vol. 50). Springer Science & Business Media, Heidelberg, Germany.

Ruth, C. E., D. Michel, M. Hirschi, and S.I. Seneviratne. 2018. Comparative study of a long-established large weighing lysimeter and a state-of-the-art mini-lysimeter. *Vadose Zone Journal*, 17:170 026. doi:10.2136/vzj2017.01. 0 0 26

Rykaart, M., D. Hockley, M. Noel and M. Paul. 2006. *Findings of international review of soil cover design and construction practices for mine waste closure. Proceeding of the 7th International Conference on Acid Rock Drainage (ICARD)*, Saint Louis, Missouri March 2730.

Hanslin, H.M., Sæbø, A., Bergersen, O. 2005. Estimation of oxygen concentration in the soil gas phase beneath compost mulch by means of a simple method. *Urban Forestry & Urban Greening*, 4(1): 37–40.

Poirier, A. 2019. *Étude du comportement thermique d'une halde à stérile en milieu nordique, département des génies géologique, civil et des mines.* MScA thesis, Polytechnique Montréal, Montréal, Québec.

Schmidt, C.D.S., F.A. de Carvalho Pereira, A.S. de Oliveira, J.F. Gomez Junior, L.M. Vellame. 2013. Design, installation and calibration of a weighing lysimeter for crop evapotranspiration studies. *Water Resources and Irrigation Management*, 2, 77–85.

Schmidt-Hattenberger, C., P. Bergmann, T. Labitzke, F. Wagner, and D. Rippe. 2016. Permanent crosshole electrical resistivity tomography (ERT) as an established method for the long-term CO2 monitoring at the Ketzin pilot site. *International Journal of Greenhouse Gas Control*, 52, 432–448.

Scott, R.L. 2010. Using watershed water balance to evaluate the accuracy of eddy covariance evaporation measurements for three semiarid ecosystems. *Agriculture and Forest Meteorology* 150, 219–225.

Shannon, C.E. 1948. A mathematical theory of communication. *Bell System Technical Journal* 27, 379–423.

Shokri, B. J., H. Ramazi, F.D. Ardejani, and A. Moradzadeh. 2014. Integrated time-lapse geoelectrical-geochemical investigation at a reactive coal washing waste pile in Northeastern Iran. *Mine Water and the Environment*, 33(3), 256–265.

Singh, D. N., and S.J. Kuriyan. 2003. Estimation of unsaturated hydraulic conductivity using soil suction measurements obtained by an insertion tensiometer. *Canadian Geotechnical Journal*, 40: 476–483.

Skaling, W. 1992. Trase: a product history. *Soil Society of America Journal*, Special publication, 30, 187–207.

Smit, A.L., A.G. Bengough, C. Engels, M. van Noordwijk, S. Pellerin, S.C. van de Geijn, 2000. *Root Methods: A Handbook.* Springer-Verlag, New York.

Smith, G.M. and W.J. Waugh. 2004. Influences of construction subtleties on the hydraulic performance of water balance. In *Proceeding of Tailings and Mine wastes 04*, Taylor & Francis Group, London, ISBN .

Smith, L. J., M. C. Moncur, M. Neuner, M. Gupton, D. W. Blowes, L. Smith, and D. C. Sego. 2013. The Diavik waste rock project: Design, construction, and instrumentation of field-scale experimental waste-rock piles. *Applied Geochemistry* 36:187–199.

SoeMoe, K.W., Biggar, K. W., Martens, S. and Hoda, R. 2013. *Application of Geotechnical Instrumentation in Monitoring in-Pit Dykes Performance.* In *GéoMontréal 2013, the 66th Canadian Geotechnical Conference and the 11th Joint CGS/IAH-CNC Groundwater Conference*, Montreal, CA; Septembre 29–Octobre 3, 2013, 9p.

Stenitzer, E. 1993. Monitoring soil moisture regimes of field crops with gypsum blocks. *Theoretical and Applied Climatology*, 48: 159–165.

Swanson, D.A., S.L. Barbour, G.W. Wilson, and M. O'Kane. 2003. Soil–atmosphere modelling of an engineered soil cover for acid generating mine waste in a humid, alpine climate. *Canadian Geotechnical Journal*, 40: 276–292.

Szymanski, M. B., & Davies, M. P. 2004. *Tailings Dams: Design Criteria and Safety Evaluations at Closure.* Paper presented at the *BC Reclamation Symposium*, British Columbia Technical and Research Committee on Reclamation, Vancouver, BC, doi:http://dx.doi.org/10.14288/1.0042456.

Tan, E., D.G. Fredlund, and B. Marjerison. 2007. Installation procedure for thermal conductivity matric suction sensors and analysis of their long-term readings. *Canadian Geotechnical Journal*, 44: 113–125

Topp, G. C. and T.P.A. Ferre. 2002. Water content. In: *Methods of Soil Analysis. Part 4.* (Ed. J.H. Dane and G.C. Topp), SSSA Book Series No. 5. Soil Science Society of America, Madison WI.

Topp, G.C., J.L. Davis, and A.P. Annan. 1980. Electromagnetic determination of soil water content: measurements in coaxial transmission lines. *Water Resources Research*, 16, 574–582.

Tso, C. H. M., O. Kuras, and A. Binley. 2019. On the field estimation of moisture content using electrical geophysics: the impact of petrophysical model uncertainty. *Water Resources Research*, 55(8), 7196–7211.

USDI. 2001. *Water Measurement Manual.* US Government Printing Office Washington, DC.

Varhola A., Wawerla J., Weiler M., Coops N.C., Bewley D. and Alila, Y., 2010. A new low-cost, stand-alone sensor system for snow monitoring. *Journal of Atmospheric and Oceanic Technology* 27 no. 12, 1973–1978.

Vigneault, B., Campbell, P.G.C., Tessier, A., De Vitre, R. 2001. Geochemical changes in sulfidic mine tailings stored under a shallow water cover. *Water Research*, 35: 1066–1076.

Vriens, B., Smith, L., Mayer, K.U., Beckie, R.D. 2019. Poregas distributions in waste-rock piles affected by climate seasonality and physicochemical heterogeneity. *Applied Geochemistry*, 100: 305–315

Waugh, W.J. 2004. *Designing Sustainable Covers for Uranium MILL Tailings. Proceedings of High Altitude Revegetation Workshop* No. 16 March 2004, Edited by Warren R. Keammerer and Jeffrey Todd.

White, I. and S.J. Zegelin 1995. Electric and dielectric methods for monitoring soil-water content. In: *Vadose Zone Characterization and Monitoring*, Chapter 22. L.G. Wilson, L.G. Everett, S.J. Cullen (eds.) CRC Press, Boca Raton, FL, pp. 343–385.

Whiteley, J. S., J.E. Chambers, S. Uhlemann, P.B. Wilkinson, and J.M. Kendall, J. M. 2019. Geophysical monitoring of moisture-induced landslides: a review. *Reviews of Geophysics*, 57(1), 106–145.

Wilkinson, P. B., P.I. Meldrum, J.E. Chambers, O. Kuras, and R.D. Ogilvy, 2006. Improved strategies for the automatic selection of optimized sets of electrical resistivity tomography measurement configurations. *Geophysical Journal International*, 167(3), 1119–1126.

Williams, D.J., Stolberg, D.J., Currey, N.A. 2006 *Long-term monitoring of Kidston's «Store/Release» cover system over potentially acid forming waste rock piles. Proceedings of the 7th ICARD*, St Louis, MO, USA, pp. 26–30.

Wilson, G.W., Fredlund, D.G., and Barbour, S.L., 1994. Coupled soil-atmosphere modelling for soil evaporation. *Canadian Geotechnical Journal*, 31(2): 151–161. doi:10.1139/t94-021.

Yanful, E.K., Riley, M.D., Woyshner, M.R., Duncan, J. 1993. Construction and monitoring of a composite soil cover for an experimental waste rock pile near Newcastle, New Brunswick, Canada. *Canadian Geotechnical Journal*, 30: 588–599.

Yolcubal, I. Brusseau, M.L., Artiola, J.F., Wierengal, P., Wilson, L.G. 2004. Environmental physical properties and processes. In *Environmental Monitoring and Characterization*, Elsevier, Academic Press, USA, pp. 207–239, https://doi.org/10.1016/B978-012064477-3/50014-X.

Zegelin, S.J., I. White, and G.F. Russell. 1992. A critique of the time domain reflectomery technique for determining field soil-water content. *Soil Sciences Society of America*, 30, 187–207.

Zhan, G., Keller, J., Milczarek, M., Giraudo, J. 2014. 11 years of evapotranspiration cover performance at the AA leach pad at Barrick Goldstrike Mines. *Mine Water and the Environment*, 33(3):195–205. doi:10.1007/s10230-014-0268-6

Zhang X., Leong E.C., H. Rahardjo 2001. *Evaluation of a thermal conductivity sensor for measurement of matric suction in residual soil slopes*. In: *Proceedings of 14th southeast Asian geotechnical conference*, Hong Kong, pp. 611–616.

Zhou, Z., Smith, J. A., Wright, D. B., Baeck, M. L., Yang, L., & Liu, S. 2019. Storm catalog-based analysis of rainfall heterogeneity and frequency in a complex terrain. *Water Resources Research*, 55(3), 1871–1889.

11 Passive Treatment of Acid Mine Drainage at the Reclamation Stage

Carmen M. Neculita, Gérald J. Zagury, and Bruno Bussière

11.1 INTRODUCTION

The main objectives of mine site reclamation are to control the flux of contaminants and to limit the potential adverse effects on the environment – including the impairment of water quality resources and sediments – and human health, plant life, and aquatic species (Simate and Ndlovu 2014; Candeias et al. 2015). Sometimes, the reactions leading to water contamination (see Chapter 1) have started long before the implementation of the reclamation strategy, and some contaminated water will eventually flow out of the mine site. Some reclamation technologies (see Chapters 4 to 9) significantly reduce the level of contamination but occasionally fail to respect the imposed environmental regulations. It is then important to collect water and temporarily treat residual contamination from reclaimed mine sites to avoid any significant impact on the environment. At the reclamation stage, it is preferable to use passive approaches – that is, water treatment technologies that do not involve active treatment, such as the continuous addition of chemicals or substrates, and electricity consumption. Moreover, passive treatment is usually less expensive in terms of investment and operation, does not necessitate daily supervision and maintenance, and uses locally available materials and natural biochemical phenomena to treat the contamination. The objectives of passive treatment are to increase the pH and alkalinity of acid mine drainage (AMD) and to facilitate metal/metalloid and sulfate removal. Notably, the treatment of other major contaminants specific to mining effluents (e.g., cyanides and its derivatives, and ammonia nitrogen) is not discussed in this chapter; we focus instead on the aforementioned contaminants in AMD.

The AMD passive treatment is based on several abiotic and biogeochemical processes, which involve neutralization and redox reactions, dissolution and precipitation of solids, complex formation, ion exchange, and sorption on solid surfaces (Figure 11.1) (Johnson and Hallberg 2005a; Neculita et al. 2007; Genty 2012). Passive technologies are mainly used as polishing steps on reclaimed mine sites; however, as they were developed to fit a wide variety of water conditions, in the United States, many are now also being used at active mine sites as well (Skousen et al. 2017). The choice of a passive treatment considers water chemistry, flow rate, local topography, and site characteristics. The construction of a passive treatment requires the collection of all drainage water from a mine site at the point where the treatment system will be installed. A thorough hydrological study must, therefore, be conducted before its installation (Younger et al. 2002).

This chapter presents the major mechanisms involved in the treatment processes, the classification of the technologies, and the main factors of influence for the three categories of passive treatment: passive chemical, passive biochemical, and passive multistep systems. A methodological approach for the design of such systems is then described. Case studies that used passive treatment technologies are identified, and their main advantages and limits are summarized. Finally, research needs are identified. It is worth mentioning that the terms *AMD* and *CND* (contaminated neutral

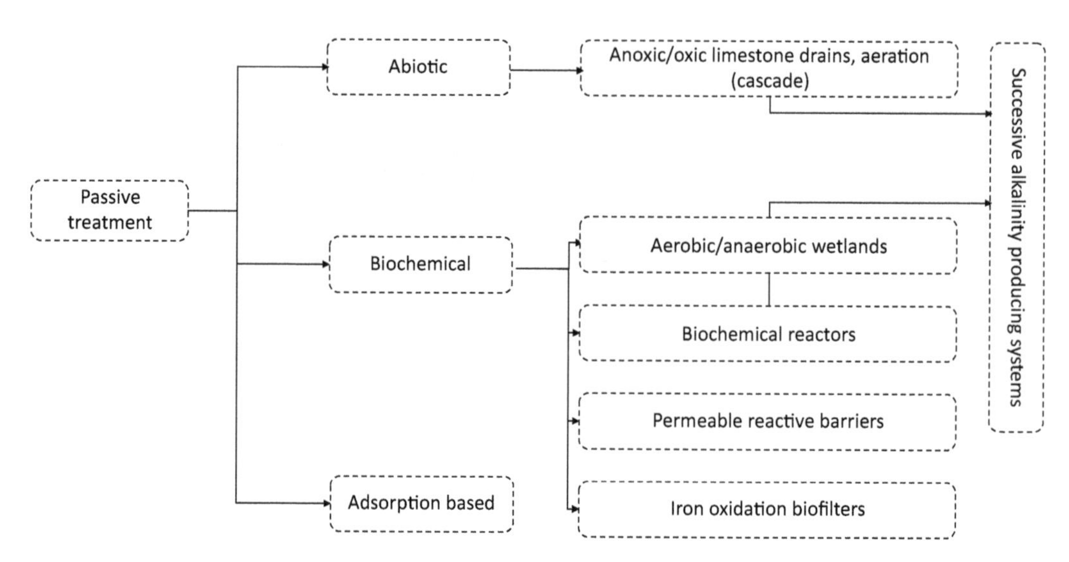

FIGURE 11.1 AMD treatment options (Adapted from Johnson and Hallberg 2005a; Neculita et al. 2007; Genty 2012).

drainage) will be used to characterize the mine effluents because water from reclaimed mine sites can evolve with time, after reclamation, from AMD to CND.

11.2 PRINCIPLE, CLASSIFICATION, AND FACTORS OF INFLUENCE

In this chapter, the efficiency and the overall performance of a passive treatment system are defined as follows: (1) efficiency (%) corresponds to contaminant removal, calculated from the initial (inlet) and final (outlet) concentrations; (2) performance includes, in addition to efficiency, residual contamination (excess of untreated contaminant and the contaminants leached by the materials used), compliance with prescribed discharging limits of the treated water (criteria), and quantity/quality (biogeochemical stability) of sludge (or metal-rich solids).

11.2.1 Passive Chemical Systems

11.2.1.1 Principle

Passive chemical systems are constructed in the form of shallow trenches (drains) or ponds with the bottom layered by natural neutralizing material (limestone or dolomite) or natural sorbent materials (Figure 11.2). The treatment principle of AMD (characterized by low pH, high acidity, and variable concentrations of dissolved metals and sulfates; see Chapter 1) relies on pH and alkalinity ($CO_3^{2-} + HCO_3^- + OH^-$) increase, together with acidity consumption, by neutralizing materials (carbonates, silicates, and hydroxides) dissolution, followed by metal and sulfate removal.

The acidity of mine drainage (mg/L as $CaCO_3$) consists of free acidity (due to the low pH) and residual acidity (due to the presence of often high concentrations [mg/L] of common metals such as Al, Fe, and Mn; see Chapter 1), as in the following equation (Kirby and Cravotta 2005a):

$$\text{Acidity}_{\text{calculated}} = 50\left\{1000\left(10^{-\text{pH}}\right)+\left[2\left(\text{Fe}^{\text{II}}\right)+3\left(\text{Fe}^{\text{III}}\right)\right]/56+2\left(\text{Mn}\right)/55+3\left(\text{Al}\right)/27\right\} \quad (11.1)$$

These common metals in AMD are acidogenic (i.e., source of residual acidity following their hydrolysis and precipitation in the form of hydroxides), as shown in the following reaction:

$$\text{Me}^{\text{n+}} + \text{H}_2\text{O} \leftrightarrow \text{Me}\left(\text{OH}\right)_{\text{n}} + \text{nH}^{+} \quad (11.2)$$

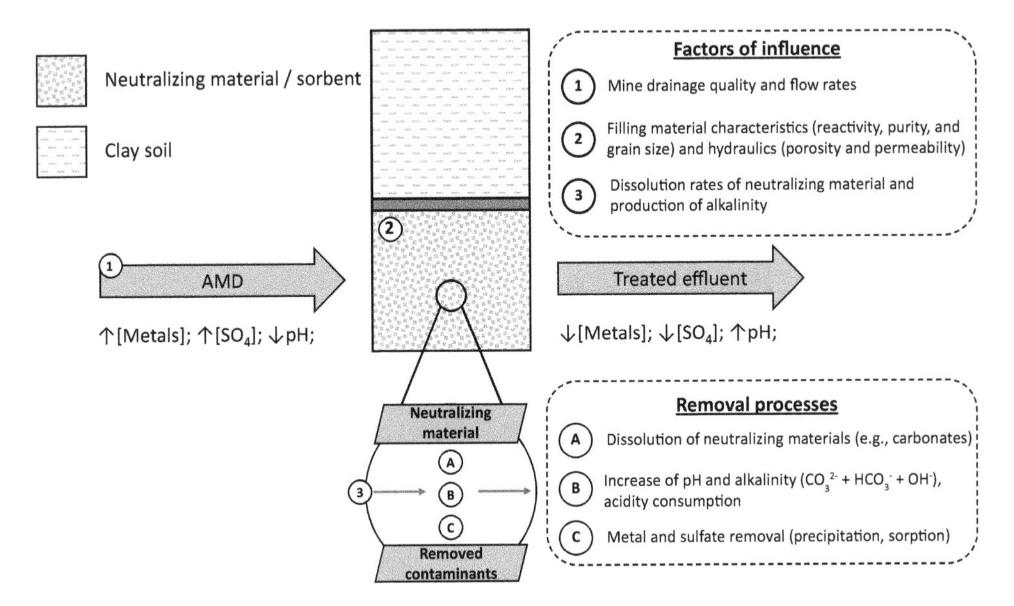

FIGURE 11.2 Scheme of a passive chemical system for the treatment of AMD.

The dissolution of calcite (limestone principal component) involves the following reactions (Cravotta 2003); the reaction 11.3 occurs at pH < 6.3, whereas reaction 11.4 occurs at pH > 6.3:

$$CaCO_3(s) + 2H^+ \leftrightarrow Ca^{2+} + H_2CO_3* \quad (11.3)$$

$$CaCO_3(s) + H_2CO_3* \leftrightarrow Ca^{2+} + 2HCO_3^- \quad (11.4)$$

where $[H_2CO_3*] = [CO_2(aq)] + [H_2CO_3^0]$ (Stumm and Morgan 1981).

The contaminant removal involves reverse precipitation (to the reactions leading to mine drainage generation; see Chapter 1) in the form of oxyhydroxides, hydroxyl-sulfates, and carbonates (following the neutralization) or in the form of carbonates and sulfides (following the sulfate reduction), and sorption (Johnson and Hallberg 2005a; Potvin 2009; USEPA 2014). Precipitation is the main mechanism of metal removal in AMD by passive chemical systems mainly because of the predominance of trivalent metals (Al^{3+} and Fe^{3+}), which precipitate at acidic pH (3 to 3.5 and 4 to 5, respectively, with respect to total concentrations) (see Figures 11.3 and 11.4 for a typical AMD quality) and remains stable with pH increase up to an optimal value (Macías et al. 2017a). On the contrary, bivalent metal removal occurs at neutral to alkaline pH. Sorption mechanisms are also responsible for partial removal of mostly bivalent metals in AMD, following the pH increase and trivalent metal removal. Passive treatment can also remove solid materials in suspension driven by water flow feeding the system, via physical mechanisms. In the long term, common issues include the reduction in the material mass (by dissolution), grain coating (by secondary minerals, especially iron oxides-hydroxides and gypsum), particle segregation, and settlement/compaction (due to time or heavy machinery movement).

11.2.1.2 Classification

The basic classification of chemical treatment systems for surface mine drainage includes open limestone channels (OLCs), oxic/anoxic limestone drains (OLDs/ALDs), limestone leach beds, steel slag leach beds, diversion wells, limestone sand, and low pH Fe oxidation channels (Johnson and Hallberg 2005a; USEPA 2014; Skousen et al. 2017). The most common systems (OLCs and OLDs/ALDs) are discussed below (Figure 11.2).

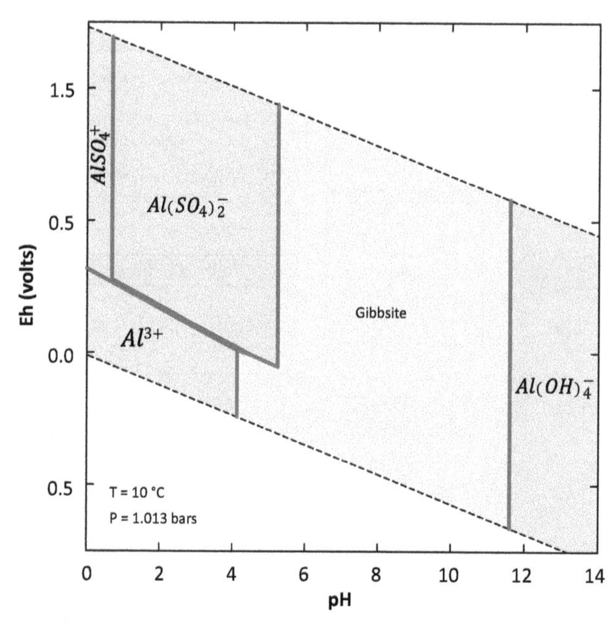

FIGURE 11.3 Pourbaix diagram for Al.

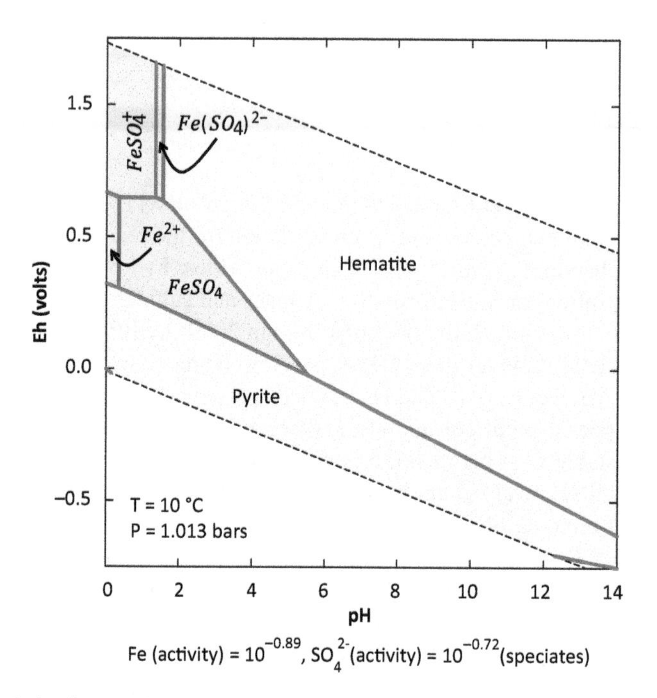

FIGURE 11.4 Pourbaix diagram for Fe.

The OLC (water flows on the surface of neutralizing material) and OLD (water flows through it) allow the pH and alkalinity increase, as well as metals and sulfate removal, and are used for polishing the mine drainage, which contains > 1 mg/L dissolved oxygen (DO) (Hedin and Watzlaf 1994). In ALDs, the limestone is overlaid by a solid cover (a buried bed to minimize/exclude DO contact with the AMD). These systems are used for the treatment of water unexposed to air (e.g., from the underneath of a pile of reactive waste). Inside the ALD, the pH and alkalinity are expected to increase, whereas, at the exit, metal oxidation and precipitation are expected to occur. In principle, the residual acidity entailed by the oxidation of Fe^{2+} to Fe^{3+} (Figure 11.4), followed by the precipitation of the $Fe(OH)_3$, is buffered by the excess of alkalinity generated in the ALD. For all systems, the treatment efficiency is controlled by the behavior and specific reactions of the inorganic carbon in the carbonate system (solubility and pH in anaerobic vs. aerobic; i.e., presence vs. absence of a CO_2 reservoir, and unusual higher solubility at low vs. higher temperature), hydraulic retention time (HRT; i.e., contact time between mine drainage and reactive material), and neutralizing/filling material characteristics (Cravotta 2003, 2015; Genty et al. 2012a).

The main reactions of calcite/dolomite with sulfuric acid are presented below; other neutralizing reactions in AMD are presented elsewhere (see Chapter 1):

$$2CaCO_3 + H_2SO_4 \rightarrow 2Ca^{2+} + 2HCO_3^- + SO_4^{2-} \tag{11.5}$$

$$CaMg(CO_3)_2 + H_2SO_4 \rightarrow Ca^{2+} + Mg^{2+} + 2HCO_3^- + SO_4^{2-} \tag{11.6}$$

A few examples of passive chemical treatment are presented in Table 11.1, while the findings of available compilations of hundreds of field systems are discussed in Section 11.4. The main factors of influence on efficiency and long-term performance, as identified in these studies, are briefly discussed below.

11.2.1.3 Factors of Influence

The short- and long-term performance of limestone/dolomite drains is significantly influenced by several factors, including the (1) mine drainage quality and flow rates; (2) filling material characteristics (reactivity, purity, and grain size); (3) hydraulics (porosity and permeability); (4) dissolution rates of neutralizing material and production of alkalinity as function of water chemistry, HRT, material characteristics, and environmental conditions (temperature and CO_2 concentrations) (Figure 11.2).

The flow rates and mine drainage quality greatly influence the performance of limestone/dolomite drains. The acidity of mine drainage controls the speciation of contaminants and the order of their removal (Kirby and Cravotta 2005a, 2005b; Maree et al. 2013). The design of these systems must therefore consider the total acidity, together with concentrations of trivalent and bivalent metals, sulfate, and DO, as well as CO_2 pressure. The CO_2 can produce alkalinity in the reaction with limestone, as shown in Equations 11.3 and 11.4 (Stumm and Morgan 1981). The flow rates should also maintain little variation as inconsistent flow of mine drainage may lead to variable performance. The construction of settling ponds, which is a common practice in many water treatment systems for ensuring constant inflow, could be a potential solution but might prove impractical in the case of drains, especially the ALDs, with minimal DO design requirements.

In addition, the reactivity, purity, and grain size of the material also control the drain's performance. Hence, the selection of a calcareous material (e.g., dolomite vs. calcite) should be based on several criteria, including availability, reactivity, and vulnerability to gypsum coating (Huminicki and Rimstidt 2008; Genty et al. 2012a). Precisely, the more rapid dissolution of calcite produces more gypsum, whereas the slower dissolution of dolomite produces less gypsum and soluble magnesium sulfate. Therefore, sulfate treatment efficiency is limited with dolomite, but the lifespan of the treatment system could be prolonged because less gypsum can coat the limestone rocks.

In principle, the hydraulics of OLDs and OLCs are less influenced by clogging because they are not prone to such operational issues. On the contrary, drain geometry (slope, length, width, and profile) could influence the hydraulics and the performance of ALDs (Bernier et al. 2001, 2002). The ALDs' performance could, however, deteriorate over time, as they are adversely impacted by the presence of variable concentrations of Al, Fe, and Mn in the mine drainage as well as fine particles

TABLE 11.1

Reported Efficiency of Passive Treatment Systems for Mine-Drainage-Impacted Waters

Type system /Reference	pH		Acidity (mg/L $CaCO_3$)		Alkalinity (mg/L $CaCO_3$)		Fe_{total} (mg/L)		Al^{3+} (mg/L)		Mn^{2+} (mg/L)		Ni^{2+} (mg/L)		Zn^{2+} (mg/L)		Cd (mg/L)		SO_4^{2-} (mg/L)	
	I	E	I	E	I	E	I	E	I	E	I	E	I	E	I	E	I	E	I	E
Howe Bridge ALD, PA, USA (Cravotta 2003)	5.8	6.3	495	369	36	158	255	254	<0.2	<0.2	40	40	–	–	–	–	–	–	1240	1230
Morisson ALD, PA, USA (Cravotta 2003)	5.3	6.4	434	76	30	291	209	165	0.5	<0.2	47	39	–	–	–	–	–	–	1200	1010
Buck Mountain ALD, PA, USA (Cravotta 2003)	4.9	6.4	19	–61	2	82	10	10	0.5	<0.2	1	1	–	–	–	–	–	–	61	56
Rid-2L ALD, Appalachia, USA (Hedin and Watzlaf 1994)	2.3	6.2	–	–	0	469	1416	202	486	<0.1	23	11	–	–	–	–	–	–	6719	2227
REM-R ALD, Appalachia, USA (Hedin and Watzlaf 1994)	4.3	5.5	–	–	0	69	589	507	5	3	136	132	–	–	–	–	–	–	2825	2655
Lorraine dolomite drains, QC, Canada (Potvin 2009)	3–4	4–6	–	–	0	0–40	3000–7000	200–5000	–	–	–	–	1–11	<0.3–1	–	–	–	–	6000–13500	2100–12000
Laboratory PBR (Waybrant et al. 2002)	5.5–6.0	6.5–7.0	–	–	<50	300–1300	300–1200	<0.01–220	–	–	8.6–19.8		0.8–12.8	<0.01	0.6–1.2	<0.01–0.15	–	–	1010–3894	500–1000

Laboratory PBR 10 days vs. 7.3 days HRT (Neculita et al. 2008a)	2.9–5.7	6–9.2	–	–	–	0–3500	1066±78 (12wks) 504±83 (31wks)	201 vs 91	–	–	10.1±2.6	Leaching	13.7±1.0	0.096 vs. 0.08	14.5±2.1	0.01	9.8±1.8		4022±583	2200–3700
Laboratory PBR 4 days vs. 2 days HRT (Vasquez et al. 2016b)	3.0–3.7	7.0	–	-	–	From 1500±54 to 800±28 vs. From 2400±32 to 1200±60	200±44	Up to 40	–	–	30±2	77±4 vs. 60±2	–	–	19±2	Up to 6	–	–	2500±170 vs 3382±31	1300±32 to 1700±98 vs. 1000±50
Laboratory limestone-DAS (Rötting et al. 2008b)	2.8	6–8	1727	127–827	–	230–300	250	BDL	106	BDL	22	NR	0.84	MR	365	MR	0.38	MR	3510	3000–4000
Bench multi-unit (Fe pretreatment, peat biofilter, PBR, ALD) (Champagne et al. 2005)	3.2	7.0	–	–	–1140±159	1340±94	189±17	< 1	91.3±5	< 1	25.2±1.4	< 1	20.2±0.9	< 1	101±7.6	< 1	6±0.3	2	3140±721	948±108
Bench multi-unit (ash-, calcite /dolomite-DAS, PBR) (Rakotonimaro et al. 2017b)	3–4	6.3–8.3	2300–4000	5–4000	–	5–280 (2d HRT) vs. 10–2230 (3 or 5 d HRT)	2500±171	0–1550	1.6±0.6	BDL	8.2±1.0	MR to leaching	0.7±0.4	BDL to 0.6	0.2±0.3	BDL	–	–	5395±988	1000–7500
Wheal Jane multi-unit, UK (OLD/ALD, aerobic/anaerobic cells, rock filter) (Johnson and Hallberg 2005b)	3.7	4.2	–	–	–	–	253 (171–298)	0.4 (0–0.6)	–	–	–	–	–	–	–	–	–	–	760.8 (542–1093)	678 (282–1019)

Table 11.1 (Continued)

Reported efficiency of passive treatment systems for mine-drainage-impacted waters

Type system / reference	pH		Acidity (mg/L $CaCO_3$)		Alkalinity (mg/L $CaCO_3$)		Fe_{total} (mg/L)		Al^{3+} (mg/L)		Mn^{2+} (mg/L)		Ni^{2+} (mg/L)		Zn^{2+} (mg/L)		Cd (mg/L)		SO_4^{2-} (mg/L)	
	I	E	I	E	I	E	I	E	I	E	I	E	I	E	I	E	I	E	I	E
Cadillac Molybdenite bi-unit (PBR, OLD), QC, Canada (Kuyucak et al. 2006)	3.5	6.7	–	–	–	–	13.5 (32–max)	0.1	43	9.4	5.8	3.5	0.6	0.01	1.4	0.01	–	–	887	360 (lowest)
RAPS, Spain (Ayora et al. 2013)	2.9	6.2	4546	39	–	–	663.6	10.7	125.5	<0.1	–	–	–	–	18.4	<0.03	–	–	3180	3108
Mina Esperanza, multistep (limestone-/ magnesia-DAS) Spain (Caraballo et al. 2011)	2.4–3	6.5	2500	1000	-	0–400	755–1100	135–930	128–167	BDL	4–6	–	0.15–0.25	TR to leaching	19–33	TR to leaching	0.07–0.1	BDL	3324–4515	–
Lorraine pilot tri-unit (2PBRs and a wood ash filter), QC, Canada (Genty et al. 2016)	4.3–6.9	5.8–7	74	–	2670	–	1799	411	0.5	–	5	5	0.6	0.06	0.26	0.07	–	–	4750	2070

I – Influent; E – Effluent; BDL – below detection limit; NR – No removal; MR – Minor removal; TR – total removal; HRT – Hydraulic Retention Time; DAS – Dispersed Alkalinity Substrate; PBR – Passive Biochemical Reactor; OLD – Oxic Limestone Drain; ALD – Anoxic Limestone Drain; RAPS – Reducing and Alkalinity-Producing Systems, i.e. an ALD overlain by an organic matter layer, with the role of the consumption of dissolved O_2 and reduction of Fe(III) to Fe(II).

transported during the water flow. The coating (passivation) by newly formed precipitates (secondary minerals) must be periodically assessed and potentially controlled, via monitoring—that is, visual observations and the periodic sampling and analysis of treated water.

11.2.2 Passive Biochemical Systems

11.2.2.1 Principle

Most passive biochemical systems are designed to promote the activity of sulfate-reducing bacteria (SRB) to increase pH and alkalinity, reduce sulfates to soluble sulfides (H_2S, HS^-, and S^{2-}, proportions pH-related), and entail precipitation (and therefore removal) of dissolved metals as solid metal sulfides. To do so, organic carbon (various solid organic natural or residual materials) must be added to the system to serve as substrate (an electron donor and energy source) (Equations 11.7 and 11.8).

$$2CH_2O + SO_4^{2-} \rightarrow 2HCO_3^- + H_2S \quad (11.7)$$

$$H_2S + M^{2+} \rightarrow MS \downarrow +2H^+ \quad (11.8)$$

where CH_2O represents a simple organic carbon source and M is a cationic metal.

Optimal basic conditions for SRB development and growth include a pH of 5 to 8, Eh <-100 mV, and temperature around 20°C, although some SRB can be adapted to sometimes extreme environments (e.g., highly alkaline and saline environments, and volcanic lakes) (Postgate 1984). Depending on carbonate content, surface characteristics, and sorption capacities of organic substrates added to the system, other metal removal mechanisms such as hydroxides and carbonates precipitation, as well as sorption, are involved in AMD treatment (Johnson and Hallberg 2005b; Zagury et al. 2006; Neculita et al. 2007, 2008a, 2008b; Vasquez et al. 2016a, 2016b; Calugaru et al. 2017, 2018a, 2018b). Since sulfate is used as the electron acceptor, passive biochemical systems based on sulfate reduction are most suitable for treatment of AMD rather than CND (which usually contains low sulfate concentrations). Sorption is the main removal mechanism of metals/metalloids in CND (Westholm et al. 2014; Calugaru et al. 2018b).

Sorption is highly influenced by surface characteristics, such as specific surface, cation exchange capacity (CEC), anion exchange capacity, and point of zero charge (pH_{PZC} – pH at which there are equal amounts of positive and negative charges at the surface). At near-neutral pH, with respect to surface functional groups (carboxyl, sulfhydryl, hydroxyl, and amino groups), surface ionization is negative or positive, and the substrates might perform better for the sorption of anions or cations. As a surface-driven mechanism, sorption efficiency in mine drainage treatment depends on the surface characteristics of the sorbent material. Substrates with low pH_{PZC} would be best suited to treat effluents contaminated with cations, whereas substrates with high pH_{PZC} would be more appropriate to immobilize anions (Bakatula et al. 2018). As a large variety of inorganic and organic materials is available, maximum sorption capacity and required contact time to reach equilibrium conditions are the most important parameters to be determined in laboratory. These parameters can be obtained through sorption equilibrium and kinetics, prior to the upscaling and construction of full-scale treatment systems (Calugaru et al. 2016, 2017, 2018a, 2018b; Bakatula et al. 2018).

11.2.2.2 Classification

The basic classification of passive biochemical treatment systems for AMD includes passive biochemical reactors (PBRs), permeable reactive barriers (PRBs), and constructed wetlands (Johnson and Hallberg 2005a; Neculita et al. 2007; USEPA 2014; Skousen et al. 2017).

A PBR is a reactor that uses a simple flow-through design, with an AMD feed over a solid reactive mixture acting as a carbon source for SRB and as a physical support for microbial attachment

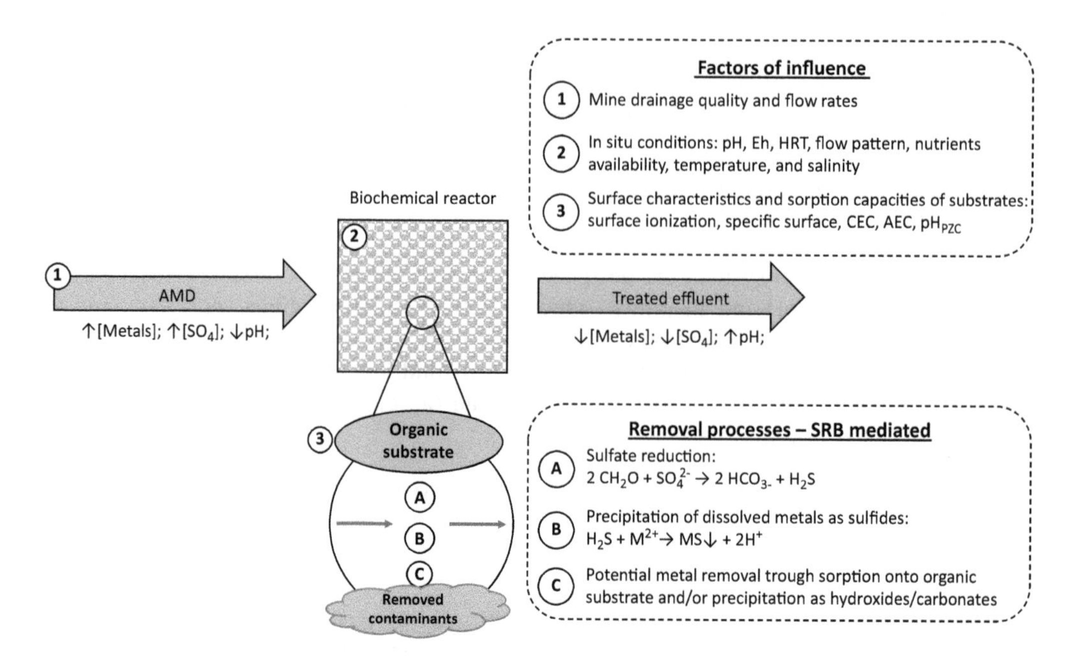

FIGURE 11.5 Scheme of a passive biochemical system for the treatment of AMD.

and metal sulfide precipitation (Figure 11.5) (Johnson and Hallberg 2005a; Neculita et al. 2007). The mixture generally contains different carbon sources for SRBs, a porous support (sand or gravel), and a neutralizing agent such as limestone to help AMD treatment and the SRB during their acclimation period (Cocos et al. 2002; Zagury et al. 2006; Neculita et al. 2007).

The PRBs have been found efficient for the treatment of dissolved metals, AMD, and dissolved nutrients (Blowes et al. 2000). The PRB consists of a reactive cell installed below the ground surface in the path of AMD-contaminated groundwater. Reactive materials include organics (compost, wood shavings, and compost) and inorganics (zero-valent iron [ZVI], limestone, zeolites, activated carbon, apatite, etc.) (USEPA 2014). Moreover, a mixing of organics (peat) and inorganics (ZVI) could extend the PRB lifespan relative to sole materials (Jeen et al. 2014). The reactive mixtures used, and the mechanisms of sulfate and metal removal, are similar in organic-based PBRs and PRBs (Waybrant et al. 1998, 2002; Gibert et al. 2003; Neculita and Zagury 2008). Once properly designed and implemented, PRBs could involve minimal maintenance costs for at least 5 to 10 years; clogging (by biofouling and mineral precipitation) is the main documented cause of performance decline (USEPA 2014).

Typical constructed/engineered wetlands (aerobic and anaerobic) are shallow, surface-flow ponds often planted with cattails (*Typha* sp.) that may or may not be lined with a synthetic liner or clay barrier. They generally contain a bed of limestone beneath or mixed with an organic carbon source, which favors alkalinity generation. Compared to PBRs and PRBs, constructed wetlands allow less system control and are more subject to seasonal variations, while involving several combinations of physicochemical and biological treatment processes (Johnson and Hallberg 2005a; USEPA 2014).

11.2.2.3 Factors of Influence

The main factors of influence for the passive biochemical treatment are (1) mine drainage quality and flow rates, (2) in situ conditions, and (3) surface characteristics and sorption capacities of substrates (Figure 11.5).

Similar to chemical treatment, water quality and flow rates are basic influence factors for the long-term performance of biochemical treatment systems, too. An essential component of mine drainage composition is sulfate concentration (the electron acceptor in the sulfate-reducing process). To maximize the bio-precipitation of metals as sulfides in the matrix of passive biochemical systems, enough soluble metals must be present in the mine water. Otherwise, H_2S could have toxic effects on SRB, may entail a sharp decline in their counts and their activity, and lead eventually to treatment efficiency deterioration (Neculita et al. 2007; Vasquez et al. 2016b).

In field conditions, the reactive mixture composition and thickness: ensure the availability of dissolved organic carbon (DOC) (and nutrients); controls the HRT, the temperature, and salinity of the effluent; and allow for a thriving microflora (Kuyucak and St-Germain 1994; Waybrant et al. 2002; Neculita et al. 2007; Neculita and Zagury 2008; Ben Ali et al. 2019a, 2019b, 2020). The components and proportions of reactive mixture composition are crucial for the short- and long-term performance of passive biochemical treatment (Waybrant et al. 1998; Zagury et al. 2006; Neculita and Zagury 2008; Neculita et al. 2011). There is consensus that single organic substrates do not significantly promote the activity of SRB, and higher sulfate-reduction rates have been obtained with reactive mixtures containing more than one organic carbon source (Waybrant et al. 1998; Cocos et al. 2002; Gibert et al. 2003; Zagury et al. 2006). Effective reactive mixtures contain relatively biodegradable DOC sources (poultry manure, cow manure, or municipal sludge) as short-term substrates, as well as more recalcitrant ones (sawdust, hay, alfalfa, woodchips, peat moss etc.), with high C/N ratios, as long-term substrates (Neculita et al. 2007). Since SRB cannot degrade complex organic carbon (carbohydrates, proteins, lipids, cellulose, and hemicellulose) (Postgate 1984), the synergism between groups of microorganisms (acidogens, methanogens, and SRB), in mixed communities, is also required to provide short-chain organic carbon, available for SRB in the longer term (Kuyucak and St-Germain 1994; Vasquez et al. 2018). The thickness of reactive mixture could also greatly influence treatment efficiency, with around 1 m deemed optimal, as based on documented case studies (URS 2003; Yim et al. 2015).

The HRT is another important influence factor (see Table 11.1). Longer HRT could lead to a better efficiency but reduce the hydraulic properties of the reactive mixture, the lifespan, and long-term efficiency of PBR (Neculita et al. 2008a; Vasquez et al. 2016a, 2016b, 2018). A longer HRT may also lead to the depletion of either the available organic matter source or the sulfate source for SRB. Hence, although the quality of treated AMD (highly contaminated) can be significantly improved with a longer HRT compared to a shorter one, the porosity and the permeability of the reactive mixture can drastically be reduced with the longer HRT (Neculita et al. 2008a). Typically, an HRT of four or more days is often required, especially for efficient treatment of highly contaminated AMD as it is accepted that bio-precipitation of metal sulfides occurs in at least three to five days (Younger et al. 2002; URS 2003). However, a too short HRT may sometimes not allow adequate time for SRB activity to neutralize acidity and precipitate metals or may result in biomass being washed out of the bioreactor (Neculita et al. 2007). Therefore, the HRT must balance the need to respect specified discharge limits for metals and acidity against the need to limit problems related to the deterioration of hydraulic properties of reactive mixtures (Neculita et al. 2008a).

Among the in situ factors of influence on sulfate reduction in passive biochemical systems, temperature is one of the most crucial. High temperatures – around 30°C – could partially explain the greater concentrations of sulfate removal (up to 2.5 g/L) observed by Vasquez et al. (2016b).

More recent research findings also show that, in laboratory testing, the detrimental impact on the microbial microflora of high salinity was lesser than low temperature for AMD/CND efficient treatment (Ben Ali et al. 2019b, 2020). However, microflora composition was reported stable over the 14-month operating period despite temperature shifts from 17°C to 5°C in a semi-passive in situ pilot-scale PBR, at a closed mine in the Yukon Territory, Canada (Nielsen et al. 2018).

Sorption mechanisms (adsorption, surface precipitation, cation exchange, and polymerization on inorganic support, solid organic matter, bacteria, and metal precipitates), in addition to precipitation, which occur during the AMD passive biochemical treatment, are also factors of influence (Neculita et al. 2007). Sorption is an important removal process for metals and sulfate from AMD, especially upon the startup of the passive bioreactor, whereas metal sulfides precipitation becomes the predominant mechanism once sulfate-reducing conditions are established (USEPA 2014; Genty et al. 2017). Main parameters describing the sorption are isotherms and kinetics. Given the large variation of AMD chemistry and substrates' composition, a literature review revealed that it is very hard to normalize the data on these parameters reported from the different available studies (Westholm et al. 2014). For example, organic residual materials (manures, municipal wastewater sludge, and compost) showed higher sorption capacity than cellulosic waste (maple chips and sawdust) for Fe removal from Fe-rich AMD (4 g Fe/L) in PBR batch testing (Genty et al. 2017). Various physicochemical properties of the filter materials tested (e.g., pH, specific surface, and particle grain size) and experimental testing conditions (e.g., solid to liquid ratio, AMD chemistry, hydraulic loading, and temperature) may also entail the exposure of original solid material to precipitation and lead to a new solid surface with different characteristics. The new layer of iron precipitates on the surface of solids from reactor filling materials could also have different sorption characteristics relative to the original surface.

11.2.3 Passive Multistep Systems

11.2.3.1 Principle

Research and field experience show that passive treatment systems perform better when used to complement mine drainage prevention measures (USEPA 2014) (see Chapters 4 to 9). However, despite the application of reclamation methods, AMD on several mine sites is often highly contaminated with metal/metalloids (Fe, Ni, Mn, Zn, As, etc.) (Genty et al. 2016, 2018, 2020; Macías et al. 2017a, 2017b). Iron is the most common contaminant in highly contaminated AMD, with concentrations >0.5 g/L and often up to several hundred g/L (Nordstrom et al. 2015). Combinations of two or more passive treatment units (multistep systems or trains) have been developed (Ayora et al. 2013; Genty et al. 2016; Macías et al. 2017a; Rakotonimaro et al. 2018a) to treat such AMD. They consist of chemical and biochemical, and aerobic and anaerobic/anoxic units, in which mine drainage treatment is performed via various processes, for maximal Fe removal in pre-treatment units, followed by progressive residual contamination polishing in the subsequent units (Figures 11.6 and 11.7).

11.2.3.2 Classification

A combined, compact system consisting of an ALD (minimum depth of 0.5 to 0.6 m of a limestone layer) and a permeable organic substrate (depth of 0.5 to 1 m, overlaying the limestone) is known as vertical flow pond (VFP) or successive alkalinity-producing system or reducing and alkalinity-producing system. The common substrate is mushroom compost, consisting of a mixture of manure, wood shavings, and limestone. In this design, the substrate's main purpose, in addition to increasing the dissolution of organic material in the mine drainage and DO elimination, is to favor Fe^{3+} to Fe^{2+} reduction. In a VFP, the input mine drainage is fed at the top and flows down through a mushroom compost layer (for dissolution of organic matter and creation of reducing conditions), then flows through the limestone below (and gains alkalinity), and finally exits the system at the base, into a settling pond. The pH raises result in Al, Cu, and Fe precipitation at the base of the VFP and in the settling pond (USEPA 2014).

Moreover, configuration as individual, separate units has the advantage of facilitating long-term maintenance, such as the replacement of a deficient or clogged unit (Rakotonimaro et al. 2018a). Hence, the multistep passive treatment system consisting of individual units known as dispersed alkalinity substrate (DAS), which is a mixture of a coarse organic material (e.g., wood shavings) and a neutralizing agent (e.g., calcite-$CaCO_3$ or magnesia-MgO), separated by aeration ponds, showed promising results for metal removal from highly contaminated AMD, including Fe (up to 1.5 g/L) and Zn (up to 0.6 g/L) (Rötting et al. 2008a, 2008b; Ayora et al. 2013; Genty et al. 2016, 2017; Rakotonimaro et al. 2016, 2017a, 2017b, 2018a; Macías et al. 2017a).

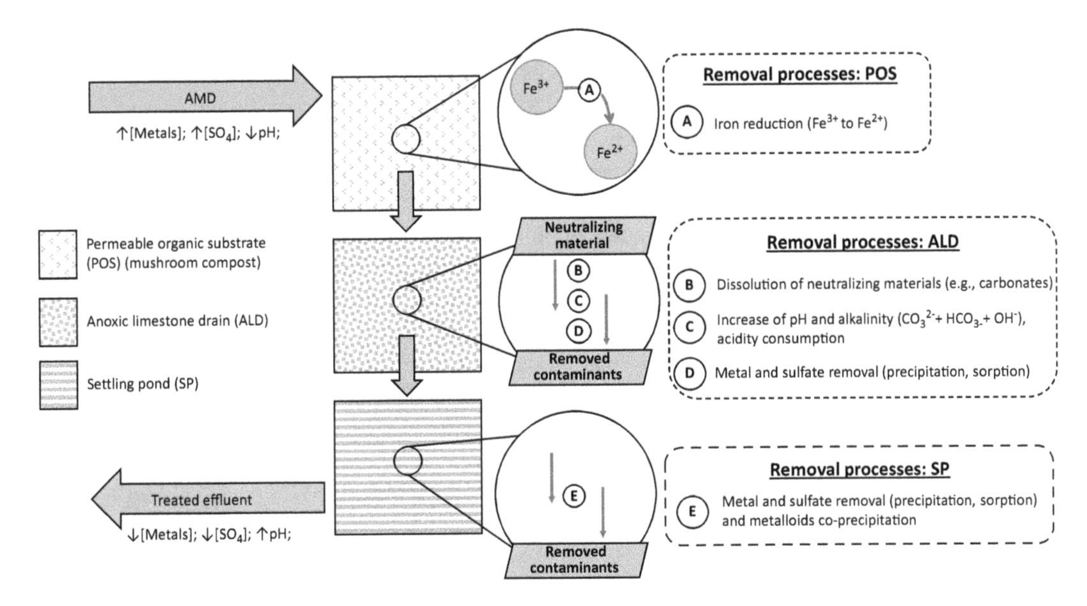

FIGURE 11.6 Scheme of a passive multistep system for the treatment of AMD.

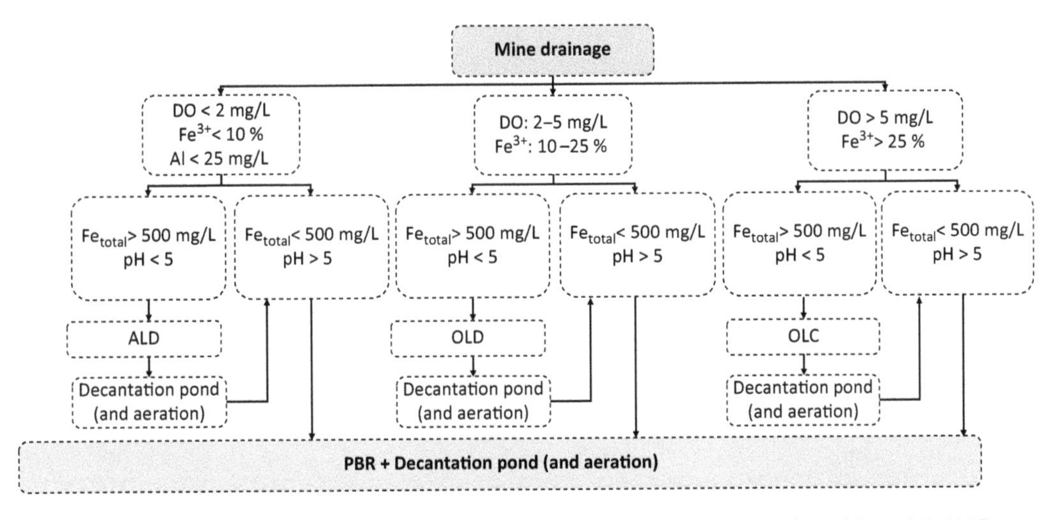

FIGURE 11.7 Selection criteria for units of a treatment system for highly contaminated iron-rich AMD.

The treatment steps aim to control contaminants' concentration at the final effluent by using the following principles and sequences of processes and removal mechanisms (Figures 11.6 and 11.7):

a) Metals are removed by sorption (at neutral pH, until the saturation of available sites) and precipitation (trivalent metals, i.e., Fe^{3+} and Al^{3+} at acidic pH, whereas bivalent metals are precipitated at neutral to alkaline pH). In addition, chemical (Genty et al. 2012b; Ayora et al. 2013; Macías et al. 2017a; Rakotonimaro et al. 2018a) or biological (Florence et al. 2016) Fe pre-treatment (down to around 0.5 g/L), using DAS units or natural Fe-oxidizing lagoons is advantageous to improve the efficiency of the subsequent treatment steps (e.g., PBRs). Moreover, As, which is a common metalloid in mine drainage, is usually removed by co-precipitation with Fe, even by natural attenuation processes (Ayora et al. 2013).
b) Sulfate is removed by sorption, precipitation (e.g., gypsum, schwertmannite), and bacterial reduction.

c) Among bivalent metals, Mn is particularly difficult to remove because of its complex chemistry, including seven oxidation states, three of which (+2, +3, and +4) are common in mine drainage (Neculita et al. 2007; Neculita and Rosa 2019). Hence, Mn treatment in mine drainage is prone to variation in efficiency, which depends on several factors, including pH, Mn, Fe, and DO concentrations, Fe to Mn ratio, catalysis by Mn or Fe solids, presence of ligands, and bacteria. Some causes of limited efficiency of Mn removal in PBRs are the high solubility of Mn minerals relative to other metals; inhibition of Mn precipitation at a molar ratio Fe/Mn >4; reductive dissolution of Mn oxides by organics, sulfides, and Fe(II); and the high pH (>8) required for oxidation and precipitation (Song et al. 2012a, 2012b; Calugaru et al. 2018a; Neculita and Rosa 2019).
d) Aeration cascades (mostly for the oxidation of iron and residual organic material) and decantation ponds of precipitates must be included as intermediate and/or last polishing steps.

11.2.3.3 Factors of Influence

The factors of influence on the efficiency of passive multistep treatment systems depend on whether chemical or biochemical units are used (Figure 11.6). The main factors in those types of units have been identified and discussed above (see Sections 11.2.1 and 11.2.2).

11.3 DESIGN APPROACH – PRACTICAL METHODOLOGY

The methodology for the design of a passive system, chemical or biochemical, for AMD treatment consists of progressive upscaling, from small (a few mL) to medium (a few L) and then to higher pilot scale (a few m^3) and full scale (hundreds to thousands m^3) (Figure 11.8). The objective of small-scale testing is to conduct parametric studies with as many variables (water chemistry, material(s) mixtures, environmental conditions, including temperature, salinity, etc.) as possible to assess their relative influence on treatment efficiency and contaminant removal mechanisms (precipitation vs. sorption). The most promising scenarios are then tested in upscaled experiments to select and confirm the optimal design parameters. As general practice, the initial testing usually begins with synthetic AMD, as surrogate of real AMD, and, if possible, a final testing with actual effluents is conducted (in laboratory or field pilot testing), before the construction of the full-scale passive treatment system.

Once the mine site is identified, the design phases can be summarized as follows: (1) mine water and materials characterization, and (2) laboratory and field testing (Figure 11.8).

11.3.1 Mine Water and Materials Characterization

The first step of the design involves characterizing the AMD quality and flow rates, and performing a materials' characterization. Flow rates must be thoroughly measured over several seasons and various weather conditions, from normal to extreme conditions. Information on field parameters (morphology, slope, water table level, weather, hydrological balance, etc.) must also be collected. Then, the proximity and availability of a source of reactor construction materials should be assessed. Available materials are sampled and characterized for their reactivity, purity, and composition; physicochemical (e.g., pH, elemental composition, DOC, metals, and sulfates); hydrogeochemical and hydrological (e.g., porosity, permeability, and saturated hydraulic conductivity); microbiological (e.g., total aerobic/anaerobic heterotrophic/autotrophic bacteria, and SRB and iron-reducing bacteria); and mineralogical (e.g., scanning electron microscopy, X-ray diffraction) characteristics.

11.3.1.1 Limestone/Dolomite Drains

The rates of acidity and metal removal, as well as of alkalinity generation, were developed empirically (Watzlaf et al. 2004). Based on the collected data from more than 20 limestone drains (oxic

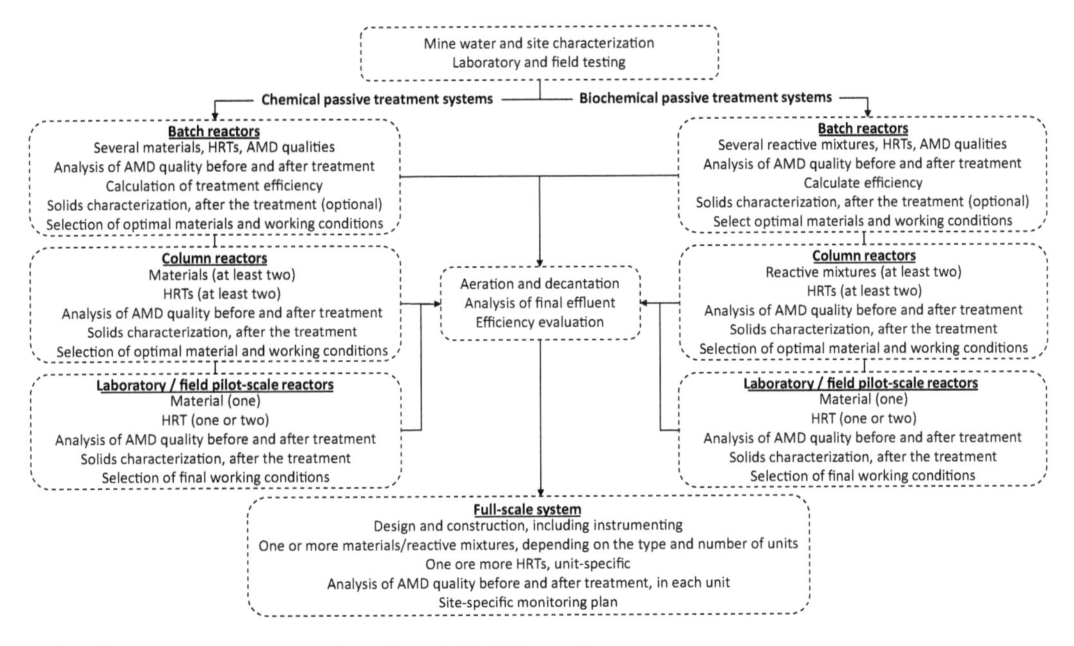

FIGURE 11.8 Design methodology for a field AMD passive treatment system.

and anoxic), with respect to HRT and the chemistry of influent and effluent water, Equation 11.9 was derived to estimate the necessary initial mass of limestone (Hedin and Watzlaf 1994).

$$M_0 = Q\left[\left(t_L C_M / X_{CaCO_3}\right) + \left(t_d \rho_B / n\right)\right] \tag{11.9}$$

where Q is the volumetric flow rate, t_L is the longevity or duration of treatment, C_M is the maximum alkalinity concentration, X_{CaCO_3} is the limestone weight fraction as $CaCO_3$, t_d is the HRT, ρ_B is the bulk density, and n is the porosity. Using this equation, reasonable limits on the size and cost for the treatment system were found for a 20-year longevity of limestone, HRT of 15 h, and porosity of 0.49 (Hedin et al. 1994a, 1994b). Later, the cubitainer test for designing ALDs and predicting their performance was proposed (Cravotta 2003). The cubitainer test (4 kg of crushed limestone, completely inundated with 2.8 L AMD), conducted for 11 to 16 days, could provide comparable estimates to long-term field data (5 to 10 years) on the rates of alkalinity production and limestone dissolution. Given the slower dissolution of dolomite relative to calcite (Busenberg and Plummer 1982), HRTs longer than 15 h should be designed when dolomite is used as a neutralizing agent in drains, compared to about 7 to 8 h for calcite (Bernier et al. 2001). Nevertheless, although the dolomite reactivity is slower than calcite, uncoated dolomite reactivity was found higher than gypsum-coated calcite after only one month of gypsum growth (Huminicki and Rimstidt 2008).

11.3.1.2 Passive Biochemical Systems

In the case of PBRs, the design is based on the following equation, derived from Darcy's law:

$$Q = \frac{V_v}{HRT} = \frac{V_t \times n_{eff}}{HRT} \tag{11.10}$$

where V_t is the total volume (m^3), V_v is the void volume (m^3), n_{eff} is the effective porosity (interconnected voids), and Q is the mean flow rate (m^3/s) (HRT, in s). While Q is measured on site, n is estimated at the design stage (on the selected construction materials), and the HRT is selected among at least two testing values, based on the results of the comparative performance in column and/or pilot reactors. As a guide, the porosity of established compost-filled reactors ranges from 0.15 to 0.35;

however, it is recommended that the saturated hydraulic conductivity and porosity of specific materials prior to final selection and system sizing are tested (Younger et al. 2002). Finally, the V_t of the final treatment system is calculated. In principle, there is no allowable limit on HRT, as longer HRT allows for the efficient treatment of lower quality AMD. Nevertheless, as the precipitation of metal sulfides usually occurs for at least three to five days (URS 2003), field treatment systems consisting of one unit rarely exceed the five days of HRT (Younger et al. 2002).

11.3.1.3 Multistep Treatment Systems

In the case of multistep treatment systems, the same principles apply as with individual units, while the total HRT is variable and site dependent and AMD quality dependent. As an example, the total HRT of a tri-unit field-pilot passive treatment system constructed and successfully operated on the Lorraine mine site, QC, Canada, was 12 days (Genty et al. 2016; Rakotonimaro et al. 2018a).

11.3.2 Laboratory and Field Testing Prior to the Construction of a Field Treatment System

Regardless of the number of units and type of passive treatment (chemical, biochemical, or mixed), once the materials and water to be treated have been characterized, the next stage is to scale up the system, from the laboratory to field testing (Figure 11.8). The design considers the rates of metal and acidity removal and of alkalinity generation, which are developed empirically (Watzlaf et al. 2004). The main design steps are as follows:

Step 1: The first step is batch testing with different variables, such as source and characteristics of the materials to be used, solid to liquid ratio, temperature, and chemistry of mine water. The tests are performed in small to medium reactors (up to 1 L) to treat synthetic AMD under no-flow conditions. This allows for the selection of the most efficient materials under given conditions (Waybrant et al. 1998; Cocos et al. 2002; Zagury et al. 2006; Neculita and Zagury 2008; Neculita et al. 2011; Rakotonimaro et al. 2016; Vasquez et al. 2016a; Genty et al. 2017). The cubitainer method could be used to predict the alkalinity within an ALD and estimate the mass of limestone required for its construction. For a given quality of limestone and AMD quality (and flow rates), the HRT is determined in laboratory (by assessing the comparative performance with two to three tested HRTs). Limestone porosity is also measured, while a 20-year lifespan of the system is usually considered.

Step 2: Column testing of the most efficient materials identified in step 1 are performed with synthetic AMD under continuous flow conditions, in reactors from a few liters to a few hundreds of liters, at different HRTs and configurations. This testing allows for the selection of optimal operating conditions (Waybrant et al. 2002; Johnson and Hallberg 2005b; Neculita et al. 2008a, 2008b; Rötting et al. 2008b; Song et al. 2012a, Genty et al. 2017; Rakotonimaro et al. 2017b). Some studies also assessed the comparative toxicity of AMD vs. treated effluent from laboratory column testing (Neculita et al. 2008c) or field-scale PBRs (Butler et al. 2011).

Step 3: The third step is laboratory and/or field pilot testing, with larger reactors (a few m^3) and use of the most promising operating conditions selected in step 2, to improve the understanding of the influence of scaling up and/or feeding with real AMD on the treatment system behavior (Song et al. 2012b; Yim et al. 2015; Genty et al. 2016; Rakotonimaro et al. 2018a). Site-specific conditions, including periodic events such as a wet period (heavy rains and snow melting), a dry period (low water flow rates in the summer), and freeze-thaw cycles, may influence the selection and sizing of passive treatment systems and must be considered prior to the final design of the full-scale systems.

Step 4: The use of modeling tools is beneficial to finalize the design (Ben Ali et al. 2019a). Numerical modeling effectively predicted the hydrogeochemical behavior of passive treatment at different scales: batch reactors (Hemsi et al. 2005), column reactors (Amos et al. 2004; Viggi et al. 2011; Pagnanelli et al. 2014), and field reactors (two years; Drury 2000; over three years, Benner et al. 2002; Mayer et al. 2006). Models were able to simulate organic solids degradation by SRB and metal precipitation (Drury 2000; Amos et al. 2004; Hemsi et al. 2005), contaminants sorption (Viggi et al. 2011; Pagnanelli et al. 2014), and reactive transport modeling (e.g., MIN3P; Benner et al. 2002; Mayer et al. 2006). These studies showed the importance of the kinetics used to describe the decomposition of organic solids and metal precipitation (Hemsi et al. 2005) and the major contribution of sorption to contaminants removal (Viggi et al. 2011; Pagnanelli et al. 2014). The temperature dependence of the sulfate reduction rate can also be simulated (Benner et al. 2002).

Step 5: The last step is to estimate the treatment costs. Estimates can be made by using the AMDTreat computer application (https://amd.osmre.gov/). This application considers several passive systems and provides financial and scientific tools, including a long-term financial forecasting module and calculators for acidity and sulfate reduction, Langelier saturation index, mass balance, passive treatment alkalinity, abiotic and biotic homogeneous Fe^{2+} oxidation, and metric conversion. A specific techno-economical study can also be performed (Rakotonimaro 2017; Calugaru 2019).

As a final remark related to the design of passive treatment systems, it is worth mentioning that better efficiency is usually achieved in laboratory testing (under controlled conditions, including the synthetic AMD chemistry used in laboratory settings) than in the field (Neculita et al. 2007). Moreover, due to the simplistic setup, more optimistic performance is predicted from batch testing compared to column-type testing.

11.4 CASE STUDIES

Compilations of case studies can be found in the literature (details on some cases are presented in Table 11.1). The main findings of these compilations (mainly from coal mines) are summarized below.

a. Watzlaf et al. (2000) assessed the long-term performance of ten ALDs based on the findings of a 10-year monitoring period of influent and effluent water data. Results showed that alkalinity in the effluent ranged from 80 to 320 mg/L $CaCO_3$, with near maximum levels reached after about 15 h of HRT and consistent alkalinity over time when influent contained less than 1 mg/L of Fe^{3+} and Al.
b. Rose and Dietz (2002) compiled data on 30 VFPs, including inflow and outflow chemistry, flow rates, dimensions, design features, and problems encountered. The results show that all systems produced a positive increase in net alkalinity (even though ranging widely from 7 to 686 mg/L $CaCO_3$), with better performance for less contaminated inflow (acidity loading less than 40 g/m^2/day produced net alkalinity) and increased HRT. A value of 25 g/m^2/day acidity was suggested as a design criterion for VFP, instead of HRT.
c. Ziemkievicz et al. (2003) evaluated 83 systems of five types (ALD, OLC, limestone leach bed, slag leach bed, and VFP), with age from 1 to 12 years, for their efficiency in acidity treatment. The results showed average acidity treated ranging from 9.9 t/year for OLCs to 22.2 t/year for ALDs, with ALDs and anaerobic wetlands as the least and the most expensive systems, respectively. Acidity reduction occurred at HRT higher than 15 h (ALD and LSB), while removal (ALD, OLC, and LSB) ranged from 20 to 50 g/day/t.
d. Watzlaf et al. (2004) empirically developed the rates of metal and acidity removal and alkalinity generation ("convenient pre-construction rule-of-thumb") in the most common types of

passive technologies: aerobic wetlands (Fe removal from alkaline water: 10 to 20 $g/m^2/day$), ALDs (add 150 to 300 mg/L of alkalinity in about 15 h of HRT, imparting 5 to 20 mg/L of alkalinity per hour of contact), and VFPs (add 15 to 60 $g/m^2/day$ of alkalinity).

e. Skousen and Ziemkiewicz (2005) assessed 116 systems of eight types (ALD, OLC, limestone leach beds, slag leach beds, aerobic and anaerobic wetlands, sulfate-reducing bioreactors, and VFP) in eight states, with ages between 2 and 12 years. The results showed that 105 systems reduced the acidity (90%). Average acid load reductions were: 0.8 t/year for aeration/decantation ponds; about 9 t/year for OLC, anaerobic and aerobic wetlands, and vertical flow wetlands; 76 t/year for SLB; and about 15 t/year for LSB and ALD. Average acidity removal rates varied from 18 to 2.3×10^3 g/day/t for the limestone systems and from 1.7 to 87 $g/m^2/day$ for the ponds and wetlands. Some systems showed large declines in efficiency, while most passive systems were effective for more than five years.

f. Skousen et al. (2017) reviewed the current state of development of passive treatments systems, described results for various types of treatment systems, and provided guidance for their design and effective operation. Their compiled design factors for the main passive systems are as follow: aerobic wetland (10 g $Fe/m^2/day$; 1 g $Mn/m^2/day$), anaerobic wetland (3.5 g acidity/m^2/day; 10 g $Fe/m^2/day$), VFP (35 g acidity/m^2/day), Mn removal beds (2 to 10 g $Mn/m^2/day$), bioreactors (low flow rates; readily degradable organics), ALD (15 h of HRT; 50 g acidity/m^2/day), OLC (30 g acidity/m^2/day), limestone leach bed (2 h of HRT; 10 g acidity/m^2/day), and steel slag leach bed (1000 g acidity/m^2/day).

g. Multistep treatment system: The Wheal Jane mine, located in the Carnon Valley, Cornwall, UK, was exploited for tin and other metals for several hundred years prior to its closing in 1991 (Johnson and Hallberg, 2005b). Then, in 1992, an adit plug failed and released about 50,000 m^3 of moderately contaminated AMD (Table 11.1) to the Fal Estuary and Carnon River. An active treatment plant was constructed and became operational in 2000. Then, a multistep treatment system was constructed and operated temporarily from 2000 to 2002. The passive system consisted of three alternative pretreatments to increase the pH (lime dosing, ALD, and lime free). Water then flowed through aerobic wetlands (for Fe and As removal), PBRs (for other metals and sulfate removal), and, finally, an aerobic rock filter to promote Mn removal. The system increased the pH from 3 to up to 6.8 and decreased metals and sulfate concentrations efficiently (see Table 11.1).

h. Passive multistep treatment system DAS type: The Iberian Pyrite Belt, in southwest Spain and Portugal, has about 100 abandoned mine waste impoundments and galleries that release high volumes of AMD to the Tinto and Odiel Rivers (Ayora et al. 2013; Macías et al. 2017a, 2017b). To avoid coating, passivation, and clogging of traditional passive treatments with highly contaminated AMD, after more than 10 years of laboratory and field testing, the DAS-type systems were developed. They consist of a mixture of highly porous materials (such as woodchips and shavings) and neutralizing agents (calcite or caustic magnesia for trivalent and bivalent metal removal, respectively; Table 11.1). A limestone-DAS system proved efficient for 20 months of operation under field conditions (480 m^3 volume; designed to treat 1 L/s at 5.5 days of HRT for AMD with pH 2.9, 126 mg/L Al, 664 mg/L Fe, 17 mg/L Cu, 18 mg/L Zn, 3.2 g/L sulfate, and 4.6 g/L acidity), during which more than half of the initial reactive substrate was consumed without any significant armoring problems. This design showed better performance than other passive treatment (e.g., VFP) for highly contaminated AMD. Nevertheless, due to the lack of previous experience and ongoing DAS testing at full scale, construction engineering guidelines are not yet available for these systems.

i. Passive multistep system at the Lorraine mine site (QC, Canada; Genty et al. 2016; Rakotonimaro et al. 2018a). The rehabilitation of this mine site started with AMD prevention using a cover with capillary barrier effect (CCBE; see Chapter 7). To treat the already contaminated pore water, different passive treatment systems have been implemented since the site reclamation in 1999. The first treatment system consisted of one limestone drain and three anoxic dolomitic drains (DOL-1 to DOL-3), which were installed to treat the Fe-AMD that was flowing, by gravity, from underneath the already CCBE-covered tailings (Genty et al. 2016; Rakotonimaro et al. 2018a). The drains DOL-1 and DOL-2 were found efficient in polishing the iron-rich AMD even almost 20 years (Rakotonimaro et al. 2017b). Two of the drains (DOL-1 and DOL-2) showed efficient results for pH increase (from 3.6 to up to 6.5), alkalinity production (up to 200 mg/L), and metal and sulfate removal (up to 100% Al, 89% Cd, 59% Co, 91% Cr, 89% Fe, 96% Ni, 95% Pb, 85% Zn, and 77% sulfate) (Potvin et al. 2009; Potvin 2009). However, over time, high Fe and sulfate concentrations were observed in the exit of the DOL-3 drain. Moreover, Fe-AMD began to accumulate on the top of the drain, as an indication of its progressive clogging (Potvin 2009). Factors of influence for the limited performance of the Lorraine treatment system include dolomite-rich limestone particles coating, porosity reduction by precipitation of secondary minerals, and unsaturated portions of the drains affecting their hydrogeochemical behavior. In 2011, to remediate this issue, the DOL-3 drain was replaced by a tri-step passive treatment system composed of two PBRs separated by a wood ash (WA) filter (Genty 2012). The first unit, PBR1, was designed to reduce the redox potential, neutralize the acidity, and partially remove metals. The second unit, WA, was aimed for Fe treatment. Finally, the PBR2 was intended to act as a polishing unit to remove the residual metals. Based on five-year survey data, Fe and sulfate removal exceeded 69% and 79%, respectively, while the pH remained at ~5.8 (Genty et al. 2016; Rakotonimaro et al. 2018a).

11.5 ADVANTAGES AND LIMITS

Regardless of their design and the biogeochemical processes involved, passive treatment systems are acknowledged as convenient and efficient for AMD mitigation and control, especially when used as a complement of prevention measures to limit AMD generation. The main reason is that the lifespan of such systems is reduced when they are used for the treatment of AMD characterized by high concentrations (hundreds to thousands of mg/L) of Al and Fe. Their main advantages include the use of underused or residual materials, satisfactory performance (under specific conditions), aesthetics, and lower costs compared to active technologies.

However, despite the undeniable advantages, passive treatment has several drawbacks, including the difficulty of handling large flows without occupying large areas; the difficulty of treating high concentrations of Al, Fe, and Mn; the unpredictability of long-term performance (various factors of influence: flow and quality, and site- and zone-related specificity); as well as the physical alteration and loss of reactivity (clogging and coating). The geochemical and environmental behavior of the contaminant-rich residues at the end of the lifespan of passive systems is also an issue. More recently, it was found that there is inconsistency among regulation with respect to the classification of these solids, which could regenerate contaminated mine drainage; furthermore, the freezing/thawing cycles could enhance the rates of contaminants leaching, and a solidification/stabilization treatment is required before their disposal in landfills (Jouini et al. 2019a, 2019b, 2019c, 2020a, 2020b, 2020c).

11.6 PERSPECTIVES AND RESEARCH NEEDS

To limit coating/passivation (loss of reactivity) and clogging (loss of permeability) caused by secondary minerals (e.g., gypsum and metal oxides-hydroxides) precipitated during the treatment, the innovative approach known as DAS has been developed (Rötting et al. 2008a, 2008b, 2015; Macías et al. 2012a, 2012b, 2012c, 2017a, 2017b; Ayora et al. 2013). More research is needed on DAS systems in the remediation of iron-rich AMD in closed and abandoned mine sites worldwide (Ayora et al. 2013; Rakotonimaro et al. 2018a). Multistep treatment may involve the grouping of DAS units with cascade aeration (Rötting et al. 2008a) or with decantation pond steps (Caraballo et al. 2009,

2011; Macías et al. 2012b, 2012c), which would extend the lifespan and treatment efficiency of the DAS with highly contaminated iron-rich AMD (Rakotonimaro et al. 2016, 2017a, 2017b).

Another option to prevent, limit, and treat water pollution by mine drainage involves modified materials (natural and residual). The modification (e.g., charring, alkaline fusion, hydrothermal treatment, and grafting of functional groups) enhances the removal capacity of metals and metalloids, as well as the physical and chemical stability of the materials and the sludge. Charring of dolomite was identified as one of the promising avenues for enhancing its surface characteristics (porosity and CEC), its alkaline character and neutralizing capacity, and its sorption capacity for several contaminants in AMD and CND (Calugaru et al. 2016, 2017, 2018b, 2019, 2020). Element recovery, material regeneration, water reuse, evaluation of treatment efficiency for real effluents as well as the applicability of these materials in active and passive treatment systems are thoroughly discussed in an extensive literature review (Calugaru et al. 2018b).

Despite the temporarily satisfactory performance of the multistep treatment systems on some mine sites (e.g., Iberian Pyrite Belt, Spain; Lorraine mine site, Quebec), more research and systematic compilation of available data at the pilot-scale level is necessary before design criteria can be established for DAS-based systems. New knowledge is also required for the prediction of their long-term efficiency (Macías et al. 2017b; Rakotonimaro et al. 2017b, 2018a). The stability of metal-rich solids produced during passive treatment and their sustainable long-term storage is relatively rarely studied (Macías et al. 2012a, 2017b; Rakotonimaro et al. 2017a; Jouini et al. 2019a, 2019b, 2019c, 2020a, 2020b, 2020c). The presence of untreated thiosalts (source of delayed acidity; see Chapter 1) in treated mine drainage might deteriorate treatment efficiency and sludge stability. The geochemical behavior and toxicity of thiosalts in mixed mine effluents is also poorly understood (Neculita et al. 2020). The low temperature and salinity effects on the efficiency of passive treatment is another area where more research is needed, including the modeling (Ben Ali 2019), as data show enhanced adverse effects of cold temperatures with salinity on AMD treatment in PBRs (Ben Ali et al. 2019a, 2019b, 2020).

Modified materials might prove efficient not only in the treatment of mine drainage but also in the prevention of mine drainage generation through amendments by immobilizing the metals and metalloids in pore water and limiting their leaching (Calugaru et al. 2018b; Rakotonimaro et al. 2018b). Further research is also necessary to evaluate their field-scale performance and to complete cost calculations (Calugaru 2019).

Finally, mixes of contaminants from several sources within the same effluents (e.g., leaching water from waste rock dumps containing mine-drainage-generating minerals and derived from the release of undetonated explosives and blasting residual minerals, such as ammonia nitrogen, chloride, and perchlorate) is one of the emerging challenges in mine drainage treatment.

REFERENCES

Amos, R., U. Mayer, D. W. Blowes, and C. J. Ptacek. 2004. Reactive transport modeling of column experiments for the remediation of acid mine drainage. *Environmental Science and Technology* 38, no. 11 (April): 3131–3138. doi:10.1021/es0349608.

Ayora, C., M. A. Caraballo, F. Macías, T. S. Rötting, J. Carrera, and J. M. Nieto. 2013. Acid mine drainage in the Iberian Pyrite Belt: 2. lessons learned from recent passive remediation experiences. *Environmental Science and Pollution Research* 20, no. 11 (November): 7837–7853. doi:10.1007/s11356-013-1479-2.

Bakatula, E., D. Richard, C. M. Neculita, and G. J. Zagury. 2018. Determination of point of zero charge of natural organic substrates. *Environmental Science and Pollution Research* 25, no. 8 (March): 7823–7833. doi:10.1007/s11356-017-1115-7.

Ben Ali, H. E. 2019. Traitement passif du drainage minier à faible température et forte salinité. PhD diss., Université du Québec en Abitibi-Témiscamingue, Canada, 256, https://depositum.uqat.ca/id/eprint/810.

Ben Ali, H. E., C. M. Neculita, J. W. Molson, A. Maqsoud, and G. J. Zagury. 2019a. Performance of passive systems for mine drainage treatment at low temperature and high salinity: A review. *Minerals Engineering* 134 (April): 325–344. doi:10.1016/j.mineng.2019.02.010.

Ben Ali, H. E., C. M. Neculita, J. W. Molson, A. Maqsoud, and G. J. Zagury. 2019b. Efficiency of batch biochemical reactors for mine drainage treatment at low temperature and high salinity. *Applied Geochemistry* 103 (April): 40–49. doi:10.1016/j.apgeochem.2019.01.014.

Ben Ali, H. E., C. M. Neculita, J. W. Molson, A. Maqsoud, and G. J. Zagury. 2020. Salinity and low temperature effects on the performance of column biochemical reactors for the treatment of acidic and neutral mine drainage. *Chemosphere* 243 (March): 125303. doi:10.1016/j.chemosphere.2019.125303.

Benner, S. G., D. W. Blowes, C. J. Ptacek, and K. U. Mayer. 2002. Rates of sulfate reduction and metal sulfide precipitation in a permeable reactive barrier. *Applied Geochemistry* 17, no. 3 (March): 301–320. doi:10.1016/S0883-2927(01)00084-1.

Bernier, L. R., M. Aubertin, A. M. Dagenais, B. Bussière, L. Bienvenu, and J. Cyr. 2001. Limestone drain design criteria in AMD passive treatment: Theory, practice and hydrogeochemistry monitoring at Lorraine mine site, Témiscamingue. CIM Minespace 2001, Annual Meeting Proceedings, Technical Paper no. 48, 9p.

Bernier, L. R., M. Aubertin, C. Poirier, and B. Bussière. 2002. On the use of limestone drains in the passive treatment of acid mine drainage (AMD). In *Proceedings of the 1st Symposium on the Environment and Mines*, Rouyn-Noranda, QC, Canada, November, 19p.

Blowes, D. W., C. J. Ptacek, S. G. Benner, C. W. T. McRae, T. A. Bennett, and R. W. Puls. 2000. Treatment of inorganic contaminants using permeable reactive barriers. *Journal of Contaminant Hydrology* 45, no. 1–2 (September): 123–137. doi:10.1016/S0169-7722(00)00122-4.

Busenberg, E., and L. N. Plummer. 1982. The kinetics of dissolution of dolomite in CO_2–H_2O systems at 1.5–65°C and 0–1 atm P_{CO2}. *American Journal of Science* 282, no. 1: 45–78.

Butler, B. A., M. E. Smith, D. J. Reisman, and J. M. Lazorchak. 2011. Metal removal efficiency and ecotoxicological assessment of field-scale passive treatment biochemical reactors. *Environmental Toxicology and Chemistry* 30, no. 2 (February): 385–392. doi:10.1002/etc.397.

Calugaru, I. L. 2019. Amélioration de l'efficacité du traitement du drainage minier par les matériaux naturels et résiduels modifiés. PhD diss., Université du Québec en Abitibi-Témiscamingue, Canada, 232, https://depositum.uqat.ca/id/eprint/812.

Calugaru, I. L., T. Genty, and C. M. Neculita. 2018a. Treatment of manganese in acid and neutral mine drainage using modified dolomite. *International Journal for Environmental Impacts* 1, no. 3: 323–333. doi:10.2495/EI-V1-N3-323-333.

Calugaru, I. L., C. M. Neculita, T. Genty, B. Bussière, and R. Potvin. 2016. Performance of thermally activated dolomite for the treatment of Ni and Zn in contaminated neutral drainage. *Journal of Hazardous Materials* 310 (June): 48–55. doi:10.1016/j.jhazmat.2016.01.069.

Calugaru, I. L., C. M. Neculita, T. Genty, B. Bussière, and R. Potvin. 2017. Removal of Ni and Zn in contaminated neutral drainage by raw and modified wood ash. *Journal of Environmental Science and Health, Part A* 52, no. 2 (January): 117–126. doi:10.1080/10934529.2016.1237120.

Calugaru, I. L., C. M. Neculita, T. Genty, and G. J. Zagury. 2018b. Metals and metalloids treatment in contaminated neutral effluents using modified materials. *Journal of Environmental Management* 212 (April): 142–159. doi:10.1016/j.jenvman.2018.02.002.

Calugaru, I. L., C. M. Neculita, T. Genty, and G. J. Zagury. 2019. Removal efficiency of As(V) and Sb(III) in contaminated neutral drainage by Fe-loaded biochar. *Environmental Science and Pollution Research 26*, no. 9 (March): 9322–9332. doi:10.1007/s11356-019-04381-1.

Calugaru, I. L., C. M. Neculita, T. Genty, and G. J. Zagury. 2020. Removal and recovery of Ni and Zn from neutral mine drainage by thermally activated dolomite and hydrothermally activated wood ash. *Water, Air, and Soil Pollution* (in corrections).

Candeias, C., P. Freire Ávila, E. Ferreira da Silva, A. Ferreira, N. Durães, and J. P. Teixeira. 2015. Water–rock interaction and geochemical processes in surface waters influenced by tailings impoundments: Impact and threats to the ecosystems and human health in rural communities (Panasqueira Mine, Central Portugal). *Water, Air, and Soil Pollution* 226, no. 23: 1–30. doi:10.1007/s11270-014-2255-8.

Caraballo, M. A., F. Macías, T. S. Rötting, J. M. Nieto, and C. Ayora. 2011. Long term remediation of highly polluted acid mine drainage: A sustainable approach to restore the environmental quality of the Odiel river basin. *Environmental Pollution* 159, no. 12 (December): 3613–3619. doi:10.1016/j.envpol.2011.08.003.

Caraballo, M. A., T. S. Rötting, F. Macías, J. M. Nieto, and C. Ayora. 2009. Field multi-step limestone and MgO passive system to treat acid mine drainage with high metal concentrations. *Applied Geochemistry* 24, no. 12 (December): 2301–2311. doi:10.1016/j.apgeochem.2009.09.007.

Champagne, P., P. Van Geel, and W. Parker. 2005. A bench-scale assessment of a combined passive system to reduce concentrations of metals and sulphate in acid mine drainage. *Mine Water and the Environment* 24 (September): 124–133. doi:10.1007/s10230-005-0083-1.

Cocos, I. A., G. J. Zagury, B. Clement, and R. Samson. 2002. Multiple factor design for reactive mixture selection for use in reactive walls in mine drainage treatment. *Water Research* 36, no. 1 (January): 167–177. doi:10.1016/S0043-1354(01)00238-X.

Cravotta, A. C., III. 2003. Size and performance of anoxic limestone drains to neutralize acid mine drainage. *Journal of Environmental Quality* 32, no. 4 (July–August): 1277–1289. doi:10.2134/jeq2003.1277.

Cravotta, A. C., III. 2015. Monitoring, field experiments, and geochemical modeling of Fe(II) oxidation kinetics in a stream dominated by net-alkaline coal-mine drainage, Pennsylvania, USA. *Applied Geochemistry* 62 (November): 96–107. doi:10.1016/j.apgeochem.2015.02.009.

Drury, W. J. 2000. Modeling of sulfate reduction in anaerobic solid substrate bioreactors for mine drainage treatment. *Mine Water and the Environment* 19 (March): 19–29. doi:10.1007/BF02687262.

Florence, K., D. J. Sapsford, D. B. Johnson, C. M. Kay, and C. Wolkersdorfer. 2016. Iron-mineral accretion from acid mine drainage and its application in passive treatment. *Environmental Technology* 37, no. 11 (June): 1428–1440. doi:10.1080/09593330.2015.1118558.

Genty, T. 2012. Comportement hydro-bio-géo-chimique de systèmes passifs de traitement du drainage minier acide fortement contaminé en fer. PhD diss., Université du Québec en Abitibi-Témiscamingue, Canada, 271, https://depositum.uqat.ca/id/eprint/269.

Genty, T., B. Bussière, M. Benzaazoua, C. M. Neculita, and G. J. Zagury. 2017. Efficiency of iron removal during the treatment of highly contaminated acid mine drainage in biochemical reactors. *Water Science and Technology* 76, no. 7–8 (October): 1833–1843. doi:10.2166/wst.2017.362.

Genty, T., B. Bussière, M. Benzaazoua, C. M. Neculita, and G. J. Zagury. 2018. Changes in efficiency and hydraulic parameters during the passive treatment of ferriferous acid mine drainage in biochemical reactors. *Mine Water and the Environment* 37 (January): 686–695. doi:10.1007/s10230-018-0514-4.

Genty, T., B. Bussière, M. Benzaazoua, C. M. Neculita, and G. J. Zagury. 2020. Treatment efficiency of iron-rich acid mine drainage in a tri-unit pilot system. *Environmental Science and Pollution Research* 27 (January): 8418–8430. doi:10.1007/s11356-019-07431-w.

Genty, T., B. Bussière, M. Paradis, and C. M. Neculita. 2016. Passive biochemical treatment of ferriferous mine drainage: Lorraine mine site, Northern Québec, Canada. In *Proceedings of the International Mine Water Association (IMWA)*, Leipzig, Germany, pp. 790–795. http://www.mwen.info/docs/imwa_2016/IMWA2016_Genty_10.pdf.

Genty, T., B. Bussière, R. Potvin, M. Benzaazoua, and G. J. Zagury. 2012a. Dissolution of calcitic marble and dolomitic rock in high iron concentrated acid mine drainage: Application to anoxic limestone drains. *Environmental Earth Sciences* 66 (December): 2387–2401. doi:10.1007/s12665-011-1464-3.

Genty, T., B. Bussière, R. Potvin, M. Benzaazoua, and G. J. Zagury. 2012b. Capacity of wood ash filters to remove iron from acid mine drainage: Assessment of retention mechanism. *Mine Water and the Environment* 31 (August): 273–286. doi:10.1007/s10230-012-0199-z.

Gibert, O., J. de Pablo, J. L. Cortina, and C. Ayora. 2003. Evaluation of municipal compost/limestone/iron mixtures as filling material for permeable reactive barriers for *in-situ* acid mine drainage treatment. *Journal of Chemical Technology and Biotechnology* 78, no. 5: 489–496. doi:10.1002/jctb.814.

Hedin, R. S., R. W. Nairn, and R. L. P. Kleinmann. 1994a. *Passive treatment of coal mine drainage*. Information Circular IC 9389. U.S. Bureau of Mines, Pittsburgh, PA.

Hedin, R. S., and G. R. Watzlaf. 1994. The effects of anoxic limestone drains on mine water chemistry. In *Proceedings of the International Land Reclamation and Mine Drainage Conference and the 3rd International Conference on the Abatement of Acidic Drainage*, Pittsburgh, PA. doi:10.21000/JASMR94010185.

Hedin, R. S., G. R. Watzlaf, and R. W. Nairn. 1994b. Passive treatment of acid mine drainage with limestone. *Journal of Environmental Quality* 23: 1338–1345. doi:10.2134/jeq1994.00472425002300060030x.

Hemsi, P. S., C. D. Shackelford, and L. A. Figueroa. 2005. Modeling the influence of decomposing solids on sulfate reduction rates for iron precipitation. *Environmental Science and Technology* 39, no. 9 (May): 3215–3225. doi:10.1021/es0486420.

Huminicki, D. M. C., and J. D. Rimstidt. 2008. Neutralization of sulfuric acid solutions by calcite dissolution and the application to anoxic limestone drain design. *Applied Geochemistry* 23, no. 2 (February): 148–165. doi:10.1016/j.apgeochem.2007.10.004.

Jeen, S. W., J. G. Bain, and D. W. Blowes. 2014. Evaluation of mixtures of peat, zero-valent iron and alkalinity amendments for treatment of acid rock drainage. *Applied Geochemistry* 43 (April): 66–79. doi:10.1016/j.apgeochem.2014.02.004.

Johnson, D. B., and K. B. Hallberg. 2005a. Acid mine drainage remediation options: A review. *Science of the Total Environment* 338, no. 1–2 (February): 3–14. doi:10.1016/j.scitotenv.2004.09.002.

Johnson, D. B., and K. B. Hallberg. 2005b. Biogeochemistry of the compost bioreactor components of a composite acid mine drainage passive remediation system. *Science of the Total Environment* 338, no. 1–2 (February): 81–93. doi:10.1016/j.scitotenv.2004.09.008.

Jouini, M., M. Benzaazoua, C. M. Neculita, and T. Genty. 2020a. Performances of stabilization/solidification process of acid mine drainage passive treatment residues: Assessment of the environmental and mechanical behaviors. *Journal of Environmental Management* (pending acceptance).

Jouini, M., C. M. Neculita, T. Genty, and M. Benzaazoua. 2019a. *Environmental assessment of residues from field multi-step passive treatment of Fe-AMD: Case study of the Lorraine mine site, Quebec, Canada.* In *Proceedings of Tailings and Mine Waste*, Vancouver, Canada, pp. 137–150.

Jouini, M., C. M. Neculita, T. Genty, and M. Benzaazoua. 2020b. Environmental behavior of metal-rich residues from the passive treatment of acid mine drainage. *Science of the Total Environment* 712 (April): 136541. doi:10.1016/j.scitotenv.2020.136541.

Jouini, M., C. M. Neculita, T. Genty, and M. Benzaazoua. 2020c. Freezing-thawing effects on geochemical behavior of residues from acid mine drainage passive treatment systems. *Journal of Water Process Engineering* 33 (February): 101087. doi:10.1016/j.jwpe.2019.101087.

Jouini, M., T. V. Rakotonimaro, C. M. Neculita, T. Genty, and M. Benzaazoua. 2019b. Stability of metal-rich solids from laboratory multi-step treatment system for ferriferous acid mine drainage. *Environmental Science and Pollution Research* 26, no. 35 (December): 35588–35601. doi:10.1007/s11356-019-04608-1.

Jouini, M., T. V. Rakotonimaro, C. M. Neculita, T. Genty, and M. Benzaazoua. 2019c. Prediction of the environmental behavior of residues from the passive treatment of acid mine drainage. *Applied Geochemistry* 110 (November): 104421. doi:10.1016/j.apgeochem.2019.104421.

Kirby, C. S., and C. A. Cravotta III. 2005a. Net alkalinity and net acidity 1: Theoretical considerations. *Applied Geochemistry* 20, no. 10 (October): 1920–1940. doi:10.1016/j.apgeochem.2005.07.002.

Kirby, C. S., and C. A. Cravotta III. 2005b. Net alkalinity and net acidity 2: Practical considerations. *Applied Geochemistry* 20, no. 10 (October): 1941–1964. doi:10.1016/j.apgeochem.2005.07.003.

Kuyucak, N., F. Chabot, and J. Martschuk. 2006. Successful implementation and operation of a passive treatment system in an extremely cold climate, northern Quebec, Canada. In *Proceedings of the 7th International Conference on Acid Rock Drainage ICARD*, St. Louis, MO, pp. 980–992. http://mwen.info/docs/imwa_2006/0980-Kuyucak-ON.pdf.

Kuyucak, N., and P. St-Germain. 1994. In situ treatment of acid mine drainage by sulfate reducing bacteria in open pits: Scale-up experiences. In *Proceedings of the International Land Reclamation and Mine Drainage Conference and the 3rd International Conference on the Abatement of Acidic Drainage*, vol. 2, pp. 303–310. doi:10.21000/JASMR94020303.

Macías, F., M. A. Caraballo, and J. M. Nieto. 2012a. Environmental assessment and management of metal-rich wastes generated in acid mine drainage passive remediation systems. *Journal of Hazardous Materials* 229–330 (August): 107–114. doi:10.1016/j.jhazmat.2012.05.080.

Macías, F., M. A. Caraballo, J. M. Nieto, T. S. Rötting, and C. Ayora. 2012b. Natural pretreatment and passive remediation of highly polluted acid mine drainage. *Journal of Environmental Management* 104 (August): 93–100. doi:10.1016/j.jenvman.2012.03.027.

Macías, F., M. A. Caraballo, T. S. Rötting, R. Pérez-López, J. M. Nieto, and C. Ayora. 2012c. From highly polluted Zn-rich acid mine drainage to non-metallic waters: Implementation of a multi-step alkaline passive treatment system to remediate metal pollution. *Science of the Total Environment* 433 (September): 323–330. doi:10.1016/j.scitotenv.2012.06.084.

Macías, F., R. Pérez-López, M. A. Caraballo et al. 2017a. A geochemical approach to the restoration plan for the Odiel River basin (SW Spain), a watershed deeply polluted by acid mine drainage. *Environmental Science and Pollution Research* 24 (February): 4506–4516. doi:10.1007/s11356-016-8169-9.

Macías, F., R. Pérez-López, M. A. Caraballo, C. R. Canovas, and J. M. Nieto. 2017b. Management strategies and valorization for waste sludge from active treatment of extremely metal-polluted acid mine drainage: A contribution for sustainable mining. *Journal of Cleaner Production* 141 (January): 1057–1066. doi:10.1016/j.jclepro.2016.09.181.

Maree, J. P., M. Mujuru, V. Bologo, N. Daniels, and D. Mpholoane. 2013. Neutralisation treatment of AMD at affordable cost. *Water SA* 39, no. 2 (January): 245–250. doi:10.4314/wsa.v39i2.7.

Mayer, K. U., S. G. Benner, and D. W. Blowes. 2006. Process-based reactive transport modeling of a permeable reactive barrier for the treatment of mine drainage. *Journal of Contaminant Hydrology* 85, no. 3–4 (May): 195-211. doi:10.1016/j.jconhyd.2006.02.006.

Neculita, C. M., L. Coudert, E. Rosa, and C. Mulligan. 2020. Future prospects for treating contaminants of emerging concern in water and soils/sediments. In *Advanced nano-bio technologies for water and soil treatment*, ed. J. Filip, T. Cajthaml, P. Najmanová, M. Černík, and R. Zbořil, 589–605. Cham, Switzerland: Springer. doi:10.1007/978-3-030-29840-1_29.

Neculita, C. M., and E. Rosa. 2019. A review of the implications and challenges of manganese removal from mine drainage. *Chemosphere* 214 (January): 491–510. doi:10.1016/j.chemosphere.2018.09.106.

Neculita, C. M., B. Vigneault, and G. J. Zagury. 2008c. Toxicity and metal speciation in acid mine drainage treated by passive bioreactors. *Environmental Toxicology and Chemistry* 27, no. 8 (August): 1659–1667. doi:10.1897/07-654.1.

Neculita, C. M., G. J. Yim, G. Lee, S. W. Ji, J. W. Jung, H. S. Park, and H. Song. 2011. Comparative effectiveness of mixed organic substrates to mushroom compost for treatment of mine drainage in passive bioreactors. *Chemosphere* 83, no. 1 (March): 76–82. doi:10.1016/j.chemosphere.2010.11.082.

Neculita, C. M., and G. J. Zagury. 2008. Biological treatment of highly contaminated acid mine drainage in batch reactors: Long-term treatment and reactive mixture characterization. *Journal of Hazardous Materials* 157, no. 2–3 (September): 358–366. doi:10.1016/j.jhazmat.2008.01.002.

Neculita, C. M., G. J. Zagury, and B. Bussière. 2007. Passive treatment of acid mine drainage in bioreactors using sulfate-reducing bacteria: Critical review and research needs. *Journal of Environmental Quality* 36, no. 1 (January): 1–16. doi:10.2134/jeq2006.0066.

Neculita, C. M., G. J. Zagury, and B. Bussière. 2008a. Effectiveness of sulfate-reducing passive bioreactors for treating highly contaminated acid mine drainage: I. Effect of hydraulic retention time. *Applied Geochemistry* 23, no. 12 (December): 3442–3451. doi:10.1016/j.apgeochem.2008.08.004.

Neculita, C. M., G. J. Zagury, and B. Bussière. 2008b. Effectiveness of sulfate-reducing passive bioreactors for treating highly contaminated acid mine drainage: II. Metal removal mechanisms and potential mobility. *Applied Geochemistry* 23, no. 12 (December): 3445–3560. doi:10.1016/j.apgeochem.2008.08.014.

Nielsen, G., I. Hatam, K. A. Abuan et al. 2018. Semi-passive *in-situ* pilot scale bioreactor successfully removed sulfate and metals from mine impacted water under subarctic climatic conditions. *Water Research* 140 (September): 268–279. doi:10.1016/j.watres.2018.04.035.

Nordstrom, D. K., D. W. Blowes, and C. J. Ptacek. 2015. Hydrogeochemistry and microbiology of mine drainage: An update. *Applied Geochemistry* 57 (June): 3–16. doi:10.1016/j.apgeochem.2015.02.008.

Pagnanelli, F., C. C. Viggi, F. Beolchini, L. Grieco, F. Veglio, and L. Toro. 2014. Bioactive and passive mechanisms of pollutant removal in bioreduction processes in fixed bed columns: Numerical simulations. *Environmental Progress & Sustainable Energy* 33, no. 1 (April): 70–80. doi:10.1002/ep.11753.

Postgate, J. R. 1984. *The sulfate-reducing bacteria*. 2nd ed. Cambridge, UK: Cambridge University Press.

Potvin, R. 2009. Évaluation à différentes échelles de la performance de systèmes de traitement passif pour des effluents fortement contaminés par le drainage minier acide. PhD diss., Université du Québec en Abitibi-Témiscamingue, Canada, 366, https://depositum.uqat.ca/id/eprint/51.

Potvin, R., B. Bussière, M. Benzaazoua, G. J. Zagury, M. St-Arnold, and J. Cyr. 2009. Efficacité des drains dolomitiques installés au site de l'ancienne mine Lorraine, Témiscamingue, Québec. In *Proceedings of the CLRA (Canadian Land Reclamation Association) Workshop*, Québec, QC, Canada, August, 10p.

Rakotonimaro, T. V. 2017. Prétraitement et traitement passif du drainage minier acide ferrifère. PhD diss., Université du Québec en Abitibi-Témiscamingue, Canada, 251, https://depositum.uqat.ca/id/eprint/725.

Rakotonimaro, T. V., M. Guittonny, and C. M. Neculita. 2018b. Stabilization of hard rock mines tailings with organic amendments: Pore water quality control and revegetation: A review. *Desalination and Water Treatment* 112 (April): 53–71. doi:10.5004/dwt.2018.22395.

Rakotonimaro, T. V., C. M. Neculita, B. Bussière, M. Benzaazoua, and G. J. Zagury. 2017a. Recovery and reuse of sludge from active and passive treatment of mine-drainage impacted waters: A review. *Environmental Science and Pollution Research* 24 (January): 73–91. doi:10.1007/s11356-016-7733-7.

Rakotonimaro, T. V., C. M. Neculita, B. Bussière, T. Genty, and G. J. Zagury. 2018a. Performance assessment of laboratory and field-scale multi-step passive treatment of iron-rich acid mine drainage for design improvement. *Environmental Science and Pollution Research* 25, no. 18 (June): 17575–17589. doi:10.1007/s11356-018-1820-x.

Rakotonimaro, T. V., C. M. Neculita, B. Bussière, and G. J. Zagury. 2016. Effectiveness of various dispersed alkaline substrates for the pretreatment of ferriferous acid mine drainage. *Applied Geochemistry* 73 (October): 13–23. doi:10.1016/j.apgeochem.2016.07.014.

Rakotonimaro, T. V., C. M. Neculita, B. Bussière, and G. J. Zagury. 2017b. Comparative column testing of three reactive mixtures for the bio-chemical treatment of iron-rich acid mine drainage. *Minerals Engineering* 111 (September): 79–89. doi:10.1016/j.mineng.2017.06.002.

Rose, A. W., and J. M. Dietz. 2002. Case studies of passive systems: Vertical flow systems. In *Proceedings of the 2002 National Meeting of the American Society of Mining and Reclamation*, Lexington, KY, pp. 776–797. doi:10.21000/JASMR02010776.

Rötting, T. S., M. A. Caraballo, J. A., Serrano, C. Ayora, and J. Carrera. 2008a. Field application of calcite dispersed alkaline substrate (calcite-DAS) for passive treatment of acid mine drainage with high Al and metal concentrations. *Applied Geochemistry* 23, no. 6 (June): 1660–1674. doi:10.1016/j.apgeochem.2008.02.023.

Rötting, T. S., L. Luquot, J. Carrera, and D. J. Casalinuovo. 2015. Changes in porosity, permeability, water retention curve and reactive surface area during carbonate rock dissolution. *Chemical Geology* 403 (May): 86–98. doi:10.1016/j.chemgeo.2015.03.008.

Rötting, T. S., R. C. Thomas, C. Ayora, and J. Carrera. 2008b. Passive treatment of acid mine drainage with high metal concentrations using dispersed alkaline substrate. *Journal of Environmental Quality* 37, no. 5 (August): 1741–1751. doi:10.2134/jeq2007.0517.

Simate, G. S., and S. Ndlovu. 2014. Acid mine drainage: Challenges and opportunities. *Journal of Environmental Chemical Engineering* 2, no. 3 (September): 1785–1803. doi:10.1016/j.jece.2014.07.021.

Skousen, J., and P. F. Ziemkiewicz. 2005. Performance of 116 passive treatment systems for acid mine drainage. In *Proceedings of American Society of Mining and Reclamation*, Breckenridge, CO, pp. 1100–1133. doi:10.21000/JASMR05011100.

Skousen, J., C. E. Zipper, A. Rose et al. 2017. Review of passive systems for acid mine drainage treatment. *Mine Water and the Environment* 36 (March): 133–153. doi:10.1007/s10230-016-0417-1.

Song, H., G. J. Yim, S. W. Ji, I. H. Nam, C. M. Neculita, and G. Lee. 2012a. Performance of mixed organic substrates during treatment of acidic and moderate mine drainage in column bioreactors. *Journal of Environmental Engineering ASCE* 138, no. 10 (October): 1077–1084. doi:10.1061/(ASCE)EE.1943-7870.0000567.

Song, H., G. J. Yim, S. W. Ji, C. M. Neculita, and T. Hwang. 2012b. Pilot-scale passive bioreactors for the treatment of acid mine drainage: Efficiency of mushroom compost vs mixed substrates for metal removal. *Journal of Environmental Management* 111 (November): 150–158. doi:10.1016/j.jenvman.2012.06.043.

Stumm, W., and J. J. Morgan. 1981. *Aquatic chemistry: An introduction emphasizing chemical equilibria in natural waters*. 2nd ed. New York: John Wiley and Sons.

URS (United Registrar of Systems Corporation). 2003. *Passive and semi-active treatment of acid rock drainage from metal mines-state of the practice*. Final draft report. Portland, ME: URS.

USEPA (United States Environmental Protection Agency). 2014. *Reference guide to treatment technologies for mining-influenced water*. EPA 542-R-14-001. Washington, DC: EPA.

Vasquez, Y., M. C. Escobar, C. M. Neculita, Z. Arbeli, and F. Roldan. 2016a. Selection of reactive mixture for biochemical passive treatment of acid mine drainage. *Environmental Earth Sciences* 75 (March): 576. doi:10.1007/s12665-016-5374-2.

Vasquez, Y., M. C. Escobar, C. M. Neculita, Z. Arbeli, and F. Roldan. 2016b. Biochemical passive reactors for treatment of acid mine drainage: Effect of hydraulic retention time on the evolution of efficiency, composition of reactive mixture and microbial activity. *Chemosphere* 153 (June): 244–253. doi:10.1016/j.chemosphere.2016.03.052.

Vasquez, Y., M. C. Escobar, J. Saenz et al. 2018. Effect of hydraulic retention time on microbial community in biochemical passive reactors during treatment of acid mine drainage. *Bioresource Technology* 247 (January): 624–632. doi:10.1016/j.biortech.2017.09.144.

Viggi, C. C., F. Pagnanelli, and L. Toro. 2011. Sulphate reduction processes in biological permeable reactive barriers: Column experimentation and modeling. *Chemical Engineering Transactions* 24: 1231–1236. doi:10.3303/CET1124206.

Watzlaf, G. R., K. T. Schroeder, and C. L. Kairies. 2000. Long-term performance of anoxic limestone drains. *Mine Water and the Environment* 19 (September): 98–110. doi:10.1007/BF02687258.

Watzlaf, G. R., K. T. Schroeder, R. L. P. Kleinmann, C. L. Kairies, and R. W. Nairn. 2004. *The passive treatment of coal mine drainage*. United States Department of Energy/National Energy Technology, DOE/NETL-2004/1202, 72 p.

Waybrant, K. R., C. J. Ptacek, and D. W. Blowes. 1998. Selection of reactive mixtures for use in permeable reactive walls for treatment of acid mine drainage. *Environmental Science and Technology* 32, no. 13 (May): 1972–1979. doi:10.1021/es9703335.

Waybrant, K. R., C. J. Ptacek, and D. W. Blowes. 2002. Treatment of mine drainage using permeable reactive barriers: Column experiments. *Environmental Science and Technology* 36, no. 6 (February): 1349–1356. doi:10.1021/es010751g.

Westholm, L. J., E. Repo, and M. Sillanpää. 2014. Filter materials for metal removal from mine drainage – a review. *Environmental Science and Pollution Research* 21, no. 15 (May): 9109–9128. doi:10.1007/s11356-014-2903-y.

Yim, G. J., S. W. Ji, Y. W. Cheong, C. M. Neculita, and H. Song. 2015. The influences of the amount of organic substrate on the performance of pilot-scale passive bioreactors for acid mine drainage treatment. *Environmental Earth Sciences* 73 (April): 4717–4727. doi:10.1007/s12665-014-3757-9.

Younger, P. L., S. A. Banwart, and R. S. Hedin. 2002. *Mine water: Hydrogeology, pollution, remediation*. Dordrecht, The Netherlands: Kluwer Academic Publ.

Zagury, G. J., V. I. Kulnieks, and C. M. Neculita. 2006. Characterization and reactivity assessment of organic substrates for sulphate reducing bacteria in acid mine drainage treatment. *Chemosphere* 64, no. 6 (August): 944–954. doi:10.1016/j.chemosphere.2006.01.001.

Ziemkiewicz, P. F., J. G. Skousen, and J. Simmons. 2003. Long-term performance of passive acid mine drainage treatment systems. *Mine Water and the Environment* 22 (November): 118–129. doi:10.1007/s10230-003-0012-0.

12 Revegetation of Mine Sites

Marie Guittonny

12.1 INTRODUCTION

Industrial activities such as mining replace natural ecosystems with human-made substrates called technosols, which are often devoid of any organisms. Technosols comprise anthropized soils created or modified by urbanization and industrialization (Chesworth and Spaargaren 2008). After cessation of the industrial activity, technosols become biologically active (Leguédois et al. 2018) through colonization by organisms – in particular, plants – despite sometimes adverse living conditions (Bradshaw 1997). Plant establishment on these substrates occurs through natural processes called plant primary succession (Clements 1916). However, such processes can take several hundred years to grow a stable plant cover that will sustain a diverse natural food web (Walker and del Moral 2009). Revegetation concerns the techniques that aim to help and accelerate plant establishment on substrates. Knowledge and observation of natural succession of plant communities (plant ecology) as well as the knowledge of plant organisms' adaptations to environmental constraints (ecophysiology) are important to design revegetation.

Mine activities generate particular technosols for plant establishment that consist of mine solid wastes such as overburden, waste rocks, and residues from the treatment of ore to extract commercial value (tailings) and sometimes from the treatment of contaminated waters (sludges). Revegetation or natural plant establishment on mine wastes has been extensively studied and reviewed for coal mines, oil sands, clay pits, as well as metalliferous, hard rock mines all over the world and their climatic zones and social contexts (e.g., Winterhalder 1996; Bradshaw 1997, 2000; Tordoff et al. 2000; Cooke and Johnson 2002; Burke 2003; Wong 2003; Li 2006; Sheoran et al. 2010; Zipper et al. 2011; Mosseler et al. 2014; MacDonald et al. 2015; Sebelikova et al. 2019). The revegetation goals – that is, the final desired plant community and the means to create it – evolve from renaturalization, rewilding, reuse, valorization, and conservation, depending on communities' needs, culture, density, and land use (agricultural, urban, natural, and pristine).

At the closure of hard rock mines, surface materials that need to be revegetated are mainly non-acid-generating waste rocks and mill tailings (see Chapter 1), or materials used in the surface layers of reclamation covers (see Chapters 4 to 9) if the wastes are acid generating. Even if they are not acid generating, mine waste materials present common constraints and challenges to revegetation, which will be discussed in this chapter and viewed within the larger context of the ecological knowledge of plant establishment on mineral substrates. The aim is to provide reclamation engineers and practitioners with useful knowledge of plant and soil processes and factors that influence the success of revegetation. This knowledge will help them select appropriate strategies and techniques to reach a wide range of revegetation objectives. To guide them in the development of their revegetation plan, a general methodological approach is proposed. Finally, possible avenues to increase the value of revegetated areas are identified, as well as research and development needs.

Revegetation can also be used temporarily to physically or chemically stabilize (phytostabilization) or to extract metals (phytoextraction) in acid-generating or metal-rich mine wastes. These phytoremediation techniques can be useful on abandoned mine sites awaiting final reclamation, but they won't be deeply investigated in this chapter. Indeed, in the long term, acid- or metal-contaminated wastes will need to be reclaimed with a cover that controls contaminant generation at the source (see Chapters 4 to 9) to confine the problematic mine wastes before final revegetation can be implemented.

12.2 FUNDAMENTALS RELATED TO PLANT ESTABLISHMENT ON MINE SUBSTRATES

To select realistic objectives of revegetation, define adapted revegetation strategies, and realize a functional revegetation design, it is necessary to better understand how plants establish on mineral substrates such as mine wastes.

12.2.1 Plant Succession

Plant succession describes the colonization, growth, and death process of a pool of species on a site, which induces changes in space and time of the composition of the plant community (Clements 1916). Plant primary succession occurs on mineral substrates, such as after glacier retreat (Reiners et al. 1971) or on China clay deposits (Bradshaw 1983). These substrates are devoid of any organisms and associated organic matter (OM). Resources needed by plants and provided by the substrate, such as water and nutrients, are limited, which shapes plant colonization and development as well as further ecosystem processes (Walker and del Moral 2009). Plant succession is composed of several stages until reaching a stable state without disturbance. Each stage corresponds to different plant communities. As stages succeed, the level of some resources provided by the substrate generally increases with OM enrichment, such as increased nitrogen storage capacity (Chapin et al. 2011; Young et al. 2013). At the beginning of primary succession, resource limitations filter the establishment of plants on mineral substrates, and only plants that are able to overcome these limitations (also called ecological filters) can successfully establish (Keddy 1992).

12.2.2 Ecological Filters Associated with Mineral Substrates

Table 12.1 provides examples of the three levels of limitations to plant establishment associated with mineral substrates, and in particular mine wastes.

12.2.2.1 Physical Filters

At the beginning of plant succession, mineral substrates such as mine wastes are often physically unstable due to lack of structure and susceptibility to fine particle movement. The substrate topography and grain-size distribution will condition its ability to provide favorable stable microsites, with fine particles accumulation allowing water retention and seed germination (Cooper 1923). Topography influences local microclimate (temperature and humidity) through wind (altitude) and solar radiation (slope aspect) exposure, as well as through drainage, runoff, and fine particle loss or deposition (slope angle and position). It also controls snow accumulation, which influences water reserve at snowmelt, as well as freeze-thaw exposure (MacDonald et al. 2015). Figure 12.1 illustrates how a sloping terrain influences natural plant colonization.

Moreover, mineral substrates devoid of any sheltering plants or roughness are characterized by excessive variations in temperature at the soil level, which expose small and fragile seedlings to

TABLE 12.1
Examples of Limitations Associated to Mine Substrates

Level	Substrate Limitation
Physical	Soil loss or deposition by water or wind, too high bulk density, too low or too high hydraulic conductivity, inappropriate structure and aggregate stability, compaction
Chemical	High: acidity, alkalinity, salinity, toxicity, nutrient leaching No organic matter
Biological	No humus, vegetation, or microbiome

FIGURE 12.1 Photograph of waste rock slope with vegetation naturally established down slope, close to a water collection ditch.

desiccation or frost damages. As soon as a first plant succeeds in surviving, it shelters temperature variations and collects fine particles at its feet, which constitutes a fertile island (defined by Walker and del Moral 2009) for the colonization of other plants. These positive interactions among plants (i.e., facilitation) in the first stages of plant succession are particularly important for the plant community to evolve toward more stable and productive stages where inhibiting interactions (i.e., competition) become preponderant (Grime 1977; Bertness and Callaway 1994). Plants that facilitate the establishment of other plant species are called nurse plants (Connell and Slatyer 1977). These nurse plants can also help other plants' establishment by modifying soil chemical conditions in their vicinity, for example, by providing OM or fixing atmospheric nitrogen by symbiosis occurring in their roots.

12.2.2.2 Chemical Filters

Some substrates characterized by fine grain size (silts and clays), such as mine tailings, can provide enough water, but their slow drainage can induce lack of oxygen availability to plant roots. Oxygen and water are essential resources for plant development (see Table 12.2). Mineral substrates can also be chemically unstable due to weathering of the minerals exposed to water and air in surface conditions. This weathering can induce salt excess in the substrate, which decreases water absorption of plants and causes a pH change (see Chapter 1). Weathered cations may be rapidly leached and unavailable for plant nutrition due to low cation exchange capacity associated with a lack of OM. Finally, mineral substrates are often characterized by nutrient deficiencies, especially nitrogen, while phosphorus availability is less problematic since the weathering of minerals often provides phosphorus in available forms (Chapin et al. 2011). Nitrogen and phosphorus are the main macronutrients needed by plants (see Table 12.2). Nitrogen deficiency in plants directly results in a decrease in productivity since photosynthetic enzymes represent a major sink of leaves' N.

12.2.2.3 Biological Filters

Plant establishment cannot rely on a seed bank (i.e., the reserve of seeds contained in the soil) in human-made substrates, and adult plants producing propagules (i.e., plant-regenerating and

TABLE 12.2
Resources Needed by Plants And Factors that Influence Their Availability

Resource	Quantity in Plants	Absorption Organ	Source	Availability Range	Factors Limiting Availability
Red (400–500 nm) and blue (350 nm) radiations	–	Leave	Light exposure	Not limiting if not shaded	Shading
CO_2	+ 90% of dry matter = CHO	Leave	Ambient air	Often non-limiting	
Water	80% to 95% of plants (in volume)	Fine root (diameter <2 mm), mycorrhizae	Soil water	Suction lower than the permanent wilting point (theoretical value: 1500 kPa)	Excessive drainage, evaporation, salinity
O_2	–	All plant cells	Ambient air and soil atmosphere	Minimum = 1% to 2% of gaseous O_2	Flooding, compaction
Nutrients	Major nutrients (in terms of % dry mass) N = 0.25%–7.5% P = 0.02%–1.4% K = 0.1%–12% Ca = 0.04%–7% Mg = 0.05%–2% Micronutrients: Fe, Zn, Cu, Mn, B, and sometimes Na, Cl	Fine root Symbiotic organisms	Mineral ions in soil solution (roots) Organic or gaseous forms (symbiotic organisms like mycorrhizae and N_2 fixing bacteria)	Targeted total nitrogen stock in the soil: 1000 kg/ha; annual N consumption/fertilization: 50–100 kg/ha for temperate ecosystems. To be increased in colder climates due to lower decomposition rates	Excessive leaching, soil drying, insoluble forms, competition among ions
Space and support (not a resource)	–	Root	Soil pores >15 µm	Bulk density <1.4 Macroporosity (pores >30–75 µm) = 10% to 30% of total volume	Compaction, crusting

plant-disseminating organs like seeds) can be far from mine waste disposal areas. Once a seed germinates on a mineral substrate, organisms such as symbiotic bacteria, mushrooms, and nurse plants (which help seedlings thrive in resource-limited environments) are absent. At adult stages, pollinator populations and dispersion of seeds by animals may be limited in mine waste environments due to incomplete food web or habitats, and may limit plant reproduction and spreading.

12.2.3 Useful Plant Characteristics to Overcome Filters on Mineral Substrates

Pioneer plants are the first plants to establish on mineral substrates. They show a large dispersion capacity through very mobile dissemination organs (e.g., seeds). Seed dispersion capacity often shapes the community of the first stage of succession and can be more crucial than resource availability. Species with a small seed, which are wind-dispersed over large distances, will preferentially reach the site to colonize. However, many seeds reach the sites but do not successfully germinate (Walker and del Moral 2009). Revegetation can rely on natural colonization of a site by pioneer plants if the site is small, if the soil allows plant establishment, and if the natural ecosystem is close enough (Prach and Pysek 2001). In forested environments of Central Europe, species' richness reached greater levels when sites were left to spontaneous colonization compared to assisted afforestation (Sebelikova et al. 2019). To accelerate plant establishment, seeding or planting are used to overcome the lack of dissemination organs that reach the site. The selection of native plants will insure that they are adapted to local climatic conditions and variations, especially regarding water availability (Bradshaw 1997; Tordoff et al. 2000).

Once a plant reaches a site, it must find on the spot what it needs to grow, reproduce, and survive because plants are immobile organisms. Small plants that take low resources, with short life duration and an ability to quickly reproduce, will dominate the communities established on mineral substrates (Chapin et al. 2011). The larger the size of a plant, the greater its resource needs. For example, the mean size of vegetation increases with water availability in differing climatic areas (Chapin et al. 2011). Moreover, long-lived plants must keep up in periods of resource shortages, whereas short-lived plants can concentrate their life period when resource availability is maximal. Long-lived plants thus adapt with difficulty to habitats where the mean level of some resources is low. For example, long-lived, large-seeded, and large-sized plants, such as trees, will dominate later succession stages when resource levels are improved and adapted to their greater needs. They can colonize mineral substrates early in the succession, but they only reach a satisfying development later in the succession (Cooper 1923; Bradshaw 1983), unless they are adapted to elevated stress levels.

Yet, when facing a stress due to a lack of resources, some plants can decrease their needs or optimize their supplying through morphological or physiological adaptations (Lambers et al. 2008). Drought-adapted plants, for example, often have small, thick-walled leaves to decrease water losses. In mine revegetation, it is important to select plants that are adapted to the availability and spatio-temporal distribution of resources associated with reclaimed mine wastes. For example, fertilization with quick release of nutrients, such as mineral fertilizers, will better benefit plants that have a high growth rate and a strategy of quick resource consumption (Aerts and Van Der Peijl 1993; Urbas and Zobel 2000). By contrast, a slow-release fertilizer will better benefit plants with a conservative use of resources, such as plants that produce long-lived tissues with low concentrations of nutrients and that retranslocate nutrients from leaves to shoots before shedding (Rundel 1988; Aerts 1995). Table 12.3 summarizes the characteristics of pioneer plants that can help to overcome filters associated with mineral substrates like mine solid wastes.

To accelerate the establishment of later successional plant species like trees, resource limitations can be attenuated by modifying the substrate until reaching the characteristics of a more evolved soil found in later succession.

TABLE 12.3

Characteristics of Pioneer Plants Adapted to Mineral Substrates Like Mine Solid Wastes

Possible filters Associated with Mineral Substrates	Characteristics of Pioneer Plants
No plant propagules (e.g., no seed bank)	Small, wind-dispersed seeds with large dispersion capability
Physical instability due to lack of structure: soil loss and deposition	Short-lived plants, small plants
Chemical instability due to weathering	Stress tolerant (e.g., to drought, nutrient deficiencies, salinity, frost, flooding)
Excessive variations in temperature at the soil surface	
Low nitrogen concentrations	
No positive interactions with other organisms (e.g., no nurse plants)	

12.2.4 Characteristics of a Functional Soil

It is difficult to compare technosols to natural soils since the study of technosol pedogenesis is still in its infancy (Leguédois et al. 2018). Mine wastes represent a young, rapidly evolving soil (Leguédois et al. 2018) that could be compared to the already fragmented C horizon (bedrock) of a generic profile of soil (a generic soil profile can be found in Chapin et al. 2011). When deposited, they have not yet encompassed surface chemical or biological processes that change rock pieces along time. If the mine wastes have already been exposed to surface conditions, as they would be in abandoned waste storage facilities, chemical processes may have begun. Weathering creates fine mineral particles, which aggregate to form bigger units and in turn create structure (Ciarkowska et al. 2016).

Aggregates are the result of the cementation of mineral particles that crack into bigger units under frost or desiccation action. Cementation relies on several agents, including OM, iron and aluminum oxides, polyvalent cations, clay minerals, silica, and earthworms' action. Soil aggregates create pores, especially macropores, which quickly drain and allow air to flow. Macropore size varies due to differences in measurement techniques but is generally greater than 30, 60, or 75 μm (Beven and Germann 1982). Macroporosity influences fluids' movement and creates space for the colonization of elongating roots. It thus influences water and oxygen uptake by, and growth of, roots (Table 12.2). Thinner pores store available water for plant roots until the permanent wilting point is reached (Table 12.2), beyond which soil suction does not allow plant roots to extract water bounded to soil particles.

In mine revegetation, mineral substrate improvement often aims at reconstructing organic (O) and organo-mineral (A) horizons (sensu Chapin et al. 2011) (see Figure 12.2) because they support the maximal activity of plants and soil organisms and provide the greatest availability of nutrients. Both horizons are characterized by the presence of OM. Soil OM, provided by colonizing organisms, plays a very important role in improving soil ability to provide resources to plants. Since

FIGURE 12.2 Evolution of gold mine waste deposits toward a more evolved soil through pedogenic processes: (a) and (c) desulfurized tailings seven years after revegetation with agronomic herbaceous plants; (b) acid-generating waste rocks three years after willow planting.

OM is negatively charged, it forms electrostatic bonds with other sites negatively charged at the surface of mineral particles (e.g., clays), thanks to divalent cations, which participate in aggregate formation. Moreover, it increases soil cation exchange capacity and allows a progressive release of cationic nutrients (Table 12.2). Finally, it is mineralized by decomposers, and this process of mineralization releases nutrients in an available form for plants (Berg and Laskowski 2005). A functional soil is also a welcoming habitat for organisms that can be useful to plants.

Most importantly, decomposers fragment and oxidize OM. Their alive or dead tissues, and their excretions, participate in stable aggregate formation and substrate structure creation (Oades 1984). Other organisms can establish symbiotic relations with plant roots and participate in water and nutrient uptake in exchange for being fed with photosynthetic products. These relations are reciprocally beneficial. They occur between roots of the majority of higher plants (mainly angiosperms and gymnosperms) and mycorhizal fungi (mushrooms), which greatly increase the soil volume exploited by roots for water uptake (1 to 15 m of fungi length for 1 cm of root, Chapin et al. 2011). Plants, however, do not form mycorhizal symbiosis when nutrients are easily available, and their occurrence is limited under excessive fertilization. Some plant species associate with N_2-fixing bacteria, which develop in root nodules (Chapin et al. 2011), or more generally, with plant growth-promoting bacteria (Ullah et al. 2015). These symbiotic bacteria allow N uptake by plants from a non-limiting N source. However, N_2 fixing by these bacteria consumes a lot of plant energy. It thus occurs more frequently in habitats where light availability is great and soil N availability is low, like in little-evolved soils such as mine wastes.

12.3 METHODOLOGICAL APPROACH TO DESIGNING A REVEGETATION PLAN

Figure 12.3 summarizes the four main steps of designing a revegetation plan. Each step of the methodological approach proposed in Figure 12.3 is detailed in the following sections. This approach is inspired and developed from those proposed in the literature reviews of Cooke and Johnson (2002), as well as Tordoff et al. (2000), which deal with revegetation specific to metalliferous mine wastes.

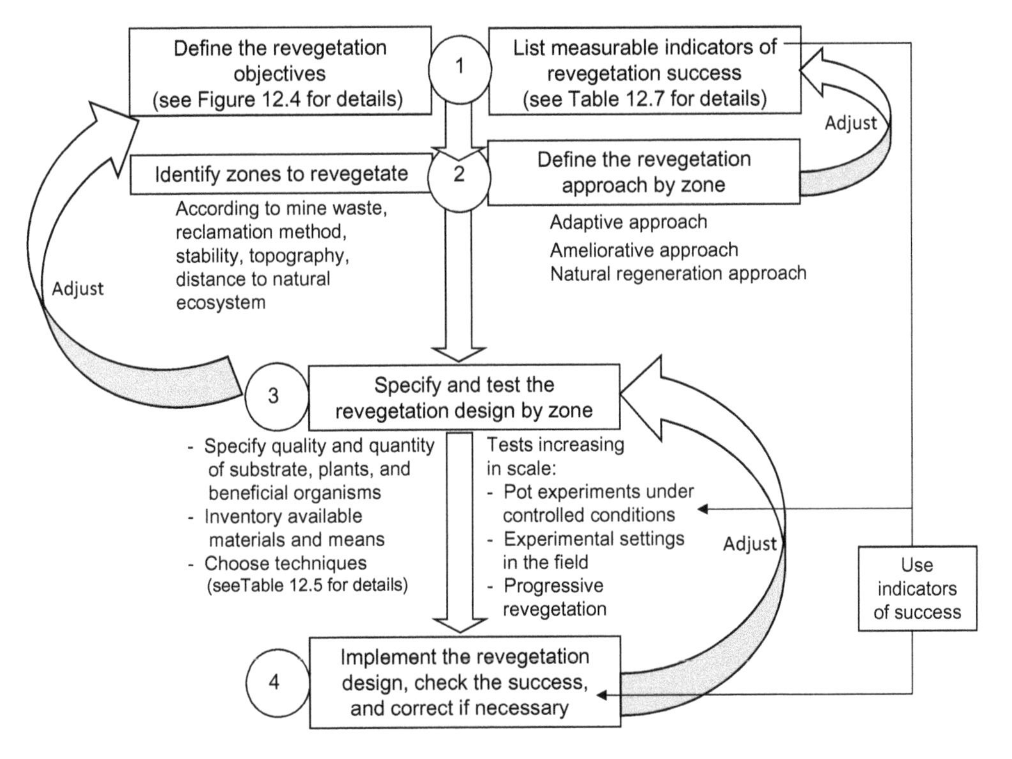

FIGURE 12.3 Design approach for the revegetation plan.

12.4 DEFINING THE REVEGETATION OBJECTIVES AND ASSOCIATED INDICATORS OF REVEGETATION SUCCESS

12.4.1 Revegetation Objectives

Revegetation objectives can be very diverse – from restoring the historical ecosystem and greening to increasing evapotranspiration, producing wood or biomass, attracting fauna, and sometimes controlling vegetation establishment. To define a robust revegetation objective, several elements must be considered and integrated. Figure 12.4 summarizes four main areas to consider in determining revegetation objectives: (1) regulation framework, (2) reclamation plan, (3) community's expectations of land use, and (4) compensation or added value.

Revegetation objectives must comply with the regulatory framework of mine reclamation. The main requirements of reclamation vary with the country of mine operations but generally concern the control of contaminant generation (in particular acid mine drainage, described in Chapter 1) and mobility (exportation in surface or drainage water or in plants exposed to fauna browsing, erosion control). Regarding the established vegetation itself, regulations can demand that it is self-sustaining (with no maintenance), native, and allows biodiversity conservation. Some regulations also require that a reclamation plan and associated costs estimation are provided and evaluated at the beginning of a mine project. Such a plan commits the mine operators to specific revegetation objectives and budget. Finally, regulations can require that mine operators consult with and involve communities in the mine reclamation process, including revegetation.

Since mine companies nowadays cannot act without social license to operate, they have to take into account the community's desired final land use of closed mine sites. Revegetation is the most visible component of reclamation as it aims to repair what the community often perceives as a degraded mine site. Degradation can be defined as a change or disturbance of the environment that induces a loss of value and is perceived as detrimental (Johnson et al. 1997). Mine waste disposal areas constitute a disturbance, which destroys plants that were there before mine operations and changes the substrate on which revegetation will take place. A quantitative or qualitative change in established plant communities compared to the surrounding environment can be perceived as a degradation. Moreover, as the final step of mine site reclamation, revegetation significantly shapes the land use and value of the reclaimed site. Thus, the revegetation objective must be connected to the perceptions and perspectives of community users (individuals, society, and governments) depending on their socioeconomic context, temporal scale of vision, values, and interests. For example, final land use desired by the communities can be recreational, productive, or patrimonial; or they may want a return to a historical ecosystem.

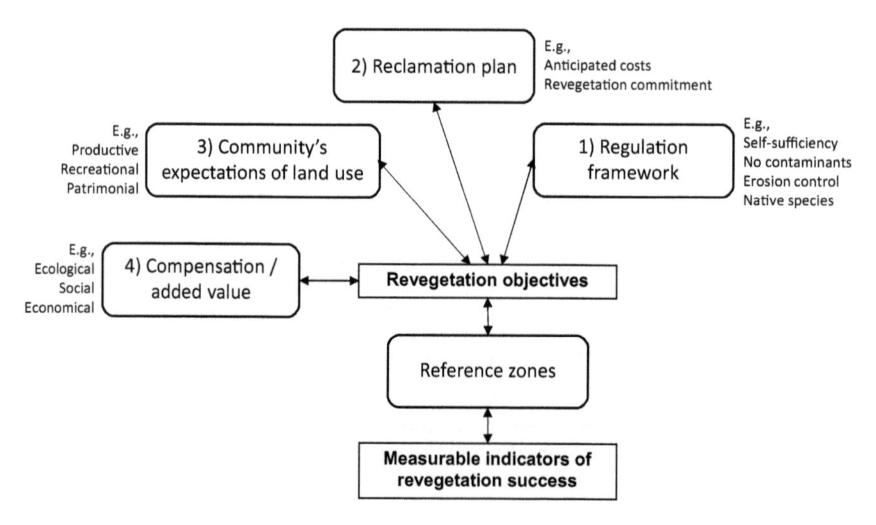

FIGURE 12.4 Revegetation objectives: Summary of elements to consider.

The revegetation objective must also deal with ecological constraints and available budget (e.g., cost of material transport, thicknesses of soil layers, and landscaping) to be realistic. At the ecological level, mine waste deposition corresponds to a disturbance of high intensity because it destroys former vegetation and drastically changes the substrate, sometimes on hundreds of hectares. Restoring the former ecosystem is possible when the applied disturbance level does not exceed ecological resiliency of the ecosystem (or inertia, Westman 1986), defined as the capacity of disturbance absorption of the ecosystem before it changes its evolutionary trajectory. Cooke and Johnson's (2002) review provides a useful conceptual framework for understanding the different objectives of mine site revegetation – from reconstructing to repairing and restoring the ecosystem – as well as examples of ecological indicators of success.

12.4.2 Indicators of Revegetation Success

Once a realistic revegetation objective is identified, it is important to define measurable indicators of success and their targeted levels to determine when the objective is reached (Figure 12.4). Useful indicators can be, for example, the total number of species, the number of rare species, or other diversity indexes; the percentage of cover, the aboveground biomass, or other plant productivity measurements; and the quantity of transpired water, the quantity of sequestered carbon, or other ecosystem services measurements. Reference zones, where ecosystems similar to the one desired on the revegetated mine site are present, are useful to define appropriate reference levels of these indicators.

Sometimes, regulatory obligations, community point of view, and ecological and technical constraints regarding reclamation and revegetation are difficult to reconcile on certain zones of mine sites. Ecological or other forms of compensation can then be considered outside the mine site.

12.5 IDENTIFYING ZONES AND THE REVEGETATION APPROACH USED ON EACH ZONE

Revegetation can be left to natural regeneration processes on mine wastes, but some mine wastes, even if not geochemically problematic, remain devoid of vegetation for decades (e.g., Figure 12.5). To accelerate the establishment and development of vegetation, sites are actively revegetated. Two main assisted revegetation approaches exist: the ameliorative approach, in which the physical, chemical, and/or biological

FIGURE 12.5 Photograph of a gold mine tailings facility still devoid of vegetation 58 years after closure.

nature of the site are improved to meet the requirement of target or native surrounding plant species; and the adaptive approach, in which introduced plant species, cultivars, or ecotypes are adapted to the constraints present on the site (Tordoff et al. 2000). In the first approach, preparing the substrate and letting natural regeneration of plants occur allow local and diverse species and genotypes to establish. However, revegetation success is less predictable and slower than with introduced vegetation. Moreover, on large waste storage facilities, it is possible that the sources of indigenous seeds and other propagules are too far to ensure adequate dispersal on the surfaces to revegetate. The distance to seed sources is thus an important criterion to consider in defining zones where natural colonization is expected to take place after amelioration of the receiving substrate. On certain zones, islands of desired vegetation could be strategically established to serve as seed sources for the recolonization of the whole site.

Constraints limiting vegetation establishment are specific to mine waste type and their corresponding physicochemical characteristics. For example, bottomland, fine-grained tailings storage facilities, and coarse-grained waste rock piles will constitute zones that may need differing revegetation approaches (see Table 12.4). On other parts of the mine site, it is possible that vegetation is not desirable due to technical constraints regarding the chemical or physical stability of reclaimed areas. For example, biopore creation associated with deep root colonization is not desired in oxygen barrier or low k_{sat} covers used to control acid mine drainage of reactive mine wastes (see Chapter 14 for more details). Similarly, vegetation on mine dikes, especially trees, is often removed to avoid their roots altering the integrity of dike materials. Moreover, some communities express concern about establishing vegetation on mine waste storage areas due to possible contaminant transfer in the food web. At the spatial scale, mine site revegetation can thus demand a combination of approaches adapted to several zones that have differing constraints, which should be identified early in the process. It should be noted that species migration between different zones can be expected, which can be desirable or not. It should also be important to define the time frame at which revegetation objectives must be reached on each zone. For these reasons, revegetation objectives must sometimes be qualified after the site zoning step (see Figure 12.3). Finally, combining several revegetation approaches and techniques on one site can decrease the amplitude of revegetation failure in case of a site disturbance.

TABLE 12.4
Examples of Factors to Consider in Identifying Differing Revegetation Zones

Factor to consider for zoning	Example
Mine waste type	Waste rock, tailings, water treatment sludge
Reclamation method	Oxygen barrier cover Low k_{sat} cover
Chemical and structural stability	Dikes, steep slopes, contamination
Topography	Elevation, slope exposure, and steepness
Ability of native seeds to reach the zone	Area of the revegetated zone, elevation, distance from native and mature ecosystem
Time frame	Temporary vs. final revegetation, progressive revegetation

12.6 SPECIFYING THE REVEGETATION DESIGN

On each zone, revegetation designing means specifying the quality and quantity of:

- substrates in terms of physical, chemical, and biological properties;
- target plants in terms of species, propagules, density, spacing, combinations, etc.; and
- beneficial organisms (symbiosis and nursing).

The best moment to prepare the substrates and to install and maintain the plants must be defined, as well as the techniques and means/resources used for each step (see Table 12.5). It is important that the revegetation design matches available resources close to the mine site.

TABLE 12.5
Examples of Techniques, Means, and Materials for Revegetation

Techniques	Means	Materials
Nurse plant	Machinery	Soils
Hydroseeding	Humans	Amendments
Mechanical decompaction	Natural processes	Fertilizers
Placed fertilization	Storage capacity	Plant material
Planting	Budgeted costs	Seed rain
Seeding		
Irrigating		
Inoculating		
Amending		

12.6.1 Substrate Selection and Improvement of Physical, Chemical, and Biological Properties

First, available substrates, their limitations to plant establishment and growth, and the means by which they can be modified to reach favorable properties must all be characterized. Second, it is useful to identify available inputs, their quantity, and possible application techniques to improve the substrate quality. Indications of targeted properties levels to reach are given in Table 12.3 for main soil resources, while techniques to improve substrate properties are given in Table 12.6.

TABLE 12.6
Techniques that can Improve Mine Substrate Properties at the Physical, Chemical, and Biological Levels

Level	Treatment	Effect
Physical	Scarify/subsoil/freeze and thaw	Decompact
	Create roughness/level/mulch	Stabilize the soil against particle movement, catch seeds
	Mix with OM rich amendment	Increase macroporosity
	Cover with finer material/mix with OM rich amendment	Increase water storage
	Drain	Increase aeration
	Mulch/create rugosity/use nurse plants	Buffer surface temperature
Chemical	Seed N-fixing plants	Bring nitrogen
	Fertilize/use organic amendment	Bring nutrients
	Use organic amendment/natural weathering	Decrease pH
	Liming (temporary solution)	Increase pH
	Leach/irrigate	Decrease salinity
Biological	Use nurse plants/inoculate mycorrhiza or N-fixing bacteria/ bring decomposers	Increase beneficial relations
	Vegetation control/browsing control	Decrease competition / herbivory
	Establish islands of seed stock	Create seed rain

OM: organic matter

12.6.1.1 Topsoil and Mineral Soils (Overburden)

On mine sites, at the beginning of a mine project, the natural soils present at the location of the future pits, buildings, working pads, roads, and waste or water storage areas are worth saving for revegetation, if they are not contaminated. Topsoil, which is the superficial soil colonized by organisms (roots and soil fauna) and containing OM (darker color), is a strategic material for revegetation. It brings OM, soil organisms, native plant seeds as well as nutrient stock in balance with the surrounding natural environment to the revegetated area (Marcus 1997; Cooke and Johnson 2002). Thus, topsoil use will help the establishment of plant and soil organism species from the native ecosystem (Martinez-Ruiz and Fernandez-Santos 2005). To optimize topsoil as a source of seeds and spores, only the superficial layers should be collected because seed density rapidly decreases with soil depth (Rokich et al. 2001; Singh et al. 2002). Moreover, long duration of topsoil storage can decrease seed viability, and direct placement after excavation should be considered (Rokich et al. 2001). However, topsoil is generally in limited quantity, which restricts its use on large surfaces (Whitbread-Abrutat 1997). Under the topsoil, the mineral soil (also called overburden when covering the ore) can also be of interest for the site revegetation. It contains no or low OM. Digging a trench to observe the soil profile can help separate topsoil and mineral soil to manage them separately.

12.6.1.2 Soil Transport and Storage: Avoid Compaction

It appears necessary to protect the physical, chemical, and biological properties of soils from excavation until use for revegetation. Before excavation, it is useful to characterize soil grain-size distribution, plasticity limit, structure, bulk density, water content, and OM concentration to anticipate its vulnerability to compaction during excavation, transport, and storage. Soil compaction decreases soil macroporosity and air circulation, water infiltration, as well as root penetration, while increasing water runoff in slopes. Moreover, soil OM may decompose with time during storage (Cooke and Johnson 2002). Resulting anaerobic conditions (mainly in the center of storage piles) change the redox potential and the pH and increase the ratio of ammonium to nitrate ions, the organic acid concentrations, and the availability of some metals like Cu, Zn, and Mn (Cooke and Johnson 2002). Aerobic soil organisms such as fungi (mushrooms) and soil fauna will eventually die during storage, even if some recent study suggest that arbuscular mycorrhizae fungi may re-establish after 10 years of stockpiling (Birnbaum et al. 2017). OM loss will alter the links between soil particles and aggregate stability. It is recommended to store the different soils in different small piles, to avoid having machinery pass over it, and to temporarily revegetate it to control erosion of fine particles, bring OM, and keep soil organisms alive.

12.6.1.3 Substrate Placement for Revegetation: Decompaction, Thickness, Layering, and Topography

If no soil is available for revegetation, mine wastes such as tailings and waste rocks can be improved to constitute an acceptable growing substrate for plants, if they are not contaminated (Green and Renault 2008; Larchevêque et al. 2013; Guittonny-Larchevêque and Pednault 2016). Mineral soils compacted by storage can also need improvement before being used as substrates for revegetation. Several mechanical techniques are used to decrease substrate compaction, like subsoiling, scarifying, and freeze and thaw action. However, recompaction can be observed for mine wastes because the porosity created by mechanical action is not stabilized. Moreover, scarifying waste rocks can create finer particles and improve water retention capacity. But finer particles may be lost by erosion or move with water deeper into the pile if not stabilized. Mixing the substrates with organic amendments will more permanently decrease the substrate bulk density, favor structuration (i.e., aggregation and creation of macropores), and improve aeration. Finally, seeding annual plants with a strong taproot, such as alfalfa (*Medicago sativa* L.), can help to create macropores in compacted soils (also named biotilling; Yunusa and Newton 2003) before establishing other plants. In all cases, a sufficient thickness of non-compacted substrate should be provided to allow adequate root growth and anchorage of plants. The minimum required thickness will change according to root mean depth of target plants as well as with substrate quality in terms of nutrient and water availability. A thickness greater than 25 cm is recommended to allow enough moisture and nutrient reserves for adequate root development, and survival and growth of the vegetation (Meredith and Patrick 1961; Tordoff et al. 2000; Evanylo et al. 2005). It can be useful to combine several layers of materials of differing grain-size distribution to improve water storage, for example, by using capillary break effects (MacDonald et al. 2015) (see Chapter 3 for a description of the phenomenon). When surficial substrate is mechanically placed or worked, it is worth creating heterogeneous topography at the microsite scale (<1 m) (MacDonald et al. 2015), for example, to create mounds. Soil roughness associated with the presence of coarse elements (rocks, wood logs, and coarse mulch) will decrease the speed of water runoff and wind, and trap seeds and detached soil particles by erosion. Roughness and low compaction (that allows water infiltration) are particularly important in slopping areas to protect soils from erosion (Morgan and Rickson 2005). Roughness will also create a variety of microhabitats for plants with a range of surface microclimatic conditions in terms of light and wind exposure.

12.6.1.4 Substrate Fertilization

Mine wastes and mineral soils present deficiencies in nutrients that can be overcome by fertilization. The main limiting nutrient is nitrogen (N) because it is not included in soil minerals. Nitrogen can be brought by mineral fertilization (in the form of NH_4^+, NO_3^-, and urea) to revegetated zones, but it can be lost by leaching if it is not rapidly used by plants. Nitrates are very mobile in soils, and since mine wastes do not contain OM or clay, their cation exchange capacity is very low, and thus cations like NH_4^+ are also mobile. Slow-release fertilizers are thus recommended on mine wastes to avoid the need of repeated applications of mineral fertilizers.

Unlike N, phosphorus ions (P) come from the alteration of rock minerals in soils, like potassium (K), magnesium (Mg), and calcium (Ca). However, P is often fixed in soils under forms that are not available to plant roots (Bradshaw 1983). P fertilization is used to compensate for this low availability. It should be noted that excessive fertilization can stimulate the establishment of plant species that are not those targeted in the revegetation plan; for example, exotic or invasive species can compete with desired native species. If the revegetation objective is to regenerate an ecosystem similar to the one surrounding the mine site, it is recommended to investigate natural element concentrations of soils in this ecosystem (nutrient total and available concentrations) to design appropriate fertilization rates.

12.6.1.5 Organic Amendments

Regarding OM amendment, a concentration of 4% (dry matter) in mixtures with mine wastes improves the structure and can bring a nutrient stock for several years (Marcus 1997; Bendfeldt et al. 2001). This effect, however, varies with the OM quality and with the climate. For example, the greater the C to N ratio of the OM, the lower its decomposition rate and the more stable its structural effect, while nutrients will be released more slowly (Berg and Laskowski 2005). Moreover, the colder the climate, the lower the decomposition rate of OM (Bradshaw 1983). The salinity of organic amendments should be checked with care before use in mine revegetation. An electrical conductivity higher than 1.4 $mS.cm^{-1}$ in the mixture could limit some plant root growth and survival (Larchevêque et al. 2013). Tailings especially contain chemical additives used in the metal extraction process that can increase their salinity, in addition to the salinity coming from the natural weathering of minerals exposed to atmospheric conditions. The pH of organic amendments is variable and can be used to increase or decrease the pH of mine wastes (Pichtel et al. 1994), like lime amendments used to increase the pH of acidic wastes (Bradshaw 1997; Tordoff et al. 2000). A pH value of the growing substrate lower than 8 and greater than 4 may not decrease the growth of plants, while at pH 3, plants stop growing (Arnon and Johnson 1942).

OM amendment generally brings a nitrogen reserve to the system, but since plant roots can only absorb nitrogen under mineral forms (NH_4^+ and NO_3^-), decomposers must breakdown organic nitrogen to mineral forms before being used by plants. If ectomycorrhiza are associated with plant roots, they could help their host plants to obtain usable amino acids from N-rich organic compounds

(Wu 2011). The use of N-fixing plants, like those of the *Fabaceae* family, can help increase the N stock of the soil. However, this N stock comes from plant litter and must be decomposed to be recycled in an available form to plants.

12.6.1.6 Beneficial Organisms

Finally, spores of soil organisms (e.g., mycorrhiza) can be inoculated to plant seeds or seedlings to obtain a synergistic effect. A little amount of natural topsoil can also be used as inoculant (Bradshaw 2000). Soil organisms (bacteria, mushrooms, and soil fauna) play a crucial role in the self-sufficiency of revegetated areas in the long term because important ecosystem processes such as recycling of nutrients rely on them. Soil macrofauna, such as earthworms, improve the soil structure by bringing fine earth to the soil surface, mixing organic and mineral matter, and creating biopores that aerate the soil (Bradshaw 1983). It has been observed that they are able to recolonize the revegetated zones from adjacent ecosystems depending on several factors, in particular the quantity of available organisms in the adjacent ecosystem, the type of organisms (e.g., earthworms have little mobility), the nature of the soil and vegetation of the revegetated ecosystem, the distance from the adjacent ecosystem, and the presence of barriers to colonization (e.g., large mining surfaces or inappropriate vegetation). For more information on the presence of soil organisms in post-mining sites, see book by Frouz (2013). Pollinators are also important contributors to the reproduction of flowering plants and their long-term survival on revegetated mine sites.

12.6.2 Selection of Plants and Revegetation Techniques

Once the substrate properties are designed, adapted plant combinations should be selected and coupled with appropriate establishment techniques.

12.6.2.1 Adaptive Approach: Plants Adapted to the Mine Substrate Constraints

On difficult substrates like chemically unstable, acid-generating, and/or metal-rich wastes, a plant cover can be temporarily established to stabilize the surface against erosion and to intercept and transpire precipitation, decreasing the amount of contaminated water. It requires plants tolerant to elevated metal concentrations and salinity, and also repeated lime or OM amendments to improve the pH, as well as fertilization to provide nutrients. Some tolerant cultivars are available for phytoremediation (Tordoff et al. 2000) but remain rare and often specific to particular climatic zones or particular metal mixtures. Their growth can be slow, which decreases their efficiency against erosion. More information on the use of phytoremediation plants can be found in review papers by Tordoff et al. 2000; Mendez and Maier 2008; Ullah et al. 2015; and Wang et al. 2017. Bradshaw's (2000) review also provides extensive knowledge on the reclamation of mine soils using natural processes.

On substrates with low toxicity levels, one of the most used revegetation techniques is herbaceous seeding coupled with fertilization because its costs are low when hay agronomic species are used. Moreover, some agronomic species are easily found all over the world, and they are adapted to a wide range of soil conditions and their fertilization requirements are well known. Finally, herbaceous seeding is very efficient to control water and wind erosion of mine wastes (Haigh and Sansom 1999). The seeding should be planned well before or after the annual periods of intense climatic events (freeze, rain, and drought). Agronomic herbaceous seeds generally need two weeks of favorable humidity conditions to germinate. Seeding rates vary around several hundreds of kilogram of seeds by hectare. Optimal seed mixtures generally contain three main types of plants: a nurse plant, graminoids (*Poaceae*), and legumes (*Fabaceae*) (Fields-Johnson et al. 2012). The nurse plant, usually an annual cereal, quickly germinates and stabilizes the soil, preventing the seeds of species that germinate more slowly from being removed from the site or buried. Perennial graminoids are used to quickly provide a persistent cover, while legumes progressively enrich the soil in nitrogen taken from the atmosphere thanks to symbiotic bacteria in root nodules. The seeding mixture can also include plant species with a strong taproot to help decompaction of the soil (e.g., alfalfa). Thanks to legumes, the nitrogen fertilization requirements may decrease in the medium term, but phosphorus needs may remain longer (Tordoff et al. 2000). Seed mixture can contain commercial spores of symbiotic bacteria and/or mushrooms to facilitate the establishment of the desired symbiosis.

Revegetation with agronomic herbaceous seeds relies on several techniques, depending on the site characteristics:

- Hand seeding on small surfaces.
- Seeding with agriculture equipment on flat and well-drained surfaces. The soil is prepared and seeds are buried. This technique requires fewer seeds than the two others.
- Hydraulic seeding on slopping or uneven terrains where equipment access is difficult. Seeds are spread on the terrain in a mixture containing water, mulching and/or binding agents, and fertilizers (Brofas et al. 2007).

Agronomic herbaceous plants initiate plant soil processes and progressively decrease resources constraints for the establishment of new species. In this way, they can favor native plant succession. However, they can also be very competitive to native plants' establishment, (e.g., forest trees) (Rizza et al. 2007; Franklin et al. 2012; Bouchard et al. 2018). Moreover, graminoids generally form a dense cover that creates habitats for rodents, which attack planted trees (Skousen et al. 2007). To insure agronomic plants do not gain advantage over native plants, they can be chosen sterile or unable to complete their life cycle (until reproduction) under the local climate.

Shrubs tolerant to water, nutrient, and salt stresses, such as N-fixing *Alnus* and *Acacia* species, can also be used to revegetate mine wastes directly (Boriŝev et al. 2018). However, woody species are usually less efficient than herbaceous species to control soil water erosion. Their cover is less complete at the early stages of establishment, and their distance from the soil surface is greater. Potentially invasive species should be excluded from revegetation designs, even if they are sometimes very performant on mine sites, to avoid their spread in natural ecosystems surrounding mine sites.

In regions where the climatic conditions are unfavorable to plant growth part of the year (e.g., frost, drought), the use of native plant species adapted to local climatic stresses may decrease the risks of revegetation failure. However, plant material production is mainly developed for important commercial species in agriculture or forestry, less for non-commercial native species. If seed rain (i.e., wind-dispersed seeds falling on the ground) is occurring on the revegetated site, natural recolonization of native plant species could then be favored by treatments that improve microclimatic conditions like annual or low-density nurse plants, mulches, and watering.

12.6.2.2 Improvement Approach: Trees Established on Improved Substrates

Most tree species cannot tolerate soil compaction, extreme pH, and elevated salinity (Angel et al. 2006; Emerson et al. 2009), whereas most graminoids and legumes can (Burger and Zipper 2002). Thus, for the majority of mine substrates, the return by natural processes to a plant community dominated by trees is very slow (hundreds to thousands years) (Walker and del Moral 2009). If the mine substrate is of low quality, tree plantation after a few years of growth of agronomic herbaceous mixtures could help tree establishment. However, care should be given to choose a herbaceous mixture compatible with trees at an appropriate density (25% to 50% cover, Buckley and Franklin 2008) to decrease competition with establishing trees. Once trees are present on revegetated mine sites, they can catalyze the return to forested ecosystems by favoring an understory of native species tolerant to shade over agronomic species intolerant to shade, especially at canopy closure (Strong 2000).

The presence of trees can also change the microclimatic conditions at the soil level and favor natural recolonization by native tree seedlings (Bouchard et al. 2018). Broadleaved species are generally better catalysts of the forested ecosystem restoration than coniferous species (Singh et al. 2002).

More generally, nurse plants can be used to improve the establishment of trees. Nurse plants were shown to increase tree establishment through seed trapping, less temperature variations, more available water, more nutrients in the soil, protection against erosion, herbivory or stamping, physical support, less soil compaction (Flores and Jurado 2003). However, the effect of nurse plants toward large plants like trees would be more often positive on survival and emergence than on their later growth (Gomez-Aparicio 2009). In consequence, to insure satisfying tree growth in the long term, improved substrates such as topsoils and mine wastes or mineral overburden amended with OM are usually required.

12.6.2.3 Tree Planting

Regarding the techniques used to establish trees on mine surfaces, direct seeding was rarely found successful (Hutnik and McKee 1990). Tree planting is preferred. Trees can be planted in plantation holes filled with improved soil or mine wastes, or in favorable soil pockets (Guittonny-Larchevêque and Pednault 2016), to decrease the quantity of material needed and the revegetation costs. However, trees have a larger root system than herbaceous plants, which varies with the tree species (Schenk and Jackson 2002), and need enough soil surface and thickness to develop an adequate root system. Tree mechanical stability on revegetated surfaces depends on the appropriate architecture of root systems, in particular, the symmetry, spread, and number of lateral roots (Khuder et al. 2007) as well as the presence of sink and deep roots binding a large volume of soil (Danjon et al. 2005). Moreover, limited thickness of favorable substrate may decrease tree aboveground development (Larchevêque et al. 2014).

Cuttings (shoot pieces) as well as seedlings or rooted cuttings can be used as tree planting material. Cuttings (e.g., from *Salix* sp. and *Populus* sp.) allow the quick establishment of trees (Marcus 1997) but require favorable humidity conditions for root formation during several weeks after planting and low levels of competition with herbaceous plants (Hutnik and McKee 1990). Cuttings, however, are often clonal (i.e., same genetics for all cuttings), which decreases the diversity of the planted population. Seedlings and rooted cuttings, which have roots before planting, can suffer important planting shock if water availability is reduced during the first steps of establishment (Hutnik and McKee 1990). Planting stock of greater size can help planted trees to overcome competition with herbaceous plants (Jobidon et al. 1998). Planting material with a greater root/shoot biomass ratio will decrease tree water losses and will be more suitable for planting on water-stressed sites. Planting material can also be exposed to treatments at the nursery to accumulate nutrients and more easily overcome nutrient limitations associated with mine wastes (MacDonald et al. 2015).

Tree material conservation and manipulation until planting should be adequate. The planting time can vary according to the tree species from the beginning to the end of the growing season. The planting of cuttings must be deep enough to allow appropriate root/shoot ratio and satisfying tree anchorage. The topography of planting microsites will condition the water availability for the planted tree. Trees more tolerant to water stress, such as conifers (Bugmann 1996; Gao et al. 2002), could be planted in slopping and wind-exposed terrains, while broadleaved trees may be installed in flat basins or bottom slope areas. Fast-growing trees with higher nutrients requirements may benefit from OM-richer parts of the site to revegetate. Planting density will control the time to reach canopy closure as well as the competition intensity among planted trees (Remaury et al. 2019). Localized fertilization at the foot of each tree (Van den Driessche 1999) can be used to fulfill nutrient needs of more demanding species. More information on mine afforestation can be found in MacDonald et al.'s (2015) review (especially specific to surface mines).

Finally, whatever the selected plants for revegetation, diversity of species and genotypes is important to increase the chances of plant establishment in variable living conditions on reclamation sites and to increase resistance to stresses and diseases. Moreover, diverse planting or seeding mixtures will provide differing habitats for recolonization by a diverse flora and fauna.

12.7 VALIDATING EXPERIMENTALLY AND ADJUSTING THE FINAL REVEGETATION DESIGN

Since tree survival and growth on mine wastes vary a lot according to the tree species (Larchevêque et al. 2013; Larchevêque et al. 2014; Guittonny-Larchevêque and Pednault 2016), it is recommended to experimentally test tree establishment on improved substrates at a small scale before applying it at a large scale. More specifically, tests at several scales are needed to prepare and adjust the revegetation plan before implementing it in the field: tests under controlled conditions at the small scale (pot experiments in the greenhouse or in the laboratory), tests at the small (experimental cells of 100 m^2) or medium (1 ha experimental settings) scale in the field, or tests through progressive revegetation. Choices of design then rely on demonstrated facts and can be explained based on concrete results. In revegetation, it is difficult to rely on mechanistic models to test the designs since the evolution of a revegetated zone in relationship with the natural surrounding environment is both deterministic (known pattern of plant growth and succession) and stochastic (unplanned events) (Cooke and Johnson 2002). Success indicators used to validate the tests belong to several categories of measurements (see Table 12.7). These success indicators, and the scale needed to test them, must comply with the defined revegetation objectives (see 12.4.1). However, they can be limited by the spatiotemporal scale of the tests, especially under controlled conditions. In situ tests are important to guarantee that the revegetation plan and its objectives match the specific characteristics of the site and its natural variability. Progressive revegetation can help to test which approaches and associated techniques are efficient in specific zones and how much time it takes to reach desired levels of success indicators. It also offers possibilities of organizational learning regarding the operationalization of revegetation technics to be ready at mine closure.

12.8 PLANNING SITE MAINTENANCE AND MEASUREMENT OF REVEGETATION SUCCESS

Once the final revegetation design is applied, the follow-up of revegetation success should include several important elements:

1) Validate if the design was respected.
2) Measure if revegetation objectives are reached thanks to the associated success indicators.
3) Check that no undesirable effect occurred at the larger ecological scale of the landscape surrounding the mine site.
4) Maintain the follow-up at the medium term to validate the self-sufficiency of the revegetated zone, meaning that structural and functional attributes remain without any inputs (e.g., propagules, fertilizers).

Some indicators of revegetation success are listed in Section 12.4.2 and techniques of monitoring are given in Chapter 10.

Since the risk of failure is real, revegetation managers may define alternative revegetation objectives in their plan, as well as corrective procedures if the results of the follow-up are unsatisfying. This means that any revegetated site should remain accessible for further work if needed (Cooke and Johnson 2002).

TABLE 12.7
Some Categories of Measurements Performed to Test a Revegetation Design

Substrate characterization	Element concentrations in plants
Viability of plant propagule	Water budget
Plant survival and productivity	Diversity
Efficiency of natural recolonization	Mechanical stability of trees
Soil erosion	Plant reproduction

12.9 REVEGETATION WITH ADDED VALUE

The ecosystems created on mine sites by revegetation can be designed to provide useful services and produce goods. Ecosystem services (see Fisher et al. 2008) rely on ecosystem properties maintaining the environment quality for living organisms (e.g., cleaning soil, water, and air; storing and recycling nutrients, and regulating climate). Soils of urban, industrial, traffic, mining, and military areas (SUITMAs) can sustain ecosystem services (Morel et al. 2014). These services can provide "nature-based solutions" (Song et al. 2019) for the remediation of contaminated soils. Revegetation can be designed to reduce ecotoxicological risks associated with contaminated mine soils. In the mining context, where contaminants are persistent metallic elements, it is mainly phytostabilization that is of interest. Phytostabilization reduces contaminant fluxes and transfers through soil stabilization against erosion by plant cover and root networks, through the decrease of contaminated water percolation by increasing evapotranspiration, and through the decrease of contaminant mobility by adsorption on root surfaces (Mendez and Maier 2008). On mine waste facilities, phytoextraction of residual commercial metals can also be used. It relies on plants accumulating metals in their aboveground parts. Metals in leaves of selected plants can reach concentrations greater than 1% of dry matter (e.g., for Ni, Zn, and Mn) (Baker and Whiting 2002). Hyperaccumulator plants are especially identified for Ni and are used for the development of agromining (Nkrumah et al. 2016; Kidd et al. 2018) in order to produce metals for industrial applications. Metal yields can reach up to 100 kg of Ni by hectare in R&D experiments in Europe (Nkrumah et al. 2016). However, access to the sites bearing hyperaccumulator plants may need to be controlled to avoid the contamination of browsing fauna.

Revegetated mine sites can also be used for CO_2 fixation in plant biomass to contribute to climate change mitigation. For such objectives, trees are more efficient since they produce a large biomass, a deep rooting, and lignified tissues that decompose slowly. The soil can also act as a CO_2 sink in revegetated areas, especially under cold climates of boreal regions (Bolin et al. 2000). Mine soils were demonstrated to have a greater potential of C accumulation compared to non-disturbed soils (Stahl et al. 2003). This accumulation was quantified between 0.7 and 4 t C/ha/year in US mine soils (Shestra and Lal 2006) compared to 2.6 t C/ha/year in a forest ecosystem (Sperow 2006).

The use of organic wastes to revegetate mine sites can help increase C storage in the soil concomitantly to improve soil quality. Several types of organic wastes were tested to stabilize and revegetate mine wastes (Rakotonimaro et al. 2018)—for example, composts (green wastes, municipal solid waste, olive mill waste, sewage sludge, and spent mushroom), biosolids (sewage sludge, food wastes, sanitary wastes, and anaerobic digestate), manures (cow, cattle, pig, and poultry), biochars (rice straw, hardwood, oak tree, fir tree, pruning residues, and manure pellets), and slurry (pig and cattle). The valorization of such organic wastes in the mining context allows their diversion from landfills and favors circular economy. However, some studies show that in the long term, using poor substrates favors the stability of the established plant community, while the use of organic substrates richer in nutrients allows a rapid vegetation development but with a greater risk of vegetation regression in the long term (Tordoff et al. 2000).

The valorization of organic wastes like sewage sludges, domestic waste composts, and paper sludges can be advantageously coupled to the production of plant biomass for bioenergy purposes. It can create economic value associated with revegetated mine sites. And it proposes an alternative to the use of agricultural soils for biofuel production. For example, the Green Mines Green Energy initiative from Natural Resources Canada (Hargreaves et al. 2012) tested annual (canola, safflower, and corn) and perennial (willow and switchgrass) cultures on tailings facilities covered by paper sludges or municipal wastes compost. However, the quality of organic wastes is variable; some could bring contaminants to the revegetated site, increase the mobility of metalloids (Rakotonimaro et al. 2018), or increase the mobility of metals in secondary minerals of oxidized tailings (Beauchemin et al. 2018). Thus, the potential accumulation of contaminants by the cultivated plants should be surveyed, as well as the surface and groundwater water quality.

12.10 CHALLENGES AND R&D NEEDS

This chapter illustrates that basic knowledge on plant and soil processes underlying the success of revegetation is available. Strategies and techniques have been proposed to establish diverse types of plants on mine substrates. Avenues were explored to increase the value of revegetated sites. However, knowledge is still missing on several aspects of mine site revegetation, some of which are listed below.

- The revegetation of mine sites is important to return the reclaimed mine site to its surrounding natural environment but also to shape future land use by the communities. More research is needed to better integrate the communities' vision in the design of the revegetation plan.
- When sites are reclaimed with engineered covers to control the chemical stability of the waste storage facilities (see Chapters 4 to 9), care should be taken to install a vegetation that is not harmful to the performance of the cover. The relationships of plants and engineered covers have to date been partially studied, especially for oxygen barrier covers. For more details on this subject, see Chapter 14.
- Plants' effects on slope stability and erosion control (see, e.g., Morgan and Rickson 2005) were studied in civil engineering contexts (e.g., in roadsides stabilization) or bioengineering contexts (e.g., riverside restoration). The knowledge developed in these fields could be transferred to improve the geotechnical stability of reclaimed mine sites. For example, research could be developed to quantify revegetated surfaces' contribution to the water budget of reclaimed mine sites and tailings' consolidation (water storage) or water volumes to collect (runoff and infiltration).
- When the revegetation objective aims to re-establish a natural, mature ecosystem, or a high level of biodiversity, only long-term follow-up can document if the revegetation design succeeded in reaching the objective. Indeed, the natural processes underlying mature ecosystems and diverse communities take time to become established. Moreover, unplanned disturbances can change or delay the evolutive trajectory of the revegetated system. Vegetation regression was also observed with time on some revegetated sites. Thus, more long-term follow-up of revegetation case studies are needed. Since climate changes are expected to modify the distribution of plant species (Price et al. 2013), they should be included in the long-term prediction of vegetation influence on the performance of engineered covers (see Chapter 14).
- Establishing plants is the first step to favor ecosystem reconstruction on mine sites. Once established, they will attract fauna and microbes to participate in ecosystem processes and evolution. The relationships between plants and communities of other organisms and how these interactions improve the success of ecosystem restoration should be better studied. Recent advances in metagenomics are providing researchers with tools that will advance knowledge on these subjects.

REFERENCES

Aerts, R. 1995. The advantages of being evergreen. *Tree* 10, no. 10: 402–407.

Aerts, R., and M. J. Van Der Peijl. 1993. A simple model to explain the dominance of low-productive perennials in nutrient-poor habitats. *Oïkos* 66, no. 4: 144–147.

Angel, P., D. H. Graves, C. Barton, R. C. Warner, P. W. Conrad, R. G. Sweigard, and C. Agouridis. 2006. Surface mine reforestation research: Evaluation of tree response to low compaction reclamation techniques. In *7th International Conference on Acid Rock Drainage (ICARD), 26–30 March 2006, St. Louis, MO*, ed. R. I. Barnhisel, 45–58. Lexington, KY: American Society of Mining and Reclamation (ASMR).

Arnon, D. I., and C. M. Johnson. 1942. Influence of hydrogen ion concentration on the growth of higher plants under controlled conditions. *Plant Physiology* 17, no. 4: 525–539.

Baker, A. J. M., and S. N. Whiting. 2002. In search of the Holy Grail: A further step in understanding metal hyperaccumulation? *New Phytologist* 155: 1–7.

Beauchemin, S., J. S. Clemente, Y. Thibault, S. Langley, E. G. Gregorich, and B. Tisch. 2018. Geochemical stability of acid-generating pyrrhotite tailings 4 to 5 years after addition of oxygen-consuming organic covers. *Science of the Total Environment* 645: 1643–1655.

Bendfeldt, E. S., J. A. Burger, and W. L. Daniels. 2001. Quality of amended mine soils after sixteen years. *Soil Science Society of America Journal* 65: 1736–1744.

Berg, B., and R. Laskowski. 2005. *Litter decomposition: A guide to carbon and nutrient turnover*. Amsterdam: Elsevier.

Bertness, M. D., and R. Callaway. 1994. Positive interactions in communities. *Trends in Ecology and Evolution* 9: 191–193.

Beven, K., and P. Germann. 1982. Macropores and water flow in soils. *Water Resources Research* 18, no. 5: 1311–1325.

Birnbaum, C., L. E. Bradshaw, K. X. Ruthrof, and J. B. Fontaine. 2017. Topsoil stockpiling in restoration: Impact of storage time on plant growth and symbiotic soil biota. *Ecological Restoration* 35, no. 3: 237–245.

Bolin, B., R. Sukumar, P. Ciais, et al. 2000. Global perspective. In *Land use, land-use change and forestry: A special report of the IPCC*, ed. R. T. Watson, I. R. Noble, B. Bolin, N. H. Ravindranath, D. J. Verardo, and D. J. Dokken, 23–51. Cambridge: Cambridge University Press.

Borišev, M., S. Pajević, N. Nikolić, A. Pilipović, D. Arsenov, and M. Župunski. 2018. Chapter 7: Mine site restoration using silvicultural approach. In *Bio-geotechnologies for mine site rehabilitation*, ed. M. N. V. Prasad, P. J. de Campos Favas, and S. K. Maiti, 115–130. Amsterdam: Elsevier.

Bouchard, H., M. Guittonny, and S. Brais. 2018. Early recruitment of boreal forest trees in hybrid poplar plantations of different densities on mine waste rock slopes. *Forest Ecology and Management* 429: 520–533.

Bradshaw, A. 1983. The reconstruction of ecosystems: Presidential address to the British Ecological Society. *Journal of Applied Ecology* 20, no. 1: 1–17.

Bradshaw, A. 1997. Restoration of mined lands—using natural processes. *Ecological Engineering* 8, no. 4: 255–269.

Bradshaw, A. 2000. The use of natural processes in reclamation—advantages and difficulties. *Landscape and Urban Planning* 51, no. 2: 89–100.

Brofas, G., G. Mantakas, K. Tsagari, M. Stefanakis, and C. Varelides. 2007. Effectiveness of cellulose, straw and binding materials for mining spoils revegetation by hydro-seeding, in Central Greece. *Ecological Engineering* 31, no. 3: 193–199.

Buckley, D. S., and J. A. Franklin. 2008. Early tree and ground cover establishment as affected by seeding and fertilization rates in Tennessee. In *National Meeting of the American Society of Mining and Reclamation, Richmond, VA, New Opportunities to Apply Our Science*, 14–19 June 2008, ed. R. I. Barnhisel, 180–191. Lexington, KT: ASMR.

Bugmann, H. 1996. Functional types of trees in temperate and boreal forests: Classification and testing. *Journal of Vegetation Science* 7, no. 3: 359–370.

Burger, J. A., and C. E. Zipper. 2002. How to restore forests on surface-mined land: Reclamation guidelines for surface mined land in Southwest Virginia. Virginia Cooperative Extension, publication 460–123.

Burke, A. 2003. Practical measures in arid land restoration after mining: A review for the southern Namib. *South African Journal of Science* 99, no. 9: 413–417.

Chapin III, F. S., P. A. Matson, and P. Vitousek. 2011. *Principles of terrestrial ecosystem ecology*. New York: Springer-Verlag.

Chesworth, W., and O. Spaargaren. 2008. Technosols. In *Encyclopedia of soil science*, ed. W. Chesworth, 765–766. New York: Springer.

Ciarkowska, K., L. Gargiulo, and G. Mele. 2016. Natural restoration of soils on mine heaps with similar technogenic parent material: A case study of long-term soil evolution in Silesian-Krakow Upland Poland. *Geoderma* 261: 141–150.

Clements, F. E. 1916. Plant succession. Publication N. 242, Carnegie Institute, Washington.

Connell, J. H., and R. O. Slatyer. 1977. Mechanisms of succession in natural communities and their role in community stability and organization. *American Naturalist* 111: 1119–1144.

Cooke, J. A., and M. S. Johnson. 2002. Ecological restoration of land with particular reference to the mining of metals and industrial minerals: A review of theory and practice. *Environmental Reviews* 10, no. 1: 41–71.

Cooper, W. S. 1923. The recent ecological history of Glacier Bay, Alaska: The present vegetation cycle. *Ecology* 4, no. 3: 223–246.

Danjon, F., T. Fourcaud, and D. Bert. 2005. Root architecture and wind-firmness of mature Pinus pinaster. *New Phytologist* 168, no. 2: 387–400.

Emerson, P., J. Skousen, and P. Ziemkiewicz. 2009. Survival and growth of hardwoods in brown versus gray sandstone on a surface mine in West Virginia. *Journal of Environmental Quality* 38, no. 5: 1821–1829.

Evanylo, G. K., A. O. Abaye, C. Dundas, et al. 2005. Herbaceous vegetation productivity, persistence, and metals uptake on a biosolids-amended mine soil. *Journal of Environmental Quality* 34, no. 5: 1811–1819.

Fields-Johnson, C., C. Zipper, J. Burger, and D. Evans. 2012. Forest restoration on steep slopes after coal surface mining in Appalachian USA: Soil grading and seeding effects. *Forest Ecology and Management* 270: 126–134.

Fisher, B., K. Turner, M. Zylstra, et al. 2008. Ecosystem services and economic theory: Integration for policy relevant research. *Ecological Applications* 18, no. 8: 2050–2067.

Flores, J., and E. Jurado. 2003. Are nurse-protégé interactions more common among plants from arid environments? *Journal of Vegetation Science* 14, no. 6: 911–916.

Franklin, J. A., C. E. Zipper, J. A. Burger, J. G. Skousen, and D. F. Jacobs. 2012. Influence of herbaceous ground cover on forest restoration of eastern US coal surface mines. *New Forest* 43: 905–924.

Frouz, J., ed. 2013. *Soil biota and ecosystem development in post mining sites*. Boca Raton, FL: CRC Press.

Gao, Q., P. Zhao, X. Zheng, X. Cai, and W. Shen. 2002. A model of stomatal conductance to quantify the relationship between leaf transpiration, microclimate and soil water stress. *Plant Cell and Environment* 25, no. 11: 1373–1381.

Gomez-Aparicio, L. 2009. The role of plant interactions in the restoration of degraded ecosystems: A meta-analysis across life-forms and ecosystems. *Journal of Ecology* 97, no. 6: 1202–1214.

Green, S., and S. Renault. 2008. Influence of papermill sludge on growth of Medicago sativa, Festuca rubra and Agropyron trachycaulum in gold mine tailings: A greenhouse study. *Environmental Pollution* 151, no. 3: 524–531.

Grime, J. P. 1977. Evidence for the existence of three primary strategies in plants and its relevance to ecological and evolutionary theory. *American Naturalist* 111: 1169–1194.

Guittonny-Larchevêque, M., and C. Pednault. 2016. Substrate comparison for short-term success of a multispecies tree plantation in thickened tailings of a boreal gold mine. *New Forests* 47, no. 5: 763–781.

Haigh, M. J., and B. Sansom. 1999. Soil compaction, runoff and erosion on reclaimed coal-lands (UK). *International Journal of Surface Mining, Reclamation and Environment* 13, no. 4: 135–146.

Hargreaves, J., A. Lock, P. Beckett, et al. 2012. Suitability of an organic residual cover on tailings for bioenergy crop production: A preliminary assessment. *Canadian Journal of Soil Science* 92: 203–211.

Hutnik, R. J., and G. W. McKee. 1990. Revegetation. In *Surface mining*, ed. B. A. Kennedy, 811–817. Littleton, Colorado: Society for Mining, Metallurgy, and Exploration Inc.

Jobidon, R., L. Charette, and P. Y. Bernier. 1998. Initial size and competing vegetation effects on water stress and growth of Picea mariana (Mill.) BSP seedlings planted in three different environments. *Forest Ecology and Management* 103, no. 2–3: 293–305.

Johnson, D. L., S. H. Ambrose, T. J. Bassett, et al. 1997. Meanings of environmental terms. *Journal of Environmental Quality* 26, no. 3: 581–589.

Keddy, P. 1992. Assembly and response rules: Two goals for predictive community ecology. *Journal of Vegetation Science* 3, no. 2: 157–164.

Khuder, H., A. Stokes, F. Danjon, K. Gouskou, and F. Lagane. 2007. Is it possible to manipulate root anchorage in young trees? *Plant and Soil* 294, no. 1: 87–102.

Kidd, P. S., A. Bani, E. Benizri, et al. 2018. Developing sustainable agromining systems in agricultural ultramafic soils for nickel recovery. *Frontiers in Environmental Science* 6, article 44.

Lambers, H., F. S. Chapin, and T. L. Pons. 2008. *Plant physiological ecology*. New York: Springer.

Larchevêque, M., A. Desrochers, B. Bussière, H. Cartier, and J.-S. David. 2013. Revegetation of non-acid-generating, thickened tailings with boreal trees: A greenhouse study. *Journal of Environmental Quality* 42, no. 2: 351–360.

Larchevêque, M., A. Desrochers, B. Bussière, and D. Cimon. 2014. Planting trees in soils above non-acid generating wastes of a boreal gold mine. *Ecoscience* 21, no. 3–4: 217–231.

Leguédois, S., G. Séré, A. Auclerc, et al. 2018. Modelling pedogenesis of technosols. *Geoderma* 262: 199–212.

Li, M. S. 2006. Ecological restoration of mineland with particular reference to the metalliferous mine wasteland in China: A review of research and practice. *Science of the Total Environment* 357: 38–53.

MacDonald, S. E., S. M. Landhaüsser, J. Skousen, et al. 2015. Forest restoration following surface mining disturbance: Challenges and solutions. *New Forests* 46: 703–732.

Marcus, J. J. 1997. Chapter 6: Technologies for environmental protection. In *Mining Environmental Handbook: Effects of mining on the environment and American environmental controls on mining*, ed. J. J. Marcus, 190–282. London: Imperial College Press.

Martinez-Ruiz, C., and B. Fernandez-Santos. 2005. Natural revegetation on topsoiled mining-spoils according to the exposure. *Acta Oecologica* 28, no. 3: 231–238.

Mendez, M. O., and R. M. Maier. 2008. Phytoremediation of mine tailings in temperate and arid environments. *Reviews in Environmental Science and Bio/Technology* 7, no. 1: 47–59.

Meredith, H. L., and W. H. Patrick. 1961. Effects of soil compaction on subsoil root penetration and physical properties of three soils in Louisiana. *Agronomy Journal* 53, no. 3: 163–167.

Morel, J. L., C. Chenu, and K. Lorenz. 2014. Ecosystem services provided by soils of urban, industrial, traffic, mining, and military areas (SUITMAs). *Journal of Soils and Sediments* 15: 1659–1666.

Morgan, R. P. C., and R. J. Rickson. 2005. *Slope stabilization and erosion control: A bioengineering approach*, 1st ed. London: Chapman and Hall.

Mosseler, A., J. E. Major, and M. Labrecque. 2014. Growth and survival of seven native willow species on highly disturbed coal mine sites in eastern Canada. *Canadian Journal of Forest Research* 44, no. 4: 340–349.

Nkrumah, P. N., A. J. M. Baker, R. L. Chaney, et al. 2016. Current status and challenges in developing nickel phytomining: An agronomic perspective. *Plant and Soil* 406: 55–69.

Oades, J. M. 1984. Soil organic matter and structural stability: Mechanisms and implications for management. *Plant and Soil* 76: 319–337.

Pichtel, J. R., W. A. Dick, and P. Sutton. 1994. Comparison of amendments and management practices for long-term reclamation of abandoned mine lands. *Journal of Environmental Quality* 23, no. 4: 766–772.

Prach, K., and P. Pysek. 2001. Using spontaneous succession for restoration of human-disturbed habitats: Experience from Central Europe. *Ecological Engineering* 17, no. 1: 55–62.

Price, D. T., R. I. Alfaro, K. J. Brown, et al. 2013. Anticipating the consequences of climate change for Canada's boreal forest ecosystems. *Environmental Reviews* 21, no. 4: 322–365.

Rakotonimaro, T., M. Guittonny, and C. Neculita. 2018. Stabilization of hard rock mines tailings with organic amendments: Pore water quality control and revegetation—a review. *Desalination and Water Treatment* 112: 53–71.

Remaury, A., M. Guittonny, and J. Rickson. 2019. The effect of tree planting density on the relative development of weeds and hybrid poplars in revegetated mine slopes vulnerable to erosion. *New Forests* 50, no. 4: 555–572.

Reiners, W. A., I. A. Worley, and D. B. Lawrence. 1971. Plant diversity in a chronosequence at Glacier Bay, Alaska. *Ecology* 52, no. 1: 55–69.

Rizza, J., J. Franklin, and D. Buckley. 2007. The influence of different ground cover treatments on the growth and survival of tree seedlings on remined sites in eastern Tennessee. In *National Meeting of the American Society of Mining and Reclamation, Gillette, WY, 30 Years of SMCRA and Beyond, 2–7 June 2007*, ed. R. I. Barnhisel, 663–677. Lexington, KY: ASMR.

Rokich, D. P., K. W. Dixon, K. Sivasithamparam, and K. A. Meney. 2001. Topsoil handling and storage effects on woodland restoration in western Australia. *Restoration Ecology* 8, no. 2: 196–208.

Rundel, P. W. 1988. Leaf structure and nutrition in Mediterranean-climate sclerophylls. In *Mediterranean-type ecosystems: A data source book*, ed. R. L. Specht, P. Rundel, W. E. Westman, P. C. Catling, J. Majer, and P. Greenslade, 157–167. New York: Kluwer.

Schenk, H. J., and R. B. Jackson. 2002. Rooting depths, lateral root spreads and below-ground/above-ground allometries of plants in water-limited ecosystems. *Journal of Ecology* 90, no. 3: 480–494.

Sebelikova, L., G. Csicsek, A. Kirmer, et al. 2019. Spontaneous revegetation versus forestry reclamation: Vegetation development in coal mining spoil heaps across Central Europe. *Land Degradation and Development* 30, no. 3: 348–356.

Sheoran, V., A. S. Sheoran, and P. Poonia. 2010. Soil reclamation of abandoned mine land by revegetation: A review. *International Journal of Soil, Sediment and Water* 3, no. 2: article 13.

Shestra, R. K., and R. Lal. 2006. Ecosystem carbon budgeting and soil carbon sequestration in reclaimed mine soil. *Environment International* 32, no. 6: 781–796.

Singh, A. N., A. S. Raghubanshi, and J. S. Singh. 2002. Plantations as a tool for mine spoil restoration. *Current Science* 82, no. 12: 1436–1441.

Skousen, J., J. Gorman, and P. Emerson. 2007. Tree recruitement and growth on 20-year-old, unreclaimed surface mined lands in West Virginia. In *National Meeting of the American Society of Mining and Reclamation, Gillette, WY, 30 Years of SMCRA and Beyond, 2–7 June 2007*, ed. R. I. Barnhisel, 771–787. Lexington, KY: ASMR.

Song, Y., N. Kirkwood, C. Maksimović, et al. 2019. Nature based solutions for contaminated land remediation and brownfield redevelopment in cities: A review. *Science of the Total Environment* 663: 568–579.

Sperow, M. 2006. Carbon sequestration potential in reclaimed mine sites in seven east-central states. *Journal of Environmental Quality* 35, no. 4: 1428–1438.
Stahl, P. D., J. D. Anderson, L. J. Ingram, G. E. Schuman, and D. L. Mummey. 2003. Accumulation of organic carbon in reclaimed cola mine soils of Wyoming. Paper presented at the *National Meeting of the ASMR*, June 3–6. ASMR, Lexington, KY.
Strong, W. L. 2000. Vegetation development on reclaimed lands in the Coal Valley Mine of western Alberta, Canada. *Canadian Journal of Botany* 78, no. 1: 110–118.
Tordoff, G. M., A. J. M. Baker, and A. J. Willis. 2000. Current approaches to the revegetation and reclamation of metalliferous mine wastes. *Chemosphere* 41, no. 1–2: 219–228.
Ullah, A., S. Heng, M. F. H. Munis, S. Fahad, and X. Yang. 2015. Phytoremediation of heavy metals assisted by plant growth promoting (PGP) bacteria: A review. *Environmental and Experimental Botany* 117: 28–40.
Urbas, P., and K. Zobel. 2000. Adaptive and inevitable morphological plasticity of three herbaceous species in a multi-species community: Field experiment with manipulated nutrients and light. *Acta Oecologica* 21, no. 2: 139–147.
Van den Driessche, R. 1999. First-year growth response of four Populus trichocarpa × Populus deltoïdes clones to fertilizer placement and level. *Canadian Journal of Forest Research* 29, no. 5: 554–562.
Walker, L. R., and R. del Moral. 2009. Lessons from primary succession for restoration of severely damaged habitats. *Applied Vegetation Science* 12, no. 1: 55–67.
Wang, L., B. Ji, Y. Hu, R. Liu, and W. Sun. 2017. A review on in situ phytoremediation of mine tailings. *Chemosphere* 184: 594–600.
Westman, W. E. 1986. Resilience: Concepts and measures. In *Resilience in Mediterranean-type ecosystems*, ed. B. Dell, A. J. M. Hopkins, and B. B. Lamont, reprint of Tasks for vegetation science, vol. 16. Dordrecht: Springer.
Whitbread-Abrutat, P. H. 1997. The potential of some soil amendments to improve tree growth on metalliferous mine wastes. *Plant and Soil* 192: 199–217.
Winterhalder, K. 1996. Environmental degradation and rehabilitation of the landscape around Sudbury, a major mining and smelting area. *Environmental Reviews* 4: 185–224.
Wong, M. H. 2003. Ecological restoration of mine degraded soils, with emphasis on metal contaminated soils. *Chemosphere* 50: 775–780.
Wu, T. 2011. Can ectomycorrhizal fungi circumvent the nitrogen mineralization for plant nutrition in temperate forest ecosystems? *Soil Biology and Biochemistry* 43, no. 6: 1109–1117.
Young, I. W. R., C. Naguit, S. J. Halwas, S. Renault, and J. H. Markham. 2013. Natural revegetation of a boreal gold mine tailings pond. *Restoration Ecology* 21, no. 4: 498–505.
Yunusa, I. A. M., and P. J. Newton. 2003. Plants for amelioration of subsoil constraints and hydrological control: The primer-plant concept. *Plant and Soil* 257, no. 2: 261–281.
Zipper, C. E., J. A. Burger, J. G. Skousen, et al. 2011. Restoring forests and associated ecosystem services on Appalachian coal surface mines. *Environmental Management* 47, no. 5: 751–765.

13 Alternative and Innovative Integrated Mine Waste Management Approaches

Isabelle Demers and Thomas Pabst

13.1 INTRODUCTION

Chapters 4 to 9 presented different approaches to prevent geochemical instability related to mine waste disposal. Cover systems aim to reduce and limit the interactions between reactive mine waste and oxygen and/or water. These approaches are usually applied at mine closure – that is, after deposition and consolidation in the final storage facility is completed. However, choices made earlier in the mine design process can alter the ease and cost of reclamation and usually dictate the type of reclamation strategy that will be required at the end of operations. Geotechnical stability concerns (not discussed in this book) also need to be taken into account in the selection of the reclamation scenario to apply for a given tailings storage facility or waste rock pile.

The "designing for closure" concept (Steffen Robertson and Kirsten (BC) Inc. 1989; Aubertin et al. 2002a, 2016) proposes to integrate closure and reclamation into the mining production cycle, ideally concurrently to mine planning. Designing for closure opens up to alternative and innovative mine waste management strategies to reach an optimal mine waste management plan. The main objective of these mine waste integrated management strategies is to improve the performance of reclamation, but also to make reclamation easier, faster, and more cost effective than the conventional approaches.

This chapter concisely presents several approaches to manage mine waste and advantageously improve reclamation, addressing both geochemical or geotechnical stability issues at the same time. These approaches by themselves are usually not sufficient to reclaim mine sites and must be combined with the methods presented in the previous chapters. In this chapter, some selected integrated mine waste management approaches are presented sequentially, and their advantages and limits for subsequent reclamation are exposed. The intent is to provide a toolbox of options for mine operators and designers to rethink and integrate mine waste management and reclamation strategies.

13.2 DENSIFIED TAILINGS

13.2.1 General Principle

Tailings dam stability has raised an increasing concern among mine operators, scientists, and communities in the past few decades. Mine tailings are typically transported hydraulically, at a high water content (typically 20–45% solid content; Bussière 2007; Blight 2010; MEND 2017), and disposed of at the surface in tailings storage facilities (TSF). Engineered dikes are necessary to retain the tailings slurry and limit the footprint of the TSF. These can be built with locally available natural materials (e.g., sand, gravel, clay, till, or moraine), but are more commonly made with mine wastes. Consolidation in impoundments is usually slow because of the relatively low hydraulic conductivity of tailings (Vick 1990), which therefore remain prone to liquefaction and large deformations

(James et al. 2011; Pépin et al. 2012). Despite recent technical developments (e.g., dam design and construction techniques, monitoring systems) and the publication of construction guides and guidelines (MAC 2017), geotechnical instabilities still regularly occur (Caldwell 2017). Events at the Mount Polley Mine in 2014 (British Columbia, Canada), the Bento Rodrigues mine (Mariana, Minas Gerais, Brazil) in 2015 (Morgenstern et al. 2015, 2016), and the Brumadinho mine (Brazil) in 2019, tragically demonstrated the catastrophic consequences of dam failures. Although reasons for failures are numerous, they typically involve water (e.g., overtopping, internal or external erosion, or liquefaction; Rico et al. 2008; Strachan and Goodwin 2015).

Dewatering, which consists of reducing the water content (i.e., increasing the solid content) of tailings, aims to improve the geotechnical properties of tailings and reduces the need for confinement structures, thus decreasing the risk for geotechnical instabilities (Robinsky 1975, 1999; Simms 2017). Densified tailings can be characterized by their gravimetric or volumetric pulp density, and/or their yield stress (Boger 2013). Three types of densified tailings are generally distinguished, based on their properties upon deposition: thickened, paste, and filtered tailings (Bussière 2007; MEND 2017; Table 13.1, Figure 13.1).

TABLE 13.1
Main characteristics of Conventional and Densified Tailings (Adapted from Martin 2018)

B100Tailings	% Solids (w/w)	Yield Stress (Pa)	Critical Flow Velocity	Dewatering Technique	Transport Mode
Slurry (conventional)	<45	<5–20	Yes	Conventional thickeners	Centrifuge pumps
Thickened	45–70	20–100	Yes	HDS	Centrifuge or volumetric pumps
Paste	70–85	100–800	No	UHDS	Volumetric pumps
Filtered	>85	>800	No	Vacuum or press filters	Conveyor or trucks

Notes: The values provided are typical ranges and/or rough estimates and can vary depending on the mineralogy and particle grain size. HDS: high-density sludge, (U)HDS: (ultra) high-density sludge

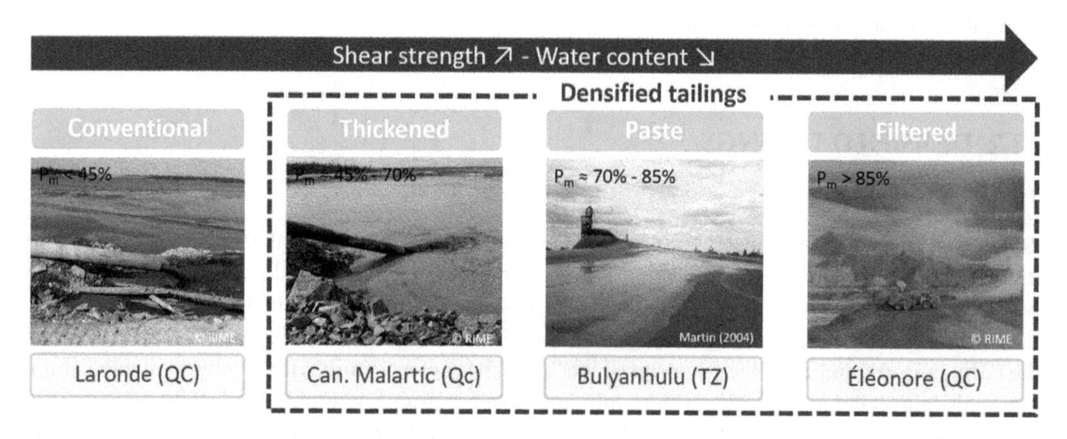

FIGURE 13.1 Comparison of conventional slurry tailings and densified tailings with example of applications. P_m: gravimetric solid content. (Adapted from Martin 2018).

13.2.2 Thickened Tailings

Thickened tailings approach was first proposed by Robinsky (1975) with the objective to create a homogeneous self-supporting tailings mass that would contribute to reducing the requirements for confining dams and eliminate settling ponds. Thickened tailings are generally produced using thickeners, which have a diameter ranging from a few meters to over 100 m. Thickeners are based on sedimentation and usually rely on the addition of flocculants to accelerate tailings settling rate. High-rate or high-capacity thickeners can increase tailings solid content (also called gravimetric pulp density) up to 60%, while it can reach 70% with high-density or high-compression thickeners (Jewell and Fourie 2015). Thickened tailings exhibit a critical flow velocity (i.e., flow can become turbulent for high gradients), but do not segregate during sedimentation (Barbour et al. 1993). Bleed water can be observed from thickened tailings following a slump test or after disposal in a TSF. The solid content is usually comprised between 45% and 70%, corresponding to a gravimetric water content between 40% and 120%. Thickened tailings have a small yield stress (generally lower than 50 kPa; Simms 2017) and can be transported with centrifuge pumps. They are often deposited from a central discharge point to form a conical stack (typical average slope is around 1–2%; Simms 2017), and containing dams are required (MEND 2017). Thickened tailings were or are produced for example at Kidd Creek mine (Canada; Al and Blowes 2000; Kam et al. 2009); Canadian Malartic mine (Canada; Doucet et al. 2015), and Musselwhite mine (Canada; Kam et al. 2011).

13.2.3 Paste Tailings

Paste tailings, which have a lower water content than thickened tailings, are inspired from underground mines backfill developed in the 1970s (Hassani and Archibald 1998) and were later proposed as an alternative to conventional slurry disposal in surface TSF (Bussière 2007). Ultra high-density thickeners can be employed to increase solid content over 70%; they resemble conventional thickeners but have a greater height/diameter ratio to enhance particle settlement (Wills and Finch 2015). More generally, however, paste tailings are prepared by addition of a small amount of water (or re-pulping) to filtered tailings (see Section 13.2.4) (Shuttleworth et al. 2005). Paste tailings do not have a critical flow velocity but can be pumped using volumetric or positive displacement pumps, provided the applied yield stress exceeds the shear yield stress (between 100 and 800 kPa; Simms 2017). Paste tailings are often disposed of from central points (generally deposition towers) in impoundments. The tailings slope can be as high as 4–6% close to the deposition point (Daliri et al. 2014), but values around 2% are more often observed in practice (MEND 2017; Simms 2017). Small dams may be required to contain the spreading of tailings in the short term but can be considered temporary structures. The solid content of paste tailings is typically comprised between 70% and 85%, corresponding to a gravimetric water content of 20–40% (Bussière 2007). Paste tailings are or have been produced at Myra Falls mine (Canada; MEND 2017), Neves-Corvo mine (Portugal; Verburg and Oliveira 2016), and Bulyanhulu mine (Tanzania; Simms et al. 2007; Martin et al. 2010), among others.

13.2.4 Filtered Tailings

Filtered tailings are generally characterized by a solid content above 85% (Bussière 2007), and a gravimetric water content less than approximately 20%. They were first developed to reduce the risks of liquefaction and find a solution to tailings transport in cold climate (Davies and Rice 2001). Filtered tailings are often produced by press or disc filters and form a filter cake, which is transported by trucks or conveyors to the TSF (Wills and Finch 2015). Filtered tailings are compacted, which confers them a greater strength and a relatively good homogeneity (no segregation). Dams are usually not required for filtered TSFs. Filtered tailings have been quite successfully used in over 30 mine sites around the world and for various climatic conditions (Davies 2011), including La Copa mine (Chile), Raglan mine (Quebec, Canada), Green Creeks mine (Alaska, USA), Casposo mine (Argentina), and Eleonore mine (Quebec, Canada). Even if the initial investment and the operation costs for filtered tailings can be higher than for conventional slurry or other densified tailings, these factors are usually balanced by a lower geotechnical risk and reduced constraints on dam construction (Fitton and Roshdieh 2014; Carneiro and Fourie 2018).

13.2.5 Advantages and Limits of Tailings Densification

Dewatering can contribute to optimizing water recirculation and decrease the risks of liquefaction (Lara et al. 2014). However, tailings dewatering is complex, and its performance depends on the size and shape of particles, the presence of clay particles, fluid temperature, and viscosity. The process is generally divided into several successive phases, and water is removed progressively. In practice, the final solid content is often 3–5% lower than expected, mainly because of scale effects and inherent variability in tailings production (MEND 2017).

Densifying tailings provide several advantages over traditional slurry tailings, including a decreased demand for borrow materials for the construction of impoundment structures, the reduction (or even elimination in the case of filtered tailings) of ponding, a more rapid consolidation and trafficability, and the reduction of water to collect and treat (Johnson et al. 2005; Simms et al. 2005, 2007; Fourie 2012a, 2012b; Fitton 2017), thus facilitating the implementation of reclamation.

Despite these advantages, the approach (especially filtration) can be, however, poorly suited for highly reactive wastes, which would be more rapidly exposed to oxygen and water (Davies 2011), increasing the risk for contaminated drainage generation (see Chapter 1). In this case, progressive reclamation is necessary to prevent oxidation and reduce the potential for acid mine drainage (AMD) and contaminated neutral drainage (CND) generation during operations (MERN 2017). The implementation of progressive reclamation includes the construction of intermediate and temporary cover systems, frost enhancement, in situ neutralization of surface tailings, and selective disposal. Also, the evolution of the water table position in the densified TSFs and the effect on tailings geotechnical stability upon rewetting have not yet been addressed either, thus raising questions about the stability of these structures in the long-term, at the reclamation temporal scale. A well-designed cover system could help to control water balance in situ (Oldecop and Rodari 2017), but tailings may remain exposed to climatic conditions for a certain time before reclamation.

Precipitation can also erode the surface of the TSFs before reclamation, especially for slopes above 4% (Reid and Bolshoff 2015). Dust generation is also an issue and requires specific mitigation procedures (Jewell et al. 2002). Finally, effects of frost and wet-dry cycles on filtered tailings are little known. Most of these advantages and drawbacks depend, however, on the processing method and site-specific conditions. Mining operators are well aware of these issues and risks, and research is currently ongoing to improve our knowledge about long-term behavior of densified tailings.

13.3 BACKFILLING

Mine openings, both underground stopes and surface pits, can be considered as potential storage areas for tailings and waste rock. The filling of man-made openings is generally more acceptable from a social perspective than creating new storage areas that can cover large surface areas, affect the local environment, and present potential geotechnical and geochemical risks. Significant advantages in terms of site reclamation can be gained by using backfill during the operation.

13.3.1 Underground Backfilling

Backfilling of underground openings created by the mining operation is a storage method, applicable to both tailings and waste rock, that has been implemented worldwide in the past decades (Hassani and Archibald 1998). The main objective of underground backfilling is to fill openings (stopes) and provide ground support for exploitation of remaining adjacent stopes (pillar recovery), thus increasing the mine throughput and economic efficiency (Hassani and Archibald 1998). Backfill techniques were historically introduced and designed on a case-by-case basis with minimal research considerations. By the 1990s, research was initiated to better understand the geotechnical behavior of backfill (Landriault et al. 1997; Hassani and Archibald 1998), as well as the gain in mechanical strength of the backfill mass (Benzaazoua et al. 1999). The 2000s saw an increase in research to understand the environmental behavior of backfill, and more specifically cemented paste backfill (e.g., Coussy et al. 2011; Ouellet et al. 2006), its rheological behavior (e.g., Boger 2013; Mehdipour and Khayat 2018; Qi et al. 2018), and its mechanical behavior in relationship with the stope conditions (e.g., Li et al. 2005; Helinski et al. 2010; Sivakugan et al. 2015; Qi and Fourie 2019).

Three categories of backfill are generally recognized: hydraulic backfill, rock backfill, and cemented paste backfill. Rock backfill consists of waste rock (possibly crushed) being returned in empty stopes either by truck, conveyor belt, or hoist (Hassani and Archibald 1998). A fluid mixture of binder, such as cement, is usually added to the waste rock during placement to strengthen the backfill. Hydraulic and cemented paste backfill are both made with tailings, water, and binders. For hydraulic fill, tailings are deslimed – that is, the fine particles are removed (Sivakugan et al. 2006). The removal of fines helps to accelerate drainage in the stope, so the desired mechanical strength is reached faster. The tailings slurry (density of 60–70% solids) is mixed with a binder that forms cementitious bonds between particles via mineral precipitation (Benzaazoua et al. 2004), which are essential for mechanical strength acquisition. Hydraulic backfill is carried to the empty underground stopes by pumping or gravity, or both. Cemented paste backfill does not require desliming (Sivakugan et al. 2006). The slurry density is increased and binders are used to increase mechanical strength. The lower amount of water drained from paste backfill reduces the binder losses with drainage, thus allowing the optimization of the proportion of binder, which is an expensive component of backfill (Belem and Benzaazoua 2008). Because of its higher viscosity, paste backfill distribution in empty stopes often requires the use of powerful positive displacement pumps (see Figure 13.2).

Apart from ground support, another major advantage of backfill is the possibility to return tailings or waste rock underground and consequently reduce the volume stored in surface storage areas.

FIGURE 13.2 Distribution of cemented paste backfill in an empty stope. (Photo courtesy of Li Li).

From 40% to 60% of the tailings mass can be returned underground as backfill, reducing accordingly the size of the surface TSF, the environmental risks, and the reclamation costs (Belem and Benzaazoua 2008).

Underground backfill is a beneficial process not only for the operation but also for closure and reclamation of the mine site. By diverting up to 60% of tailings in the backfill stream, the volume of tailings to store in surface facilities is significantly reduced, and the reclamation efforts and costs are also reduced correspondingly. When tailings are characterized and sorted by their acid-generation potential (see Chapter 2), or their contamination potential, it is possible to selectively use problematic tailings for cemented paste backfill and keep the inert or low-risk tailings in surface storage facilities, where reclamation will be easier (Benzaazoua et al. 2008). The careful selection of backfill recipe is necessary to provide long-term stabilization of contaminants and/or sulfides. Similarly, acid-generating waste rock could be returned underground provided that the hydrological conditions favor limited and saturated flow where oxidation would be avoided. In that case, waste rock remaining in surface piles would have low environmental risk.

Specific care must be taken when preparing backfill with sulfidic tailings and process water that contains sulfate. Indeed, a concept known as "sulfate attack" was observed when sulfur-containing material is used to make concrete, which causes degradation of the concrete (Benzaazoua et al. 1999). Some studies have also shown that sulfides in tailings and sulfate in process water can have deleterious effects on the strength of cemented paste backfill (Benzaazoua et al. 1999, 2002). The sulfate attack phenomenon can be minimized, and possibly avoided, by a careful selection of binder components through laboratory tests. For example, tests by Benzaazoua et al. (2002) showed that Portland cement was preferable to slag and fly ash as a binder in the preparation of paste backfill with a high sulfide content, while another study (Benzaazoua et al. 2008) found that paste backfill made with a desulfurization concentrate (12% S) and a binder composed of Portland cement and slag gained more strength than a mixture made with low-sulfide tailings (3% S), and showed no long-term loss of strength or contaminant leaching. These examples highlight the need to experiment with several recipes to determine the optimal one for the given type of tailings. Cement addition to backfill may also be used to chemically stabilize contaminants, such as arsenic (Coussy et al. 2011, 2012), and improve its unconfined compressive strength.

The cost of backfill is highly dependent on the type and amount of binders used. Portland cement is the most usual binder. However, for economic and environmental reasons alternative binders and additives have been developed. Industrial by-products, such as fly ash, blast furnace slags, copper slags, and biomass ash can totally or in part substitute the cement (Peyronnard and Benzaazoua 2011). Laboratory tests are usually necessary to optimize the backfill recipe (type of binder, proportion of binder, and water and tailings) that will provide the required fluidity for pumping and subsequent strength development (Benzaazoua et al. 2002; Belem and Benzaazoua 2008B15).

13.3.2 In-pit Disposal

In-pit disposal of mine wastes can be an attractive approach to reduce their environmental impact in the short and long term. In-pit disposal is often considered efficient to isolate sulfide-bearing mine wastes and limit the risk for AMD generation, improve the long-term geochemical and geotechnical stability of the disposed wastes, and stabilize the walls of the pit (Castendyk and Eary 2009). The technique can also reduce the need for building TSF or waste rock piles and improve the esthetics of the sites. This approach is and has been used on many mine sites to dispose of both acid-generating and neutral mine wastes (MEND 2015), but also coal mine tailings (Park et al. 2019), oil sands tailings (Dompierre and Barbour 2016), and radioactive mine wastes from uranium mines (Donahue et al. 2000). Depending on the design, the technique can be suitable for wet and dry climates, and examples of backfilled pits can be found around the world, including Doyon-Westwood Mine

(QC, Canada; Figure 13.4), Whistle Mine and Owl Creek Mine (ON, Canada), Equity Silver Mine (BC, Canada), Solbec Mine (QC, Canada), Jundee Mine (Australia), Los Frailes (Spain), Kimheden Mine (Sweden), among others. Even if the approach seems relatively straightforward (i.e., filling a pit with mine wastes), several challenges need to be addressed to develop a successful reclamation plan, particularly regarding the geochemical stability of the disposed wastes, the reclamation of backfilled pits (see also Chapter 6 on water covers), the control of water flow and exchanges with the surrounding environment, and the operational aspects of the disposal.

For example, tailings are usually disposed of in pits as slurry (Figure 13.4a) and are therefore subjected to long-term consolidation and deformations. Settlement in a backfilled pit can be significant, especially because of the thickness of disposed tailings (sometimes several hundred meters) and the usually relatively quick rate of disposal (because of the limited surface area). The geometry of the pit and the presence of benches may also induce differential settlements (Puhalovich and Coghill 2011). Accelerating consolidation (e.g., using waste rock inclusions (WRIs); see Section 13.5.2) may reduce displacements and optimize the available space and increase the quantity of mine wastes that can be stored in pits. In-pit disposed waste rocks are somewhat less prone to large deformation but may still lead to slope instabilities during disposal, especially if waste rock is placed from the surface (Ayres et al. 2007; Figure 13.4b). The additional loads induced by backfilled materials may also compromise the stability of surface pillar in the case where the operation continues underground, not to mention the risks for water infiltration in the case of tailings slurry disposal. In-pit disposal of filtered tailings may solve parts of these problems but considerably increases the operation costs.

Reclamation of open pits often consists in filling them with a limited quantity of mine wastes and then flooding them (Figure 13.3; Schultze et al. 2011). These pit lakes are usually designed to function as "hydraulic traps" (McCullough et al. 2013; see also Chapter 6), so water from the regional water table is directed toward the pit, thus preventing it from migrating to the environment. This approach has several major limitations, however: only a fraction of the total volume of the pit is used to store waste, the concept of hydraulic trap may be altered over time (e.g., because of climate change and water balance), water exchanges with the fractured rock are usually difficult to quantify (Miller and Zègre 2014), and the water quality in the pit may deteriorate over time (Castendyk 2005).

Filling the pits over the regional water table and up to the surface would maximize the quantity of wastes that can be disposed of and reduce the need for additional storage facilities. A large part of the materials, however, may become partially saturated and exposed to oxidation and AMD generation. In theory, engineered cover systems (see Chapters 4, 5, 7, and 9) could be installed on top of reactive wastes to prevent their oxidation (Adu-Wusu and Yanful 2007), but case studies are scarce as only a limited number of pits have been reclaimed to date (MEND 2015). Cover systems may also modify the water balance inside the disposed wastes and around the pit.

Previous studies on reclaimed pits have also shown that backfilled wastes could continue to oxidize despite efficient cover systems (Villain et al. 2013). Oxygen may flow through cracks in

FIGURE 13.3 Reactive tailings from Gallen mine (Fonderie Horne-Noranda) were transported and disposed of in the flooded pit of Don Rouyn mine. The 1.2-m-thick (in average) water cover contributes to limit oxygen diffusion to the tailings (also see Chapter 6 for more details). (Photo courtesy of Akué Sylvette Awoh).

(a)

(b)

FIGURE 13.4 (a) In-pit disposal of Westwood mine (IAMGOLD) tailings and (b) Waste rock (right) in the old Doyon mine pit. The Westwood mine is located close to an abandoned mine site whose pit is being used for disposal. The slurry tailings are directly discharged from the surface of the pit by spigoting. The waste rocks were disposed of by truck in two benches and required the rehabilitation of the ramp to access the bottom of the pit.

fractured rock and reach reactive wastes from the sides (Villain et al. 2015). Large water fluxes, favored by highly permeable rock masses along blasted walls, may also bring significant amounts of dissolved oxygen, which can oxidize backfilled wastes and generate AMD. Therefore, strategic disposal approaches (e.g., sorting and placing more reactive wastes deeper in the pit; Guerin et al. 2006 – see also Chapter 2 regarding mine waste sorting based on their reactivity and contamination potential) need to be considered and planned from the start of the operations. Interactions with the environment is also a major concern, both regarding inflow of dissolved oxygen and Fe(III) ions and the outflow of contaminants (either process water or AMD; Ben Abdelghani et al. 2015). The pervious surround (or permeable envelope) technique consists of placing coarse and permeable materials (e.g., nonreactive waste rock) along the pit walls to reduce gradients and limit water flow through the reactive tailings (Belfadhel 2000). Water is diverted and avoids contacts with problematic masses. The performance of such technique, however, depends on regional gradients, pit geometry, and material relative hydraulic conductivities, among other site-specific properties (Rousseau and Pabst 2018).

Finally, the practical aspects of in-pit disposal require site-specific investigations. Resource sterilization (i.e., preventing access to future resources) is a major concern among operators. Also, the disposal of waste rock (either as backfilled material or for co-disposal purposes) is complex as surface dumping may lead to geotechnical instabilities, while bottom-up disposal would prohibitively increase costs. In-pit disposal may also require temporary storage facilities while the pit is being mined, thus increasing environmental risks and operation costs.

13.4 ENVIRONMENTAL DESULFURIZATION

Environmental desulfurization aims to remove sulfides from tailings prior to disposal in a TSF, making these tailings non-acid generating and simpler to reclaim. The removed sulfides, being concentrated, occupy a much smaller volume than the initial tailings. Environmental desulfurization has gained significance for tailings management mainly in the past 20 years. It evolved from work in Canada and Europe in the 1990s (e.g., McLaughlin and Stuparyk 1994; Bussière et al. 1995; Humber 1995; Leppinen et al. 1997). Since the late 1990s, much of the research work has been conducted in Canada and South Africa (e.g., Benzaazoua et al. 2000; Bussière et al. 2004; Yalcin et al. 2004; Hesketh et al. 2010; Kazadi Mbamba et al. 2012; Broadhurst and Harrison 2015). A few industrial applications are reported (e.g., Martin and Fyfe 2007; Dobchuk et al. 2013; Ethier et al. 2018b), and more extensive use of environmental desulfurization is expected in the coming years.

13.4.1 Environmental Desulfurization Process

The primary objective of environmental desulfurization is to reduce the acid-generation potential of tailings to prevent AMD once deposited in a TSF and after the site's closure (Humber 1997; Leppinen et al. 1997; Benzaazoua et al. 2000). An evaluation of the acid-generating potential must be performed on the tailings to determine potential acidity (AP), neutralization potential (NP), net neutralization potential (NNP), and neutralization potential ratio (NPR; see Chapter 2); this information is then used to identify the desulfurization target. Indeed, the maximal allowable sulfide content may be obtained using the applicable acid-base accounting criteria; equations 13.1 and 13.2 calculate the maximum % S considering the NNP criterion and the NRP criterion, respectively.

$$\%S = \frac{NP - 20}{31.25} \tag{13.1}$$

$$\%S = \frac{NP}{3 \times 31.25} \tag{13.2}$$

For example, tailings with a NP of 30 kg $CaCO_3$/t would require desulfurization below 0.32% S using equations 13.1 and 13.2, while tailings with a NP of 50 kg $CaCO_3$/t would require a maximum % S of 0.96% using eq. 13.1 and 0.53% using eq. 13.2. Safety factors (case specific) are necessary to account for possible variation in sulfur grade in the tailings, and heterogeneity in neutralization potential.

Froth flotation is the main process used to remove sulfide minerals from the tailings. The flotation process is a century-old mineral processing method used on sulfide base metal and gold ores for coal and industrial minerals (Wills and Finch 2015). Gravity-based processes were also tested and provided good environmental desulfurization results in some specific cases; however, flotation is readily applicable to a wide range of tailings composition and sulfide content (Humber 1997; Bois et al. 2005; Hesketh et al. 2010). Flotation involves the use of surfactants as reagents to modify the surface properties of minerals to facilitate their separation (Wills and Finch 2015). Sulfide minerals, and most often pyrite, are made hydrophobic (tendency to repel water) by reagents known as collectors, while the remaining minerals in the tailings are hydrophilic (attracted to water). The tailings slurry is agitated in a flotation cell, where air bubbles are inserted. Hydrophobic minerals (sulfides in this case) attach to air bubbles and rise to the top of the cell by buoyancy, whereas hydrophilic minerals remain in the bulk of the slurry. Loaded air bubbles form a froth that is removed from the cell. The froth constitutes the sulfide concentrate, and the remaining slurry constitutes the desulfurized tailings (see Figure 13.5).

The environmental desulfurization process differs from the typical flotation process in several aspects:

1) The focus of desulfurization is the production of a non-acid-generating tailings, while typical flotation targets a concentrate grade and recovery for economical benefits.
2) Environmental desulfurization aims to recover sulfides indiscriminately, that is, as a bulk sulfide concentrate, while typical flotation targets a specific metal bearing mineral.
3) Environmental desulfurization is often a process carried out at the end of beneficiation before disposal in the TSF. The feed material is the leftover from the previous processes, therefore not optimized for flotation, and may require particular reagents to reactivate surfaces for an optimal sulfide separation (Benzaazoua et al. 2000; Mermillod-Blondin 2005).

FIGURE 13.5 Environmental desulfurization in a pilot flotation bank. (Photo courtesy of I. Demers and M. Lopez).

13.4.2 Concentrate Management

The sulfide concentrate contains the sulfide minerals that were floated and other minerals that ended up in the concentrate by entrainment or that were hydrophobic. It may constitute between 10% and 40% of the original tailings' mass. The sulfide content in the concentrate varies depending on the initial feed, the flotation operating parameters, flotation performance, and other factors related to the tailings' characteristics. However, most sulfide concentrates are characterized as acid generating and should be managed as such (Benzaazoua et al. 2000). In the integrated tailings management concept, several management options are suggested for the concentrate. Benzaazoua et al. (2008) proposed to integrate the sulfide concentrate into paste backfill (see Section 13.3.1) for underground storage, while controlling degradation by sulfate expansion. Skandrani et al. (2019) showed that residual gold may be recovered through desulfurization when gold particles are associated with sulfides; in such cases, desulfurization could fulfill both an environmental and economic objectives. Broadhurst et al. (2018) identified several options for coal-sulfide concentrates, including their reuse as soil enhancement. Research by Stander et al. (2018) and Kazadi Mbamba et al. (2012), among others, highlighted opportunities to repurpose the sulfide concentrate and advantages to perform integrated tailings management via environmental desulfurization.

13.4.3 Desulfurized Tailings Management

Desulfurized tailings are the bulk of the desulfurization process output. They have a low sulfide content (<0.5%) and are typically characterized by particle size between 10 and 100 μm (silt), and a relatively low saturated hydraulic conductivity (10^{-4}–10^{-6} cm s^{-1}; Bussière et al. 2004). These properties make desulfurized tailings an alternative material well suited for the construction of some reclamation cover systems. The use of desulfurized tailings for reclamation purposes, instead of natural soil, is advantageous in terms of environmental and economic performance, particularly when desulfurization is integrated at the appropriate life-of-mine stage (Benzaazoua et al. 1998; Bussière et al. 2004; Demers et al. 2008b). The production of desulfurized tailings during the last years of operation should be planned to produce sufficient material to reclaim the TSF

TABLE 13.2
Selected Examples of Integrated Tailings Management Involving Environmental Desulfurization

Type of Mine	Type of Desulfurization Study (scale)	Sulfur (S) in Desulfurized Tailings (%)	Desulfurized Tailings Management	Concentrate Management	References
Gold mine	Laboratory and intermediate field scale investigation	0.1–0.5	Cover material for AMD generating TSF	Paste backfill and/or in-pit disposal	Rey et al. (2020) Demers et al. (2008, 2009) Benzaazoua et al. (2008) Blier et al. (2012)
Coal mine	Laboratory two-stage flotation, laboratory repurposing study	0.38	Disposal as benign waste	Coal concentrate as valuable product, sulfide concentrate for further processing	Kazadi Mbamba et al. (2012) Broadhurst and Harrison (2015)
Nickel mine	Plant scale desulfurization by hydrocyclone (overflow is desulfurized, underflow is sulfide concentrate)	0.4–1	Single-layer cover (min. 1.5 m) placed on tailings pond, lime addition to increase cover NP	Cyclone underflow (2–3% S) used with tailings in underground backfill	INAP (2009)

(Bois et al. 2004). The use of desulfurized tailings as cover material is discussed in more detail in Section 13.6.1. Desulfurized tailings may also be used for paste backfill preparation, especially if sulfate attack is a concern. Finally, desulfurized tailings may also be deposited in a TSF, but without the risk of generating contaminated drainage, reclamation methods would be less stringent (Benzaazoua et al. 2008).

Table 13.2 presents three cases of integrated tailings management involving environmental desulfurization at different scales of applications. These three cases show that environmental desulfurization can be applied to different types of ore (coal, gold, base metal), and different management options for desulfurized tailings and sulfide concentrate can be considered. The desulfurized tailings management options are driven by their sulfur content; when the % S is high enough to potentially generate AMD, such as in the nickel mine case, addition of a neutralizing compound is required to use these tailings as reclamation cover. The coal mine example showcases the recovery of valuable components (coal) in the concentrate during the desulfurization process.

Overall, environmental desulfurization provides significant advantages to the mine site reclamation. By removing the sulfidic portion of tailings and managing this sulfidic portion correctly (by valorization or paste backfill), the remaining tailings are non-acid generating, which facilitates the TSF reclamation and reduces the reclamation costs. The use of desulfurized tailings as cover material for existing TSF reclamation is also a major driver in the development and application of environmental desulfurization at operating mine sites.

13.5 CO-DISPOSAL OF WASTE ROCK AND TAILINGS

Co-disposal of waste rock and tailings is a concept that involves a common disposal facility for both types of mine waste, as opposed to separate waste rock piles and TSF, to ultimately make reclamation easier. As a reminder, tailings are a fine-grained, slowly consolidating material with

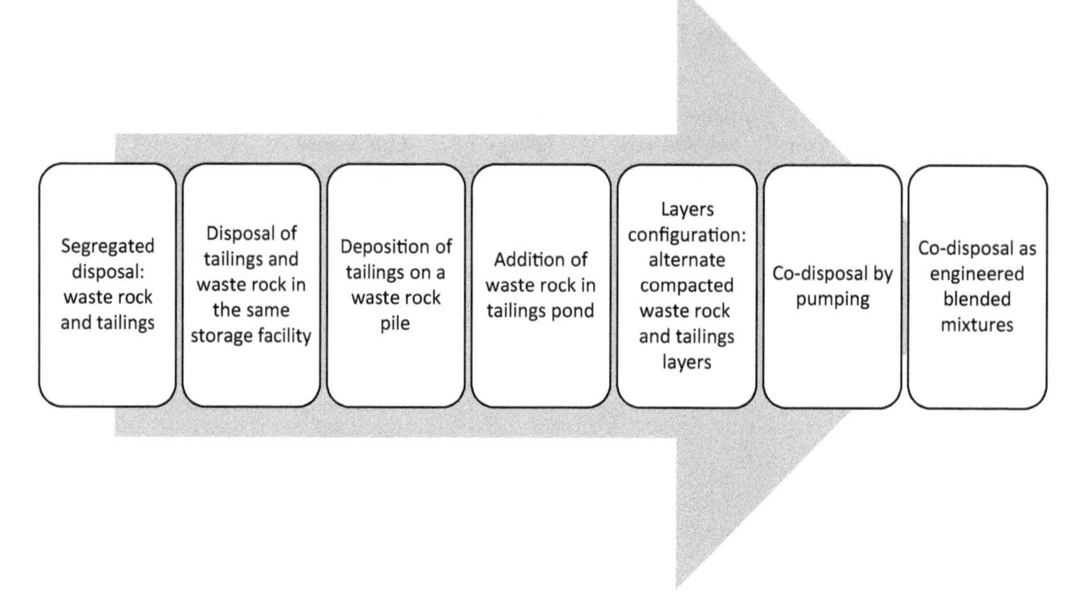

FIGURE 13.6 Methods of co-disposal of waste rock and tailings from lowest to highest degree of mixing. (Adapted from Wickland et al. 2006).

low saturated hydraulic conductivity, while waste rock has high hydraulic conductivity, high porosity, and wide particle size range. The objective of co-disposal is to take the advantages of one type of material to compensate for the limitations of the other, in terms of hydrogeological and geomechanical properties. Figure 13.6 presents the spectrum of co-disposal methods, from the completely segregated state to the most mixed configuration. The development of co-disposal emerged from pumped co-disposal of fine tailings and coarse waste rock in collieries (e.g., Williams 1997; Wickland et al. 2006). Then, other configurations were experimented in field, laboratory, and numerical simulations (e.g., Lamontagne et al. 2000; Leduc et al. 2004). The selection of mixture is based on the reclamation objectives sought. This section briefly describes the co-mingling and co-mixing approaches, followed by the recent application of WRI in tailings impoundments.

13.5.1 Co-mingling and Co-mixing

Co-mingling and co-mixing approaches aim to produce a physically and chemically stable structure by combining waste rock and tailings. Layered co-mingling refers to placement of alternating layers of waste rock and tailings to reduce the AMD generation potential of waste rock piles, while co-mixing consists of creating a new binary material by mixing crushed waste rock with tailings.

With co-mingling, capillary barrier effects can be created in unsaturated conditions because of particle size distribution contrast between the waste rock and tailings layers, which maintains the tailings layers at a high degree of saturation, and therefore limits oxygen flux and diverts a large portion of the infiltrating water (Lamontagne et al. 2000; Lefebvre et al. 2001). For a waste rock pile in which tailings layers are placed, the gas movement within the pile and from the atmosphere to the pile's interior becomes limited by diffusion, as opposed to a typical waste rock pile where convection is the major gas transport mechanism (see Chapter 3). Laboratory and numerical modeling work by Lamontagne et al. (1999, 2000) demonstrated that sulfide oxidation was reduced in a reactive waste rock pile with layers of compacted tailings; however, oxygen diffusion remained too high for layered co-mingling to be considered a definitive reclamation method for waste rock piles.

The co-mixing approach, also called paste rock (Wilson et al. 2003), involves a more homogeneous mixing of waste rock and tailings, at the end of the co-disposal spectrum presented in Figure 13.6. When tailings are mixed with waste rock particles, the resulting mixture has a low permeability (controlled by the tailings fraction) and a high shear strength (provided by the waste rock particles/blocks; Wickland et al. 2006, 2010). Particle packing theory can be used to determine the ideal waste rock to tailings ratio to maximize structural stability and low oxygen diffusion (Wickland et al. 2006). An ideal mixture contains just enough tailings to fill completely the voids between waste rock particles, while waste rock particles remain in contact (Figure 13.7). A higher ratio of tailings makes the waste rock particles isolated in the tailings mass, while a higher ratio of waste rock leaves empty (air-filled) porosity. Pumped mixtures of tailings and crushed waste rock have been used in collieries for some decades (Williams 1997); however, the high water content favors fine particle segregation once deposited (Kotsiopoulos and Harrison 2018). The "just-filled" mixture of tailings in waste rock is therefore desired for optimal performance. Several parameters, other than waste rock to tailings ratio, also influence the properties of the mixtures, such as porosity and packing, water content, density of tailings and waste rock (Wickland et al. 2006; Charbonneau et al. 2018; Mjonono 2019).

Several studies evaluated the hydraulic conductivity of paste rock mixtures at different waste rock to tailings ratios. Table 13.3 presents sample results, where waste rock to tailings ratio is expressed as dry mass. Mixtures have hydraulic conductivities close to the tailings' conductivities.

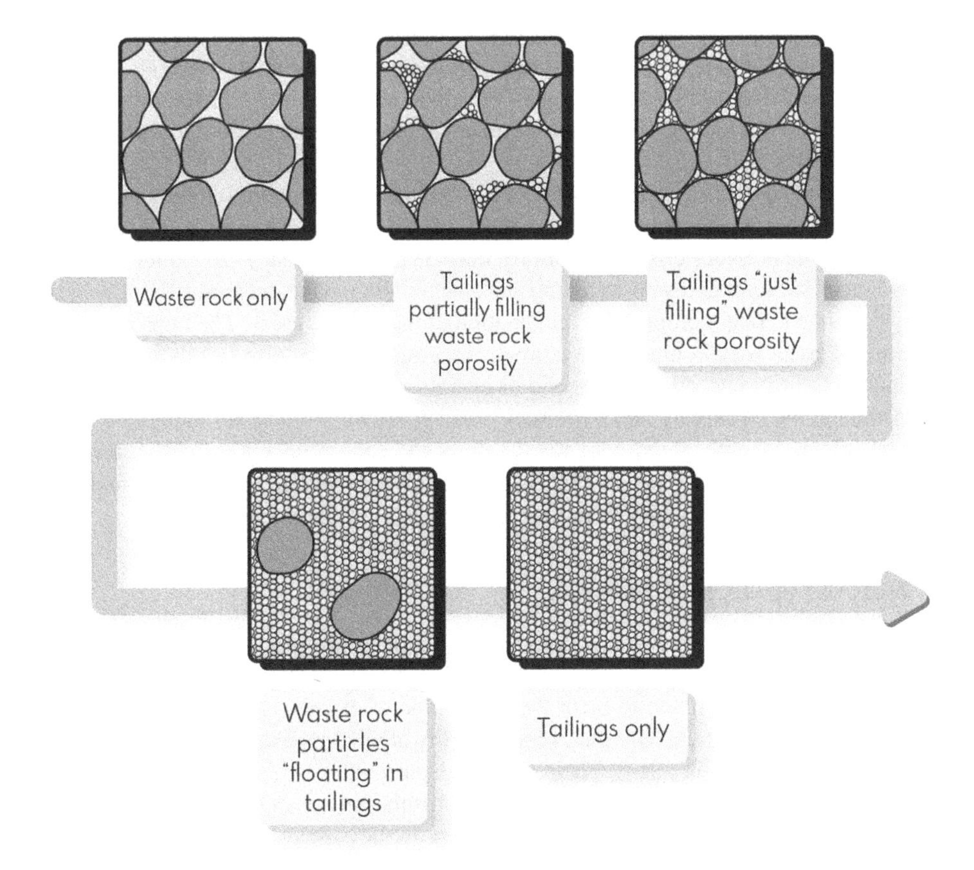

FIGURE 13.7 Fine and coarse particle arrangement in mixtures. (Modified from Vallejo 2001).

TABLE 13.3
Hydraulic Conductivity of Tailings and Waste Rock Mixtures

Type of Mixture	Saturated Hydraulic Conductivity (m s^{-1})	References
Tailings (alone)	1.8×10^{-9}–4.2×10^{-7}	Wickland et al. (2010)
Waste rock (alone)	0.3	
4.2: 1	4.9×10^{-9}–7.6×10^{-8}	
5: 1	1.0×10^{-8}–3.9×10^{-8}	
6: 1	5.4×10^{-8}	
Tailings (alone)	3.9×10^{-8}–5.7×10^{-8}	Pouliot et al. (2018)
Waste rock (alone)	1.6×10^{-5}–7.1×10^{-6}	
4.3: 1	2.7×10^{-8}	
Tailings (alone)	7.1×10^{-7}	Charbonneau et al. (2018)
Waste rock (alone)	8.1×10^{-3}	
2.4: 1	1.4×10^{-7}	
3: 1	1.0×10^{-7}	
3.6: 1	9.3×10^{-8}	

The compression index (related to the volume change after loading) of the mixtures was less than one to one order of magnitude greater than the waste rock's, while the tailings' compression index was one to two orders of magnitude higher than the mixture's (Wickland et al. 2010; Bareither et al. 2018). These results show that the mixtures have hydraulic conductivities similar to tailings, and compressibility closer to waste rock. Other geotechnical parameters were also evaluated in the context of physical stability of paste rock (Wickland and Wilson 2005; Bareither et al. 2018; Burden et al. 2018). For example, internal friction angle of paste rock is close to that of waste rock (Charbonneau 2019).

Co-mingling and co-mixing strategies deserve to be investigated more in terms of their benefits for mine site reclamation. Apart from the reduction in surface area taken by mine waste disposal, advantages in physical and/or chemical stability may be gained. Even if the sulfide oxidation rate cannot be reduced sufficiently by co-mingling and co-mixing to make the mixture non-acid generating, it may buy time to begin progressive reclamation of the storage areas. The improved geotechnical properties of paste rock compared to tailings alone may also help to install reclamation cover systems, including in slopes, either at the end of operations or as part of a progressive reclamation strategy. Although the co-mixing and paste rock concept shows promise for integrated mine waste management, more research is required for its application in metal mines and as a reclamation strategy (INAP 2009; Pouliot et al. 2018). Issues related to preparation, transport, and long-term behavior of co-mixed materials require further research and development.

13.5.2 WRIs in Tailings Impoundments

The concept of WRIs is inspired by wick drains and consists of placing waste rock linear inclusions within the tailings storage facility (James and Aubertin 2012; James et al. 2013, 2017; Figure 13.8). WRIs act as drains and contribute to favor the dissipation of excess pore pressures, accelerate the consolidation of tailings, and improve the overall stability of tailings dikes during static and dynamic loadings (Bolduc and Aubertin 2014). Their rigidity also provides structural reinforcement, and, in the case of an earthquake, WRI can significantly limit displacements at the dam crest (Ferdosi et al. 2015a, 2015b), sometimes even preventing dam failure. In the event of a dam breach, the compartmentation of the tailings impoundment would also help to significantly limit the volumes of released tailings to the surroundings, thus reducing the consequences of a dam failure. Finally, they can be

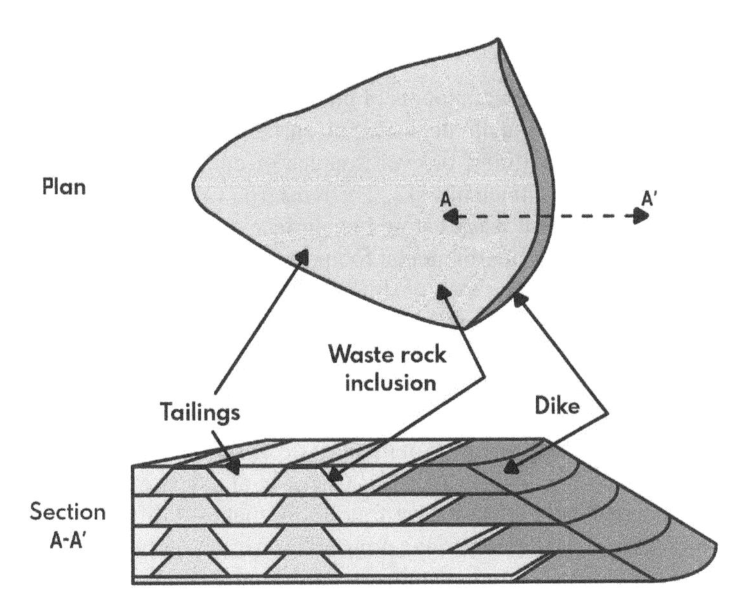

FIGURE 13.8 Waste rock inclusions (WRI) provide structural reinforcement to tailings dams and accelerate tailings consolidation. (Adapted from James and Aubertin 2012).

used for circulation on site to optimize deposition and plan progressive reclamation, thus favoring geotechnical and geochemical stability both during operations and after closure.

WRIs are usually 12–15 m wide and raised from the foundation of the TSF at the same rate as the tailings (Figure 13.8). Waste rocks are disposed of by trucks and compacted by the repeated traffic on the surface (Figure 13.9). Their design (disposition, geometry, orientation, and spacing) depends on the size of the impoundment, the properties of the waste materials, and the specific operational constraints on site (James et al. 2017). The WRI should be parallel to external dams to maximize mechanical reinforcement in case of an earthquake (Figure 13.8). The selection (e.g., based on mineralogy and particle size distribution), preparation (e.g., sieving or crushing), and placement (e.g., raising thickness, compaction) of the waste rock is also critical to avoid internal erosion and clogging, so the draining capacity of WRI remains sufficient during the entire lifetime of the impoundment (Essayad et al. 2018). Recent work has shown that if the WRI is well designed, internal erosion of WRI and migration of tailings at the contact with WRI should not significantly alter their drainage capacity in the short or long term (Essayad et al. 2018).

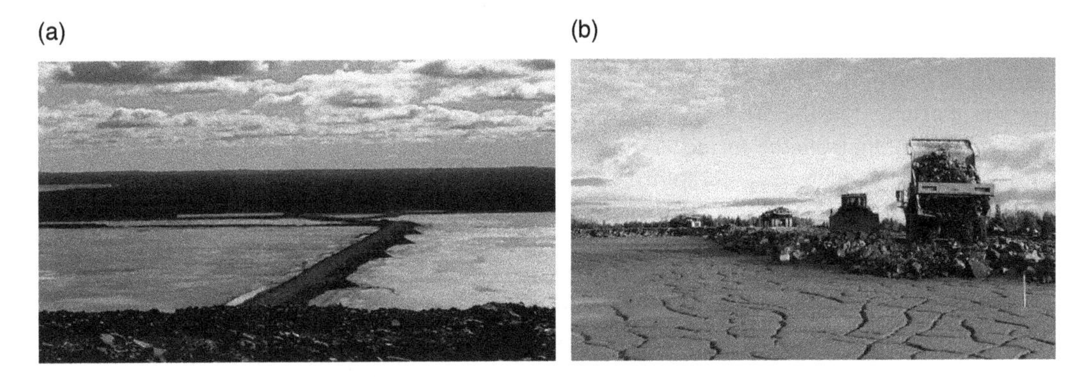

FIGURE 13.9 (a) Waste rock inclusion (WRI) on the Canadian Malartic mine tailings storage facility and (b) Construction of WRI during the construction of the tailings storage facility. (Credit: RIME).

13.6 VALORIZATION OF MINE WASTE

Mining operations produce significant amounts of mine wastes annually in the form of waste rock and tailings. These mine wastes, usually deposited on surface, represent potential sources of geotechnical and geochemical instabilities, or even sources of contaminants to surface and ground water, and therefore cause environmental risks. The reuse (or valorization) of mine wastes could decrease the amount of these mine wastes stored on surface and consequently reduce the cost of management and reclamation, the environmental footprint, and risks associated with storage facilities. Valorization can also be used for the construction of required various infrastructures on site (e.g., roads, dams, cover systems), thus reducing the quantities of natural material that need to be borrowed in the vicinity of the site or transported over long distances. In some cases, the reuse of mine waste can even help improve the performance of the infrastructure (compared to natural materials).

Given the value and repurposing potential of mine waste, the mining industry is starting to adhere to the concept of circular economy (Stewart 2001; Cheng et al. 2016; Segura-Salazar et al. 2019). However, in many jurisdictions, the regulatory framework makes it difficult to "declassify" mine waste and allow its reuse in a non-industrial setting. This section discusses the main valorization approaches applicable on mine sites that can utilize a significant portion of the mine waste produced. Note that backfilling was discussed separately (see Section 13.3) since it is considered both an operational process and mine waste valorization.

13.6.1 Reclamation and Cover Systems

The potential of using tailings as cover materials to reclaim mine sites was proposed by Nicholson et al. (1989) and Aubertin and Chapuis (1991). Low-sulfide tailings, and desulfurized tailings, generally have adequate hydrogeological properties (see Section 13.4.3) for cover systems where fine-grained material is required, such as monolayer covers with elevated water table (see Chapter 8), covers with capillary barrier effects (see Chapter 7), and store and release (SR) covers (see Chapter 5). Investigations on the use of desulfurized tailings as part of reclamation scenarios were performed at the laboratory scale, at the intermediate field scale, using numerical modeling, and at the full field scale. Several authors investigated the influence of design parameters on the performance of the cover systems to prevent AMD and/or limit oxygen diffusion (Aachib et al. 1994; Mbonimpa et al. 2003; Romano et al. 2003; Bussière et al. 2004; Demers et al. 2008a; Demers et al. 2009; Dobchuk et al. 2013). The residual sulfide can improve the reclamation performance when desulfurized tailings and low-sulfide tailings are used as oxygen barrier cover material. Indeed, the low sulfide content has been shown to react with oxygen diffusing through the cover material; therefore, oxygen is consumed by the cover before reaching the reactive tailings placed under the cover (Mbonimpa et al. 2003; Romano et al. 2003). For example, Demers et al. (2009) calculated that less than 10% of the oxygen flux entering the top of the cover reached the reactive tailings placed underneath, as most of the oxygen was consumed in the 50–100 cm cover made of 0.25–1.25% S desulfurized tailings, with water table ranging from the surface of reactive tailings to 1 m below the surface (cover material with a high degree of water saturation). Mbonimpa et al. (2003) used numerical modeling and analytical solutions to Fick's first and second laws to obtain the oxygen fluxes at the top and the base of a moisture-retaining layer of a cover with capillary barrier effects (CCBE) made with low-sulfide tailings. For a 80 cm layer at a porosity of 0.44 and a degree of saturation of 85%, unreactive cover material showed an oxygen flux of 4.79×10^{-4} kg m^{-2} day^{-1}, while a slightly reactive cover material reduced the oxygen flux reaching the bottom of the cover by one order of magnitude. This beneficial effect of desulfurized (or low-sulfide) tailings is transient, however; sulfide oxidation will occur until all sulfides are depleted. Rey et al. (2020) observed that sulfide depletion in monolayer covers is dependent on cover thickness and degree of saturation (or oxygen availability) and calculated that sulfide depletion would occur in less than 4.7 years in the laboratory experimental setting used.

Other design parameters that were tested experimentally and numerically include water table level (particularly for monolayer covers with elevated water table), cover layer thickness and configuration, and physical properties of cover materials. For monolayer covers, regardless of the cover material, water table level was identified as a major factor influencing the cover system's performance (Demers et al. 2008a, 2008b; Dobchuk et al. 2013; Pabst et al. 2017; Ethier et al. 2018a) (see Chapter 8). Physical properties, and more specifically particle size distribution, is also an important factor for monolayer and multilayer covers. Studies by Hanton-Fong et al. (1997), Aubertin et al. (1999), Bussière et al. (2004), and Kabambi et al. (2017) showed that desulfurized tailings can maintain a high degree of saturation as part of a CCBE, when placed between coarse material layers acting as capillary breaks with the appropriate particle size difference (see Chapter 7). Rey et al. (2020) also demonstrated that monolayer covers made with desulfurized tailings as fine as the reactive tailings under the cover can be efficient oxygen barriers, while monolayer covers made with desulfurized tailings coarser than reactive tailings tend to act as an evaporation barrier, thus preventing desaturation of reactive tailings.

Finally, low-sulfide or inert waste rock has also been tested as part of reclamation scenarios (Kabambi et al. 2017; Larochelle et al. 2019). The requirements for the capillary break and drainage layers of a CCBE can be fulfilled by crushed waste rock, as long as the waste rock does not produce AMD or metal leaching.

13.6.2 Mining Infrastructures and Other Uses

Tailings dams are usually built using raised embankment design (i.e., the dam is raised progressively as the TSF is filled with tailings). Natural unbounded granular materials can be used for construction of retaining structures but, in practice, they have progressively been replaced by mine wastes since the mid of the 20th century (Bussière 2007), especially for cost reduction reasons. Starter dams, from where the dam is raised, are usually made of coarse waste rock. The dam themselves are usually built using the coarse fraction of the tailings, separated out either by cyclones or naturally during spigoting (Blight 2010). Starter and raising dams are exposed to atmospheric conditions and should be non-reactive to limit contaminated mine drainage generation.

Hard rock mine waste rocks usually have strong mechanical properties, high strength, and a good drainage capacity, and are therefore particularly suitable for road construction (Tannant and Regensburg 2001; McLemore et al. 2009). Most mines already use waste rocks, crushed or not, in their road systems (Tannant and Regensburg 2001). This approach is cheap and convenient but also needs to be optimized to improve trafficability, reduce tire punctures, and limit dust generation.

Finally, mine wastes can be used outside mine sites for various applications. Non-reactive mine tailings have, for example, been used to make bricks (Zhang 2013; Gopez 2015; Taha et al. 2017); their coarse fraction was used as aggregates for construction works, and waste rocks were used many times for road and dam construction (FHWA 2016). However, regulations can be strict for outside mine site uses, as these materials are considered wastes and must therefore be thoroughly characterized to allow their use (Tardif-Drolet et al. 2020). Also, their reuse should not be a way to postpone the environmental issues and create a secondary contamination source when these infrastructures will reach their end of life (e.g., bricks made of tailings should not be considered industrial wastes later or cause environmental issues when the building they were used to build will be demolished). Outside site reuse is particularly important for countries or regions with chronic lack of construction materials.

13.7 INTEGRATED WASTE ROCK MANAGEMENT

Reclamation of waste rock piles can represent a significant challenge for several reasons (Aubertin et al. 2015, 2016):

- Waste rock piles are often very large, and can exceed several dozens of meters in height, and several dozens of hectares in surface. Their reclamation therefore requires large quantities of cover materials. Borrowing these materials outside the site could have significant impacts on the environment.
- Pile slopes are usually close to the waste rock natural repose angle, which is often greater than the internal friction angle of fine-grained materials used in oxygen barrier cover systems.
- Waste rock properties (highly permeable, coarse, and widely graded materials) and construction methods (typically end-dumping or push-dumping methods) tend to produce highly heterogeneous piles. Water flow in piles is therefore usually rapid and largely controlled by preferential flow paths, thus making it difficult to control water balance and water table position.

The control of surface infiltration during operations could contribute to limit the generation of contaminated mine drainage, integrate reclamation in the pile design (according to the concept of "Designing for Closure"; Aubertin et al. 2016), and reduce the costs for reclamation at closure.

A flow control layer (FCL) aims to limit infiltration and deep percolation of water (Aubertin et al. 2002b, 2005, 2013). A FCL is typically composed of an inert fine-grained material, placed directly on top of waste rock, at the surface of each bench of the pile (Figure 13.10). Its slope (typically

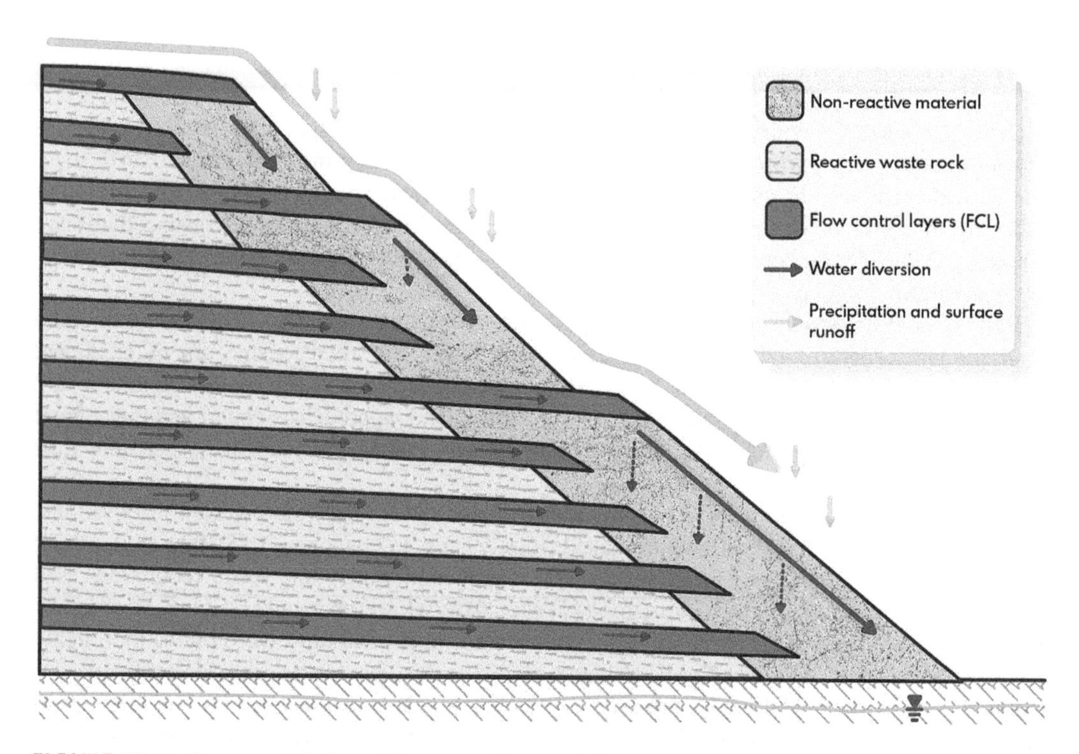

FIGURE 13.10 Conceptual design of flow control layers (FCL) in waste rock piles. FCL are slightly inclined toward the extremity of the pile, diverting water to nonreactive materials. FCL can be made with natural materials borrowed in the vicinity of the mine site, or crushed waste rock, compacted by the traffic on the surface of the pile. (Adapted from Aubertin et al. 2002b, 2013).

around 5%) toward the outside of the pile promotes runoff and favors lateral diversion. Reactive waste rock, placed in the core of the pile, is therefore protected from contact with water, reducing the risk for oxidation reactions and contaminant transport (Fala et al. 2005; Broda et al. 2013). Non-reactive waste rocks are placed on the sides of the piles, where water infiltrates. FCL should be built at the surface of each bench to control water flow and contamination during the operations (progressive reclamation). The final FCL on the surface of the pile could act as final reclamation cover system.

The principle of the FCL relies on the development of a capillary barrier effect at the interface between the coarse-grained waste rock and the fine-grained FCL. The unsaturated hydrogeological conditions within the pile induce a reduction of the hydraulic conductivity of the waste rock, which prevents deep percolation of the water (see Chapter 3). The concept is similar to that of the CCBE (see Chapter 7) or inclined SR cover systems (Chapter 5). However, the lateral diversion (or diversion capacity) increases progressively the quantity of water in the FCL. As explained in Chapter 5, when the pore water pressure in the FCL exceeds the water entry value of the underlying waste rock, water starts to infiltrate vertically. This point (or zone) is known as the down-dip limit (DDL), the effective length of the capillary barrier (L_{eff}) (Ross 1990). The L_{eff} should therefore be maximized to avoid infiltration in the reactive waste rock and reach at least the location where non-reactive waste rocks are placed (Figure 13.10).

The FCL concept was developed and validated using physical model tests (Bussière et al. 2003), numerical simulations (Fala et al. 2005, 2006; Molson et al. 2005; Broda et al. 2014) and in situ large-scale tests on an experimental waste rock pile (Martin et al. 2017, 2019). Results indicated that water diversion was mainly influenced by the contrast between waste rock and FCL hydrogeological properties. FCL material needs to be finer than the underlying waste rock to create a capillary barrier effect and prevent infiltration, but also sufficiently permeable to allow a rapid lateral transport of the water and limit water accumulation at the interface. The FCL can be composed of sand, but can also integrate mine waste materials, such as crushed non-reactive waste rock. The slope angle, compaction, and thickness of the FCL also showed, to a lesser extent, some influence on the waste rock pile hydrogeological behavior.

The properties of the FCL (materials, slope, thickness) therefore need to be adapted to the site conditions (waste rock properties, available material for FCL, and, more importantly, climate conditions, see Chapter 14) and to consider other operating limitations. For instance, using a sand layer with a higher void ratio might enhance lateral flow in the FCL by increasing its hydraulic conductivity, but a less dense sand layer could also affect the development of the capillary break above the waste rock. A steeper slope could increase not only runoff and the diversion length of the FCL but also the risk of erosion on the surface. Slope stability may also be affected by the presence and configuration of FCL in piles. Finally, the application of the technique in situ would require sorting (or segregating) reactive and non-reactive waste rocks (see Chapter 2 for characterization methods) to dispose them in either the center or sides of the pile.

The main advantages of FCL include the progressive reclamation of the pile (preventing contamination during operations, thus reducing costs for water treatment) and the use of materials produced at the mine for its reclamation (thus reducing the necessity to borrow natural materials in the environment).

13.8 CONCLUSION AND RESEARCH NEEDS

Integrated mine waste management approaches aim to propose alternative solutions to conventional deposition in tailings impoundments and waste rock dumps. The objective of these techniques is to improve both the short- and long-term geotechnical and geochemical stability of mine waste storage facilities. Integrated mine waste management can also contribute to reduce costs (for water treatment and mine closure) and reduce the environmental footprint (e.g., by reducing the amount of natural material borrowed in the environmental to build cover systems and mining infrastructures).

TABLE 13.4
Summary of Conventional and Alternative Approaches to Mine Waste Management and Reclamation

	Tailings	Waste rock
Conventional	• Slurry deposition in TSF • Engineered cover systems • Revegetation	• Waste rock piles • Engineered cover systems • Revegetation

Alternative	• Densified tailings deposition in TSF
	• Progressive reclamation
	• Co-disposal with waste rock
	• Backfill (underground and in pit)
	• Integrated tailings management
	  • Environmental desulfurization
	  • Value recovery from concentrate
	  • Use of desulfurized tailings as cover material
		• Co-disposal with tailings
		• Backfill (underground and in pit)
		• Used as cover material for reclamation
		• Used for infrastructure construction
		• Integrated waste rock management

They also facilitate and improve the performance of the reclamation. These techniques, summarized in Table 13.4, are relatively recent and will continue to be improved in the coming years, while new approaches will most probably be developed. Technological watch and research and development are therefore an important part of integrated mine waste management technologies. These approaches should be used in combination with the reclamation techniques presented in previous chapters, since they are not reclamation methods. The combination of integrated mine waste management technologies and efficient reclamation strategies results in optimal technical and environmental performance of the reclaimed mine site.

Similar to reclamation methods, application of integrated mine waste management technologies is not universal and needs to be evaluated on a case-by-case basis. Table 13.5 summarizes the advantages and limitation of the techniques described in this chapter. The limitations are currently driving the research needs, as expressed below.

- Densification: Better control (management, disposal, and especially progressive reclamation) is required to extend the technique to acid-generating tailings.
- Backfill: More work in alternative and innovative binders to reduce or replace cement in paste backfill is required. The potential for contaminant stabilization and development of binders and additives to promote long-term stabilization of toxic elements within the paste backfill matrix is a promising avenue to manage AMD-generating and metal-leaching tailings. Practical approaches to dispose of mine wastes (and especially waste rocks) in pit should also be developed to integrate the method to operation constraints.
- Co-mixing: Test application of co-mixing of waste rock and tailings in metal mines, under different climatic conditions, as reclamation material for TSF and waste rock piles. We need a better understanding of water and gas movement in such a material, and also its durability under weathering conditions.
- Environmental desulfurization: The application of desulfurization to tailings with very low neutralization potential, and/or with unusual sulfide minerals, is still challenging. Adaptation to the typical bulk sulfide flotation is required for these cases to ensure the production of inert tailings. Recovery of valuable elements concurrently to desulfurization is also a promising field of research. The long-term monitoring of covers made with desulfurized tailings is also warranted to evaluate the performance of such materials in a reclamation scenario.

TABLE 13.5
Main Advantages and Limitations of Integrated Mine Waste Management Approaches Compared to Conventional Management Methods

Technique	Advantages Over Conventional Management Approaches	Limitations
Densification	• Improved geotechnical stability (reduced risks for dam failure) • Lower requirements on dam design/construction • Little or no liquefaction risks • Recirculation of water • Tailings uniformity (reduced segregation) • Reduction of surface footprint • Facilitated progressive reclamation	• High investment costs • Poorly suited for reactive AMD-generating tailings • Higher transport and disposal costs • Higher risks for dust generation
Backfilling	• Pillar recovery (higher extraction efficiency) • Less wastes to dispose of in surface facilities • Smaller visual impact • Facilitated reclamation, and physical isolation of reactive wastes (including sulfide concentrate)	• Cost of binders and additives • Operational constrains and technical limitations for pit backfilling
Environmental desulfurization	• Reduction of volumes of problematic tailings • Production of benign tailings	• Costly • Limited large-scale operation examples • Application difficult for tailings with very low NP
Co-disposal	• Improve both hydrogeological and geotechnical properties of mine wastes • Lower requirements on dam design/construction • Reduction of liquefaction risks • Possibility to have a single waste storage facility, waste rock piles can become unnecessary in some cases • Easier management and reclamation of reactive waste rock (encapsulated in saturated tailings)	• Technical limitations because of waste rock size (depending on the mixing method) • Performance highly dependent on waste characterization
Valorization	• Reduced quantities of mine waste in storage facilities • Improved performance of structures compared to the use of natural materials (e.g., dams, roads, cover systems) • Reduced costs • Reduced impact on environment (less borrowed materials required) • Circular economy approach	• Not applicable for reactive mine waste • Trace elements may limit potential applications • Limited guidelines and application examples • Regulation can hinder applications
Integrated waste rock management	• Reclamation is integrated to the design, no final cover required • Progressive reclamation, reduces contaminated mine drainage during operations • No additional material needed for reclamation	• Require a method to sort reactive and non-reactive waste rock

- Valorization: Current research aims to optimize the generation, preparation, and sorting of mine wastes to maximize their valorization in mining and civil infrastructures.
- Impact of climatic conditions and climate change: Most of the integrated mine waste management approaches described in this chapter have been developed for specific mine sites and have to be adapted depending on site conditions. In particular, further research would be required to apply these approaches in cold climates where water balance, thermal amplitudes, and operational constrains are very particular. Also, impact of climate change has to be studied to validate the long-term efficiency of these methods (also see Chapter 14).

REFERENCES

Aachib, M., M. Aubertin, and R. Chapuis. 1994. Column test investigation of milling wastes properties used to build cover systems. Paper presented at the *International Land Reclamation and Mine Drainage Conference*. Pittsburgh, PA. *Proceedings America Society of Mining and Reclamation* 2: 128–137. https://doi.org/10.2100/JASMR94020128.

Adu-Wusu, V., and E. K. Yanful. 2007. Post-closure investigation of engineered test covers on acid-generating waste rock at Whistle Mine, Ontario. *Canadian Geotechnical Journal* 44, no. 4: 496–506.

Al, T. A., and D. W. Blowes. 2000. Identification of preferential flow effects on hydraulic conductivity measurements using a fluorescent tracer. *Canadian Geotechnical Journal* 37, no. 2: 479–484.

Aubertin, M., and R. P. Chapuis. 1991. Considérations hydro-géotechniques pour l'entreposage des résidus miniers dans le nord-ouest du Québec. In *2nd International Conference on Acid Rock Drainage (ICARD)*, vol. 3, 1–22.

Aubertin, M., M. Aachib, M. Monzon, A.-M. Joanes, B. Bussière, and R. P. Chapuis. 1999. *Étude de* laboratoire sur l'efficacité de recouvrements construits à partir de résidus miniers. In *MEND 2.22.2b*, Édit 2.22.2b MR.

Aubertin, M., B. Bussière, and L. Bernier. 2002a. *Environnement et gestion des rejets miniers*. Montréal: Presses Polytechnique Internationales.

Aubertin, M., O. Fala, B. Bussière, V. Martin, D. Campos, A. Gamache-Rochette, M. Chouteau, and R. P. Chapuis. 2002b. Analyse des écoulements de l'eau en conditions non saturées dans les haldes à stériles. In *Défis et perspectives: Symposium 2002 sur l'environnement et les Mines*, Rouyn-Noranda, QC, Canada.

Aubertin, M., O. Fala, J. Molson, et al. 2005. Évaluation du comportement hydrogéologique et géochimique des haldes à stériles. In *Symposium 2005 sur l'Environnement et les Mines*, Rouyn-Noranda, QC, Canada.

Aubertin, M., M. Maknoon, and B. Bussière. 2013. Recommandation pour améliorer le comportement hydrogétechnique des haldes à stériles. Paper presented *at GeoMontréal.*

Aubertin, M., T. Pabst, B. Bussière, et al. 2015. Rebue des meilleures pratiques de restauration des sites d'entreposage de rejets miniers générateurs de DMA. In *Symposium 2015 sur l'environnement et les mines*, Rouyn-Noranda, QC, Canada.

Aubertin, M., B. Bussière, T. Pabst, M. James, and M. Mbonimpa. 2016. Review of the reclamation techniques for acid-generating mine waste upon closure of disposal sites. In *Proceedings of the Geo-Chicago 2016 Conference*, 343–358. https://doi.org/10.1061/9780784480137.034.

Ayres, B. K., L. Lanteigne, Q. Smith, and M. O'Kane. 2007. Closure planning and implementation at CVRD Inco's Whistle Mine, Ontario, Canada. *Mine Closure*. http://bc-mlard.ca/files/supporting-documents/2009-19a-AYRES-ETAL-closure-planning-implementation-whistle-mine.pdf.

Barbour, S. L., G. W. Wilson, and L. C. St. Arnaud. 1993. Evaluation of the saturated-unsaturated groundwater conditions of a thickened tailings deposit. *Canadian Geotechnical Journal* 30, no. 6: 935–946.

Bareither, C., M. Gorakhki, J. Scalia, and M. Jacobs. 2018. Compression behavior of filtered tailings and waste rock mixtures: GeoWaste. Paper presentations at Tailings and Mine Waste '18, Keystone, CO, October 1–3, 2018.

Belem, T., and M. Benzaazoua. 2008. Underground mine paste backfill technology: Applications and design methods. *Geotechnical and Geological Engineering* 26, no. 2: 147–174.

Belfadhel, M. B. 2000. Méthodes de gestion et de déclassement des résidus miniers d'uranium au Canada. In *Proceedings of 53rd Canadian Geotechnical Conference*, 675–682.

Ben Abdelghani, F., M. Aubertin, R. Simon, and R. Therrien. 2015. Numerical simulations of water flow and contaminants transport near mining wastes disposed in a fracture rock mass. *International Journal of Mining Science and Technology* 25, no. 1: 37–45.

Benzaazoua, M., B. Bussière, R. V. Nicholson, and L. Bernier. 1998. Geochemical behavior of multilayered cover made of desulphurized mine tailings. In *Proceedings of Fifth International Conference on Tailings and Mine Waste '98*, 389–398. Fort Collins, CO: Rotterdam BE.

Benzaazoua, M., J. Ouellet, S. Servant, P. Newman, and R. Verburg. 1999. Cementitious backfill with high sulfur content: Physical, chemical, and mineralogical characterization. *Cement and Concrete Research* 29, no. 5: 719–725.

Benzaazoua, M., B. Bussière, M. Kongolo, J. McLaughlin, and P. Marion. 2000. Environmental desulphurization of four canadian mine tailings using froth flotation. *International Journal of Mineral Processing* 60: 57–74.

Benzaazoua, M., T. Belem, and B. Bussière. 2002. Chemical factors that influence the performance of mine sulphidic paste backfill. *Cement and Concrete Research* 32, no. 7: 1133–1144.

Benzaazoua, M., M. Fall, and T. Belem. 2004. A contribution to understanding the hardening process of cemented pastefill. *Minerals Engineering* 17, no. 2: 141–152.

Benzaazoua, M., B. Bussière, I. Demers, M. Aubertin, E. Fried, and A. Blier. 2008. Integrated mine tailings management by combining environmental desulphurization and cemented paste backfill: Application to mine Doyon, Quebec, Canada. *Minerals Engineering* 21, no. 4: 330–340.

Blier, A., I. Demers, M. Gagnon, and G. Bois. 2012. Eliminating acid rock drainage (ARD) at Mine Doyon: Sustained operations allow for new possibilities. In *9th International Conference on Acid Rock Drainage*, Ottawa, ON, May 20–26, 2012.

Blight, G. E. 2010. *Geotechnical engineering for mine waste storage facilities*. London: Taylor & Francis Group.

Boger, D. V. 2013. Rheology of slurries and environmental impacts in the mining industry. *Annual Review of Chemical and Biomolecular Engineering* 4: 239–257.

Bois, D., P. Poirier, M. Benzaazoua, and B. Bussière. 2004. A feasability study on the use of desulphurized tailings to control acid mine drainage. In Proceedings 2004 – 36th Annual Meeting of the Canadian Mineral Processors, Ottawa, Canada, 361–80.

Bois, D., M. Benzaazoua, B. Bussière, M. Kongolo, and P. Poirier. 2005. A feasibility study on the use of desulphurized tailings to control acid mine drainage. *CIM Bulletin* 98, no. 1087: 74–94.

Bolduc, F. L., and M. Aubertin. 2014. Numerical investigation of the influence of waste rock inclusions on tailings consolidation. *Canadian Geotechnical Journal* 51, no. 9: 1021–1032.

Broadhurst, J., and S. T. L. Harrison. 2015. A desulfurization flotation approach for the integrated management of sulfide wastes and acid rock drainage risks. In *10th International Conference on Acid Rock Drainage and IMWA Annual Conference*, ed. A. Brown et al. Chile: Gecamin Ltd.

Broadhurst, J., H.-M. Stander, J. Amaral Filho, and S. T. L. Harrison. 2018. From liability to opportunity: Developing resource-efficient approaches for the management of sulfidic mine waste. In *Symposium on Mines and Environment*, Rouyn-Noranda, Canada, June 17–20, 2018.

Broda, S., E. Hirthe, D. Blessent, M. Aubertin, and T. Graf. 2013. Using random discrete fractures for representing preferential flow in waste rock piles with compacted layers. In *Proceedings of the 66th CGS Conference GeoMontreal 2013*, Montreal, QC, Canada.

Broda, S., M. Aubertin, D. Blessent, T. Graf, and E. Hirthe. 2014. Improving control of contamination from waste rock piles. *Environmental Geotechnics* 4, no. 4: 274–283.

Burden, R., D. J. Williams, G. W. Wilson, and M. Jacobs. 2018. The shear strength of filtered tailings and waste rock blends. In *Proceedings of Tailings and Mine Waste '18*, Keystone, CO, USA, October 1–3, 2018.

Bussière, B. 2007. Colloquium 2004: Hydro-geotechnical properties of hard rock tailings from metal mines and emerging geo-environmental disposal approaches. *Canadian Geotechnical Journal* 44, no. 9: 1019–1052.

Bussière, B., J. Lelièvre, J. Ouellet, and D. Bois. 1995. Utilisation de résidus miniers désulfurés comme recouvrement pour prévenir le DMA: Analyse technico-économique sur deux cas réels. In *Sudbury '95 Conference on Mining and the Environment*, Sudbury, ON, May 28–June 1, 1995, 59–68.

Bussière, B., M. Aubertin, and R. P. Chapuis. 2003. The behavior of inclined covers used as oxygen barriers. *Canadian Geotechnical Journal* 40: 512–535.

Bussière, B., M. Benzaazoua, M. Aubertin, and M. Mbonimpa. 2004. A laboratory study of covers made of low sulphide tailings to prevent acid mine drainage. *Environmental Geology* 45: 609–622.

Caldwell, J. 2017. Tailings history: 2016 and 2017. In *Proceedings of Tailings and Mine Waste Conference Banff*, Alberta, November 5–8, 2017.

Carneiro, A., and A. B. Fourie. 2018. A conceptual cost comparison of alternative tailings disposal strategies in Western Australia. In *Paste 2018: Proceedings of the 21st International Seminar on Paste and Thickened tailings*, Perth, Australia, April 11–13, 2018, 439–454.

Castendyk, D. N. 2005. *An interdisciplinary approach to the prediction of pit lake water quality, Martha Mine pit lake, New Zealand*. PhD diss., University of Auckland.

Castendyk, D. N., and Eary, L. E., eds. 2009. *Mine pit lakes: Characterstics, predictive modeling, and sustainability*. Littleton, CO: Society for Mining, Metallurgy, and Exploration.

Charbonneau, E. 2019. *Caractérisation hydrogéotechnique des mélanges paste rock.* Maîtrise sur mesure en mines et environnement, Université du Québec en Abitibi-Témiscamingue, Rouyn-Noranda, QC.

Charbonneau, E., S. Pouliot, B. Bussière, and G. W. Wilson. 2018. Assessment of paste rock as cover material to control AMD. In *Symposium Rouyn-Noranda on Mines and the Environment*, Rouyn-Noranda, QC, June 17–20, 2018.

Cheng, T. C., F. Kassimi, and J. M. Zinck. 2016. A holistic approach of green mining innovation in tailings reprocessing and repurposing. In *Tailings and Mine Waste '16*, Keystone, CO, October 2–5, 2016, 17.

Coussy, S., M. Benzaazoua, D. Blanc, P. Moszkowicz, and B. Bussière. 2011. Arsenic stability in arsenopyrite-rich cemented paste backfills: A leaching test-based assessment. *Journal of Hazardous Materials* 185, no. 2: 1467–1476.

Coussy, S., M. Benzaazoua, D. Blanc, P. Moszkowicz, and B. Bussière. 2012. Assessment of arsenic immobilization in synthetically prepared cemented paste backfill specimens. *Journal of Environmental Management* 93, no. 1: 10–21.

Daliri, F., H. Kim, P. Simms, and S. Sivathayalan. 2014. Impact of desiccation onmonotonic and cyclic shear strength of thickened gold tailing. *Journal of Geotechnical and Geoenvironmental Engineering* 140, no. 9. https://doi.org/10.1061/(ASCE)GT.1943-5606.0001147.

Davies, M. 2011. *Filtered dry stacked tailings – The fundamentals.* In *Proceedings Tailings and Mine Waste 2011*, Vancouver, BC, November 6–9, 2011, 9p.

Davies, M. P., and S. Rice. 2001. An alternative to conventional tailings management – "dry stack" filtered tailings. In *Proceedings of the 8th International Conference on Tailings and Mine Waste*, Fort Collings, CO, January 15–18, 2001, 411–420.

Demers, I., B. Bussière, M. Benzaazoua, M. Mbonimpa, and A. Blier. 2008a. Column test investigation on the performance of monolayer covers made of desulphurized tailings to prevent acid mine drainage. *Minerals Engineering* 21, no. 4: 317–329.

Demers, I., B. Bussière, M. Benzaazoua, M. Mbonimpa, and A. Blier. 2008b. Optimisation of single-layer cover made of desulphurized tailings: Application to the Doyon mine tailings impoundment. In *SME Annual Meeting 2008*, Salt Lake City, February 24–27.

Demers, I., B. Bussière, M. Mbonimpa, M. Aubertin, and M. Benzaazoua. 2009. Oxygen diffusion and consumption in low sulphide tailings covers. *Canadian Geotechnical Journal* 46: 454–469.

Dobchuk, B., C. Nichol, G. W. Wilson, and M. Aubertin. 2013. Evaluation of a single-layer desulphurized tailings cover. *Canadian Geotechnical Journal* 50, no. 7: 777–792.

Dompierre, K. A., and S. L. Barbour. 2016. Characterization of physical mass transport through oil sands fluid fine tailings in an end ipt lake: A multi-tracer study. *Journal of Contaminant Hydrology* 189: 12–26.

Donahue, R., M. J. Hendry, and P. Landine. 2000. Distribution of arsenic and nickel in uranium mine tailings, Rabbit Lake, Saskatchewan, Canada. *Applied Geochemistry* 15: 1097–1119.

Doucet, K., N. Pepin, M. Kissiova, and C. Pednault. 2015. Thickened tailings characterization program for a tailings storage facility design update – Case Study. In *Proceedings of 19th International Conference on Tailings and Mine Waste*, Vancouver, Canada, October 25–28, 2015.

Essayad, K., T. Pabst, M. Aubertin, and R. P. Chapuis. 2018. An experimental study of the movement of tailings through waste rock inclusions. In *Proceedings of the 71st Canadian Geotechnical Conference GeoEdmonton 2018*, Edmonton, AB.

Ethier, M.-P., B. Bussière, M. Aubertin, A. Maqsoud, I. Demers, and S. Broda. 2018a. In situ evaluation of the performance of the reclamation measures implemented on an abandoned reactive tailings disposal site. *Canadian Geotechnical Journal* 55, no. 12: 1742–1755.

Ethier, M.-P., B. Bussière, M. Aubertin, A. Maqsoud, I. Demers, and S. Broda. 2018b. In situ evaluation of performance of reclamation measures implemented on abandoned reactive tailings disposal site. *Canadian Geotechnical Journal* 55, no. 12: 1742–1755.

Fala, O., J. Molson, M. Aubertin, and B. Bussière. 2005. Numerical modelling of flow and capillary barrier effects in unsaturated waste rock piles. *Mine Water and the Environment* 24, no. 4: 172–185.

Fala, O., J. Molson, M. Aubertin, B. Bussière, and R. P. Chapuis. 2006. *Numerical simulations of long term unsaturated flow and acid mine drainage at waste rock piles*. In *7th International Conference on Acid Rock Drainage (ICARD)*, March 26–30.

Ferdosi, B., M. James, and M. Aubertin. 2015a. Investigation of the effect of waste rock inclusions configuration on the seismic response of a tailings impoundment. *Geotechnical and Geological Engineering* 33, no. 6: 1519–1537.

Ferdosi, B., M. James, and M. Aubertin. 2015b. Effect of waste rock inclusions on the seismic stability of an upstream raised tailings impoundment: A numerical investigation. *Canadian Geotechnical Journal* 52, no. 12: 1930–1944.

FHWA. 2016. *User guidelines for waste and byproduct materials in pavement construction*. Federal Highway Administration.

Fitton, T. 2017. Avoiding large tailings dams without going underground – Robinsky's thickened tailings concept. In *Paste 2017: Proceedings of the 20th International Seminar on Paste and Thickened Tailings*, 243–249. Beijing, China: Australian Center for Geomechanics.

Fitton, T., and A. Roshdieh. 2014. Filtered tailings versus thickened slurry: Four case studies. In *Proceedings of the 16th International Conference on Paste and Thickened Tailings*, 275–288.

Fourie, A. 2012a. Paste and thickened tailings: Has the promise been fulfilled? In *Proceedings of GeoCongress 2012: State of the Art and Practice in Geotechnical Engineering*, ed. R. D. Hryciw, A. Athanasopoulos-Zekkos, and N. Yesiller.

Fourie, A. 2012b. Perceived and realised benefits of paste and thickened tailings for surface deposition. In *Paste 2012: Proceedings of the 15th International Seminar on Paste and Thickened Tailings*, ed. R. J. Jewell, A. B. Fourie, and A. Paterson, 53–64.

Gopez, R. G. 2015. Utilizing mine tailings as substitute construction material: The use of waste materials in roller compacted concrete. *Open Access Library Journal* 2, no. 12: 22199. https://doi.org/10.4236/oalib.1102199.

Guerin, F., S. Wilson, and R. Nicholson. 2006. Optimizing In-pit disposal of problematic waste rock using leaching tests, portable XRF, block and mass transport models. In *Proceedings of 13th Annual BC MEND Workshop*.

Hanton-Fong, C. J., D. W. Blowes, and R. Stuparyk. 1997. Evaluation of low sulphur tailings in the prevention of acid mine drainage. In *4th ICARD*, Vancouver, Canada, May–June 1997, 835–847.

Hassani, F., and J. Archibald. 1998. *Mine backfill*. CIM.

Helinski, M., M. Fahey, and A. Fourie. 2010. Coupled two-dimensional finite element modelling of mine backfilling with cemented tailings. *Canadian Geotechnical Journal* 47, no. 11: 1187–1200.

Hesketh, A. H., J. L. Broadhurst, and S. T. L. Harrison. 2010. Mitigating the generation of acid mine drainage from copper sulfide tailings impoundments in perpetuity: A case study for an integrated management strategy. *Minerals Engineering* 23, no. 3: 225–229.

Humber, A. J. 1995. Separation of sulphide minerals from mill tailings. In *Sudbury '95, Conference on Mining and the Environment*, Sudbury, ON, May 28–June 1, 1995, 149–158.

Humber, A. J. 1997. Separation of sulphide minerals from mill tailings. *Land Contamination and Reclamation* 5, no. 2: 109–116.

INAP. 2009. *The global acid rock drainage guide*. www.gardguide.com.

James, M., and M. Aubertin. 2012. The use of waste rock inclusions to improve the seismic stability of tailings impoundments. In *GeoCongress 2012: State of the art and practice in geotechnical engineering*, ed. R. D. Hryciw, A. Athanasopoulos-Sekkos, and N. Yesiller, 4166–4175. https://doi.org/10.1061/9780784412121.

James, M., M. Aubertin, D. Wijewickreme, and G. W. Wilson. 2011. A laboratory investigation of the dynamic properties of tailings. *Canadian Geotechnical Journal* 48, no. 11: 1587–1600.

James, M., M. Aubertin, and B. Bussière. 2013. On the use of waste rock inclusions to improve the performance of tailings impoundments. In *18th International Conference on Soil Mechanics and Geotechnical Engineering*, Paris, France, 735–738.

James, M., M. Aubertin, B. Bussière, C. Pednault, N. Pépin, and M. Limoges. 2017. A research project on the use of waste rock inclusions to improve the performance of tailings impoundments. In *Proceedings of GeoOttawa 2017, 70th Canadian Geotechnical Conference*, 8p.

Jewell, R. J., and A. B. Fourie. 2015. *Paste and thickened tailings: A guide*. Nedlands: Australian Centre for Geomechanics, University of Western Australia.

Jewell, R. J., A. B. Fourie, and E. R. Lord. 2002. *Paste and thickened tailings: A guide*. Nedlands: Australian Center for Geomecanics.

Johnson, J. M., J. Vialpando, and C. Lee. 2005. Paste tailings management alternative: Study results for Molycorp's lanthanide group operations in Mountain Pass, California. *Mining Engineering* 57, no. 2: 50–56.

Kabambi, A. K., B. Bussière, and I. Demers. 2017. Hydrogeological behaviour of covers with capillary barrier effects made of mining material. *Geotechnical and Geological Engineering* 35, no. 3: 1199–1220.

Kam, S., D. Yaschyshyn, M. Patterson, and D. Scott. 2009. Thickened tailings disposal at Xstrata Copper Canada Kidd Metallurgical Site. In *Proceedings 8th ICARD Conference*, Skelleftea, Sweden, June 23–26, 2009.

Kam, S., J. Girard, N. Hmidi, Y. Mao, and S. Longo. 2011. Thickened tailings disposal at Musselwhite Mine. In *Proceedings of the 14th International Seminar on Paste and Thickened Tailings, Australian Centre for Geomechanics*, Perth, 225–236.

Kazadi Mbamba, C., S. T. L. Harrison, J. P. Franzidis, and J. L. Broadhurst. 2012. Mitigating acid rock drainage risks while recovering low-sulfur coal from ultrafine colliery wastes using froth flotation. *Minerals Engineering* 29: 13–21.

Kotsiopoulos, A., and S. T. L. Harrison. 2018. Co-disposal of benign desulfurised tailings with sulfidic waste rock to mitigate ARD generation: Influence of flow and contact surface. *Minerals Engineering* 116: 62–71.

Lamontagne, A., R. Lefebvre, R. Poulin, and S. Leroueil. 1999. Modelling of acid mine drainage physical processes in a waste rock pile with layered co-mingling. In *52nd Canadian Geotechnical Conference, Regina*, Saskatchewan, Canada, October 24–27, 1999, 479–485.

Lamontagne, A., S. Fortin, R. Poulin, N. Tassé, and R. Lefebvre. 2000. Layered co-mingling for the construction of waste rock pile as a method to mitigate acid mine drainage: Laboratory investigations. In *5th International Conference on Acid Rock Drainage (ICARD 2000), Denver, CO*, 779–788. Littleton, CO: Society of Mining Engineers.

Landriault, D. A., R. Verburg, W. Cincilla, and D. Welch. 1997. Paste technology for underground backfill and surface tailings disposal applications. In *Short course notes, Canadian Institute of Mining and Metallurgy, Technical workshop*, Vancouver, BC, Canada, p 120.

Lara, J. L., E. U. Pornillos, and Munoz, H. E. 2014. Geotechnical-Geochemical and operational considerations for the application of dry stacking tailings deposits – State of the art. In *Proceedings of the 16th International Conference on Paste and Thickened Tailings*, ed. R. Jewell and A. Fourie, 249–260. Belo Horizonte, Brazil.

Larochelle, C. G., B. Bussière, and T. Pabst. 2019. Acid-generating waste rocks as capillary break layers in covers with capillary barrier effects for mine site reclamation. *Water, Air, and Soil Pollution* 230, no. 3, p. 57–72.

Leduc, M., M. Backens, and M. E. Smith. 2004. Tailings co-disposal at the Esquel Gold mine, Patagonia, Argentina. In *SME annual meeting. Denver, Colorado, February 23–25, 2004*. Society for Mining, Metallurgy, and Exploration Inc., p 1–5.

Lefebvre, R., D. Hockley, J. Smolensky, and A. Lamontagne. 2001. Multiphase transfer processes in waste rock piles producing acid mine drainage: 2. Applications of numerical simulation. *Journal of Contaminant Hydrology* 52, no. 1–4: 165–186.

Leppinen, J. O., P. Salonsaari, and V. Palosaari. 1997. Flotation in acid mine drainage control: Beneficiation of concentrate. *Canadian Metallurgical Quarterly* 36, no. 4: 225–230.

Li, L., M. Aubertin, and T. Belem. 2005. Formulation of a three dimensional analytical solution to evaluate stresses in backfilled vertical narrow openings. *Canadian Geotechnical Journal* 42, no. 6: 1705–1717.

MAC (The Mining Association of Canada). 2017. *A Guide to theManagement of Tailings Facilities*, 3rd ed. Ottawa, ON: The Mining Association of Canada, 86p.

Martin, V. 2018. Evolution of the hydrogeotechnical properties of paste tailings deposited on the surface. PhD diss., École Polytechnique de Montréal.

Martin, J., and J. Fyfe. 2007. Innovative closure concepts for the Xstrata Nickel Onaping operations. In *Mining and the Environment IV Conference*, Sudbury, ON, October 19–27, 2007, p. 119.

Martin, V., M. Aubertin, M. Benzaazoua, and G. Zhan. 2010. Investigation of near-surface exchange processes in reactive paste tailings. In *Proceedings of the 13th International Seminar on Paste and Thickened Tailings*, Australian Center for Geomechanics, Perth, Australia, 265–278.

Martin, V., B. Bussière, B. Plante, et al. 2017. Controlling water infiltration in waste rock piles: Design, construction, and monitoring of a large-scale in situ pilot study. In *Proceedings of 70th Canadian Geotechnical Conference GeoOttawa*, Ottawa, ON, Canada.

Martin, V., T. Pabst, B. Bussière, B. Plante, and M. Aubertin. 2019. A new approach to control contaminated mine drainage generation from waste rock piles: Lessons learned after 4 years of field monitoring. In *Proceedings of 18th Global Joint Seminar on Geo-Environmental Engineering*, Montreal, QC, Canada.

Mbonimpa, M., M. Aubertin, M. Aachib, and B. Bussière. 2003. Diffusion and consumption of oxygen in unsaturated cover materials. *Canadian Geotechnical Journal* 40: 916–932.

McCullough, C. D., G. Marchand, and J. Unseld. 2013. Mine closure of pit lakes as terminal sinks: Best available practice when options are limited? *Mine Water and the Environment* 32: 302–313.

McLaughlin, J., and R. Stuparyk. 1994. Evaluation of low sulphur rock tailings production at Inco's Clarabelle mill. In *Innovation in Mineral Processing Conference*, Sudbury, Ontario.

McLemore, V. T., A. Fakhimi, D. van Zyl, et al. 2009. Literature review of other rock piles: Characterization, weathering, and stability. New Mexico Bureau of Geology and Mineral Resources, Socorro, NM.

Mehdipour, I., and K. H. Khayat. 2018. Understanding the role of particle packing characteristics in rheo-physical properties of cementitious suspensions: A literature review. *Construction and Building Materials* 161: 340–353.

MEND. 2015. In-Pit disposal of reactive mine wastes: Approaches, update and case study results. MEND Report 2.36.1b. Mine Environment Neutral Drainage (MEND) Program, 250p.

MEND. 2017. Study of tailings management technologies. MEND Report 2.50.1. Mine Environment Neutral Drainage (MEND) Program, 164p.

Mermillod-Blondin, R. 2005. Influence des propriétés superficielles de la pyrite sur la rétention de molécules organiques soufrées: Application à la désulfuration des résidus miniers. PhD diss., Institut National Polytechnique de Lorraine, Université de Montréal.

MERN (2017). Guide de préparation du plan de réaménagement et de restauration des sites miniers au Québec. Ministère de l'Énergie et des Ressources naturelles (Ministry of Energy and Natural Resources), Quebec, Canada, 80 p.

Miller, A. J., and N. P. Zègre. 2014. Mountaintop removal mining and catchment hydrology. *Water* 6, no. 3: 472–499.

Mjonono, D. 2019. Development of co-disposal methods for coal discards and fine waste for the prevention of acid mine drainage. MSc thesis, University of Cape Town.

Molson, J. W., O. Fala, M. Aubertin, and B. Bussière. 2005. Numerical simulations of pyrite oxidation and acid mine drainage in unsaturated waste rock piles. *Journal of Contaminant Hydrology* 78: 343–371.

Morgenstern, N. R., S. G. Vick, and D. van Zyl. 2015. *Report on Mount Polley Tailings Storage Facility breach.* Independent Expert Engineering Investigation and Review Panel. Province of British Columbia, Canada.

Morgenstern, N. R., S. G. Vick, C. B. Viotti, and B. D. Watts. 2016. *Report on the Immediate Causes of the Failure of the Fundão Dam.* Cleary Gottlieb Steen and Hamilton LLP.

Nicholson, R. V., Gillham, R. W., Cherry, J. A., Reardon, E. J. 1989. Reduction of acid generation through the use of moisture-retaining cover layers as oxygen barriers. *Canadian Geotechnical Journal* 26: 1–8.

Oldecop, L. A., and G. J. Rodari. 2017. Unsaturated soil mechanics in mining. In *PanAm Unsaturated Soils Conference 2017*, Dallas, TX, November 12–15, 2017, 257–280.

Ouellet, S., B. Bussière, M. Mbonimpa, M. Benzaazoua, and M. Aubertin. 2006. Reactivity and mineralogical evolution of an underground mine sulphidic cemented paste backfill. *Minerals Engineering* 19, no. 5: 407–419.

Pabst, T., M. Aubertin, B. Bussière, and J. Molson. 2017. Experimental and numerical evaluation of single-layer covers placed on acid-generating tailings. *Geotechnical and Geological Engineering* 35, no. 4: 1421–1438.

Park, J. H., M. Edraki, and T. Baumgartl. 2019. A practical testing approach to predict the geochemical hazards of in-pit coal mine tailings and rejets. *Catena* 148: 3–10.

Pépin, N., M. Aubertin, and M. James. 2012. A seismic table investigation of the effect of inclusions on the cyclic behaviour of tailings. *Canadian Geotechnical Journal* 49, no. 4: 416–426.

Peyronnard, O., and M. Benzaazoua. 2011. Estimation of the cementitious properties of various industrial by-products for applications requiring low mechanical strength. *Resources Conservation and Recycling* 56: 22–33.

Pouliot, S., B. Bussière, A. M. Dagenais, G. W. Wilson, and Y. Letourneau. 2018. Evaluation of the use of paste rock as cover material in mine reclamation. In *International Conference on Acid Rock Drainage (ICARD)*, Pretoria, South Africa, ed. C. S. Wolkersdorfer et al., 559–565.

Puhalovich, A. A., and M. Coghill. 2011. Management of mine wastes using pit/underground void backfilling methods: Current issues and approaches. Mine Pit lakes: Closure and Management. Published by Australian Centre for Geomechanics.

Qi, C., and A. Fourie. 2019. Numerical investigation of the stress distribution in backfilled stopes considering creep behaviour of rock mass. *Rock Mechanics and Rock Engineering* 52, no. 9: 3353–3371.

Qi, C., Q. Chen, A. Fourie, J. Zhao, and Q. Zhang. 2018. Pressure drop in pipe flow of cemented paste backfill: Experimental and modeling study. *Powder Technology* 333: 9–18.

Reid, D., and J. Boshoff. 2015. Stability of a proposed steepened beach. In *Proceedings of the 18th International Seminar on Paste and Thickened Tailings*, Cairns, Australia, May 5–7, 2015, ed. R. Jewell and A. Fourie, 181–184.

Rey, N., I. Demers, B. Bussière, M. Mbonimpa, and M. Gagnon. 2020. A geochemical evaluation of a monolayer cover with elevated water table for the reclamation of Doyon-Westwood tailings ponds, Canada. *Environmental Earth Sciences* 79, no. 2: 58.

Rico, M., G. Benito, A. R. Salgueiro, A. Díez-Herrero, and H. G. Pereira. 2008. Reported tailings dam failures: A review of the European incidents in the worldwide context. *Journal of Hazardous Materials* 152, no. 2: 846–852.

Robinsky, E. I. 1975. Thickened discharge – A new approach to tailings disposal. *CIM Bulletin* 68, no. 764: 47–53.

Robinsky, E. I. 1999. *Thickened tailings disposal in the mining industry*. Toronto, ON: E. I. Robinsky Associates Ltd.

Romano, C. G., K. U. Mayer, D. R. Jones, D. A. Ellerbroek, and D. W. Blowes. 2003. Effectiveness of various cover scenarios on the rate of sulfide oxidation of mine tailings. *Journal of Hydrology* 271: 171–187.

Ross, B. 1990. The diversion of capillary barriers. *Water Resources Research* 26, no. 10: 2625–2629.

Rousseau, M., and T. Pabst. 2018. 3D numerical assessment of the permeable envelope concept for in-pit disposal of reactive mine wastes. In *Proceedings of 71st Canadian Geotechnical Conference GeoEdmonton 2018*, Edmonton, Canada.

Schultze, M., K.-H. Pokrandt, E. Scholz, and P. Jolas. 2011. Use of mine water for filling and remediation of pit lakes. In Proceedings of 11th IMWA Congress, Aachen, Germany, September 4–11.

Segura-Salazar, J., F. M. Lima, and L. M. Tavares. 2019. Life cycle assessment in the minerals industry: Current practice, harmonization efforts, and potential improvement through the integration with process simulation. *Journal of Cleaner Production* 232: 174–192.

Shuttleworth, J. A., B. J. Thomson, and J. A. Wates. 2005. Surface paste disposal at Bulyanhulu: Practical lessons learned. In *Proceedings of the 8th International Seminar on Paste and Thickened Tailings*, Santiago, Chile: Australian Centre for Geomechanics, 207–218.

Simms, P. 2017. 2013 Colloquium of the Canadian Geotechnical Society: Geotechnical and geoenvironmental behaviour of high-density tailings. *Canadian Geotechnical Journal* 54, no. 4: 455–468.

Simms, P., M. W. Grabinsky, and G. Zhan. 2005. Laboratory evaluation of evaporative drying from surface deposited thickened tailings at the Bulyanhulu gold mine. In *Proceedings of the 58th Canadian Geotechnical Conference and 6th Joint CGS/IAH-CNC Conference*, Saskatoon.

Simms, P., M. W. Grabinsky, and G. Zhan. 2007. Modelling evaporation of paste tailings from the Bulyanhulu mine. *Canadian Geotechnical Journal* 44, no. 12: 1417–1432.

Sivakugan, N., R. M. Rankine, K. J. Rankine, and K. S. Rankine. 2006. Geotechnical considerations in mine backfilling in Australia. *Journal of Cleaner Production* 14, no. 12: 1168–1175.

Sivakugan, N., R. Veenstra, and N. Naguleswaran. 2015. Underground mine backfilling in Australia using paste fills and hydraulic fills. *International Journal of Geosynthetics and Ground Engineering* 1, no. 2. https://doi.org/10.1007/s40891-015-0020-8.

Skandrani, A., I. Demers, and M. Kongolo. 2019. Desulfurization of aged gold-bearing mine tailings. *Minerals Engineering* 138: 195–203.

Stander, H.-M., S. T. L. Harrison, and J. Broadhurst. 2018. Re-purposing of acid generating fine coal waste: An assessment and analysis of opportunities. In *11th International Conference on Acid Rock Drainage and IMWA Annual Conference*, Johannesburg, South Africa, September 10–14, 2018.

Steffen Robertson and Kirsten (BC) Inc. 1989. *Draft acid rock drainage technical guide*, vol. 1. Vancouver: British Columbia Acid Mine Drainage Task Force.

Stewart, M. 2001. The application of life cycle assessment to mining, minerals and metals. MMSD: Mining, minerals and sustainable development. Report.

Strachan, C., and S. Goodwin. 2015. The role of water management in tailings dam incidents. In *Proceedings of Tailings and Mine Waste*, Vancouver, BC, Canada, October 26–28, 2015. http://hdl.handle.net/2429/59672.

Taha, Y., M. Benzaazoua, R. Hakkou, and M. Mansori. 2017. Coal mine wastes recycling for coal recovery and eco-friendly bricks production. *Minerals Engineering* 107: 123–138.

Tannant, D., and B. Regensburg. 2001. *Guidelines for mine haul road design*.

Tardif-Drolet, M., L. Li, T. Pabst, G. Zagury, R. Memillod-Blondin, and T. Genty. 2020. Revue de la réglementation sur la valorisation des résidus miniers hors site au Québec. *Environmental Reviews* 28, no. 1: 32–44.

Vallejo, L. E. 2001. Interpretation of the limits in shear strength in binary granular mixtures. *Canadian Geotechnical Journal* 38(5): 1097–1104.

Verburg, R., and M. Oliveira. 2016. Surface paste disposal of high sulphide tailings at Neves Corvo – Evaluation of environmental stability and operational experience. In *Proceedings of IMWA2016*, Freiberg, Germany. https://www.imwa.info/docs/imwa_2016/IMWA2016_Verburg_31.pdf.

Vick, S. G. 1990. *Planning, Design and Analysis of Tailings Dams*. Vancouver, BC: BiTech Publishers Ltd.

Villain, L., L. Alakangas, and B. Öhlander. 2013. The effects of backfilling and sealing the waste rock on water quality at the Kimheden open-pit mine, northern Sweden. *Journal of Geochemical Exploration* 134: 99–110.

Villain, L., N. Sundström, N. Perttu, et al. 2015. Evaluation of the effectiveness of backfilling and sealing at an open-pit mine using ground penetrating radar and geoelectrical surveys, Kimheden, northern Sweden. *Environmental Earth Sciences* 73, no. 8: 4495–4509.
Wickland, B. E., and G. W. Wilson. 2005. Self-weight consolidation of mixtures of mine waste rock and tailings. *Canadian Geotechnical Journal* 42, no. 2: 327–339.
Wickland, B. E., G. W. Wilson, D. Wijewickreme, and B. Klein. 2006. Design and evaluation of mixtures of mine waste rock and tailings. *Canadian Geotechnical Journal* 43, no. 9: 928–945.
Wickland, B. E., G. W. Wilson, and D. Wijewickreme. 2010. Hydraulic conductivity and consolidation response of mixtures of mine waste rock and tailings. *Canadian Geotechnical Journal* 47, no. 4: 472–485.
Williams, D. J. 1997. Effectiveness of co-disposing coal washery wastes. In *Tailings and mine waste '97: Proceedings of the Fourth International Conference on Tailings and Mine Waste '97*, Fort Collins, CO, USA, 335–342. Rotterdam: A. A. Balkema.
Wills, B. A., and J. Finch. 2015. *Wills' mineral processing technology*. 8th ed. Elsevier. E-book.
Wilson, G. W., H. D. Plewes, D. J. Williams, and J. Robertson. 2003. Concepts for co-mixing of tailings and waste rock. In *Application of sustainability of technologies: Proceedings of the 6th International Conference on Acid Rock Drainage (ICARD)*, 437–444. Cairns: Australian Institute of Mining and Metallurgy.
Yalcin, T., M. Papadakis, N. Hmidi, and B. Hilscher. 2004. Desulphurization of Placer Dome's Musselwhite mine gold cyanidation tailings. *CIM Bulletin* 97, no. 1084.
Zhang, L. 2013. Production of bricks from waste materials—a review. *Construction and Building Materials* 47: 643–655. https://doi.org/10.1016/j.conbuildmat.2013.05.043.

14 Long-Term Evolution of Reclamation Performance

Bruno Bussière and Marie Guittonny

14.1 INTRODUCTION

Engineered covers are used worldwide to confine reactive wastes in landfills and mine waste sites at closure. In mine waste facilities, where sulfide minerals are exposed to water and atmospheric oxygen, sulfuric acid can be produced. This process can decrease the pH of drainage waters and liberate potentially toxic metals (Blowes et al. 2014), thus resulting in acid mine drainage (AMD; see Chapter 1). Ensuring the chemical stability of wastes to control AMD is one of the main issues related to mine site reclamation, as well as the physical stability of waste storage facilities (not specifically treated in this book). To this end, engineered covers (see Chapters 4 to 9) can be used to control oxygen migration, water infiltration, and/or reactive mine waste temperature to limit acid generation (Aubertin et al. 2015; the fundamental processes underlying water, oxygen, and heat transport in mine covers and wastes are described in Chapter 3).

One of the main challenges associated with the engineering of mine site reclamation is that the performance of reclamation methods must be sustained over hundreds of years. The performance of the engineered cover systems, such as low-saturated hydraulic conductivity covers (LSHCCs) (see Chapter 4), store and release (SR) covers (see Chapter 5), covers with capillary barrier effects (CCBEs) (see Chapter 7), water covers (see Chapter 6), and elevated water table (EWT) with a monolayer cover (see Chapter 8), is controlled by their water budget (WB; see Chapter 3). Insulation covers are less dependent on the WB, even if an increase in precipitation could reduce their long-term performance (because water is a source of heat). The main parameters that control insulation cover performance are associated with climatic parameters and their evolution through time (e.g., temperature, snow accumulation; see Chapter 9).

The present chapter assesses the four main factors of influence on the water budget and heat transfer, and, therefore, on the long-term performance of reclamation methods: climate changes, vegetation, evolution of cover material properties, and loss of durability. The design of reclamation methods should take these four factors into consideration, but their integration is actually limited by interdisciplinary barriers. A methodological approach is proposed at the end of the chapter to better integrate these factors in the design of reclamation methods.

14.2 EFFECTS OF CLIMATE CHANGE

14.2.1 Meteorological Definitions

It is important to begin with a few meteorological definitions. First, weather and climate have different meanings. The difference between weather and climate is a measure of time. Weather refers to the conditions of the atmosphere over a short period of time (e.g., temperature, humidity, precipitation, cloudiness, brightness, visibility, wind, and atmospheric pressure), whereas climate is related to the behavior of the atmosphere over relatively long periods of time. Climate is a long-term pattern of weather.

Different terms are also used to refer to the evolution of climate, such as *climate change* and *global warming*. Climate change (CC) refers to significant changes (compared to reference values) in global climate parameters that occur over several decades or longer. Reference climate values are related to climate normals. The average value of a meteorological parameter over 30 years for a given region in space is usually defined as a climatological normal. Climate normals are usually calculated from the observed data gathered at meteorological stations since the middle of the 19th century. Global warming is more specific. It is related to the warming of Earth's climate observed since the pre-industrial period (between 1850 and 1900) due to anthropic activities. In this chapter, the term *climate change*, and the abbreviation *CC*, will be used.

Another important concept related to CC is extreme conditions, which are defined according to the field of expertise. For climate experts, an extreme event is defined as the occurrence of a value of a weather variable above (or below) a threshold value near the upper (or lower) ends of the range (e.g., <10, 5, or 1% for a given time of the year during a specific reference period) of its observed values in a specific region (Seneviratne et al. 2012). For engineers, the notion of risk enters into the equation. Risk is often defined as the probability of the occurrence of an event multiplied by its consequences. Therefore, a climate event can be considered extreme by a climate expert because of its low probability of occurrence, but is not necessarily considered extreme by an engineer because of its relatively low impacts. As Seneviratne et al. (2012) explain, some climate extremes (e.g., droughts, floods) may also be the result of an accumulation of climate events (i.e., two or more events occurring simultaneously). These compound events can lead to high impacts and are often considered the conditions that infrastructure must be able to resist for centuries. For example, a drought can be more severe if the period preceding the drought was relatively dry, or a flood could be more problematic if the soil conditions are already saturated. In both cases, the soil moisture conditions, combined with the precipitation or evapotranspiration (ET) intensity, play a role.

14.2.2 Climate Modeling

Climate models are a mathematical representation of the climate system that integrates well-documented physical processes. These equations describe interactions between the different stratum of the Earth (atmosphere, lithosphere, hydrosphere, cryosphere, and biosphere) as well as external forcing, natural (solar radiation, volcanism) or anthropic (greenhouse gas) (Charron 2016). As Flato et al. (2013) note, climate models are the primary tools available for investigating the response of the climate system to various radiative forcings (RFs) in order to make climate predictions on seasonal, decadal, and even centurial timescales.

Many research centers around the world work on different climate models that have their own specificities (physical and numerical hypothesis, types of inputs, etc.). Depending on their resolution, two types of model can be distinguished: Global Climate Models (GCMs) and Regional Climate Models (RCMs).

- **GCMs:** GCMs depict the climate using a three-dimensional grid over the globe, and typically have a horizontal resolution of 200 km, with tens of vertical layers in the atmosphere and in the oceans. Their resolution is thus quite coarse relative to the scale needed in many applied situations. In addition, there are many other sources of uncertainties related to GCMs, including the small scale of the physical processes, the simulation of various feedback mechanisms in models, and so on. Hence, for similar initial conditions, the different GCMs may simulate different responses.
- **RCMs:** RCMs predict the evolution of climate on a regional scale (e.g., North America). Due to their finer horizontal resolution, usually at 45 km or less, RCMs allow practical planning of local issues such as hydroelectricity water resources or flood occurrence. Results from GCMs are integrated to the boundary of RCMs using reanalysis. Reanalysis is climate variable estimates obtained by treating past meteorological data.

Greenhouse gases (GHG) concentration in the atmosphere is a key parameter to simulate the evolution of climate. Real GHG measurements from different meteorological stations are available for the past climate. However, the future GHG concentrations are unknown and emission scenarios must be produced to determine the evolution of future climate. In 2013, climate scientists agreed upon a new set of scenarios that focused on the level of GHG in the atmosphere in 2100; these scenarios are known as representative concentration pathways (RCPs; Moss et al. 2010). Each RCP indicates the amount of climate RF expressed in Watts per square meter (W m^{-2}) until 2100 and after relative to preindustrial (Moss et al. 2008). These scenarios are based on different hypotheses related to socioeconomic development, population growth, and technological changes (IPCC 2007). Four RCPs were selected from the published literature (see Figure 14.1) and are used in the Fifth International Panel on Climate Change (IPCC) Assessment as a basis for the CC predictions and projections (Collins et al. 2013).

- **RCP2.6** (the lowest of the four): One pathway where RF peaks at approximately 3 W m^{-2} before 2100 and then declines to 2.6 W m^{-2} in 2100;
- **RCP4.5** (medium-low) and **RCP6.0** (medium-high): Two intermediate stabilization pathways in which RF is stabilized at approximately 4.5 W m^{-2} and 6.0 W m^{-2} after 2100;
- **RCP8.5:** One high pathway for which RF reaches greater than 8.5 W m^{-2} by 2100 on a rising trajectory.

14.2.3 Impact of CC on Meteorological Parameters

GCMs allow the prediction of climate evolution for the entire Earth, and results are compiled in IPCC reports. In the context of this book, GCMs and RCMs predictions can be used to identify future climate conditions that will apply to mine site reclamation scenarios. To illustrate the type

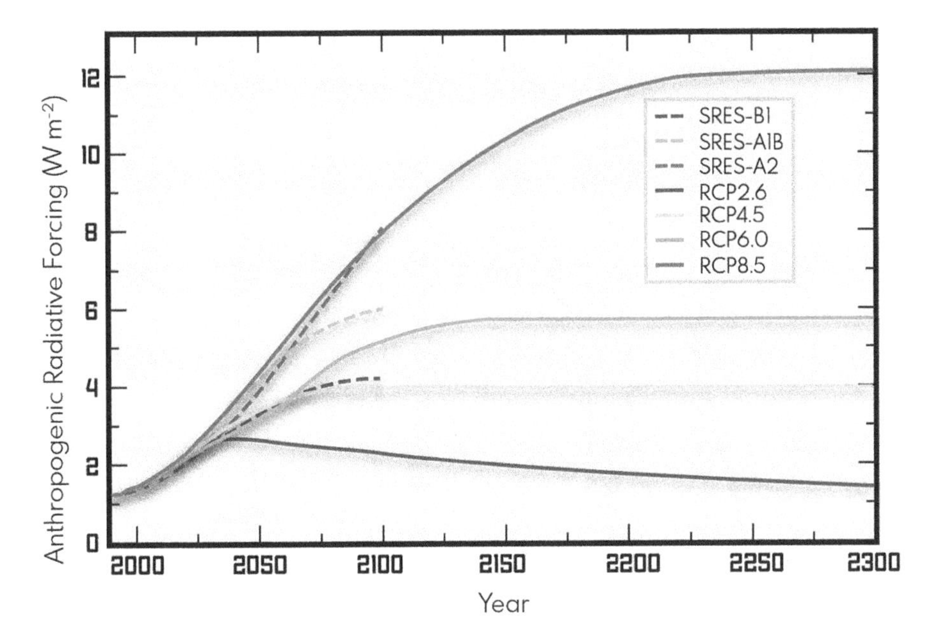

FIGURE 14.1 Time evolution of the total anthropogenic radiative forcing (RF) relative to pre-industrial (about 1765) between 2000 and 2300 for RCP scenarios and their extensions (*continuous lines*), and Special Report on Emissions Scenarios (SRES) (*dashed lines*) as computed by the Integrated Assessment Models (IAMs) used to develop those scenarios. The four RCP scenarios used in CMIP5 are as follows: RCP2.6 (*dark blue*), RCP4.5 (*light blue*), RCP6.0 (*orange*) and RCP8.5 (*red*). The three SRES scenarios used in CMIP3 are B1 (*blue, dashed*), A1B (*green, dashed*) and A2 (*red, dashed*). (Modified from Collins et al. 2013).

of information that can be extracted from GCMs and RCMs in the context of mine site reclamation, this chapter takes the province of Quebec, Canada, as a case study. Results are extracted from Bussière et al.'s (2017) report on the vulnerability of Quebec's mining sector to CC.

Roy (2016) performed simulations for two RCP scenarios (4.5 and 8.5) and for six different mining regions of Quebec (see Figure 14.2). The main objective of this study was to predict the evolution of climate parameters that can have an impact on mining development. The main parameters calculated from the simulations are mean winter temperature, mean summer temperature, minimum annual temperature, maximum annual temperature, annual precipitation, number of days with precipitation greater than 10 mm, extreme precipitation (99 percentile), accumulation during extreme precipitation, number of days without freezing, period of thawing, and maximum snow accumulation. For each RCP scenarios, 43 simulations (using 23 different models) from the Coupled Models Intercomparison Phase 5 (CMIP5) project were used. This approach allows researchers to cover a large range of possibilities and to obtain average values with realistic uncertainties for this type of exercise. Results were corrected using reference values from Natural Resources Canada (NRCAN) (Hutchinson et al. 2009). For the period of 1981–2010, differences between measured and predicted climate parameter values, called the bias, were calculated. Then, the bias for each parameter was subtracted from the predicted values for simulations (>2010). The main highlights from these simulations are as follows (Bussière et al. 2017):

- For all regions, an increase in annual mean temperature is projected, with a greater increase during winter season (see Figure 14.3).
- There is a significant decrease of the number of days of freezing per year. The freezing season will become significantly shorter, particularly in the northern regions (see Figure 14.4).
- Everywhere in Quebec, climate models predict an increase in precipitation (see Figure 14.5), with a greater increase during winter and spring seasons, compared to summer and fall.

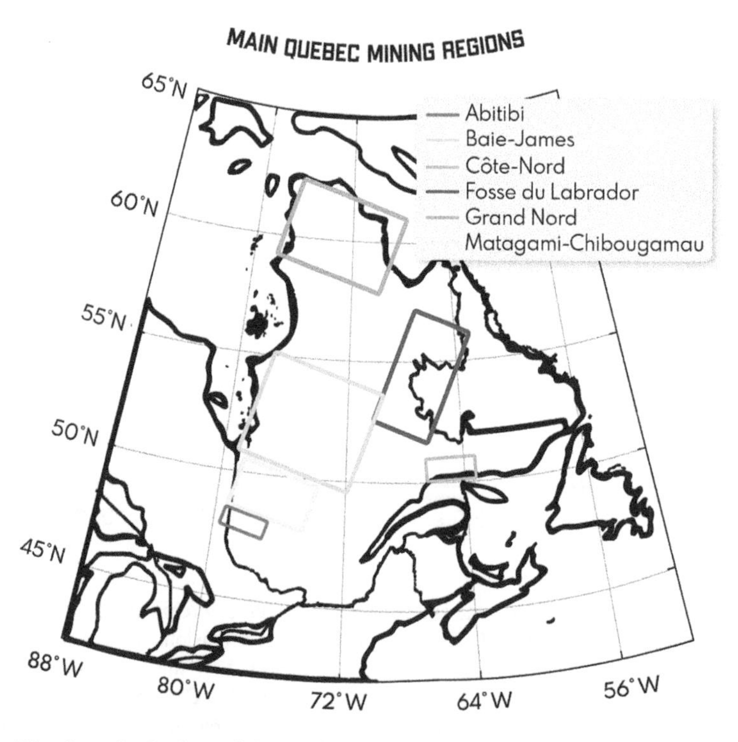

FIGURE 14.2 The six main Quebec mining regions studied by Roy (2016).

Abitibi			
Average 1981–2010	2020	2050	2080
-13.52	-11.9	-9.38	-6.84

Matagami-Chibougamau			
Average 1981–2010	2020	2050	2080
-15.22	-13.5	-10.7	-7.6

Baie-James			
Average 1981–2010	2020	2050	2080
-18.46	-16.4	-13.3	-9.81

Côte-Nord			
Average 1981–2010	2020	2050	2080
-13.47	-11.8	-9.28	-6.41

Fosse du Labrador			
Average 1981–2010	2020	2050	2080
-19.75	-17.8	-14.9	-12.0

Grand-Nord			
Average 1981–2010	2020	2050	2080
-21.88	-19.1	-14.8	-9.82

FIGURE 14.3 Average predicted winter (°C) (between December and February) temperature (RCP8.5) compared to average values measured between 1981 and 2010 for all mining regions studied.

Abitibi			
Average 1981–2010	2020	2050	2080
166	177	194	216

Matagami-Chibougamau			
Average 1981–2010	2020	2050	2080
172	184	201	223

Baie-James			
Average 1981–2010	2020	2050	2080
150	162	179	201

Côte-Nord			
Average 1981–2010	2020	2050	2080
170	181	198	219

Fosse du Labrador			
Average 1981–2010	2020	2050	2080
129	139	155	176

Grand-Nord			
Average 1981–2010	2020	2050	2080
114	126	143	167

FIGURE 14.4 Predicted number of days without freezing conditions (daily minimal temperature > 0°C) compared to average values measured between 1981 and 2010, for all mining regions studied (RCP8.5).

Abitibi			
Average 1981–2010	2020	2050	2080
853	889	940	990

Matagami-Chibougamau			
Average 1981–2010	2020	2050	2080
889	937	994	1049

Baie-James			
Average 1981–2010	2020	2050	2080
835	877	946	1014

Côte-Nord			
Average 1981–2010	2020	2050	2080
1139	1192	1262	1313

Fosse du Labrador			
Average 1981–2010	2020	2050	2080
922	974	1043	1112

Grand-Nord			
Average 1981–2010	2020	2050	2080
471	505	558	621

FIGURE 14.5 Predicted mean annual precipitation (liquid and solid) in millimeter compared to average values measured between 1981 and 2010, for all mining regions studied (RCP8.5).

- A significant increase of extreme precipitations (in mm) is expected for all Quebec regions, with a greater increase, in proportion, for the northern regions. However, climate models predict longer periods without precipitation during the summer season.
- The period of return for extreme precipitation events, which corresponds to an estimated average time between events of a similar size or intensity, will be shortened significantly for all Quebec regions. For example, an event that currently has a return period of 20 years (estimated with data obtained during the period from 1986 to 2005) will have a period of return of 10–13 years for the 2046–2065 horizon.

These generic results on the evolution of meteorological parameters for specific regions are important to assess, in a preliminary manner, the kind of impact CC could have on the different components of the mine life cycle. In the following section, links between CC and the risks associated with a performance loss of reclamation methods are presented.

14.2.4 Main CC Risks Associated with Long-Term Performance of Reclamation Methods

The main reclamation methods to control AMD from mine wastes have been presented in Chapters 4 to 9. As presented in those chapters, each method has its own advantages and limits that can be amplified by CC. This section outlines the main results of a risk assessment concerning the influence of CC on the long-term performance of AMD reclamation methods (Bussière et al. 2017); the SR cover is not considered because the Quebec climate is not suitable for that type of cover. Although it is specific to a certain geographical location (Quebec, Canada), the suggested approach can be applied to any other specific region where CC projections similar to those presented in Section 14.2.3 are available.

In engineering, the level of risk of an event is defined as a combination of the probability and the consequences of the event; the consequences are related to the potential impact on the environment, the health and safety of people, and the costs associated with the rehabilitation of the infrastructure. In a first step, different experts in the field of mine site reclamation identified the main risks that could threaten the performance of a reclamation method. In a second step, questionnaires were sent to professionals in the mining industry to qualify the risk level associated with each risk identified in the first step; the detailed results are presented in Bussière et al. (2017). The interpretation of risk level was performed using the risk matrix presented in Figure 14.6. The combination of consequences and probability allows professionals to characterize the risk: green, yellow, orange, and red cells correspond to low, moderate, high, and very high risks, respectively.

The main risks identified for the long-term performance of the different reclamation methods are presented in Table 14.1. Some of these risks are common for most of the methods. For example, the physical instability (slope instability and erosion) of inclined portions of reclaimed mine sites and/or dikes due to the increase of mean annual precipitation and extreme precipitation events magnitude and occurrence could impact the long-term performance of LSHCCs, CCBEs, EWT technique with monolayer cover, and water covers. The physical stability problem here is related to chemical stability because once the infrastructure are physically destabilized, the interactions between sulfide minerals, water, and oxygen can start again and lead to AMD generation. Some other risks are specific to one or two methods. For example, the risks associated with permafrost degradation and an increase of the thickness of the active zone are specific to insulation covers, and the loss of performance due to longer droughts during the summer can affect mainly two types of oxygen barriers: CCBEs and the EWT technique.

As presented in Table 14.1, risk level categorized as high was identified for three reclamation methods: the EWT, water covers, and insulation covers. For the EWT, the approach is sensitive to

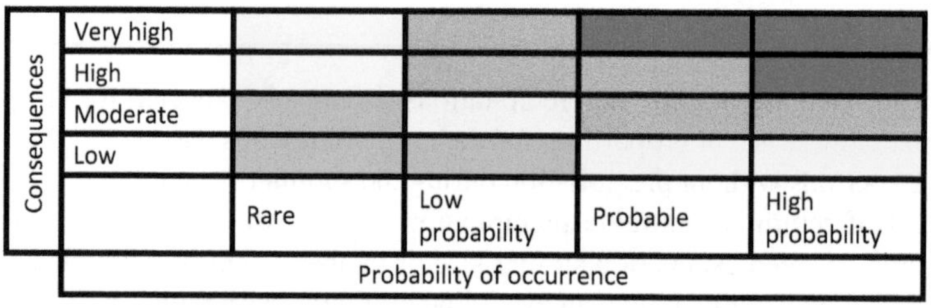

Consequences				
Very high				
High				
Moderate				
Low				
	Rare	Low probability	Probable	High probability
	Probability of occurrence			

FIGURE 14.6 Risk matrix used to assess the influence of CC on the performance of reclamation methods: green = low risk, yellow = moderate risk, orange = high risk, red = very high risk.

TABLE 14.1
Principal Risks Associated to CC for the Main Mine Site Reclamation Methods. (Modified from Bussière et al. 2017)

Risks	Risk Level
Low saturated hydraulic conductivity cover (LSHCC)	
1. Performance loss due to the increase of mean annual precipitation and extreme precipitation event and occurrence	Moderate
2. Physical instability (slope instability and erosion) of inclined portion due to the increase of mean annual precipitation and extreme precipitation event and occurrence	Moderate
Covers with capillary barrier effects (CCBE)	
1. Performance loss due to the reduction of the degree of saturation of the moisture-retaining layer (MRL) during the longer dry periods	Moderate
2. Physical instability (slope instability and erosion) of inclined portion due to the increase of mean annual precipitation and extreme precipitation event and occurrence	Moderate
Elevated water table (EWT)	
1. Greater variations of the water table position due to the change in precipitation distribution and quantity, and to longer dry periods during the summer	Moderate–High
2. Greater chances of having the AMD tailings at low degree of saturation during the dry periods	Moderate–High
3. Physical instability (slope instability and erosion) of inclined portion due to the increase of mean annual precipitation and extreme precipitation event and occurrence	Moderate
Water cover	
1. Decrease of the water cover thickness due to longer dry periods during the summer	Moderate
2. Challenges related to the physical instability of the dikes used to create the artifical reservoir due to the increase of water volume to manage	High
Insulation covers	
1. Increase of the active zone that will expose AMD tailings to above 0°C conditions	High
2. Permafrost instability that could affect retaining infrastructures	High
3. Greater infiltration due to the increase of precipitation (water is a source of heat)	Moderate

prolonged droughts that could happen more frequently due to CC, particularly when the monolayer cover is made of fine-grained material (Lieber 2019). The increase of the intensity of extreme precipitation events puts the water management structure of water covers at risk. Finally, the important increase in temperature in the northern part of Quebec will significantly impact the capacity of insulation covers to control the oxidation of the underlying AMD tailings.

Other sources of risk were not analyzed in Bussière et al. (2017). One of them is the influence of CC on the evolution of vegetation over cover systems. Indeed, if the natural evolution of vegetation (from herbaceous-shrub species to trees) on covers is modified by CC, this could affect the long-term performance of the cover systems. Hence, the effects of vegetation on covers, with or without CC, are an important issue in the long term and are discussed in more detail in the following section.

14.3 EFFECTS OF VEGETATION

To maintain long-term performance, engineered covers must be considered dynamic systems that include interactions with vegetation (Piet et al. 2005). Mine reclamation covers may include a designed vegetation component to meet end land-use objectives or may be naturally colonized by plants (see Chapter 12) even when the environment is inhospitable (McLendon et al. 1997). As soon as they are installed, individual growing plants will locally change the WB of cover systems. At the site level, evaluating the global effect of a plant community on the WB remains difficult due to the presence of multiple species and variations in plant ages and colonization intensity (Proteau et al. 2020a). Moreover, with time and under CC, plant communities will evolve, as will their effects on the WB. Finally, plant development and root colonization may be specific to the mine materials used to construct the covers, especially when mine wastes are recycled. The following sections explore these considerations further.

14.3.1 Effect of Vegetation on the Water Budget of Cover Systems

Vegetation can affect the WB of cover systems on which they grow (Albright et al. 2015). Plants pump water for their transpiration. Transpiration accounts for the water vapor lost by plants at the leaf level that is passively pumped by roots in the soil (when soil suction values are not limiting) and transported through the stems (Lambers et al. 2008); this process can change the water storage of materials. Water is pumped by fine roots (diameter <2 mm) (Lambers et al. 2008), and the fine root length by soil volume is useful to estimate water consumption (Zhang et al. 2009). In the 17-year-old CCBE at the Lorraine site in western Quebec, root colonization occurred in the first 10 cm of the moisture-retaining layer (MRL) (Guittonny et al. 2018; Proteau et al. 2020a and 2020b), and the degree of saturation (S_r) at this depth was negatively and linearly correlated with the root length in the soil volume (i.e., root length density (RLD) in cm cm^{-3}).

More generally, field studies dealing with the WB of covers, with or without vegetation, have been conducted in arid, semi-arid, and temperate climates for covers above landfills and mine wastes (Woyshner and Yanful 1995; Wilson et al. 1997; Benson et al. 2002; Madalinski et al. 2003; Yanful et al. 2003; Albright et al. 2004; Waugh 2004; Breshears et al. 2005; Stoltz and Greger 2006; Naeth et al. 2011; Benson and Bareither 2012; Apiwantragoon et al. 2014; Arnold et al. 2014; Fraser 2014; Gwenzi et al. 2014). These studies mainly analyzed evaporation or potential evapotranspiration (ET) for covers that aim to control water infiltration. In general, plant presence on covers increased ET (Wels et al. 2002; Madalinski et al. 2003) and reduced the volume of stored water that could percolate into SR covers (Shurniak and Barbour 2002; Madalinski et al. 2003; Zornberg et al. 2003; Scanlon et al. 2005; Williams et al. 2006; Barnswell and Dwyer 2011). Consequently, vegetation's contribution to the WB of low-infiltration covers is expected to improve their performance.

A case study was performed in the humid climate (around 900 mm of rainfall a year) of northwestern Quebec, where in situ lysimeter cells (see Figure 14.7) allowed for comparison between the WB of vegetated versus unvegetated mine wastes (Chevé et al. 2018). Short-term results illustrate global ET increase due to vegetation presence. Indeed, actual ET increased from 51% to 81% of total precipitation during the second growing season on a soil layer above waste rocks planted with willows compared to soil layer without vegetation (Guittonny et al. 2019). This increase in ET represented 110 mm of rain for 90 days (Guittonny et al. 2019).

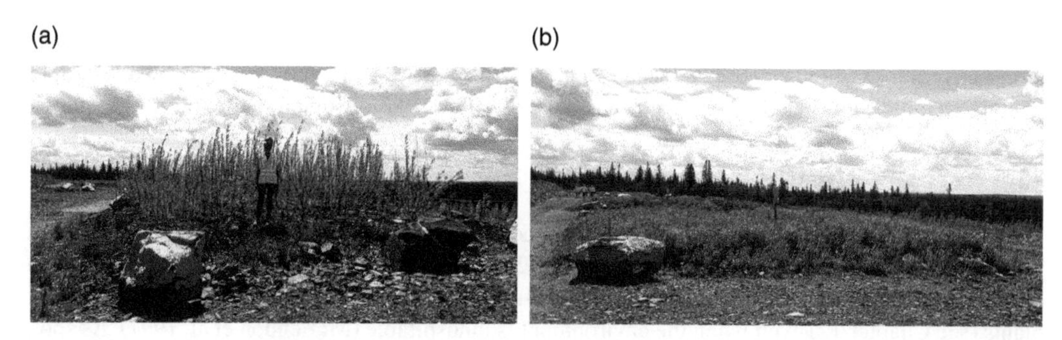

FIGURE 14.7 Photographs of two lysimeter cells constructed in situ to compare the water budget of willow planted (a) versus unplanted (b) waste rocks covered with soil layers.

However, studies have also reported that the performance of low infiltration covers decreased over time due to biological and physical processes, including root colonization (Fourie and Tibbett 2007; Traynham 2010; Traynham et al. 2012; DeJong et al. 2015), that change material properties (see Section 14.4.4). Moreover, presently, few literature data are available on the effect of vegetation on the WB of oxygen barrier covers used in humid climates. Yet vegetation is perceived as a threat on these types of covers because root colonization could desaturate the materials that need to remain close to water saturation to decrease the oxygen fluxes (see Chapters 3, 7, and 8). Research is thus needed in this particular context.

Numerical unsaturated water flow models that are used to design reclamation covers can take into account water lost by vegetation through ET and the associated decrease in the water storage in cover materials. Vegetation parameters needed as inputs are mainly the duration of the growing season, the leaf area index (LAI; i.e., the total one-sided leaf area covering a given soil surface and divided by this soil surface), the maximum rooting depth, and the distribution of RLD with depth (Botula et al. 2019). Vegetation can also reduce the quantity, intensity, and speed of runoff (Bruce and Clark 1969; Gutierrez and Hernandez 1996; Le Bissonnais et al. 2004; Zuazo and Pleguezuelo 2008; Garcia-Estringana et al. 2010). Finally, it intercepts precipitation (i.e., the parts of vegetation that are aboveground stop precipitation and prevent it to reach the soil) on engineered mine covers (Ayres and O'Kane 2013). However, the main numerical models used to design mine reclamation covers do not take into account runoff and interception effects of vegetation on the WB. The anticipated effects of vegetation on the different components of the WB and how it is taken into account in models are conceptually synthetized in Figure 14.8.

14.3.2 Taking into Account the Evolution of Vegetation in Space and Time

The influence of vegetation on the WB of cover systems is complex (Shurniak 2003), especially in multispecies environments, and it changes over time (Fourie and Tibbett 2007; Mine Environment Neutral Drainage (MEND) Program 2014; DeJong et al. 2015). The ecological process of plant succession describes the colonization of mineral substrates by plants over time (see Chapter 12). It implies successive plant communities that evolve over centuries from small mosses and lichens to large plants like forest trees. Sometimes, the surrounding forest community quickly recolonizes the site. For example, on the reclaimed Lorraine site in western Quebec, more than 10 species of temperate forest trees were present on the sandy protection layer of the CCBE in the first years after construction (Smirnova et al. 2011; Proteau et al. 2020a). However, their size (and covered surface area) was still small 16 years after construction (Guittonny-Larchevêque et al. 2016a; Proteau et al. 2020a; see Figure 14.9). Natural colonization from the surrounding ecosystem can also occur when sites are revegetated with selected species, making the plant community evolve. For example, native willows appeared in the herbaceous, agronomic plant community established by seeding 10 years ago on the desulfurized tailings monolayer at the Manitou site in northwestern Quebec (see Figure 14.10).

It is often difficult to predict the trajectory and the pace of the plant community evolution on disturbed sites (Palmer et al. 2006), such as reclaimed mine sites. Indeed, even if deterministic processes underlying plant succession are well understood, plant community evolution is also controlled by stochastic events such as disturbances or introduction of organisms from the surrounding ecosystem (Cooke and Johnson 2002). Moreover, plant succession relies on the preferential colonization of favorable microsites (Walker and del Moral 2009). This preferential colonization induces plant presence and age heterogeneities on a site. For example, on the Lorraine reclaimed site, tree age varied between 1 year for new seedlings to 11–16 years for the four dominant species, and total plant cover varied between 26% and 81% (depending on the location of measurements) 16 years after the construction of the CCBE (Proteau et al. 2020a).

Different plant species of different ages on engineered covers imply varying above- and belowground growth rates and water use (Lamoureux et al. 2016). Grasses are preferred to reduce erosion

a)

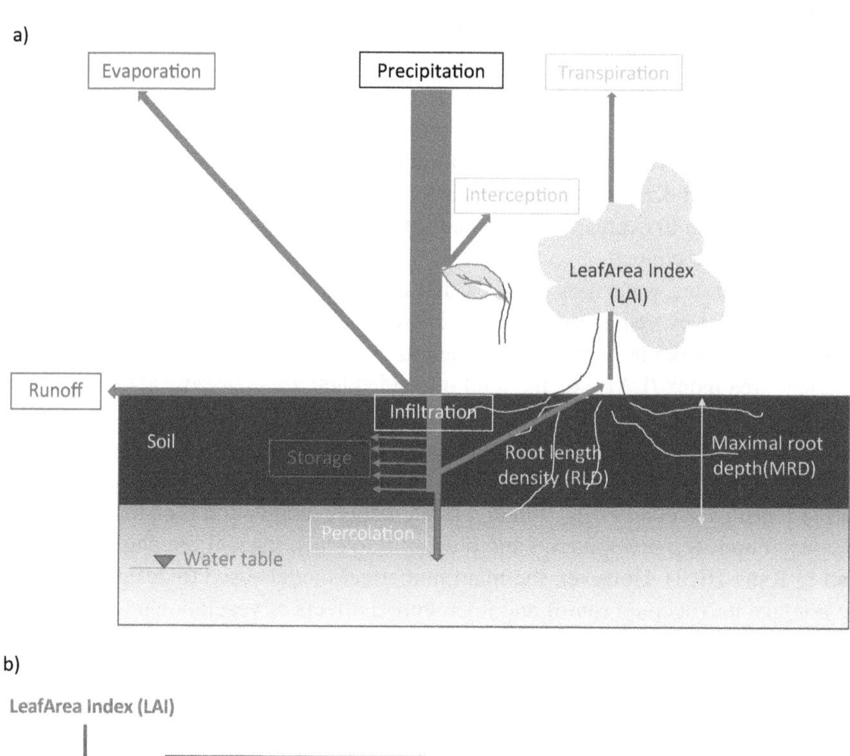

b)

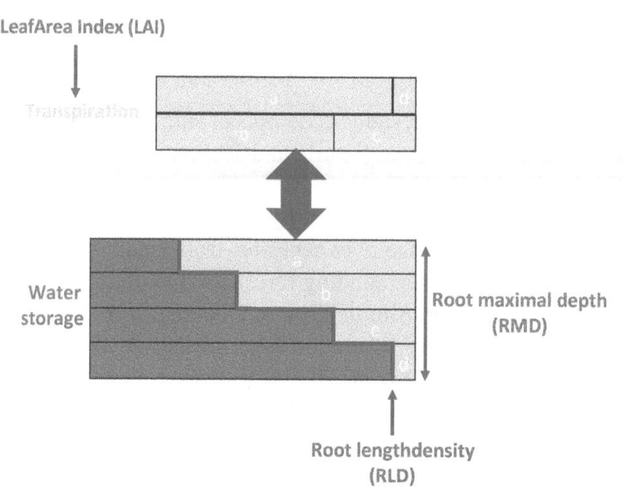

FIGURE 14.8 (a) Vegetation effects on the different components of the water budget: components in blue (Transpiration, Interception) are increased while components in red (Storage, Runoff, Evaporation) are decreased by vegetation. Components in white (Infiltration, Percolation) can be influenced in both ways. (b) Conceptual model of the effect of vegetation in numerical models of unsaturated water flow: Leaf area index (LAI) is used to calculate transpiration as a sink term. The water in this sink term is subtracted from the water storage component of the soil down to maximal root depth (MRD), being parted in the soil according to the root colonization intensity (or root length density (RLD)) by depth (*right*).

and increase the interception of precipitation, while woody plants (trees and shrubs) maximize ET. Trees especially may affect the WB of covers since they pump significant amounts of water due to their large size and can have deep, coarse rooting compared to other plants (Jackson et al. 1996; Schenk and Jackson 2002; Fourie and Tibbett 2007; Freschet et al. 2017). In particular, broadleaved trees may increase ET more than herbs or conifers (550 mm yr^{-1} compared to 200–300 mm yr^{-1},

FIGURE 14.9 Photograph (2015) of the reclaimed Lorraine site in western Quebec that shows the colonization of the sandy protection layer of the cover with capillary barrier effect by diverse woody species from the surrounding mixed forest 16 years after construction.

respectively) (Liu et al. 2003). Broadleaf trees also generally have deeper roots than coniferous trees (Picea < Pinus < Populus < Quercus) (Fan et al. 2017).

Codes used to calculate WB and to design mine covers were mainly developed for agricultural contexts, where plant communities are monocultures of homogeneous age and controlled density. New approaches are needed to take into account the diversity in species, age, and colonization intensity of plant communities on covers located in natural environments, rather than assume direct applicability from the simplified context of agriculture (MEND Program 2014). Natural analogs of cover soils (Albright et al. 2010) located in the vicinity of the mine site can help identifying and characterizing a probable plant community that may develop on the future cover.

Furthermore, CC has the potential to change the distribution of plant species as well as the structures and functions of ecosystems (McKenney et al. 2007; Price et al. 2013; Boulanger et al. 2017). For example, in Canada, a northward shift of the distribution of tree species from temperate to boreal regions is expected before 2100, especially for conifers (see Botula et al. 2019 and references therein). Thus, vegetation and CC effects on the WB of covers should be considered concomitantly. However, uncertainties remain regarding the ability of plant species to spread fast enough to keep up with the shifts in climatic zones (Loarie et al. 2009). For the Lorraine case study, it is expected that there will be an increase of deciduous tree species and a decline of coniferous tree species

FIGURE 14.10 Photograph (2019) of the reclaimed Manitou site in northwestern Quebec that shows the recolonization of the desulfurized tailings monolayer cover with elevated water table by native willows. Willows appeared in the herbaceous, agronomic plant community that was established by seeding 10 years before and that is surrounded by natural boreal forest.

(Botula et al. 2019). Broadleaf trees are expected to transpire more water pumped from greater depths (Fourie and Tibbett 2007). The range of LAI values should evolve from 0.4–9.9 to 0.4–8.1 and from 0.1–3.0 to 0.1–3.5 m for maximal rooting depth values from 2015 to 2100 (Botula et al. 2019). This change in the dominant tree species with CC is expected to have impacts on the CCBE WB and performance (Botula et al. 2019).

Finally, when clean mine wastes are reused in engineered covers (see Chapter 13), the values of vegetation input parameters may be lowered by the specific physical, chemical, and biological constraints to plant development associated with mine wastes (see Chapter 12). For example, deep root development can be delayed in recycled mined materials (Guittonny-Larchevêque and Lortie 2017), and root biomass and RLD decreased (Guittonny-Larchevêque et al. 2016b; Guittonny et al. 2018). Thus, reference values of input vegetation parameters are needed in the specific context of mine reclamation covers.

14.4 EVOLUTION OF MATERIAL PROPERTIES

14.4.1 Reactivity of Cover Layers (Sulfides, Roots, Soil Organisms)

As mentioned in the previous chapters, the use of mine wastes in cover configuration is an attractive approach to valorize their use, reduce the footprint of the reclamation, and potentially reduce covers' construction costs (see especially Chapter 13). The use of low sulfide tailings can improve the short-term performance of oxygen barriers such as CCBEs or monolayer covers combined with EWT by consuming a fraction of the diffusing oxygen (Bussière et al. 2004; Ethier et al. 2018; Rey et al. 2020). However, such consuming reaction will disappear with time as the sulfide minerals will slowly deplete. The depletion time is mainly related to the sulfide reactivity, which is a function of type of sulfide, size and surface characteristics of grains, and the effective diffusion coefficient (Mayer 1999). At the design stage, it is recommended to calculate a long-term oxygen flux worst-case scenario where oxygen consumption by the low sulfide tailings is ignored.

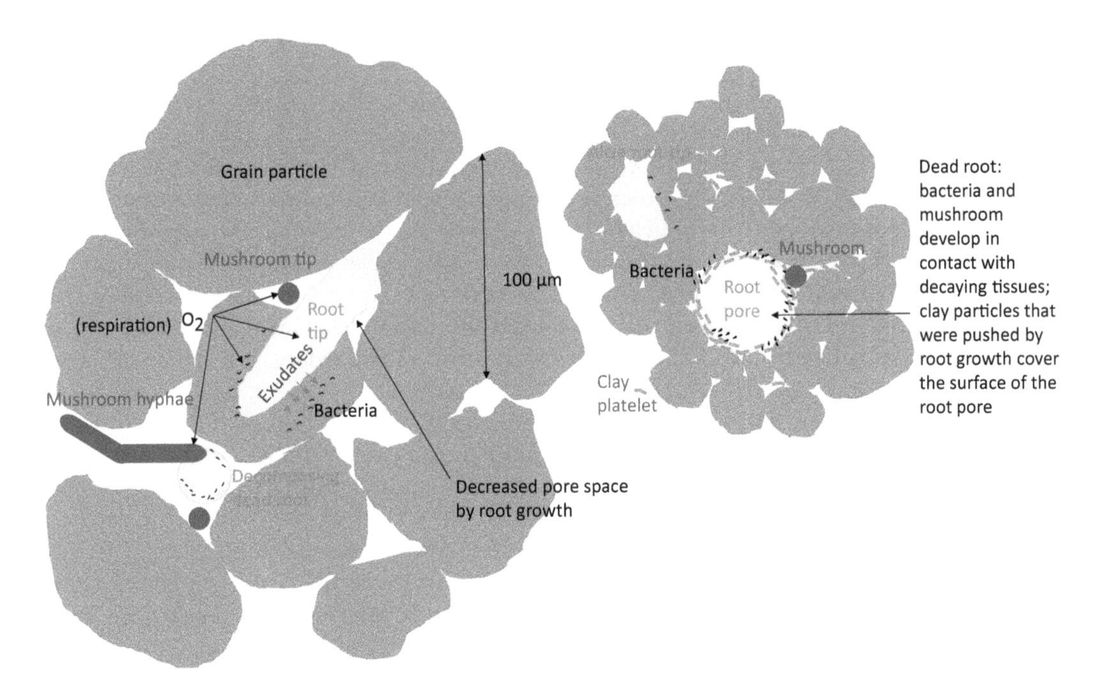

FIGURE 14.11 Conceptual representation of root presence in the soil that influences porosity, the community of organisms, and oxygen consumption for coarse-grained soils (*sand*, *left*) and fine-grained soils (*right*). Roots attract microorganisms (bacteria and mushrooms), which feed on root exudates (i.e., C-rich molecules excreted by roots) or upon decaying roots. Roots and microorganisms consume O_2 for their cellular respiration. In coarse-grained soil, roots occupy the pore space and decrease the soil void ratio, while in fine-grained soil, roots push the soil particles when they grow and reorganize clay platelets along the root axis. Root decay creates new tubular pores, which are stabilized by the remaining links between soil organisms, their excretions, and clay platelets.

The colonization of cover materials by living organisms implies oxygen consumption by these organisms through cellular respiration (see Figure 14.11). Oxygen consumption is distributed between autotrophic respiration from roots and heterotrophic respiration from microorganisms that break down organic matter (OM) (Bond-Lamberty et al. 2004; Olsson et al. 2005; Chen et al. 2017). This consumption can be considered as a biological reactivity and could contribute to increase the performance of oxygen barrier covers. In forested environments, root contribution to soil oxygen consumption is considered greater than that of heterotrophic respiration (Hanson et al. 2000). It is recognized that root respiration is reduced, but not nil, when the soil is nearly saturated (Cook and Knight 2003), such as the conditions in the MRL of a CCBE (Davidson et al. 1998).

At the Lorraine mine site, root contribution to oxygen consumption was quantified in the MRL of the 17-year-old CCBE by in situ oxygen consumption tests interpreted using numerical modeling (Proteau et al. 2020b). Oxygen consumption by roots was quantified with a reactivity coefficient (K_r) corresponding to an equivalent pyrite reactivity. RLD and K_r appeared positively and linearly related ($R^2 = 0.7$). Oxygen fluxes were decreased by 0.5–76 g m^{-2} yr^{-1} thanks to root consumption, and estimated K_r values (1.2E-8–1.0E-5 s^{-1}) were in the same range as those of desulfurized tailings (0.2–1% pyrite) used in oxygen barrier covers (Demers et al. 2009; Ethier et al. 2018). Compared to desulfurized tailings, the reactivity of roots should not exhaust with time, but could vary with the soil temperature (Rustad et al. 2000), seasonal variability of plant metabolism and phenology (Davidson et al. 1998), plant species (Reich et al. 1998; Atkin et al. 2000), as well as soil depth and root diameter (Pregitzer et al. 1998).

14.4.2 Freeze-Thaw Cycles

Results presented in Chapters 4, 5, 7, 8, and 9 showed that engineered covers can be efficient to control AMD from mine wastes. However, the long-term performance of cover systems can be affected by the evolution of cover material properties, particularly those of the fine-grained layers. One phenomenon that could affect the properties of geological materials in covers is freeze-thaw cycles. Freeze-thaw cycles can modify the structure of fine-grained plastic soils and consequently modify their properties. This section explores the impact of freeze-thaw cycles on properties of fine-grained materials once thawed. It is worth mentioning that the performance of covers to control AMD is not usually considered critical during the frozen state due to the low saturated hydraulic conductivity of frozen materials (Fredlund et al. 2012) and the low temperature that reduces the sulfide oxidation rate (see Chapter 1).

During the cold season, heat transfer between the air and the cover reduces the soil temperature to values below the freezing point. The water contained in the cover layers close to the surface will then freeze (totally or partially) and the freezing front will move downward, freezing more pore water (Phukan 1985; Benson and Othman 1993; Konrad 2001). The formation of ice in fine-grained soil pores is a complex process that involves interrelationships between the phase change of water to ice (that includes the effect of salinity), transport of water to the freezing front, and unsteady heat flow in the freezing soil (Harlan 1973; Mitchell and Soga 2005). During the process, the freezing front generates a suction that attracts water molecules from the unfrozen zone to the freezing front (Othman et al. 1994). This suction (as high as 500–3,000 kPa; Chamberlain 1981; Watanabe et al. 2011) can modify the soil structure and properties after thawing due to consolidation and the formation of cracks after ice lens melting (Nixon and Ladanyi 1978; Benson and Othman 1993). Depending on the material properties (physical and mineralogical characteristics) and in situ conditions (compaction level, earth pressure), cracks generated during the freezing process can self-heal or not. In general, it is considered that low plasticity soils (with plasticity index PI < 10–20; Othman et al. 1994; Eigenbrod 2003) and clayey soils composed mainly of swelling clay (e.g., montmorillonite) can self-heal and are less affected by freeze-thaw cycles (Eigenbrod 2003). However, plastic soils with $20 < \mathrm{PI} < 100$ can be destructured by consolidation and crack formation, and subjected to a significant permanent increase in saturated hydraulic conductivity (k_{sat}) (Konrad and Samson 2000; Eigenbrod 2003). The shape of the water retention curve (WRC) and the effective oxygen diffusion coefficient (D_e) can also change with freeze-thaw cycles. These properties (k_{sat}, D_e, and WRC) are critical for maintaining the performance of covers to control AMD generation (see Chapters 4 to 9).

Many studies can be found in the literature concerning the influence of freeze-thaw cycles on k_{sat} of fine-grained materials (with low or high plasticity), typical of those used in covers to control fluid migration. Usually, the influence of freezing and thawing arises in the first three to five cycles and then stabilizes, and the increase in k_{sat} ranges from one to three orders of magnitude (Othman et al. 1994). The main parameters that affect the magnitude of the impact of freeze-thaw cycles on k_{sat} are (Othman et al. 1994) soil type (mineralogical and geotechnical properties) and density, the supply of water, the rate of freezing, the number of cycles, and the pressure applied to the sample. It is assumed that large pores and cracks formed during the freeze-thaw cycles increase the pore size and, consequently, the k_{sat} values. Dagenais (2005) measured the WRC of clays before and after 20 freeze-thaw cycles (see Figure 14.12a). Results showed that after 20 freeze-thaw cycles, the shape of the WRC changed from a standard S shape to a bimodal shape that was best fitted with a bi-modal van Genuchten model (Burger and Shackelford 2001). Pore size distribution was quantified by mercury intrusion porosimetry (MIP; Simms and Yanful 2004) and confirmed a significant evolution of the pore size distribution with the appearance of a family of pores size between 10 and 200 microns (see Figure 14.12b).

Bentonite-based materials (geosynthetic clay liner GCL, sand-bentonite, or crushed rock-bentonite mixtures) can also be used as fine-grained materials in covers. Studies on GCL hydrated and permeated with low concentrations of cations showed that the material can self-heal and keep

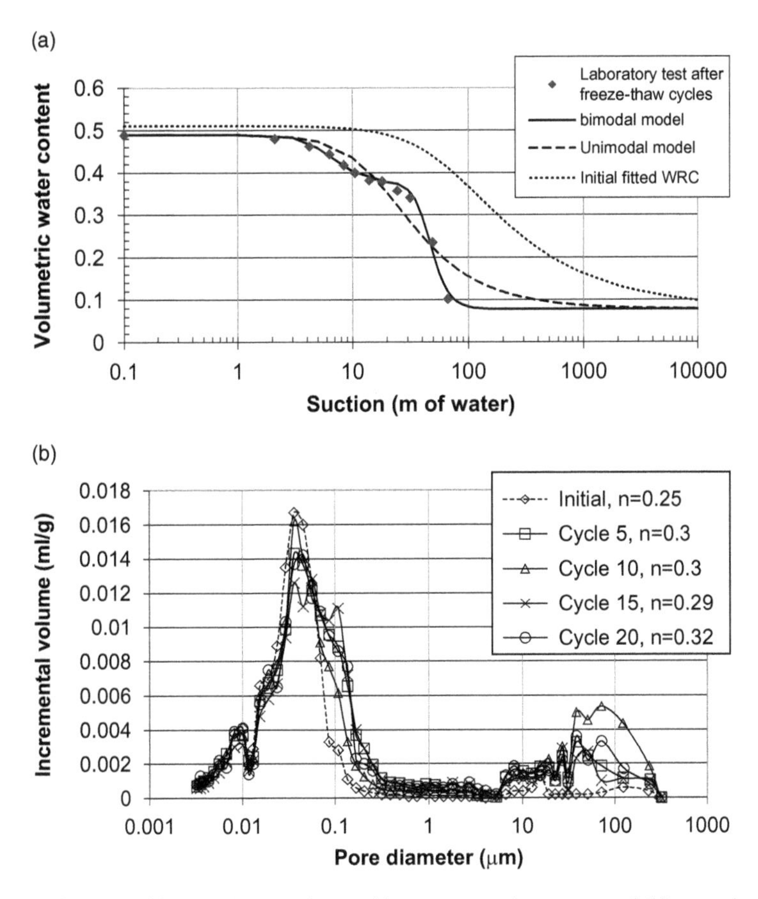

FIGURE 14.12 Influence of freeze–thaw cycles on (a) water retention curve and (b) pore size distribution.

its hydrogeological properties throughout freeze-thaw cycles (Kraus et al. 1997; Podgorney and Bennett 2006; Makusa et al. 2014; Chevé 2019). However, when the GCL is hydrated with highly contaminated AMD, k_{sat} increases by two to three orders of magnitude after 10 cycles (Chevé 2019). Soil-bentonite mixtures are recognized to be less prone to the effects of freeze-thaw cycles and maintain their k_{sat} value (Chapuis 2002). However, Boulanger-Martel et al. (2016) showed that freeze-thaw cycles can affect the microstructure of crushed rock-bentonite mixtures used as fine-grained materials in a CCBE and significantly increase the k_{sat} (two to three orders of magnitude) and D_e values.

In summary, if a fine-grained soil is subjected to freeze-thaw cycles, it could evolve over time, becoming more permeable and less able to retain water due to cracks generated by the melting of ice lenses. For covers that aim to control water infiltration and oxygen migration, this could impact their capacity to control AMD generation. It is then crucial to determine the resistance of cover materials to freeze-thaw cycles at the design stage to avoid a loss of performance in the long term.

14.4.3 Wet-Dry Cycles

Covers built to control AMD are exposed to soil–atmosphere interactions. Hence, covers are exposed to wet-dry cycles that can be more or less frequent and intense. During the drying cycle, a phenomenon called desiccation can affect the properties of fine-grained materials in a manner similar to the freeze-thaw cycles.

Desiccation occurs when a saturated fine-grained material is subjected to evaporation that induces a decrease in water content and the development of suction in the porous network of the soil. The suction that develops in the material leads to a decrease in the material volume by consolidation (see Konrad and Ayad 1997; Saleh-Mbemba et al. 2016 for more details). At the large scale, soils subjected to evaporation are restrained in term of lateral displacement. Hence, a critical level of stress concentration due to suction will initiate crack propagation to a depth controlled by both the stress state (pressure applied in all direction) and the intrinsic material properties (Konrad and Ayad 1997).

Studies performed on compacted clay covers used to control water infiltration in landfills showed that they are prone to damage from desiccation (Daniel and Wu 1993). Albrecht and Benson (2001) showed that hydraulic conductivity increased from an initial k_{sat} of 1×10^{-7} to 10^{-4} cm s^{-1} for cracked material. Cracking that occurred during the drying process induced a new pore structure with macropores that controlled the saturated hydraulic conductivity; the material moved from an unimodal to a bimodal pore size distribution with micro and macropores (Beven and Germann 1982). The main factors of influence on the magnitude of wet-dry cycles on k_{sat} are compaction properties, plasticity and clay content of the fine-grained material, and the number of cycles (with the first cycle having the most significant impact) (Albrecht and Benson 2001). Results also showed that the self-healing potential of compacted clay is limited and does not allow the maintenance of the initial k_{sat} value. Other information on the influence of wet-dry cycles on LSHCCs is presented in Chapter 4.

The influence of wet-dry cycles was also investigated for GCL. The first studies were performed in ideal conditions (full hydration with tap water) and showed no significant influence of wet-dry cycles on k_{sat} (Shan and Daniel 1991; Boardman and Daniel 1997). Other studies, however, showed that when wet-dry cycles were combined with GCL hydration by cationic solutions, the self-healing potential was reduced, and k_{sat} was significantly affected (Benson and Meer 2009).

In summary, as with freeze-thaw cycles, wet-dry cycles can significantly affect the hydrogeological properties of fine-grained cover materials and, consequently, the ability of reclamation methods to control AMD generation. It is important to prevent these cycles from occurring in plastic fine-grained materials or to select materials that are less sensitive to avoid reduced performance of cover systems in the long term.

14.4.4 Roots, Organic Matter, and Porosity

In addition to water pumping, root colonization can change cover material properties, thus impacting their ability to control fluid movement (see Chapter 3) and causing variations in water storage, infiltration, and oxygen migration (Beven and Germann 1982; Breshears et al. 2005; Melchior et al. 2010). Living roots occupy the soil pore space when they grow (see Figure 14.11), which can decrease the soil void ratio (Ng et al. 2016; Ni et al. 2019) and the saturated hydraulic conductivity (k_{sat}), and increase the air entry value (AEV; Jotisankasa and Sirirattanachat 2018). Roots can also induce cracking in materials by provoking wetting or drying conditions (Grevers and De Jong 1990; Scanlon and Goldsmith 1999; Wick et al. 2007). Root growth can also reorganize soil particles by exerting a pressure with expanding cells of the root tip (Bengough et al. 2011), e.g., from 0.3 to 1.3 MPa in pea (Clark et al. 1999), and create new pores, in particular macropores (Guittonny-Larchevêque et al. 2016c). Root diameter directly influences macropore formation and soil hydraulic conductivity (Bodner et al. 2014). Macropores larger than 0.3 mm allow rapid non-equilibrium (non-porous) water flow (Jarvis 2007), induce the evolution from a uni- to a bi-modal WRC (Beven and Germann 1982), and could impact the relationship between the WRC and k_{sat} (Connolly 1998). The magnitude of root effects on soil hydrogeological properties is influenced by soil grain-size distribution and dry density (Bodner et al. 2014; Freschet et al. 2017; Jotisankasa and Sirirattanachat 2018).

More than 50% of root biomass can die each year (Wick et al. 2007), especially fine roots (Steele et al. 1997) (from 0.015 to 2 mm diameter; Lambers et al. 2008). Root death and decay create biopores (see Figure 14.11) with high connectivity in materials (Grevers and De Jong 1990; Bodner et al. 2014). Both cracking and biopores can increase the k_{sat} of cover system materials by up to 10^3 (Albright et al. 2006; Melchior et al. 2010). Increases in the k_{sat} and effective diffusion coefficient and changes in the WRC of materials depend on the intensity of root colonization (Bodner et al. 2014; Jotisankasa and Sirirattanachat 2018), root diameter, and alive/dead status. Dead roots bring OM to cover materials, which can change their porosity and related hydrogeological properties. For example, adding OM (compost) to mine tailings can increase the total porosity of the mixture, in particular its macroporosity, and can decrease its water retention properties (Larchevêque et al. 2013; Guittonny-Larchevêque et al. 2016b). Soil detritivores feed upon this OM. Larger detritivores (e.g., earthworms, some termites, and other insects) participate in the creation of biopores in the substrate. Living roots also release C-rich molecules in adjacent soil, called exudates (see Figure 14.11), which shape the microorganism communities (mainly mushrooms and bacteria) in the soil-root environment (Shi et al. 2011). The dead or alive tissues of soil microorganisms, as well as their excretions, are involved in the formation of stable aggregates, influencing the substrate porosity (Oades 1984; see Figure 14.11).

14.5 DURABILITY OF COARSE-GRAINED COVER MATERIALS

The durability of a given geomaterial can be defined as its ability to adequately maintain a desired functionality over time in a specific working environment (Latham et al. 2006). The concept of durability is integrated in different types of design, such as in construction engineering and architecture (e.g., Chen et al. 2004; Özbek 2014), geomorphology (e.g., Hall 1999; Mackay 1999), and costal engineering (e.g., Graham et al. 2007; Ertas and Topal 2008), but has never been systematically integrated in mine cover design, despite indications that weathering of coarse-grained cover materials can occur. The main weathering phenomena that could affect the durability of coarse-grained cover materials are freeze-thaw and wet-dry cycles. The potential effects of these two phenomena on covers made of coarse-grained materials include the following (Boulanger-Martel et al. 2020): alteration of a cover's physical integrity; changes in the hydrogeotechnical properties of materials, which could, in turn, affect the interactions between cover layers and cover performance and problems related to water quality through the release of fine particles in drainage waters.

Freeze-thaw weathering of coarse-grained materials occurs when porewater freezes in pore spaces (intraparticles). If more than 90% of the pore volume is occupied by water, the 9.05% volume increase resulting from the phase change will create an increase in pore pressure. If this pressure exceeds the rock's tensile strength, new microfractures and cracks will occur and existing fractures will widen. During freeze-thaw cycles, cracking can also occur as a result of the formation of ice lenses or wedges and the excess hydraulic pressure from expelled unfrozen water (Chen et al. 2004; Boulanger-Martel et al. 2020). Wet-dry cycles can affect the durability of coarse-grained materials. The two main degradation mechanisms involved are the intraparticle volume increase of swelling minerals and minerals dissolution. In the presence of water, swelling minerals absorb water in a wet environment. The swelling will generate an increase in intergranular pressure. In the long term, these wet-dry cycles will generate cycles of loading and unloading, causing tension forces that can deconstruct the granular matrix and reduce the quality of the rock particles (Sarman et al. 1994; Özbek 2014). Wet–dry cycles can also dissolve some important minerals that bound the sedimentary rocks together. The dissolution may be associated with precipitation of secondary minerals, which can also affect the material properties. This chemical process can alter the properties of rock particles in the long term (Bell 1992; Dove 1994). It is worth mentioning that the impact of wet-dry cycles is recognized as being less aggressive than freeze-thaw cycles (Pardini et al. 1996; Özbek 2014).

Few standard test methods are available to evaluate durability of rocks. ASTM D4992-14 (ASTM 2014) suggests a series of standard tests to assess rocks to be used for erosion control, including some tests on the durability of a rock unit subjected to freeze-thaw (ASTM D5312-12; ASTM 2013). Recently, other tests for use in pavement or concrete engineering have been developed to determine the durability of coarse-grained materials to freeze-thaw cycles. Most of these durability tests are conducted in a similar manner and essentially involve exposing several specific grain-size fractions to a certain number of freeze-thaw cycles and quantifying the aggregates' weighted average mass loss due to the freeze-thaw cycles (Boulanger-Martel et al. 2020). More sophisticated techniques such as scanning electron microscopy (Martínez-Martínez et al. 2013), X-ray-computed tomography (Park et al. 2015), and nuclear magnetic resonance (e.g., Gao et al. 2017) have been used to evaluate the impacts of freeze-thaw cycles on material microstructure, which can be indicative of its degradation.

As a final remark, no specific methodology is currently available to provide an indication of the durability of coarse-grained cover materials. Therefore, the results of multiple tests must be combined and interpreted as a whole to determine the durability of a given material. The interested reader is referred to the work of Boulanger-Martel et al. (2020) for an example where multiple tests were used to characterize the durability of a coarse-grained cover material.

14.6 METHODOLOGICAL APPROACH TO INTEGRATE EVOLUTIVE PARAMETERS FOR THE PREDICTION OF LONG-TERM PERFORMANCE

The previous chapters on the different reclamation methods (Chapters 4 to 9) proposed design methodologies specific to each method. However, we can identify commonalities among these design methodologies that can be described in five steps: (1) materials characterization, (2) site characterization, (3) performance assessment, (4) design using numerical modeling, and (5) construction and monitoring (see Figure 14.13).

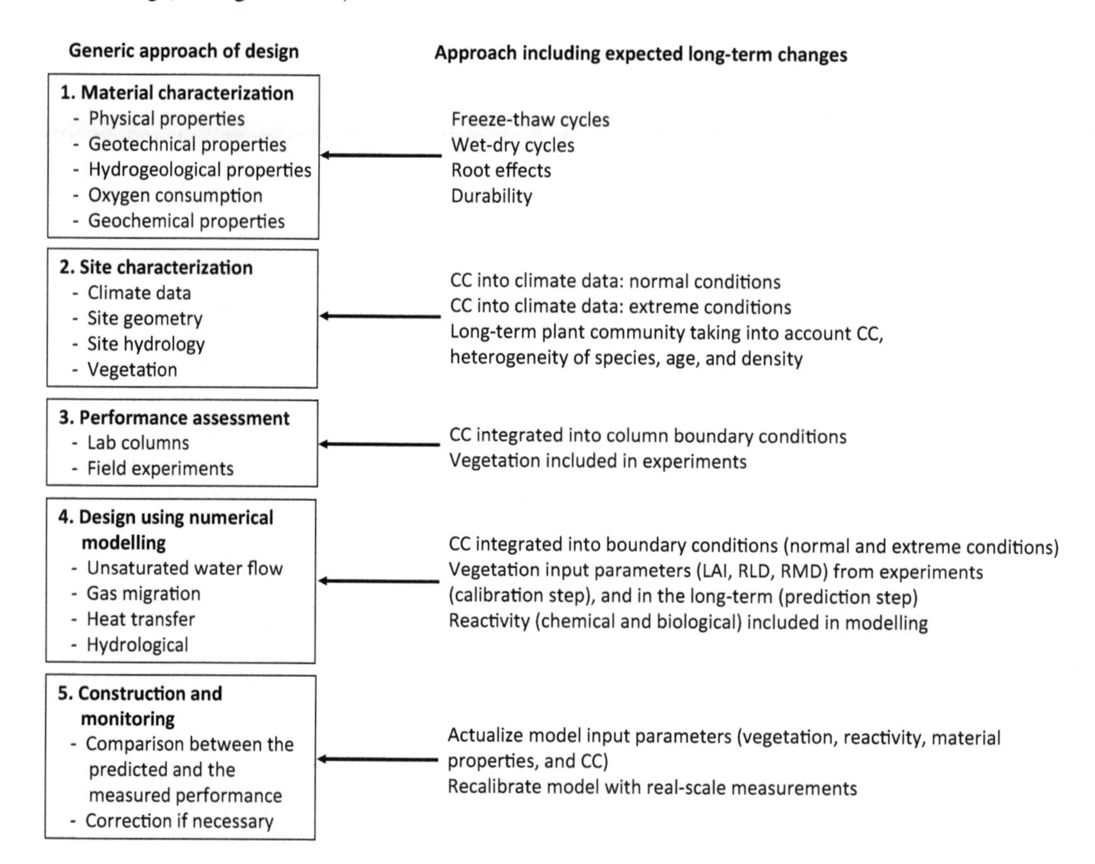

FIGURE 14.13 Integration of factors influencing the long-term performance in the design methodology of reclamation methods.

The first two steps are related to the materials and site characterization; this information is essential to perform a preselection of the most appropriate methods. For example, if no appropriate fine-grained materials are available close to the site, one or two cover options could be eliminated. Also, technologies must be appropriate for the site's climate (e.g., SR and insulation covers need arid and cold climates, respectively, to be efficient). The third step aims to validate the most promising options based on the results from steps 1 and 2 by the use of physical models such as instrumented laboratory columns and/or field experiments. The objective is to physically simulate, in the most realistic manner, a preliminary design of the reclamation methods and to validate that the tested configurations are efficient to control AMD generation. Results from this step are usually compared to numerical predictions to validate the predicted behavior. Once a numerical model is validated with the results from step 3, a new modified model is developed by designers to extrapolate the small-scale results to the real scale (step 4). At this stage, the large-scale model is used to optimize the final design. Then, the reclamation method is built with the integration of instruments to monitor the behavior of the real-size cover system. The monitoring results are then compared to those predicted at the design stage. If discrepancies exist, it is the role of the designer to understand why and to make any necessary corrections.

Based on the previous sections of this chapter, one can expect that the performance of the cover system will evolve with time, mainly because of three factors of influence: vegetation, CC, and changes in material properties. Figure 14.13 illustrates at which steps of the generic design methodology these parameters should be taken into account.

In step 1, the evolution of the material properties should be considered and/or evaluated. This evolution could be due to freeze-thaw cycles, wet-dry cycles, root effects, and a decline of durability. It is then important to integrate into the design process the short-term and long-term properties of materials if they are expected to evolve significantly. This means performing hydrogeological tests (permeability, diffusion, and suction tests) after repeated freeze-thaw and wet-dry cycles. If a protection layer with a thickness greater than the expected maximal root depth (MRD) is planned above the cover, root effects on the evolution of physical and hydrogeological properties of cover materials with time may be neglected. If not, potential changes in the porosity, k_{sat}, and WRC of cover materials with root colonization have to be integrated in the design. For fine-grained materials, the presence of roots will create macropores that will change the WRC (from uni- to bi-modal) and increase the value of k_{sat} with time. Durability tests for the top-layer material that will have to support most of the weathering cycles are also recommended (Boulanger-Martel et al. 2020).

At step 2, it is important to obtain regional climate projections from climate experts for the main meteorological parameters. Results should come from GCM and RCM models using standard approaches (Hotton et al. 2020). Once climate projections are obtained for normal conditions, it is important to determine future extreme conditions. As a reminder, extreme conditions can integrate more than one meteorological and physical parameter (e.g., soil moisture). The extreme events of importance could be different depending on the cover technology applied. For example, the extreme event of interest for water covers and water infiltration barriers are extreme precipitation events, while for CCBE and EWT with monolayer cover, the event of interest is prolonged droughts.

Another parameter related to the site that could affect the long-term performance of covers is the vegetation and the potential evolution of this vegetation with time and CC. It is important to determine plausible scenarios of mature plant communities on the reclaimed site. This scenario takes into account the constraints of cover materials as a growing substrate (especially at the surface) (see Chapter 12), the surrounding native ecosystem, and its probable evolution under CC. It is important to identify the dominant species that are expected to be present and to document their growth characteristics at the above- and belowground levels (especially MRD) in the cover context. Particular attention should be given to plants that may have a greater impact on the cover WB, such as trees, especially broadleaved trees.

In step 3, results obtained from step 2 can be integrated in the laboratory models. For example, if a prolonged drought is expected in the long term with CC, this drought could be applied to the column tests for one or two cycles as top boundary conditions (e.g., 50 days without precipitation on a system that received low volume of water in the previous cycle; see Bresson et al. 2018). This modeling will validate that the cover system will be able to resist to future extreme climate conditions. Vegetation can be implemented in column experiments or in field tests to include its effects on the WB and on oxygen consumption. Despite the fact that these vegetation effects may be limited to the first stages of vegetation establishment and growth (due to the limited duration of experiments), lab and field experiments allow the calibration of numerical models that include plants. The vegetation parameters LAI, RLD, and MRD need to be measured in step 3 experiments to be used as inputs to calibrate/validate the numerical models used to predict cover performance.

Climate projections are used in step 4 as top boundary conditions for numerical modeling. Again, both future normal and extreme conditions must be simulated to guarantee that the system can support the CC in the long term. Then, the vegetation parameters corresponding to the probable dominant plant species on the site identified at step 2 could be used for parametric studies to evaluate whether the cover performance is maintained with an evolved vegetation under CC. Another parameter that can evolve with time is cover material reactivity. On the one hand, if low reactive materials are used in oxygen barrier covers (CCBE and EWT with monolayer cover), their reactivity will diminish with time, and the benefit of having oxygen consumption in the cover will slowly disappear. On the other hand, the presence of roots and soil organisms in the cover materials can consume oxygen and improve the long-term performance of the system to control oxygen migration. It is a complicated task to precisely integrate the oxygen consumption aspect in the design. Nonetheless, a parametric numerical study with different reactivity values should be performed.

At the last step, it is important to use the in situ monitoring results from the first years to recalibrate the initial numerical model. The evolution of parameters such as k_{sat} and WRC can be assessed using in situ sampling. Vegetation parameters used in numerical models (LAI, RLD, and RMD) can be measured instead of being estimated. It is also possible to evaluate the biological and chemical reactivity using oxygen-consumption tests. If CC projections evolve due to the improvement of models and new observations (e.g., GHG emission), the top boundary conditions in the numerical model can be changed. With this recalibration, the uncertainty related to the long-term performance is reduced, and if it is estimated that the targeted performance cannot be reached in the long term, modifications will have to be applied to the cover system.

14.7 FINAL REMARKS

The main parameters that can influence the long-term performance of reclamation methods were identified in this chapter. However, other phenomena not discussed in this chapter could affect the long-term performance of mine site reclamation. Some of them are listed here.

- This handbook does not discuss the physical stability of infrastructures, even if the authors are aware that the physical integrity of the reclamation structures must be guaranteed in the long term. In fact, some geotechnical phenomena can affect the long-term geochemical stability of reclaimed mine sites. For example, conventional tailings are usually disposed of in tailings impoundments in a loose and saturated state. Because of the low k_{sat} of tailings, consolidation of the material can last for decades (see Blight 2009 for details on tailings storage geotechnical behavior). The displacement and settlement of the tailings can jeopardize the performance of the reclamation methods. Cracks and openings in covers can become preferential flow channels for water and oxygen. For materials such as GCL and geomembrane, this displacement can exceed the materials' maximum capacity and lead to significant failure of the cover systems. Hence, in addition to the parameters of influence on the long-term behavior of reclamation methods identified earlier, it is important to assess the long-term geotechnical behavior of tailings impoundment before finalizing the design of reclamation methods. This aspect is less critical for waste rock storage facilities due to the nature of materials (coarse-grained material with high k_{sat} and good mechanical strength).

- Migration of fine particles between fine-grained and coarse-grained layers can occur. This migration could modify the properties of coarse-grained materials and alter their capacity to play an effective role in the cover structure. It is important to validate, using filter criteria (e.g., McCarthy 2007), suffusion criteria (Chapuis and Saucier 2019), and laboratory column tests, that the materials used are compatible and that migration of fine particles has no significant impact on the long-term performance of the cover system.
- Passive treatment systems can be used as polishing treatment systems to finalize the reclamation of an AMD generating mine site (see Chapter 11). Indeed, sometimes, interstitial water at a mine site can be contaminated and will eventually flow out of the site, even if a reclamation method is applied. Passive treatment systems can then be used to temporarily treat this water and reduce the environmental impacts. Also, in some cases, the performance of reclamation methods does not meet environmental criteria. In these cases, passive treatment systems can be part of the reclamation scenario. As mentioned in Chapter 11, one of the main risks for these systems is clogging (by biofouling and mineral precipitation). In fact, clogging is the main documented cause of performance decline (USEPA 2014). When clogging occurs, the contaminated water cannot flow into the passive treatment system and some contaminated water can be transferred into the environment.

REFERENCES

Albrecht, B. A., and C. H. Benson. 2001. Effect of desiccation on compacted natural clays. *Journal of Geotechnical and Geoenvironmental Engineering* 127, no. 1: 67–75.

Albright, W. H., C. H. Benson, G. W. Gee, et al. 2004. Field water balance of landfill covers. *Journal of Environmental Quality* 33, no. 6: 2317–2332.

Albright, W. H., C. H. Benson, W. G. Glendon, et al. 2006. Field performance of a compacted clay landfill final cover at a humid site. *Journal of Geotechnical and Geoenvironmental Engineering* 132, no. 11: 1393–1403.

Albright, W. H., C. H. Benson, and W. J. Waugh. 2010. *Water balance covers for waste containment – Principles and practice*. Reston, VA: American Society of Civil Engineers (ASCE) Press.

Apiwantragoon, P., C. H. Benson, and W. H. Albright. 2014. Field hydrology of water balance covers for waste containment. *Journal of Geotechnical and Geoenvironmental Engineering* 141, no. 2. https://doi.org/10.1061/(ASCE)GT.1943-5606.0001195.

Arnold, S., A. Schneider, D. Doley, and T. Baumgart. 2014. The limited impact of vegetation on the water balance of mine waste cover systems in semi-arid Australia. *Ecohydrology* 8, no. 3: 355–367.

ASTM. 2013. *Standard test method for evaluation of durability of rock for erosion control under freezing and thawing conditions (ASTM D5312/5312M-12(2013))*. West Conshohocken, PA: ASTM International. https://doi.org/10.1520/D5312_D5312M-12R13.

ASTM. 2014. *Standard Practice for Evaluation of Rock to be Used for Erosion Control (ASTM D4992–14e1)*. ASTM International, West Conshohocken, PA.

Atkin, O. K., E. J. Everard, and B. R. Loveys. 2000. Response of root respiration to changes in temperature and its relevance to global warming. *New Phytologist* 147, no. 1: 141–154.

Aubertin, M., T. Pabst, B. Bussière, and M. James. 2015. Revue des meilleures pratiques de restauration des sites d'entreposage de rejets miniers générateurs de DMA. In *Proceedings of the Symposium 2015 sur l'environnement et les mines*, Rouyn-Noranda, June 14–17, 2015.

Ayres, B., and M. O'Kane. 2013. Mine waste cover systems: An international perspective and applications for mine closure in New Zealand. Paper presented at the *AusIMM New Zealand Branch Annual Conference, Nelson, New Zealand.*

Barnswell, K. D., and D. F. Dwyer. 2011. Assessing the performance of evapotranspiration covers for municipal solid waste landfills in northwestern Ohio. *Journal of Environmental Engineering-ASCE* 137: 301–305.

Bell, R. G. 1992. The durability of sandstone as building stone, especially in urban environments. *Environmental & Engineering Geoscience* 29, no. 1: 49–60.

Bengough, A. G., B. M. McKenzie, P. D. Hallett, and T. A. Valentine. 2011. Root elongation, water stress, and mechanical impedance: A review of limiting stresses and beneficial root tip traits. *Journal of Experimental Botany* 62, no. 1: 59–68.

Benson, C. H., and C. A. Bareither. 2012. Designing water balance covers for sustainable waste containment: Transitioning state-of-the-art to state-of-the practice. In *Geotechnical engineering state of the art and practice: Keynote lectures from GeoCongress 2012*, ed. K. Rollins and D. Zekkos, 1–33. Reston, VA: American Society of Civil Engineers.

Benson, C. H., and S. R. Meer. 2009. Relative abundance of monovalent and divalent cations and the impact of desiccation on geosynthetic clay liners. *Journal of Geotechnical and Geoenvironmental Engineering* 135, no. 3: 349–358.

Benson, C. H., and M. A. Othman. 1993. Hydraulic conductivity of compacted clay frozen and thawed in situ. *Journal of Geotechnical Engineering* 119, no. 2: 276–294.

Benson, C. H., W. H. Albright, A. C. Roesler, and T. Abichou. 2002. Evaluation of final cover performance: Field data from the alternative landfill cover assessment program (ACAP). In *Proceedings of the WM 2002 Conference*, Tucson, AZ, February 24–28.

Beven, K., and P. Germann. 1982. Macropores and water flow in soils. *Water Resources Research* 18: 1311–1325.

Blight, G. E. 2009. *Geotechnical engineering for mine waste storage facilities*. Boca Raton, FL: CRC Press.

Blowes, D. W., C. J. Ptacek, J. L. Jambor, et al. 2014. The geochemistry of acid mine drainage. In *Treatise on geochemistry*, 2nd ed., ed. K. Turekian and H. Holland, vol. 11, *Environmental geochemistry*, 131–190. Elsevier. E-book.

Boardman, B., and D. Daniel. 1997. Hydraulic conductivity of desiccated geosynthetic clay liners. *Journal of Geotechnical and Geoenvironmental Engineering* 122, no. 3: 204–208.

Bodner, G., D. Leitner, and H.-P. Kaul. 2014. Coarse and fine root plants affect pore size distributions differently. *Plant Soil* 380: 133–151.

Bond-Lamberty, B., C. Wang, and S. T. Gower. 2004. A global relationship between the heterotrophic and autotrophic components of soil respiration. *Global Change Biology* 10, no. 10: 1756–1766.

Botula, Y.-D., M. Guittonny, B. Bussière, and E. Bresson. 2019. Will tree colonisation increase the risks of serious performance loss of engineered covers under climate change in Québec, Canada? In *Proceedings of the 13th International Conference on Mine Closure*, ed. A. B. Fourier and M. Tibbett, 607–620. Perth: Australian Centre for Geomechanics.

Boulanger, Y., A. Taylor, D. T. Price, et al. 2017. Climate change impacts on forest landscapes along the Canadian southern boreal forest transition zone. *Landscape Ecology* 32: 1415–1431.

Boulanger-Martel, V., B. Bussière, J. Côté, and M. Mbonimpa. 2016. Influence of freeze-thaw cycles on the performance of covers with capillary barrier effects made of crushed rock-bentonite mixtures to control oxygen migration. *Canadian Geotechnical Journal* 53: 753–764.

Boulanger-Martel, V., B. Bussière, and J. Côté. 2020. Resistance of a soapstone waste rock to freeze-thaw and wet-dry cycles: Implications for use in a reclamation cover in the Canadian Arctic. *Bulletin of Engineering Geology and the Environment*, https://doi.org/10.1007/s10064-020-01930-8.

Breshears, D. D., J. W. Nyhan, and D. W. Davenport. 2005. Ecohydrology monitoring and excavation of semi-arid landfill covers a decade after installation. *Vadose Zone Journal* 4, no. 3: 798–810.

Bresson, É., I. Demers, P. Roy, T. Pabst, and Y. Chavaillaz. 2018. Efficiency of reclamation methods under climate change: Definition of a drought index. In *Proceedings of the 22th International Conference on Tailings and Mine Waste '18*, Keystone, Colorado, USA, September 30–October 2, 2018, 583–593.

Bruce, J. P., and R. H. Clark. 1969. *Introduction to hydrometeorology*. New York: Pergamon Press.

Burger, C. A., and C. D. Shackelford. 2001. Evaluating dual porosity of pelletized diatomaceous earth using bimodal soil-water characteristic curve functions. *Canadian Geotechnical Journal* 38, no. 1: 53–66.

Bussière, B., M. Benzaazoua, M. Aubertin, and M. Mbonimpa. 2004. A laboratory study of covers made of low sulphide tailings to prevent acid mine drainage. *Environmental Geology* 45: 609–622.

Bussière, B., I. Demers, P. Charron, et al. 2017. *Analyse de risque et de vulnérabilité liés aux changements climatiques pour le secteur minier québécois*. Report submitted to the Ministère de l'Énergie et des Ressources Naturelles (MERN).

Chamberlain, E. J. 1981. Overconsolidation effects of ground freezing. *Engineering Geology* 18: 97–110.

Chapuis, R. P. 2002. The 2000 RM Hardy Lecture: Full-scale hydraulic performance of soil bentonite and compacted clay liners. *Canadian Geotechnical Journal* 39, no. 2: 417–439.

Chapuis, R. P., and A. Saucier. 2019. Assessing internal erosion with the modal decomposition method for grain size distribution curves. *Acta Geotechnica* 15: 1595–1620. https://doi.org/10.1007/s11440-019-00865-z.

Charron, I. 2016. *Guide sur les scénarios climatiques: Utilisation de l'information climatique pour guider la recherche et la prise de décision en matière d'adaptation*. Montréal, QC: Ouranos.

Chen, T., M. Yeung, and N. Mori. 2004. Effect of water saturation on deterioration of welded tuff due to freeze-thaw action. *Cold Regions Science and Technology* 38, no. 2–3: 127–136.

Chen, Z., Y. Xu, J. Fan, H. Yu, and W. Ding. 2017. Soil autotrophic and heterotrophic respiration in response to different N fertilization and environmental conditions from a cropland in Northeast China. *Soil Biology and Biochemistry* 110: 103–115. http://dx.doi.org/10.1016/j.soilbio.2017.03.011.

Chevé, N. 2019. Évaluation de la performance des géocomposites bentonitiques comme barrière aux fluides dans un contexte de recouvrement minier. Master's thesis, Polytechnique Montréal, Montréal, QC.

Chevé, N., M. Guittonny, and B. Bussière. 2018. Water budget of field experimental cells with vegetated and non-vegetated soil layers placed on waste rock. In *Proceedings of the 12th International Conference on Mine Closure*, Liepzig, Germany, September 2018.

Clark, L. J., A. G. Bengough, W. R. Whalley, A. R. Dexter, and P. B. Barraclough. 1999. Maximum axial root growth pressure in pea seedlings: Effects of measurement techniques and cultivars. *Plant and Soil* 209: 101–109.

Collins, M., R. Knutti, J. Arblaster, et al. 2013. Long-term climate change: Projections, commitments and irreversibility. In *Climate change 2013: The physical science basis. Contribution of Working Group I to the fifth assessment report of the Intergovernmental Panel on Climate Change*, ed. T. F. Stocker et al. Cambridge: Cambridge University Press.

Connolly, R. D. 1998. Modelling effects of soil structure on the water balance of soil–crop systems: A review. *Soil and Tillage Research* 48: 1–19.

Cook, F. J., and J. H. Knight. 2003. Oxygen transport to plant roots: Modeling for physical understanding of soil aeration. *Soil Science Society of America Journal* 67, no. 1: 20–31. http://soil.scijournals.org/cgi/content/abstract/soilsci;67/1/20.

Cooke, J. A., and M. S. Johnson. 2002. Ecological restoration of land with particular reference to the mining of metals and industrial minerals: A review of theory and practice. *Environmental Reviews* 10, no. 1: 41–71.

Fredlund, D. G., H. Rahardjo, and M. D. Fredlund. 2012. *Unsaturated soil mechanics in engineering practice*. Hoboken, NJ: John Wiley & Sons.

Dagenais, A.-M. 2005. Techniques de contrôle du DMA basées sur les effets capillaires. PhD diss., École Polytechnique de Montréal.

Daniel, D., and Y. Wu. 1993. Compacted clay liners and covers for arid sites. *Journal of Geotechnical Engineering* 119, no. 2: 223–237.

Davidson, E. A., E. Belk, and R. D. Boone. 1998. Soil water content and temperature as independent or confounded factors controlling soil respiration in a temperate mixed hardwood forest. *Global Change Biology* 4, no. 2: 217–227.

DeJong, J., M. Tibbett, and A. Fourie. 2015. Geotechnical systems that evolve with ecological processes. *Environmental Earth Sciences* 73: 1067–1082.

Demers, I., B. Bussière, M. Mbonimpa, and M. Benzaazoua 2009. Oxygen diffusion and consumption in low sulphide tailings covers. *Canadian Geotechnical Journal* 46: 454–469.

Dove, P. M. 1994. The dissolution kinetics of quartz in sodium chloride solutions at 25 degrees to 300 degrees C. *American Journal of Science* 294, no. 6: 665–712.

Eigenbrod, K. D. 2003. Self-healing in fractured fine-grained soils. *Canadian Geotechnical Journal* 40: 435–449.

Ertas, B., and T. Topal. 2008. Quality and durability assessments of the armourstones for two rubble mound breakwaters (Mersin, Turkey). *Environmental Geology* 53, no. 6: 1235–1247.

Ethier, M.-P., B. Bussière, M. Aubertin, A. Maqsoud, I. Demers, and S. Broda. 2018. In situ evaluation of performance of reclamation measures implemented on abandoned reactive tailings disposal site. *Canadian Geotechnical Journal* 55, no. 12: 1742–1755.

Fan, Y., G. Miguez-Macho, E. G. Jobbágy, R. B. Jackson, and C. Otero-Casal. 2017. Hydrologic regulation of plant rooting depth. *PNAS* 114, no. 40: 10572–10577.

Flato, G., J. Marotzke, B. Abiodun, et al. 2013. Evaluation of climate models. In *Climate change 2013: The physical science basis. Contribution of Working Group I to the fifth assessment report of the Intergovernmental Panel on Climate Change*, ed. T. F. Stocker et al. Cambridge: Cambridge University Press.

Fourie, A. B., and M. Tibbett. 2007. Post-mining landforms—Engineering a biological system. In *Proceedings of the Second International Conference on Mine Closure*, Santiago, Chile, October 16–19, 3–12.

Fraser, S. 2014. Evaluating the influence of vegetation on evapotranspiration from waste rock surfaces in the Elk Valley, British Columbia. MSc thesis, McMaster University, Hamilton, ON.

Freschet, G. T., O. J. Valverde-Barrantes, C. M. Tucker, et al. 2017. Climate, soil and plant functional types as drivers of global fine-root trait variation. *Journal of Ecology* 105, no. 5: 1182–1196.

Gao, F., Q. Wang, H. Deng, J. Zhang, W. Tian, and B. Ke. 2017. Coupled effects of chemical environments and freeze–thaw cycles on damage characteristics of red sandstone. *Bulletin of Engineering Geology and the Environment* 76, no. 4: 1481–1490.

Garcia-Estringana, P., N. Alonso-Blázquez, M. J. Marques, R. Bienes, and J. Alegre. 2010. Direct and indirect effects of Mediterranean vegetation on runoff and soil loss. *European Journal of Soil Science* 61, no. 2: 174–185.

Graham, J., K. Franklin, M. Alfaro, and J. Wortley. 2007. Degradation of shaley limestone riprap. *Canadian Geotechnical Journal* 44, no. 11: 1265–1272.

Grevers, M. C. J., and E. De Jong. 1990. The characterization of soil macroporosity of a clay soil under ten grasses using image analysis. *Canadian Journal of Soil Science* 70, no. 1: 93–103.

Guittonny-Larchevêque, M., A. Beaulieu, A. Proteau, B. Bussière, and A. Maqsoud. 2016a. Vegetation management on tailings impoundments reclaimed with covers with capillary barrier effects. In *Proceedings of the 5th I2SM Conference*, Montréal, Canada, July 10–13.

Guittonny-Larchevêque, M., B. Bussière, and C. Pednault. 2016b. Tree-substrate water relations and root development in tree plantations used for mine tailings reclamation. *Journal of Environmental Quality* 45, no. 3: 1036–1045.

Guittonny-Larchevêque, M., Y. Meddeb, and D. Barrette. 2016c. Can graminoids used for mine tailings revegetation improve substrate structure? *Botany* 94, no. 11: 1053–1061.

Guittonny-Larchevêque, M., and S. Lortie. 2017. Above-and belowground development of a fast-growing willow planted in acid-generating mine technosol. *Journal of Environmental Quality* 46, no. 6: 1462–1471.

Guittonny, M., B. Bussière, A. Maqsoud, A. Proteau, T. Ben Khouya, and Y.-D. Botula. 2018. Root colonization of mine covers and impact on their functioning. In *Proceedings of the 6th Symposium on Mines and the environment*, Rouyn-Noranda, Canada, June 2018.

Guittonny, M., B. Bussière, N. Chevé, B. Mangane, and M. Duclos. 2019. Effects of revegetation and its supporting layers on the water budget of waste rocks. In *Proceedings of Geo-Environmental Engineering 2019*, Concordia University, Montreal, Canada, May 30–31.

Gutierrez, J., and I. I. Hernandez. 1996. Runoff and interrill erosion as affected by grass cover in a semi-arid rangeland of northern Mexico. *Journal of Arid Environments* 34: 287–295.

Gwenzi, W., C. Hinz, T. M. Bleby, and E. J. Veneklaas. 2014. Transpiration and water relations of evergreen shrub species on an artificial landform for mine waste storage versus an adjacent natural site in semi-arid Western Australia. *Ecohydrology* 7, no. 3: 965–981.

Hall, K. 1999. The role of thermal stress fatigue in the breakdown of rock in cold regions. *Geomorphology* 31, no. 1: 47–63.

Hanson, P. J., N. T. Edwards, C. T. Garten, and J. A. Andrews. 2000. Separating root and soil microbial contributions to soil respiration: A review of methods and observations. *Biogeochemistry* 48: 115–146.

Harlan, R. 1973. Analysis of coupled heat-fluid transport in partially frozen soil. *Water Resources Research* 9: 1314–1323.

Hotton, G., B. Bussiere, T. Pabst, E. Bresson, and P. Roy. 2020. Influence of climate change on the performance of a cover with capillary barrier effect to control acid generation. *Hydrogeology Journal* 28: 763–779.

Hutchinson, M. F., D. W. McKenney, K. Lawrence, et al. 2009. Development and testing of Canada-wide interpolated spatial models of daily minimum-maximum temperature and precipitation for 1961–2003. *Journal of Applied Meteorology and Climatology* 48, no. 4: 725–741.

Intergovernmental Panel on Climate Change (IPCC). 2007. *General guidelines on the use of scenario data for climate impact and adaptation assessment.* Version 2. Prepared by T. R. Carter on behalf of the Intergovernmental Panel on Climate Change, Task Group on Data and Scenario Support for Impact and Climate Assessment. http://www.ipcc-data.org/guidelines/TGICA_guidance_sdciaa_v2_final.pdf.

Jackson, R. B., J. Canadell, J. R. Ehleringer, H. A. Mooney, O. E. Sala, and E. D. Schulze. 1996. A global analysis of root distributions for terrestrial biomes. *Oecologia* 108: 389–411.

Jarvis, N. J. 2007. A review of non-equilibrium water flow and solute transport in soil macropores: Principles, controlling factors and consequences for water quality. *European Journal of Soil Science* 58: 523–546.

Jotisankasa, A., and T. Sirirattanachat. 2018. Effects of grass roots on soil-water retention curve and permeability function. *Canadian Geotechnical Journal* 54: 1612–1622.

Konrad, J. M. 2001. *Cold region engineering*. In *Geotechnical and geoenvironmental engineering handbook*, ed. R. K. Rowe. Boston: Springer.

Konrad, J. M., and R. Ayad. 1997. An idealized framework for the analysis of cohesive soils undergoing desiccation. *Canadian Geotechnical Journal* 34: 477–488.

Konrad, J.-M., and M. Samson. 2000. Hydraulic conductivity of kaolinite-silt mixtures subjected to closed-system freezing and thaw consolidation. *Canadian Geotechnical Journal* 37: 857–869.

Kraus, J. F., C. H. Benson, A. E. Erickson, and E. J. Chamberlain. 1997. Freeze-thaw cycling and hydraulic conductivity of bentonitic barriers. *Journal of Geotechnical and Geoenvironmental Engineering* 123: 229–238.

Lambers, H., F. S. Chapin III, and T. L. Pons. 2008. *Plant physiological ecology*. New York: Springer.

Lamoureux, S. C., E. J. Veneklaas, P. Poot, and M. O'Kane. 2016. The effect of cover system depth on native plant water relations in semi-arid Western Australia. In *Proceedings of the 11th International Conference on Mine Closure*, ed. A. B. Fourie and M. Tibbett, 567–578. Perth: Australian Centre for Geomechanics.

Larchevêque, M., A. Desrochers, B. Bussière, H. Cartier, and J. S. David. 2013. Revegetation of non acid generating, thickened tailings with boreal trees: A greenhouse study. *Journal of Environmental Quality* 42, no. 2: 351–360.

Latham, J. P., D. Lienhart, and S. Dupray. 2006. Rock quality, durability and service life prediction of armourstone. *Engineering Geology* 87, no. 1: 122–140. https://doi.org/10.1016/j.enggeo.2006.06.004.

Le Bissonnais, Y., V. Lecomte, and O. Cerdan. 2004. Grass strip effects on runoff and soil loss. *Agronomie* 24: 129–136.

Lieber, É. 2019. Influence des facteurs climatiques sur la performance de la nappe phréatique surélevée combinée à un recouvrement monocouche. Master's thesis, Polytechnique Montréal, Montréal, QC.

Liu, J., J. M. Chen, and J. Cihlar. 2003. Mapping evapotranspiration based on remote sensing: An application to Canada's landmass. *Water Resources Research* 39, no. 7. https://doi.org/10.1029/2002WR001680.

Loarie, S. R., P. B. Duffy, H. Hamilton, G. P. Asner, C. B. Field, and D. D. Ackerly. 2009. The velocity of climate change. *Nature* 462: 1052–1055.

Mackay, J. R. 1999. Cold-climate shattering (1974 to 1993) of 200 glacial erratics on the exposed bottom of a recently drained arctic lake, Western Arctic coast, Canada. *Permafrost and Periglacial Processes* 10, no. 2: 125–136.

Madalinski, K. L., D. N. Gratton, and R. J. Weisman. 2003. Evapotranspiration covers: An innovative approach to remediate and close contaminated sites. *Remediation* 14, no. 1: 55–67.

Makusa, G. P., S. L. Bradshaw, E. Berns, C. H. Benson, and S. Knutsson. 2014. Freeze–thaw cycling concurrent with cation exchange and the hydraulic conductivity of geosynthetic clay liners. *Canadian Geotechnical Journal* 51: 591–598.

Martínez-Martínez, J., D. Benavente, M. Gomez-Heras, L. Marco-Castaño, and M. Á. García-del-Cura. 2013. Non-linear decay of building stones during freeze–thaw weathering processes. *Construction and Building Materials* 38: 443–454.

Mayer, K. U. 1999. A numerical model for multicomponent reactive transport in variably saturated porous media. PhD diss., University of Waterloo, ON.

McCarthy, D. F. 2007. *Essentials of soil mechanics and foundations: Basic geotechnics*. 7th ed. Upper Saddle River, NJ: Pearson Prentice Hall.

McKenney, D. W., J. H. Pedlar, K. Lawrence, K. Campbell, and M. F. Hutchinson. 2007. Potential impacts of climate change on the distribution of North American trees. *BioScience* 57, no. 11: 939–948.

McLendon, T., J. Coleman, T. Shepherd, and R. E. Nelson. 1997. The inclusion of biointrusion considerations in the design of the reclamation cover for the DMC tailings impoundments. In *Proceedings of the Tailings and Mine Wastes Conference, Fort Collins, CO, USA, January 1, 1997*, 267–281. Brookfield, VT: A. A. Balkema.

Melchior, S., V. Sokollek, K. Berger, B. Vielhaber, and B. Steinert. 2010. Results from 18 years of in situ performance testing of landfill cover systems in Germany. *Journal of Environmental Engineering* 136, no. 8: 815–823.

Mine Environment Neutral Drainage (MEND) Program. 2014. *Modelling the critical interactions between cover systems and vegetation*. MEND Report 2.21.6. http://mend-nedem.org/wp-content/uploads/2.21.6.pdf.

Mitchell, J. K., and K. Soga. 2005. *Fundamentals of soil behavior*, 3rd ed. Hoboken, NJ: John Wiley & Sons.

Moss, R., M. Babiker, S. Brinkman, et al. 2008. *Towards new scenarios for analysis of emissions, climate change, impacts, and response strategies*. Geneva: Intergovernmental Panel on Climate Change.

Moss, R. H., J. A. Edmonds, K. A. Hibbard, et al. 2010. The next generation of scenarios for climate change research and assessment. *Nature* 463, no. 7282: 747–756.

Naeth, M. A., D. S. Chanasyk, and T. D. Burgers. 2011. Vegetation and soil water interactions on a tailings sand storage facility in the Athabasca oil sands region of Alberta Canada. *Physics and Chemistry of the Earth* 36, no. 1–4: 19–30.

Ng, C. W. W., J. J. Ni, A. K. Leung, and Z. J. Wang. 2016. A new and simple water retention model for root-permeated soils. *Géotechnique Letters* 6: 1–6.

Ni, J. J., A. K. Leung, and C. W. W. Ng. 2019. Modelling effects of root growth and decay on soil water retention and permeability. *Canadian Geotechnical Journal* 56, no. 7: 1049–1055.

Nixon, J. F., and B. Ladanyi. 1978. Thaw consolidation. In *Geotechnical engineering for cold regions*, ed. O. B. Andersland and D. M. Anderson, 164–215. New York: McGraw-Hill.

Oades, J. M. 1984. Soil organic matter and structural stability: Mechanisms and implications for management. *Plant and Soil* 76: 319–337.

Olsson, P., S. Linder, R. Giesler, and P. Hogberg. 2005. Fertilization of boreal forest reduces both autotrophic and heterotrophic soil respiration. *Global Change Biology* 11: 1745–1753.

Othman, M. A., C. H. Benson, E. J. Chamberlain, and T. F. Zimmie. 1994. Laboratory testing to evaluate changes in hydraulic conductivity of compacted clays caused by freeze-thaw: State-of-the-art. In *Hydraulic conductivity and waste contaminant transport in soils, ASTM STP1142*, ed. D. E. Daniel and S. J. Trautwein. Philadelphia: American Society of Testing and Materials.

Özbek, A. 2014. Investigation of the effects of wetting–drying and freezing–thawing cycles on some physical and mechanical properties of selected ignimbrites. *Bulletin of Engineering Geology and the Environment* 73, no. 2: 595–609.

Palmer, M. A., J. B. Zedler, and D. A. Falk, eds. 2006. *Foundations of restoration ecology*. Washington, DC: Island Press.

Pardini, G., G. V. Guidi, R. Pini, D. Regüés, and F. Gallart. 1996. Structure and porosity of smectitic mudrocks as affected by experimental wetting—drying cycles and freezing—thawing cycles. *CATENA* 27, no. 3–4: 149–165.

Park, J., C. U. Hyun, and H. D. Park. 2015. Changes in microstructure and physical properties of rocks caused by artificial freeze–thaw action. *Bulletin of Engineering Geology and the Environment* 74, no. 2: 555–565.

Phukan, A. 1985. *Frozen ground engineering*. Englewood Cliffs, NJ: Prentice-Hall.

Piet, S. J., R. P. Breckenridge, J. J. Jacobson, G. J. White, H. I. Inyang. 2005. Design Principles and Concepts for Enhancing Long-Term Cap Performance and Confidence. *Practice Periodical of Hazardous, Toxic, and Radioactive Waste Manage* 9, no. 4: 210–222.

Podgorney, R. K., and J. E. Bennett. 2006. Evaluating the long-term performance of geosynthetic clay liners exposed to freeze-thaw. *Journal of Geotechnical and Geoenvironmental Engineering* 132: 265–268.

Pregitzer, K. S., M. J. Laskowski, A. J. Burton, V. C. Lessard, and D. R. Zak. 1998. Variation in sugar maple root respiration with root diameter and soil depth. *Tree Physiology* 18, no. 10: 665–670.

Price, D. T., R. I. Alfaro, K. J. Brown, et al. 2013. Anticipating the consequences of climate change for Canada's boreal forest ecosystems. *Environmental Reviews* 21: 322–365.

Proteau, A., M. Guittonny, B. Bussière, and A. Maqsoud. 2020a. Aboveground and belowground colonization of vegetation on a 17-year-old cover with capillary barrier effect built on a boreal mine tailings storage facility *Minerals* 10(8): 704.

Proteau, A., M. Guittonny, B. Bussière, and A. Maqsoud. 2020b. Oxygen migration through a cover with capillary barrier effects colonized by roots. *Canadian Geotechnical Journal*. https://doi.org/10.1139/cgj-2019-0515.

Reich, P. B., M. B. Walters, M. G. Tjoelker, D. Vanderklein, and C. Buschena. 1998. Photosynthesis and respiration rates depend on leaf and root morphology and nitrogen concentration in nine boreal tree species differing in relative growth rate. *Functional Ecology* 12, no. 3: 395–405.

Rey, N. J., I. Demers, B. Bussière, M. Mbonimpa, and M. Gagnon. 2020. A geochemical evaluation of a monolayer cover with an elevated water table for the reclamation of the Doyon-Westwood tailings ponds, Canada. *Environmental Earth Sciences* 79, no. 2: 58.

Roy, P. 2016. Synthèse des changements climatiques pour le secteur minier. In *Annexe 3 de l'Analyse de risque et de vulnérabilité liés aux changements climatiques pour le secteur minier québécois*. Report submitted to the Ministère de l'Énergie et des Ressources Naturelles (MERN), Ouranos, Montréal, QC, Canada.

Rustad, L. E., T. G. Huntington, and R. D. Boone. 2000. Controls on soil respiration: Implications for climate change. *Biogeochemistry* 48: 1–6.

Saleh-Mbemba, F., M. Aubertin, M. Mbonimpa, and L. Li. 2016. Experimental characterization of the shrinkage and water retention behaviour of tailings from hard rock mines. *Geotechnical and Geological Engineering* 34, no. 1: 251–266.

Sarman, R., A. Shakoor, and D. F. Palmer. 1994. A multiple regression approach to predict swelling in mudrocks. *Environmental & Engineering Geoscience* 31, no. 1: 107–121.

Scanlon, B. R., R. C. Reedy, K. E. Keese, and S. F. Dwyer. 2005. Evaluation of evapotranspirative covers for waste containment in arid and semiarid regions of the Southwestern USA. *Vadose Zone Journal* 4: 55–71.

Scanlon, B. R., and R. S. Goldsmith. 1999. Field study of spatial variability in unsaturated flow beneath and adjacent to playas. *Water Resources Research* 33, no. 10: 2239–2252.

Schenk, H. J., and R. B. Jackson. 2002. Rooting depths, lateral root spreads and below-ground/above-ground allometries of plants in water-limited ecosystems. *Journal of Ecology* 90, no. 3: 480–494.

Seneviratne, S. I., D. Nicholls, C. M. Easterling, et al. 2012. Changes in climate extremes and their impacts on the natural physical environment. In *Managing the risks of extreme events and disasters to advance climate change adaptation: A special report of Working Groups I and II of the Intergovernmental Panel on Climate Change (IPCC)*, ed. C. B. Field et al., 109–230. Cambridge: Cambridge University Press.

Shan, H. Y., and D. E. Daniel. 1991. Results of laboratory tests on a geotextile/bentonite liner material. In *Geosynthetics 91 Conference*, vol. 2, 517–535. St. Paul, MN: Industrial Fabrics Association International.

Shi, S., A. E. Richardson, M. O'Callaghan, K. M. DeAngelis, E. E. Jones, A. Stewart, M. K. Firestone, L. M. Condron. 2011. Effects of selected root exudate components on soil bacterial communities. *FEMS Microbiology Ecology*. 77: 600–610.

Shurniak, R. 2003. Predictive modeling of moisture movement within soil cover systems for saline/sodic overburden piles. MSc thesis, University of Saskatchewan.

Shurniak, R. E., and S. L. Barbour. 2002. Modeling of water movement within reclamation covers on oil-sands mining overburden piles. In *Proceedings of the National Meeting of the American Society of Mining and Reclamation*, Lexington, KY, USA, June 9–13, 2002, 622–644. https://doi.org/10.21000/JASMR02010622.

Simms P. H., and E. K. Yanful. 2004. A discussion of the application of mercury intrusion porosimetry for the investigation of soils, including an evaluation of its use to estimate volume change in compacted clayey soils. *Géotechnique* 54, no. 6: 421–426.

Smirnova, E., B. Bussière, F. Tremblay, and Y. Bergeron. 2011. Vegetation succession and impacts of biointrusion on covers used to limit acid mine drainage. *Journal of Environmental Quality* 40, no. 1: 133–143.

Steele, S. J., S. T. Gower, J. G. Vogel, and J. M. Norma. 1997. Root mass, net primary production and turnover in aspen, jack pine and black spruce forests in Saskatchewan and Manitoba, Canada. *Tree Physiology* 17: 577–587.

Stoltz, E., and M. Greger. 2006. Root penetration through sealing layers at mine deposit sites. *Waste Management & Research* 24, no. 6: 552–559.

Traynham, B. 2010. Monitoring the long-term performance of engineered containment systems: The role of ecological processes. PhD diss., Vanderbilt University.

Traynham, B., J. Clarke, J. Burger, and J. Waugh. 2012. Engineered containment systems: Identification of dominant ecological processes for long-term performance assessment and monitoring. *Remediation* 22, no. 3: 93–103.

United States Environmental Protection Agency (USEPA). 2014. *Reference guide to treatment technologies for mining-influenced water*. EPA 542-R-14-001. Washington, DC: EPA.

Walker, L. R., and R. del Moral. 2009. Lessons from primary succession for restoration of severely damaged habitats. *Applied Vegetation Science* 12, no. 1: 55–67.

Watanabe, K., T. Kito, T. Wake, and M. Sakai. 2011. Freezing experiments on unsaturated sand, loam and silt loam. *Annals of Glaciology* 52, no. 58: 37–43.

Waugh, W. J. 2004. Design, performance, and sustainability of engineered covers for uranium mill tailings. In *Proceedings of the Workshop on Long-Term Performance Monitoring of Metals and Radionuclides in the Subsurface: Strategies, Tools, and Case Studies*. https://doi.org/10.2172/1132788.

Wels, C., S. Fortin, and S. Loudon. 2002. Assessment of store-and-release cover for Questa tailings facility, New Mexico. In *Proceedings of the Ninth International Conference on Tailings and Mine Waste*, Fort Collins, CO, USA, January 27–30, 2002, ed. R. I. Barnhisel, 459–468. Brookfield, VT: A. A. Balkema.

Wick, A. F., P. D. Stahl, S. Rana, and L. J. Ingram. 2007. Recovery of reclaimed soil structure and function in relation to plant community composition. In *Proceedings of the National Meeting of the American Society of Mining and Reclamation, Gillette, Wyoming, USA*, ed. R. I. Barnhisel, 941–957. Lexington, KY: ASMR.

Williams, D. J., D. J. Stolberg, and N. A. Currey. 2006. Long-term monitoring of Kidston's "Store/Release" cover system over potentially acid forming waste rock piles. In 7th International Conference on Acid Rock Drainage (ICARD), St Louis, MO, USA, March 26–30, 2006, ed. R. I. Barnhisel, 2385–2396.

Wilson, G. W., D. G. Fredlund, and S. L. Barbour. 1997. The effect of soil suction on evaporative fluxes from soil surfaces. *Canadian Geotechnical Journal* 34, no. 1: 145–155.

Woyshner, M. R., and E. K. Yanful. 1995. Modelling and field measurements of water percolation through an experimental soil cover on mine tailings. *Canadian Geotechnical Journal* 32, no. 4: 601–609.

Yanful, E. K., S. M. Mousavi, and M. Yang. 2003. Modeling and measurement of evaporation in moisture-retaining soil covers. *Advances in Environmental Research* 7, no. 4: 783–801.

Zhang, X., S. Chen, H. Sun, Y. Wang, and L. Shao. 2009. Root size, distribution and soil water depletion as affected by cultivars and environmental factors. *Field Crops Research* 114, no. 1: 75–83.

Zornberg, J. G., L. Lafountain, and J. A. Caldwell. 2003. Analysis and design of evapotranspirative covers for hazardous waste landfills. *Journal of Geoenvironmental and Geotechnical Engineering* 129, no. 6: 427–438.

Zuazo, V. H. D., and C. R. R. Pleguezuelo. 2008. Soil-erosion and runoff prevention by plant covers: A review. *Agronomy for Sustainable Development* 28: 65–86.

Index

Numerals in *italics* refer to tables, and those in **bold** to figures.

W

Z

For Product Safety Concerns and Information please contact our EU
representative GPSR@taylorandfrancis.com
Taylor & Francis Verlag GmbH, Kaufingerstraße 24, 80331 München, Germany

www.ingramcontent.com/pod-product-compliance
Ingram Content Group UK Ltd.
Pitfield, Milton Keynes, MK11 3LW, UK
UKHW050109250425
457818UK00014B/349